D0077372

AGRICULTURAL MARKETING AND PRICE ANALYSIS

AGRICULTURAL MARKETING AND PRICE ANALYSIS

F. Bailey Norwood and Jayson L. Lusk

Oklahoma State University

Upper Saddle River, New Jersey
Columbus, Ohio

Library of Congress Cataloging-in-Publication Data

Norwood, F. Bailey.
Agricultural marketing and price analysis/ F. Bailey Norwood and Jayson L. Lusk.
p.cm.
Includes bilbliographical references and index.
ISBN-13: 978-0-13-221121-5
ISBN-10: 0-13-221121-1
1. Farm produce—Marketing. 2. Agricultural prices. 3. Produce trade. I. Lusk, Jayson. II. Title.
HD9000.5.N62 2007
338.1'3—dc22

2007038782

Editor in Chief: Vernon Anthony
Acquisitions Editor: Vernon Anthony
Editorial Assistant: Sonya Kottcamp
Production Coordination: Rebecca K. Giusti, GGS Book Services
Project Manager: Kevin Happell
Design Coordinator: Diane Ernsberger
Cover Designer: Bryan Huber
Operations Specialist: Deidra Schwartz
Director of Marketing: David Gesell
Marketing Manager: Jimmy Stephens
Marketing Coordinator: Alicia Dysert

This book was set in Clearface Regular by GGS Book Services. It was printed and bound by Courier Westford. The cover was printed by Phoenix Color Corp.

Pearson Education Ltd.
Pearson Education Singapore Pte. Ltd.
Pearson Education Canada, Ltd.
Pearson Education—Japhan
Pearson Education Australia Pty. Limited
Pearson Education North Asia Ltd.
Pearson Education de Mexico, S.A. de C.V.
Pearson Education Malaysia Pte. Ltd.

10 9 8 7 6 5 4 3 2 1
ISBN-13: 978-0-13-221121-5
ISBN-10: 0-13-221121-1

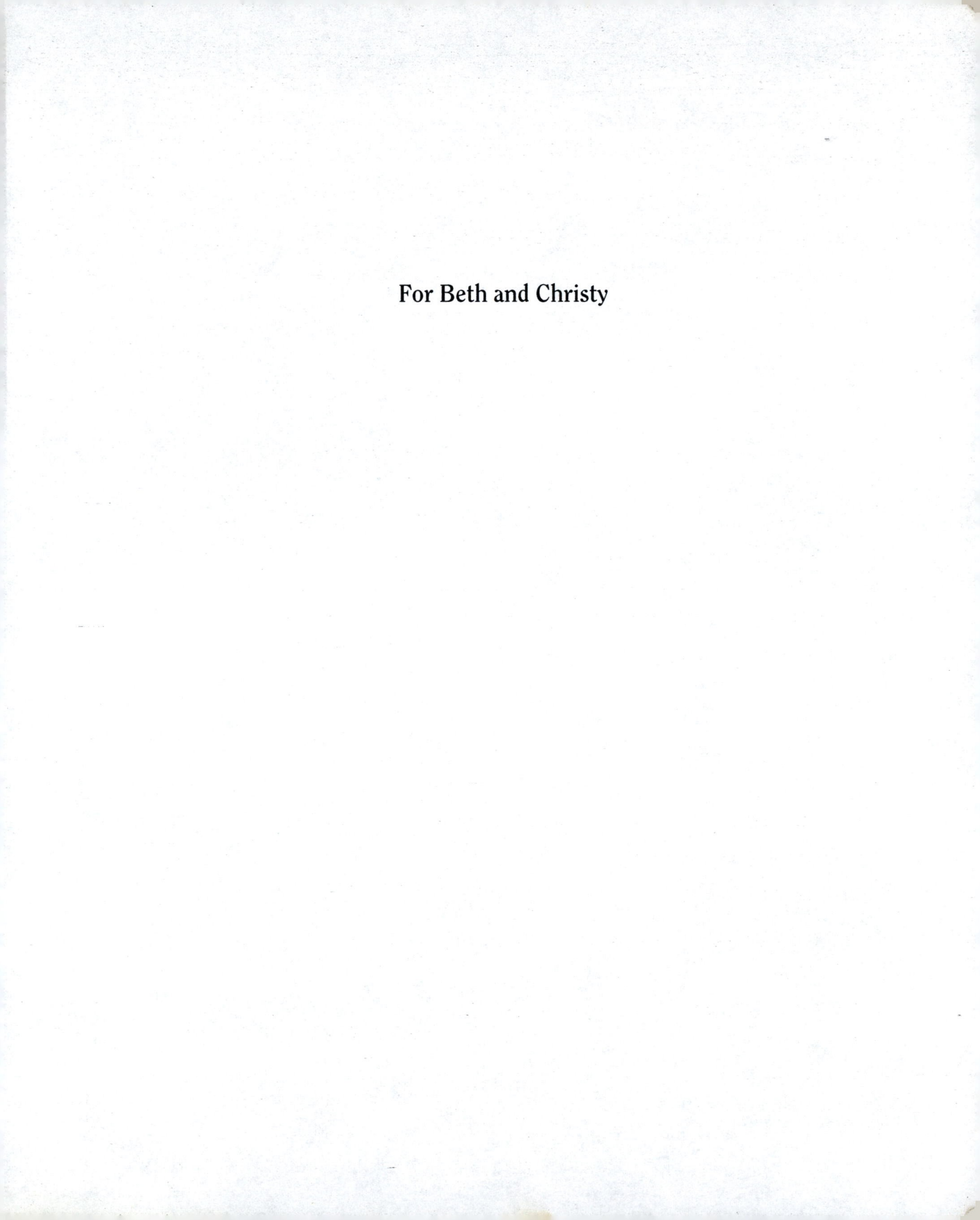

For Beth and Christy

Contents

Preface

As textbook authors go, we are relatively young. Capitalizing on our youth, this textbook brings a fresh approach to the subject of agricultural marketing and price analysis at a time when we believe a new text is particularly warranted. Agricultural economics departments have steadily evolved, but in different ways across universities. Some agricultural economics departments have merged with economics departments or changed their departmental names to include "applied economics" or "natural resource economics." For these departments, the distinction between agricultural economics and economics applied to other topics is less clear. Others have decided to place greater emphasis on agribusiness management and marketing. At these schools, lectures on economic theory are being replaced by practical agribusiness management topics. As a whole, agricultural economics departments today are less focused on the farm and instead pay more attention to the agribusiness sector and the consumer.

Yet, most schools still offer courses in basic agricultural price analysis and agricultural marketing. Instructors are expected to cover all the traditional topics in addition to emerging areas of research. An effective textbook therefore needs to reflect these changes. Refocused academic priorities are not the only changes that should be brought to bear on a textbook of this nature. With the greater prevalence of computers in the workplace, college graduates should be prepared to conduct sophisticated data analysis. Student preferences for writing styles have also changed. It is a commonly held belief that students are less inclined to read traditional textbooks; yet we have found that these same students will read less formal economics texts such as *The Armchair Economist*, *Freakonomics*, and *Naked Economics* with great enthusiasm.

This textbook discusses the topic of traditional agricultural marketing and price analysis, while being mindful of how the world has changed over the past several decades. We focus some complex topics including general equilibrium models, game theory, and econometrics; however, our aim in this book is to engage students with very little exposure to economics and with only a basic grasp of algebra.

The first section of the book, "Welcome to Economics and Price Analysis," introduces economics using a less formal—but hopefully more effective—writing style. Although the major focus and examples are agricultural in nature, we do not hesitate to use nonagricultural examples. Arbitrage in corn markets is discussed, as is arbitrage in major league baseball. This section then covers both basic supply-and-demand analysis and more advanced topics such as equilibrium displacement models and monopolistic competition. A unique feature of this section, and the book, is the attention given to equilibrium displacement models (partial and general equilibrium). These models can be of great pedagogical use but are also routinely used by industries and firms that employ our undergraduate majors.

"Understanding Agricultural Prices and Markets" includes the traditional topics of agricultural price seasonality, market adjustments, marketing margins, derived demand, and trade. In addition, an entire chapter is devoted to the use of regression analysis to study agricultural prices. Students are introduced to regression assuming no prior exposure; by the end of the chapter, they should be able to estimate short- and long-run meat demand functions, test whether advertising increases beef demand, employ time-series models to forecast prices, and conduct hedonic price analyses.

The third section, "Agribusiness Marketing Strategies," utilizes the economic concepts of the previous two sections to help the student understand and develop practical agribusiness marketing strategies. First, futures markets are described and used to illustrate how hedging, cross-hedging, and options can help managers reduce price variability. Agricultural economics students are increasingly being employed by large firms engaged in strategic competition. The chapter "Strategic Price Setting" uses game theory to uncover profitable price strategies in oligopolistic markets. Firms with market power can significantly enhance their profits via price discrimination and other pricing schemes. The chapter "Creative Pricing Schemes" covers first-, second-, and third-degree price discrimination, as well as bundling and tie-in sales.

Chapter 12 is perhaps the most unique chapter of the book. Agricultural economists are increasingly interested in the consumer end of the food marketing channel. As a result, there is an increased need to teach students about various models of consumer behavior as well as tools for eliciting consumer preferences such as experimental auctions and conjoint analysis. It is therefore prudent that these topics be included in an undergraduate textbook. Titled "Consumer Behavior and Research," this chapter allows the instructor to cover material typically found in marketing textbooks in a succinct and agriculturally relevant manner.

Some instructors may wish to cover how the firm behaves as a price taker and the history of agriculture in society. The "Additional Topics" section provides such material, containing "The Firm as a Price Taker" and "Agriculture and Society" chapters.

Due to the textbook's versatility, it could be easily adopted in a wide variety of undergraduate courses in agricultural economics. The main emphasis, however, is on agricultural price analysis, agricultural market structures, and agricultural marketing strategies. For introductory courses, we suggest relying heavily on Chapters 1 through 4, which cover core microeconomic concepts. For those teaching more advanced price analysis classes, Chapters 5 through 8 use a combination of graphs, algebra, and statistics to discuss more sophisticated models of agricultural prices and markets. Many advanced price analysis classes include futures markets, which are found in Chapter 9. A course focusing on agricultural marketing will also find Chapters 6, 8, and 9 appealing. Such courses might also cover Chapters 10 through 12 in order to prepare students for marketing in the agribusiness world. The last two chapters are added for teachers of introductory courses who wish to cover production economics and the history of agriculture.

We hope you find the friendly tone and engaging examples helpful in livening up the classroom, motivating the unmotivated, and creating a more enjoyable and effective educational experience!

Acknowledgments

Our first thanks go to our wives, who allowed us to sacrifice many weekends to complete this book. We are thankful to our bosses, including the department heads and deans at Oklahoma State University, who allowed us to momentarily stray from our research to write this book. Andrew Barkley, George Davis, and Kevin Dhuyvetter have our gratitude for the use of their data and figures.

Thanks to the reviewers of this book: Jeffrey Dorfman, University of Georgia; M. Darren Hudson, Mississippi State University; and Darrell R. Mark, University of Nebraska. Finally, our fortunate lot in life is largely due to our extraordinary teachers and mentors. These include Ted Schroeder, Michele Marra, and Sean Fox, just to name a few.

About the Authors

Bailey Norwood is an associate professor and Jayson Lusk is a professor and holds the Willard Sparles Endowed Chair of Agribusiness. Both teach in the department of agricultural economics at Oklahoma State University.

F. Bailey Norwood
Jayson L. Lusk

CHAPTER ONE

About Economics

> Did y'ever think, Ken, that making a speech on economics is a lot like pissing down your leg? It seems hot to you, but it never does to anyone else.
>
> —*Lyndon Baines Johnson,* speaking to economist J. K. Galbraith (Encarta Reference Suite 2000)

INTRODUCTION

Economists are not usually thought of as interesting people, and students rarely look forward to their first economics class. To add insult to injury, economics is often referred to as the "dismal science." Some of this reputation is deserved. Many class lectures are spent talking about interest rates, inflation, money supply, and prices. However much students love spending money, they are not crazy about studying money. Much of economics is fascinating though, and all of it is important (of course, we are a bit biased). Economists have found that seat belt laws led to more wrecks, each execution in the United States prevents approximately eight murders, you can make other people richer (though not yourself) by simply burning money, and one of the most popular girl's names in 2015 will be Clementine. All these are interesting topics, and lie within the realm of economics.

The title of this textbook is *Agricultural Marketing and Price Analysis*. Although economics is not in the title, this is an economics textbook. No economics textbook can cover everything, and we make no attempt to. The purpose of the book is to familiarize the student with how agricultural markets behave, how agricultural prices behave, and how firms in the agribusiness industry can use economics to set profitable prices and employ profitable marketing strategies. All material presented here is based on discoveries in the science called economics. It is an economics textbook for the student who wants to understand agricultural markets and work in the agriculture industry.

Because this book covers the economics of agriculture, it is necessary to discuss basic economics first. The purpose of this chapter is to introduce you to the broad and interesting field of economics by

1. defining economics and the questions economists address
2. discussing how economists think
3. presenting five important lessons of economics

Economics is broadly defined as the *study of the allocation of scarce resources*. Chances are, this doesn't mean much to you, so let us demonstrate with an example. The cow never came close to extinction but the buffalo did. Why? Both are resources in the sense that they are valuable (they both produce food, hide, and other useful things) and are scarce in the sense they are not free (they do not drop from the sky like manna from heaven). The reason buffalo almost went extinct but the cow did not is that there was clear ownership of cattle but not buffalo.

Consumers want beef today and they will want beef in the future. Ranchers understand this, and when they sell cattle they reserve some males and females for breeding. This is an investment. They forego the money they could have earned by selling the cattle today, electing instead to produce more by breeding them, earning greater profits at a later date. This is no different than you shelling out tuition money now in hopes of a higher salary after you have dressed up in a silly robe, received your degree, and thrown your graduation cap in the air. At first, it would seem that a similar argument could be posed for buffalo. Consumers in the nineteenth century wanted buffalo hides today and in the future, so it was in society's interest to not kill all the buffalo. By leaving some males and females alive, one could be assured of buffalo hides in the future. Despite this obvious truth, hunters did try to kill them all. The reason is that no one owned the buffalo, so if you decided to leave some males and females alive to breed, another person may come and kill them. Better you benefit from the kill than someone else. The same could *not* be said for cattle, because it was unlawful to steal or kill another rancher's cattle.

This is not just an interesting historical study. It has important implications for today. In many areas of Africa the elephant is endangered. Elephants are regularly hunted for their tusks, and since no one owns elephants, no one has the incentive to leave any alive for breeding. Just like the buffalo, an elephant left alive is an elephant for another hunter to kill and profit from. Lately, African countries have enacted numerous measures to protect the elephant. Zimbabwe now allows its farmers and herdsmen to own the elephants that roam their land. To hunt these elephants one must now buy a permit from the owners, and those permits can cost up to $25,000. By issuing property rights for the elephants, Zimbabwe now has too many elephants and must kill 5,000–7,000 each year just to keep the herds at sustainable levels. This is an excellent example of where the "study of the allocation of scarce resources" had a tremendous impact on the world. Economics perhaps saved the elephant from extinction! So you see, economics is more than just inflation and interest rates and is more than just a class you need for a degree. It covers a wide range of interesting issues and is perhaps one of the most important sciences today. To better describe the many faces of economics, consider the following examples.

After the Great Depression, when one quarter of Americans could not find jobs, economists sought to explain why the Depression occurred and how to prevent future depressions. We now understand much about the Depression, and although we cannot prevent all recessions, we feel we can prevent *great* depressions. These economists also study why Africa is so poor and America is so rich and what can be done about it. Those who study large economies like countries are in a field called *macroeconomics* (the "macro" meaning big economies). In the later part of the 1990s, the U.S. government took Microsoft Corporation to trial, arguing that Microsoft is a monopoly that illegally hinders product innovation. The jury had to decide whether Microsoft was indeed a monopoly and whether it illegally stymied technological innovation. To help the jury answer these questions, both prosecutors and defenders called on economists as expert witnesses. Economics applied to individual markets, a collection of markets, individual firms, and individual consumers is called *microeconomics*. Microeconomists also study individual behavior. Within microeconomics are many different fields, some of which are discussed below.

In March of 1989, the oil tanker *Exxon Valdez* briefly left its normal shipping lane to avoid icebergs but ended up running into submerged rocks and dumping 11 million gallons of oil into the Prince William Sound. In response, a class action lawsuit was filed against Exxon. But to sue Exxon for damages, one must first measure those damages. The oil spill directly harmed only a small portion of the human population, but a large portion of the population was outraged by the environmental disaster and was willing to pay some amount of money to clean the area. Chances are, you were not directly impacted by the oil spill, but you would be willing to pay one dollar to have prevented it from happening. Many people were indirectly harmed, and to measure exactly how much they were harmed—in dollars—the State of Alaska called on economists. These economists performed a nationwide study to estimate how much the country would have paid to prevent such an oil spill. They found the economic damages to the country as a whole exceeded $4.87 billion, which led lawyers to sue for $5 billion in punitive damages (Carson et al. 2003; Hirsch 2005). This same group of economists is currently calculating whether the benefits of preventing global warming outweigh the costs. As you might suspect, this area of economics is referred to as *environmental and resource economics*.

Strange as it may sound, there is even a group of economists who measure the neural activity of the brain as it undergoes economic activity. They use experiments to discover that people will extract vengeance on others for a perceived wrongdoing, even if extracting vengeance yields no tangible benefit. They fill the interface between economics and psychology—an area called *behavioral economics*. Those working in *labor economics* are busy measuring the income gap between men and women and whites and blacks, and are trying to understand what causes this gap. Finally, there is an area of economics that specializes in agriculture and food issues, an area called *agricultural economics*. The authors are agricultural economists. In 2004 a group of cattlemen filed a lawsuit against Tyson Foods alleging that Tyson used its market power to illegally lower the prices they paid for cattle. The jury had to decide whether

FIGURE 1.1 Result of the *Exxon Valdez* Oil Spill. The *Exxon Valdez* oil spill in 1989 dumped 11 million gallons of oil into Prince William Sound. As this photo shows, much of the oil was carried to shore. Economists were asked to measure in dollars how much people value cleaning up the sound.

Image courtesy of Office of Response and Restoration, National Ocean Service, National Oceanic and Atmospheric Administration.

Tyson indeed possessed such market power and whether they used this power to intentionally lower prices. Prosecutors and defenders both called on agricultural economists as expert witnesses.

As you can see, economics is a broad and relevant field. There are some areas that are strictly the domain of economics, like inflation. There are other areas where economists address the same problems as psychologists, sociologists, and political scientists. Economics is more like a set of tools than a specific object of study, which is why the line between economics and other areas like management and psychology can sometimes be blurry. Economics has many faces, and there is no doubt you will find at least one area interesting. Regardless of their diversity, economists have several things in common (besides being nerds). One is the way they think. Economists have a very specific way of thinking that has proven useful in explaining human and societal behavior. Both environmental economists and macroeconomists approach their problems with a similar frame of mind. To understand economics is to understand this way of thinking. It is impossible to fully describe the economic way of thinking in one chapter, but what follows is a sincere attempt to capture its most important features.

AN ECONOMIST'S FRAME OF MIND: THINK INCENTIVES, INTERACTIONS, AND INDIFFERENCE

The earth circles the sun every 365.25 days, the moon causes high tides every 12 hours and 25 minutes, and for every action there is an equal and opposite reaction. These physical laws govern the physical world. Economic activity is governed by its own set of laws. These laws do not strictly hold like the laws of physics—you can always count on gravity, but you cannot always count on economists being correct. But economic laws are forces that we regularly see exerted on people. Economic laws are like thermostats on heaters. The thermostat is not an absolute truth. If you set a thermostat at 72 degrees, no place in the room will it be exactly 72 degrees. But the thermostat—like economic laws—is a pervasive and important force. When the room is above 72, you can bet it will become cooler, and when below 72, you can bet it will become warmer. For this reason, we prefer to use the term *economic forces* instead of *economic laws*. Three of the most useful economic forces are the three I's of economic theory: *Incentives*, *Interactions*, and *Indifference*.

Incentives

People respond to incentives—incentives of every kind. Responding to incentives simply means people are trying to make themselves better off. If the government taxes more of people's income, people will work less. If gas prices rise, people will (eventually) drive less. Saying that people respond to incentives may not sound controversial, but where economists differ from the normal person is that, as Steven Landsburg states, economists "insist on taking it seriously at all times."[1] Politicians often raise taxes under the assumption that it will not change consumers' purchasing decisions. Economists disagree, arguing it will. Study after study has proven that the economists are right, even in seemingly bizarre cases. Matadors (Spanish bullfighters) earn money and acclaim by putting on a dangerous show. Matadors who take few chances of being mauled by the bull bore their fans. The fans boo them, and they are not asked to return to the arena. On the other hand, matadors who seem to escape death by a hair bring the fans to their feet, win the audience's applause, and get invited back to fight again.

The matador, then, faces a tough trade-off. He must weigh the benefits of the crowd's approval with the cost of being mauled. The expected cost of serious injury is the probability of being mauled times the health cost of the mauling. People respond to incentives, and the matador's performance depends on this cost. In recent years, surgeons have become more successful at treating complicated and serious gorings. As a consequence, in 2005, matadors have been more daring in their fights, leading to the bloodiest bullfighting season ever (Johnson 2005).

---

[1]Steven Landsburg is the author of the book *The Armchair Economist*, an extremely interesting read. Not written as a textbook, even people not fond of economics (or reading in general for that matter) find it hard to put down.

Higher taxes on alcohol influence people's consumption, and this has important effects on drinkers' behavior. Researchers have shown that regions with higher beer taxes also have lower rates of physical child abuse and child homicide deaths. Even one's proximity to a liquor store impacts the number of children killed. In regions with a high liquor retail density, child homicide rates rise (Sen 2006). Just making alcohol more inconvenient to purchase leads to fewer child deaths.

Economists Steven Landsburg and Charles Wheelan provide excellent examples illustrating the importance of incentives. Several of these examples are described below (Landsburg 1995; Wheelan 2002). The 1960s ushered in numerous automobile safety regulations, including the mandatory use of seat belts, padded dashboards, and penetration-resistant windshields. This made driving safer, even reckless driving. Since the risk of reckless driving decreased, economists predicted that (reacting to incentives) people will drive more recklessly. Indeed, the number of wrecks rose after these regulations. Does capital punishment deter murder? Yes, every execution in the United States deters eight murders. Even rats and pigeons eat less when the cost of procuring food rises.[2]

In East Berlin during the days of Soviet rule, car manufacturers were not rewarded for making dependable cars. The communist government told them how many cars to produce and how to produce them. The autoworkers' income remained the same no matter how well they did their job. The result? Many of the cars were of such poor quality you could get more money by destroying it and selling it for its steel than you could selling it as a drivable car. Compare this to market economies where businesses have an incentive to create high-quality cars, and that incentive is profits. Read the business section of any newspaper and there is always a merger going on. Studies have shown that corporate mergers rarely benefit stockholders; so why do CEOs pursue them? Because establishing control over a larger business means more prestige. As Charles Wheelan states, "Big companies have big offices, big salaries, and big airplanes."

Interactions

No man is an island, and often actions taken by one person affect another. In society, people interact. When people follow incentives, and in the process interact with others, this interaction often changes those incentives. Consider a checkout line in a grocery store. You see two lines. One line is shorter, so you proceed to the shorter line. Other shoppers follow this incentive, and as they do, the line becomes longer, taking away the incentive to choose that line. In 1990, Congress passed a tax on yachts in an effort to tax the rich. It did not work. The problem is that rich people do not have to buy yachts; they can spend their money on other luxury items. Another problem was that the yacht makers were not rich, and yacht factories can only produce yachts. Rich people had options to avoid paying the tax; yacht producers did not.

---

[2]For example, when rats must press a lever to dispense food, the harder the lever becomes to press, the less food they consume.

The incentive of the tax was to discourage rich people from buying boats, and that is what happened. When total yacht demand fell, yacht makers had to lower their prices to entice the rich people back. After yacht buyers followed their incentives, the interactions between yacht buyers and sellers caused the price to fall. The effect was to decrease yacht prices significantly, essentially taxing poorer yacht makers rather than richer yacht buyers.

Think back to the buffalo example. The hunter's incentive was to kill every buffalo he could. Following these incentives, hunters almost drove the buffalo to extinction. Each hunter's pursuit of their individual incentives had an effect on all other hunters and society in general. Individuals do not pursue incentives in isolation (no man is an island), since their actions affect others. To reiterate, we assume people take actions consistent with the incentives presented to them, and those actions may affect other people and prices.

Indifference

We have said that people respond to incentives and they interact. The yacht tax induced rich people to buy fewer yachts, but yacht makers needed to sell yachts so they lowered prices. If incentives lead to interactions between people, and those interactions change incentives, where does it end? Does the circle ever end? Yes, it does end, when incentives seem to disappear and people are indifferent about altering their behavior. This section describes the Indifference Principle as articulated by Steven Landsburg in his entertaining and thoughtful book *The Armchair Economist*. The Indifference Principle makes a bold statement: except when people have unusual tastes or unusual talents, all actions must be equally desirable. It describes an *equilibrium* where there are no incentives for people to modify their actions.

The Indifference Principle

Except when people have unusual tastes or unusual talents, all actions must be equally desirable. The Indifference Principle describes an *equilibrium*, where there are no incentives for people to modify their behavior.

Of course, the Indifference Principle does not always hold true. In many cases, you as a human being know exactly what you want to do. All actions are not equally desirable. If you are one of those face-painting college football fanatics, going to a college football game is the *only* thing you want to do on an autumn Saturday afternoon. You are not indifferent between going to a football game and going to a movie. But this is an unusual taste; most people do not love football this much. If you happen to be a star football player, you likely want to play professional football as a career. You are not indifferent between being an NFL linebacker and an accountant, but then, you have unusual talents. When people have roughly the same tastes and talents, the Indifference Principle does not hold perfectly, but the force striving for indifference is present and drives much economic behavior. Remember the analogy between economic laws and the thermostat. They do not hold perfectly, but they hold on average.

The best example of the Indifference Principle is a checkout line at Wal-Mart. You have filled your buggy with Wal-Mart items and now wish to check out. As you approach the checkout lines, you get lucky and spot a short line. As you proceed to that line, you and everyone else ready to check out head towards that line. Quickly, that line becomes the same length as the other lines. Now, a new shopper arrives wishing to check out but faces multiple lines of the same length. She is indifferent

People respond to **incentives.** They perform an action whenever the perceived benefit exceeds the perceived cost. For example, consumers buy a product when the perceived benefit exceeds the price.

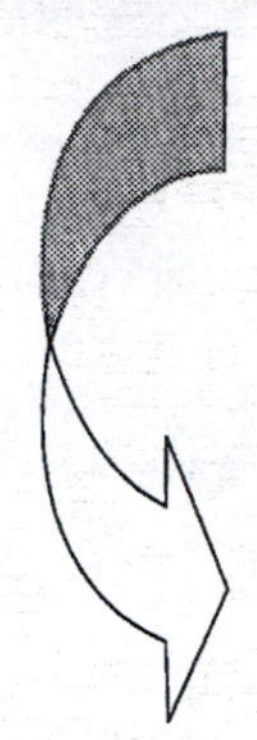

People continually respond to incentives, and their responses continually change incentives, and the circle repeats over and over until the **Indifference Principle** is satisfied.

The Indifference Principle states that all actions must be equally desirable. It is also referred to as an equilibrium. Once the Indifference Principle is satisfied, people have no desire to change their behavior.

The collection of people's actions (**interactions**) can change those benefits and costs. For example, consumers buy a product when the value exceeds the price, but greater consumer demand may drive up the price.

FIGURE 1.2 The Three I's of Economic Theory: Incentives, Interactions, and Indifference.

between which line to take, and the Indifference Principle holds. Now imagine yourself on a congested five-lane highway. The left-most lane looks the fastest, so you proceed to it. So do other drivers, further congesting the left-most lane until it is equally slow as the competing lanes. By each driver quickly moving to the lane that looks the fastest, they slow that lane down until it proceeds at the same pace as competing lanes. You are now indifferent between which lane to take. Wherever you look, the Indifference Principle is there.

Consider one more example. It is a cold day in the city. You were planning on visiting the zoo, which is outdoors, but due to the cold, indoor activities like visiting the local aquarium are more appealing. You are not indifferent between all activities; you would rather be at the aquarium than the zoo. However, other people are thinking the same thing. You all hit the streets and head towards the aquarium, only to find the crowd enormous. Seeing the wait to get in is two hours, the aquarium looks less attractive. Still you prefer going to the aquarium and head to the end of the line. As more people enter the line, the wait now becomes three hours. The line keeps growing until finally people decide the zoo is more attractive, despite the cold. But when people head to the zoo instead of the aquarium, the line at the aquarium shortens, and people start heading to the aquarium instead of the zoo. The line then becomes too long, and people lean more towards the zoo. It goes back and forth, back and forth. On average, the length of the aquarium line causes people to be indifferent between the aquarium and the zoo. The line is a thermostat: When the aquarium is preferred the line grows, causing people to prefer the zoo, and when people prefer the zoo, the line shortens, causing people to prefer the aquarium. But for the city as a whole, on average they are indifferent between the aquarium and the zoo.

Using the three I's to address real questions is not easy. Thinking like an economist is not easy at first. It takes practice, both in class and on your own. What follows are several more examples of the three I's. Hopefully, this way of thinking will become more natural, and you will find yourself thinking like an economist without even knowing it (and for that we apologize).

Does Farm Aid Benefit Farmers?

A farmer's income depends on the weather and the price farmers receive can fall greatly with little warning. Periodically, this leaves farms in financial trouble. Government has always tended to favor farmers over the average American, for a variety of reasons, and has responded by giving farmers extra money to cover these tough times. Some say the subsidies are due to the fact that farmers are a powerful lobbying group. Others say that government research aimed at an inexpensive food supply hurts the farmer by lowering the prices they receive, so to make up for it government should give them extra payments.[3] Whether it is fair that farmers get payments from the government and automobile mechanics do not is not the issue here. We are concerned with whether direct payments to farmers even benefit farmers at all.

This may sound silly. How can the government giving money directly to the farmer *not* benefit the farmer? To answer this, we must use our economic theory. Suppose a farmer named Wren rents her cropland from a landowner. Wren farms wheat, making $40,000 each year before paying land rent. After paying her rent of $5,000, Wren is left with profits of $35,000. Her best alternative to farming is to work in industry, which pays $40,000. However, Wren enjoys farming, and is willing to forego up to $5,000 (but no more) in profits to farm versus work in industry as a laborer. That is, the value of farming to her is $5,000 per year. At this point, she is indifferent between being a farmer and an industry worker. Both the value of farming and working in industry is $40,000.

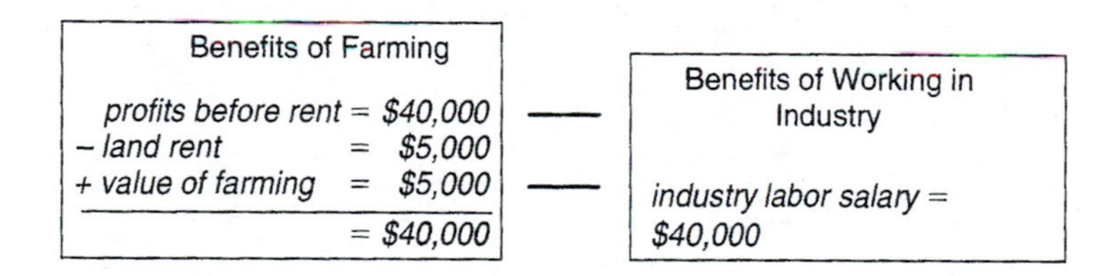

FIGURE 1.3 Wren is Indifferent Between Farming and Industry Labor.

[3]For example, the U.S. government has researched technologies for increasing crop yields for many years. As yields rise, prices fall, causing some farms to go bankrupt. If a technological advancement means it takes fewer farmers to produce the same amount of food, some farmers must be forced out of business. However, since the government policy adversely impacts farmers, some think it only right to make up for it through government payments.

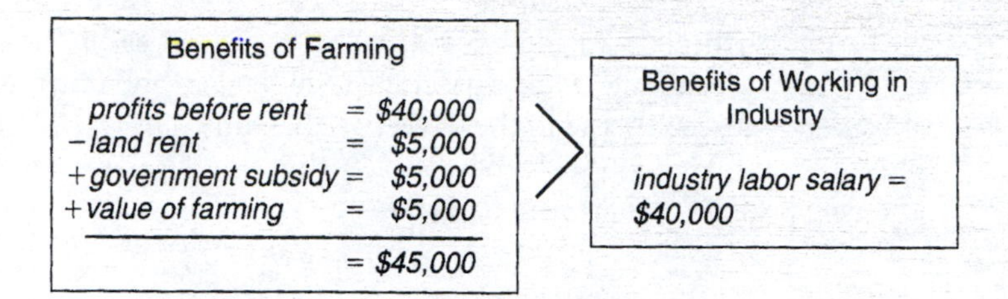

FIGURE 1.4 With the Government Subsidy, Wren Prefers Being a Farmer.

Now suppose that the government feels that Wren deserves more money and gives her a government subsidy of $5,000 per year. The benefits of farming are now clearly greater than she could earn in industry. She is no longer indifferent between the two jobs; she strictly prefers farming. However, this gives Wren's landowner negotiating power over rent. Before, if the landowner charged any more than $5,000 in rent, Wren would prefer being a laborer and therefore would not rent the land. But now, the landowner can charge more than $5,000 in rent and Wren will still rent the land. If the rent is raised to $6,000, she will still rent and the landowner will make more money. The same happens if the rent is raised to $7,000. The subsidy changes each party's incentive, and leads to interactions, which raises the price of land. The landowner will eventually charge her $10,000 in rent, because that is the maximum amount of rent that can be extracted with Wren still renting the land.

Something important happened. Every dime the government gave to Wren was passed on to the landowner in the form of greater land rents. Before the subsidy, Wren was indifferent between farming and industry labor. After the subsidy, she was still indifferent. The government program failed to achieve its goal and ended up subsidizing landowners. Of course, farmers who own all their farmland do benefit from the subsidy. But this is because they are landowners, not because they are farmers.

Steven Landsburg points out a second important lesson from the Indifference Principle: Only the owner of a fixed resource can avoid the indifference principle.

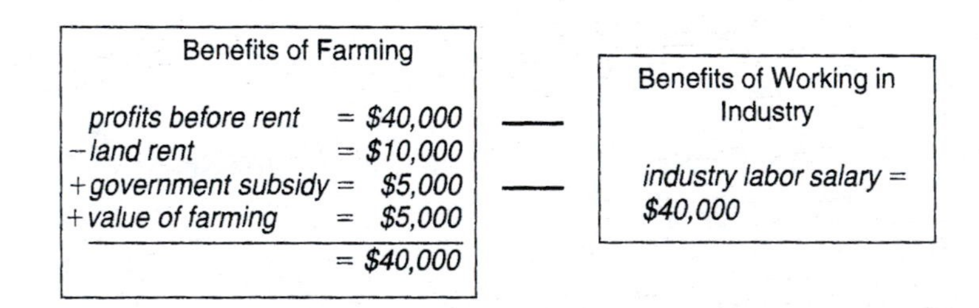

FIGURE 1.5 The Government Subsidy Raises Land Rent, Leaving Wren Indifferent Once Again.

The subsidy made the farmer no better off; she is indifferent between jobs as before. Conversely, the landowner is clearly better off by $5,000 each year and is not indifferent between being a landowner and not being a landowner. Assuming that the amount of land that can be farmed is fixed, the landowner owns a fixed resource and avoids the Indifference Principle.

Only the owner of a fixed resource can avoid the Indifference Principle.

Is this a plausible scenario? Yes, for some farmers. In Iowa, about 40% of all farmers rent the land they farm and government subsidies are responsible for about 45% of land values (Duffy and Holste 2005). Think about what this means. If you are the owner of farmland, about 45% of all the rent you receive is attributable to government payments to the people who farm the land. That's quite a lot.

Does Genetically Modified Seed Benefit Farmers?

Bt cotton is a cotton variety that has been genetically modified to produce its own pesticide. By no longer having to spray expensive pesticides, using Bt cotton could greatly lower production costs and food prices for consumers. The operative word is "could." If Bt cotton seed costs the same as regular cotton seed, farmers would undoubtedly experience lower costs. The production of Bt cotton seed is patented by Monsanto corporation. Monsanto and only Monsanto can sell it. This gives them a monopoly for Bt cotton seed, and if their seed does save farmers pesticide expenses, they likely seek to capture some of these savings by charging a higher price than regular cotton seed.

Indeed, if the three I's of economic theory holds, we would expect farmers to be indifferent between planting regular cotton seed and Bt cotton seed. Remember that only the owner of a fixed resource can avoid the Indifference Principle. Farmers will be indifferent between the two seeds, but Monsanto will clearly not be indifferent as to which they prefer selling. If farmers' pesticide expenses fall by $20 per acre from the use of Bt cotton, we would expect the price of planting Bt cotton to be $20 per acre more than regular cotton seed. By charging $20 per acre Monsanto has ensured itself of maximum profits. Farmers will pay no more than $20 for the seed, but at $20 will still plant it. In reality, Monsanto may charge a little less than $20 just to make sure that farmers slightly prefer Bt cotton seed to regular seed. However, they will not want to charge much less than $20 because that will only lower Monsanto's profits. Of course, they would not want to charge more than $20 either, because no one would purchase it.

The data support this notion, but suggest that Monsanto may have overestimated the savings in pesticide costs from Bt cotton seed (Qaim and De Janvry 2003). A study on the adoption of Bt cotton seed in Argentina reveals that the benefits to farmers adopting Bt seed is small. Surprisingly, it was found that Monsanto overcharged for Bt cotton seed. Had Monsanto lowered their seed price, its profits could have been 3.6 times higher, the farmers purchasing the seeds would have higher profits, and consumers would face lower food prices. Monsanto simply overestimated the benefits of Bt cotton seed, and in an effort to extract the maximum profits possible (leaving farmers with little or no benefit from the genetically modified seed) it priced the seed too high. Recall that the Indifference Principle does not hold perfectly. Rather, it is a force, much like a thermostat. Just like the temperature in a room will occasionally be higher than the thermostat setting, the seed price in this example is above that predicted by the Indifference Principle.

Previously we saw the price was higher than what the Indifference Principle would predict, but in other cases it is lower. In keeping with our previous example, suppose the adoption of Monsanto seed makes the farmer $20 per acre if the seed were free. The Indifference Principle suggests Monsanto would charge a price very close to $20 per acre for the seed, but because people care about fairness it might be less. There is a famous game called the ultimatum game. It consists of two people: the allocator and the receiver. The allocator is given $10 and is asked to offer the receiver a portion of this money, any portion the allocator wishes. The receiver then has the option of accepting or rejecting the offer. If the receiver accepts the offer, both keep their money. For example, if the allocator splits the $10, giving the receiver $5, and the receiver accepts the $5, both get $5. However, if the receiver rejects the offer, neither get the money. If the allocator offers the receiver only $0.25 of the $10, and the receiver rejects the offer, neither will receive any money.

If people did not care about fairness, the Indifference Principle would predict that the allocator would give the receiver one penny and the receiver would accept it. The receiver prefers one penny to no money, so the receiver would accept the penny, and knowing this, the allocator maximizes her profit by offering only one penny. However, this is not what we observe when the game is actually played. Allocators never offer one penny because no receiver would accept only a penny. Receivers feel equally deserving of the money. After all, the allocator did not earn the money. Thus, a very unequal split is considered unfair. Receivers place more value on punishing an unfair offer than receiving a small sum of money. This experiment has been conducted hundreds of times in different cultures, and on average the allocator gives 40% of the money to the receiver. While the proportion varies, allocations outside the 30% to 50% range are rare.

Now back to the Monsanto seed example. If Monsanto seed saves the farmer $20 per acre in costs but Monsanto asks $19 per acre for the seed, farmers may be quick to reject the offer. They may reject it because the benefit ($20 − $19 = $1) is too little for the hassle or because the offer is deemed unfair. For this reason, it is said (though we cannot provide any citation for this; our source requested confidentiality) that Monsanto seeks to charge a price equal to two-thirds of the benefit. If the benefit of the seed was $30 per acre, Monsanto will ask around $20 per acre because that is the maximum price farmers will pay. Does this violate the Indifference Principle? No, it just means that farmers are indifferent between purchasing the seed when the seed price equals two-thirds of its benefit. The Indifference Principle is driven by more than money. Concepts of fairness matter as well.

The ~~Law~~ *Force* of One Price

The *full price* paid for an item can be broken into two components: the transaction price and the transaction cost. The transaction price refers to the money the buyer gives to the seller, and the transaction cost refers to any other costs the buyer or seller incurs in the transaction. Suppose that you live in Raleigh, North Carolina, and wish to purchase a car. There are many car lots in Raleigh, or you could go to Charlotte, North Carolina, where there are even more. The *transaction price* of cars may or may not be the same in both towns, but the *transaction cost* is not. It costs you money and time to travel to Charlotte. Therefore, the transaction cost of purchasing a car in Charlotte is greater than in Raleigh (if you live in Raleigh).

Full Price = Transaction Price + Transaction Cost

Abstract from reality for the moment and suppose that all transaction costs were zero. In this weird world, it is just as convenient to purchase a television in San Diego, California, as in Montreal, Quebec, regardless of where you live. Consider any good, and ask yourself how the *price* of an identical good would vary across regions in this fictitious setting. If you said the price of all identical goods would be the same—brace yourself—you are starting to think like an economist! The reason is that prices should make you indifferent about where you purchase. If transaction costs are zero and the price of cars is lower in Charlotte than Raleigh, everyone will purchase cars from Charlotte. Charlotte car dealers will increase their price in response to greater demand and Raleigh dealers will lower their prices to entice their customers to return. This continues until prices are the same in all regions, and people are indifferent between purchasing in Charlotte or Raleigh. This is referred to as the *Law of One Price*.

Law of One Price: If transaction costs are zero, the transaction price of identical goods should be the same across regions.

This law holds only in fictitious worlds. In the real world transaction costs are never zero. It always takes time to shop and money to travel. Thus, the transaction cost of purchasing a product will differ depending upon where the item is located. Many prices are higher in the city than surrounding rural areas. Yet, city residents are not indifferent about where to purchase most of their goods. They purchase most of their goods within their city because the transaction cost of purchasing from rural areas is large (e.g., gas and time consumed in driving to the rural areas). Although the Law of One Price does not hold perfectly, it is a persistent force, keeping prices between regions similar. For this reason, it is better described as a *force of one price*.

Consider the following map, showing the difference in prices farmers receive for their wheat crop in selected Midwestern states. All price differences are relative to the price in the eastern portion of the states where demand is greatest. For example, farmers in western Kansas receive about −$0.89 less than their counterparts in the eastern portion. In southeast Oklahoma, farmers receive $1.01 more per bushel than farmers in western Kansas (0.30 − (−0.89) = 0.71). Clearly, the law of one price does not hold. The reason for the price difference is that most wheat is sold near river ports in eastern Kansas and eastern Oklahoma. Thus, if you purchase wheat in Dodge City (western Kansas) to resale in Kansas City (eastern Kansas), you must pay the transaction costs of transporting the wheat across the state. The Indifference Principle states that wheat buyers should be indifferent between buying in Kansas City and Dodge City. How can this occur if purchasing wheat from Dodge City incurs greater transportation costs? The answer is that the transaction price in Dodge City must fall just enough to offset the transportation costs. If prices are $0.50 lower in Dodge City compared to Kansas City, you can bet that the cost of hauling wheat across the state of Kansas is around $0.50 per bushel.

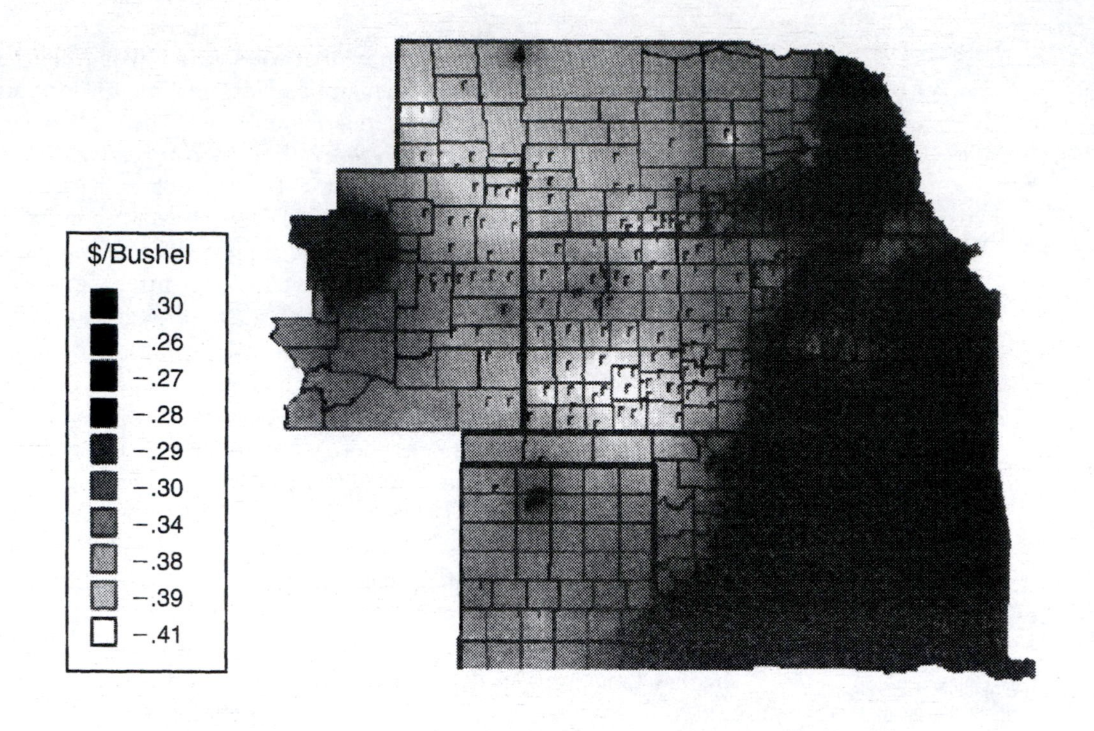

FIGURE 1.6 Price Differences for Wheat in Selected Midwestern States. All differences are relative to river ports in eastern Kansas and Oklahoma (a negative difference means the price is lower relative to river ports).

Map made available by Agmanager, Department of Agricultural Economics at Kansas State University. Available at http://www.agmanager.info. Accessed September 20, 2005.

Imagine if this was not true. Suppose that it costs $0.50 per bushel to haul wheat from Dodge City to Kansas City. The price in Dodge City is $3.00 and the price in Kansas City is $4.00. The price difference is larger than the transaction cost of buying in Dodge City and selling in Kansas City—you can profit from this. Simply purchase wheat in Dodge City for $3.00, haul it to Kansas City, and receive $4.00 per bushel. After paying the $0.50 per bushel transaction cost, you are left with $0.50 per bushel profit. This is referred to as *arbitrage*: profiting from price differences across markets.

Arbitrage: The act of profiting from price differences across markets. Arbitrage can occur across different time periods and regional markets.

Making profits from price differences is great. More money is always great. The problem is that these prices are publicly known and everyone likes to make profits. There are many businesses that seek these arbitrage profits. For example, an investment fund named Long-Term Capital Management would keep track of the price of similar investments. If investments are similar, they should sell for similar prices, at least most of the time. As soon as they saw the prices diverge, they would purchase the cheap asset. They anticipated that the prices would soon converge again, due to the Force of One Price, and the investment they purchased would rise in value, after which they would sell it to make a profit (Surowiecki 2004). There are even businesses that set up "robo-traders," which are computers that trade things like commodities and stocks. These computers search for arbitrage opportunities, and when they are found, the computers execute trades that make their programmer profits (*The Economist* 2005). They scour market information looking for price differences large enough to yield profits. They make money, lots of money at times, by betting that the Force of One Price exists.

Think about what happens when these firms and computers aggressively behave in arbitrage. Refer back to the example where it costs $0.50 per bushel to haul wheat from Dodge City to Kansas City; the price in Dodge City is $3.00, and the price in Kansas City is $4.00. Profit-seekers will buy wheat in Dodge City and sell it in Kansas City at a profit, and they will keep doing this as long as the price difference is greater than the transportation costs. As they continue to buy more and more wheat in Dodge City, they bid up the price in Dodge City, and as they sell more and more wheat in Kansas City, they will bid that price down. The Dodge City price rises and the Kansas City price falls. This continues until the price difference no longer allows arbitrage profits. At this point, the price in Dodge City must be exactly $0.50 per bushel less than in Kansas City, because the transportation cost between the two regions is $0.50 per bushel. So long as people pursue profits, the Indifference Principle ensures that the Force of One Price holds.

The Force of One Price: The price difference of identical goods in two regions must not exceed the transaction costs of buying the good in one region and selling it in another.

Thus, the price differences for wheat in Figure 1.6 are not random. The differences occur for a good reason. The Indifference Principle allows price differences to differ, but only by so much. They can never exceed the transportation cost between two regions. If they do, arbitrage opportunities arise. People follow their *incentives*, which lead to *interactions* causing prices to quickly change, and soon the price differences are less than transportation costs once again and the *Indifference* Principle holds. Evidence for this is clearly given in Figure 1.6. The further away one is from the major wheat-purchasing sites (eastern Kansas), the lower the price of wheat. The explanation is simple. Wheat buyers located in eastern Kansas must be indifferent between purchasing wheat in eastern or western Kansas.

Many agricultural economics graduates go on to become commodity traders. Before they purchase their first house, they may handle thousands of dollars of commodities for their companies. Some traders spend most of their day arbitraging. They look for price differences that should not exist if the Indifference Principle holds, and make their market transactions assuming it will hold in the near term. Meet one of these traders, one of our former students, Tim Cassidy.

The Price of Grain Between Harvests

For most crops, there is a year lag between harvests. Soybeans are harvested around November and no more soybeans will be available until the next November.[4] However much we harvest this November is how much we have to consume until the next November. Therefore, it is important that we store enough grain throughout the year to meet our food needs.

[4]We are ignoring harvests from countries south of the equator, which are small.

Tim Cassidy, a native of Wyoming, graduated with a bachelor's degree in Agricultural Economics from Oklahoma State University in 2005. Soon after graduation he went to work for the Archer Daniels Midland (ADM) Company as a commodity trader. Tim earns his salary by arbitraging for ADM. Each day at work, Tim scouts the differences in soybean, corn, wheat, and soybean meal prices across regions, and if profit opportunities present themself, he places orders to buy or sell thousands of dollars of a commodity. Not every arbitrage will make money, but if one understands market forces, significant profits can be made over time.

> "Within weeks of my first job out of college I was managing thousands of dollars of soybeans at one time. At first I was overwhelmed with the responsibility, but by holding fast to the principles I learned in agricultural economics, I have learned to do my job well and have a lot of fun in the process!"
>
> —Tim Cassidy

A Word from ADM

Archer Daniels Midland Company, an industry leader in agricultural processing and fermentation technology, recruits qualified candidates for internship and entry-level positions. Our internships are designed to give students meaningful work experience and business exposure through a global leader in the agricultural industry. We actively recruit for positions in the following areas: Accounting, Commodity Trading/Elevator Management, Engineering, Grain Terminal Operations Management, Internal Audit, Information Technology, and Specialty Areas (i.e., Finance, Marketing, R&D, Human Resources). To learn more about opportunities at ADM, please visit us at http://www.admworld.com under Careers.

FIGURE 1.7 Meet Tim Cassidy.

How are storage decisions made? Your government does not store grain to make sure we do not run out before the next harvest. In fact, no one stores grain out of kindness and a concern for the nation's food supply. Storage decisions are made based on profit motives. There are thousands of people across the country with grain storage capacity, and they will not store grain unless they feel they can profit from it. If storing grain is profitable, it will be stored. If storing grain is not profitable, the grain will be sold for current consumption.

Think back to the Indifference Principle and ask yourself: At any point in time, will people want to store grain or not store grain? If you answered they will be indifferent between storing or not storing, congratulations! Consider the following thought experiment. The price of soybeans is currently $5.00 per bushel, and it costs $0.10 per bushel to store grain one month. This storage cost includes both the physical cost of storage capacity, insurance, as well as the opportunity cost of money (you could sell now, and invest the $5.00 per bushel, earning interest for one month). Note that the physical storage costs cover the wages and salaries of those maintaining storage facilities.

If you expect the price of soybeans next month to be $5.30, you will store the grain because you expect to make $0.20 of profits. This is another example of arbitrage, where you are profiting from price differences across time. Assuming people form roughly the same expectations about future prices, everyone will begin storing soybeans. There will be less soybeans today, causing today's soybean prices to rise. Plus, when it becomes known that everyone else is storing soybeans, you realize there will be more soybeans next month and realize that next month's prices will actually be lower than $5.30. Today's price rises and next month's expected price falls until the price difference is only $0.10. That is, prices will change until the price difference between months exactly equals storage costs.

The Indifference Principle implies a very powerful statement, that *the price difference of a storable crop in different time periods should equal the cost of storing the crop between those time periods, so long as there is no new harvest*. Once a new harvest arrives, there is little incentive to store the crop any longer. Storage entails costs, and you have a brand-new harvest, so the linkage between prices across different time periods is broken once a new harvest is underway. This is not just a conjecture; it is revealed in real-world data.

The Indifference Principle can also be used to articulate how prices for crops in different time periods should change as storage costs rise or fall. Suppose storage costs rise. This immediately implies that the price difference for corn in May and June should now be larger compared to when storage costs were lower. If storage costs are larger, then prices must rise throughout the crop year by a greater amount than before to compensate people for storing the crop.

From society's point of view, these storage decisions are very important. We depend upon them for our food. Is it a good idea to rely on people seeking profits to

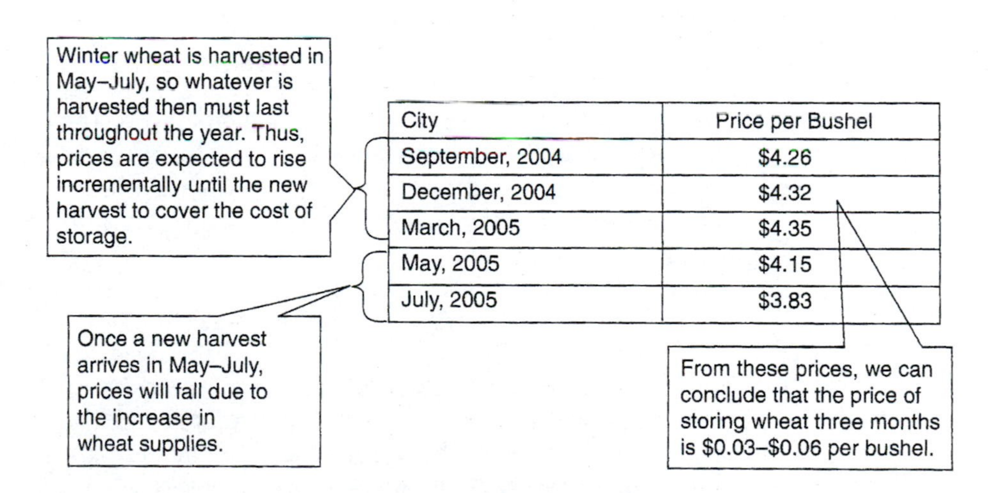

City	Price per Bushel
September, 2004	$4.26
December, 2004	$4.32
March, 2005	$4.35
May, 2005	$4.15
July, 2005	$3.83

FIGURE 1.8 Price of Winter Wheat in Kansas City.
Source: *Tulsa World.* March 31, 2004.

. . . man has almost constant occasion for the help of his brethren, and it is in vain for him to expect it from their benevolence only. . . . It is not from the benevolence of the butcher, the brewer, or the baker, that we expect our dinner, but from their regard to their own interest.

Adam Smith. *The Wealth of Nations.* Chapter 2.

Adam Smith in 1776 gave birth to economics in his book *The Wealth of Nations.* The A&E channel listed Adam Smith as the twentieth most influential person of the second millennium.

FIGURE 1.9 A Few Words from the Father of Economics.
Image made available by the Adam Smith Institute. Available at http://www.adamsmith.org/.

ensure we store enough grain? Perhaps a better question is: Can you ever recall a food shortage occurring? Not anytime recently in the United States. The reason is that the profit motive, however selfish it may be, causes price differences in different time periods to equal storage costs (so long as no new harvest arrives). This means that the price difference will pay for the wages and salaries of workers involved in maintaining storage facilities, as well as all other storage costs. Arbitrage and the seeking of profits ensure an adequate supply of food throughout the year. Markets have no heart or compassion for people, yet they do a pretty good job of providing our needs. This is summed up nicely by the father of economics, Adam Smith.

How the Red Sox Won the World Series

How do we measure a baseball player's ability? Mark McGwire and Barry Bonds are obviously good players because they seem to hit home runs every time they step up to the plate. Thus, they get paid more than other players. Professional baseball team managers must measure performance carefully, because if they offer players less than what they are worth, they will lose them to other teams, and if they overpay certain players, they have less money to hire good players for the remainder of the team.

One measure of performance is the *slugging percentage*, which is the total number of bases reached divided by the number of times at bat, not counting when players are walked. Players with a higher slugging percentage are paid considerably more. However, slugging percentage has the drawback that it does not count walks. Surely, a player with a good eye for bad pitches that enable him to make it to first without hitting the ball improves the team's winning ability, yet this is not accounted for in the slugging percentage. It is accounted for in the *on-base percentage*, which is the number of times a player makes it to first base (or further) divided by the number of times at bat.

It turns out that a high on-base percentage has a much larger impact on winning percentage than the slugging percentage. Yet, prior to 1999 a player's slugging percentage had a much larger impact on their compensation than their on-base percentage. In short, teams were overpaying players with high slugging percentages and underpaying those with high on-base percentages. One commodity was undervalued by the market and another was overvalued. There was room for arbitrage, but at the end of the 1990s, the Oakland A's were the only ones who knew.

Around 2000, the Oakland A's began recruiting players with high on-base percentages, paying them more than other teams, while focusing less on players with high slugging percentages. Essentially, they bought players with high on-base percentages and sold those with high slugging percentages. This is arbitrage, purchasing things that are undervalued and selling things that are overvalued in the market. As a result, in 2001–2003 Oakland had one of the highest winning percentages in the country, while at the same time having one of the smallest budgets. Clearly, Oakland had smart managers, in particular, two young Ivy League graduates who were hired due to their quantitative backgrounds. Other teams soon noticed Oakland's success and began arbitraging. One of the Ivy Leaguers was hired away by the Toronto Blue Jays and the other by the Los Angeles Dodgers. The Boston Red Sox hired two well-known sabermetricians (statisticians of baseball) to incorporate a similar management style. The Red Sox won the World Series in 2004.

Players with high on-base percentages began making more money, because all teams were offering them more money. Soon there were no arbitrage opportunities available. Players are now paid an amount closer to their "real worth," whereas before some were underpaid while others were overpaid. It was arbitrage that brought the player's prices closer to reality. And just as large agricultural firms such as ADM make profits by buying commodities that are undervalued and selling commodities that are overvalued, the Oakland A's increased their winning percentage (and consequently their profits) by doing the same for baseball players (Hakes and Sauer 2006).

When Will We Run Out of Oil?

The world consumes approximately 29 billion barrels of oil each year. There are about 1,278 billion barrels of proven oil reserves (Lomborg 2001). When will we run out of oil? The conventional answer is to divide 1,278 by 29 to get 44 years. That answer is wrong because it ignores how incentives change when oil runs low. The correct answer is we will never run out of oil. There will come a day when we quit extracting oil, but we will never run out. There are at least three reasons that oil will always exist. First, oil explorers respond to incentives. Although there is surely a finite amount of oil on the planet, the amount of oil reserves constantly changes. Some oil reserves are harder to find than others. We find the most noticeable oil reserves first, then move on to the more hidden reserves. It only becomes profitable to locate the harder-to-find reserves when prices rise. For example, in the mid-1970s, there were only about 650 billion barrels of proven oil reserves, but today there are more than 1.2 billion barrels of reserves. Known oil reserves have actually become more abundant over time!

Second, as oil becomes scarcer, the incentives for people to invent and invest in technology increases. Consider the fact that the average car in the United States is 60% more fuel efficient than it was in the early 1970s. In fact, the technology exists today to produce cars that get upwards of 200 miles per gallon, but these cars are expensive; people will only buy such cars when the price of oil rises to a much higher level and they become indifferent to the older-style cars and the new high-mileage cars. Finally, oil is not the only source of energy. The more we extract, the harder and more expensive it becomes to find and retrieve more oil. The more oil we consume, the more expensive it becomes to acquire more oil. Right now, alternative energy sources, such as wind energy, are 30% to 50% more expensive than oil. At some point, we will become indifferent between searching for more oil and using these alternative energy sources. Eventually, it will become so expensive to extract oil and there will be much cheaper alternatives; the result is that unused oil will remain under the earth.

SIX ECONOMICS LESSONS

The economic theory of the three I's gives you a basic idea of how economists think. It by no means describes economics fully. Economics has been an evolving science since its beginning. Although we have much yet to discover, there are several lessons that can be drawn from this 200 plus years of thinking. These lessons are from the microeconomic side of economics.

Economic Lesson 1: Think Toys, Not Dollars

Most of the readers of this book are not children, so let us qualify what we mean by "toys." Toys in this textbook refer to anything that you get pleasure from consuming. It is obvious that a Playstation, motorboat, or high-definition TV is a toy. Leisure is also a toy, so long as you enjoy it. Houses may not seem like fun, but you enjoy life more living in a house than in a tent, and so houses constitute toys as well. Even hiring someone to rake your leaves can be deemed a toy in economists' eyes because it allows you to consume more leisure. We use the word *toys* instead of "goods and services" to stress the fact that goods and services contribute nothing unless their consumption makes us happy. However, keep in mind that these *toys* are not meant to represent hedonistic desires. Mother Teresa consumed very little in her life, electing instead to give to others. All the charity work she performed and all the money she gave to charity were her toys. Some get pleasure from watching TV, while others get pleasure from giving to others. Anything that makes you happy is a toy.

As a country we often measure our standard of living according to dollars. The measurement usually used is gross domestic product, but this does not measure everything that makes us happy. Both of the authors attended school in Kansas, which is known for violent hailstorms. Homeowners often find themselves replacing their roofs after a big storm. After a particularly big hailstorm, one of the authors overheard someone say that the hailstorm was good for the state. By damaging roofs, the hail created business for construction companies. The money the roofers make would then be spent in the local community, providing economic benefits everywhere. The basic argument is that destroying property creates economic growth in an effort to repair that property, which benefits the economy as a whole. If this is true, then during the Great Depression when 25% of the workforce was unemployed, governments should have demolished neighborhoods and burnt down houses to spur economic growth.

This type of thinking is often called the *broken window fallacy* and represents an ill-founded confusion on the importance of toys and dollars. True, hail-damaged roofs may increase the number of people working and even profits made by a community. However, people will not have as many toys. They will be busy replacing the toys they once had. Suppose you owned a 2001 Honda Accord, and we came to your house and destroyed it with a steam roller, just to be jerks. You need a car, so you work overtime to purchase another, and end up buying a 2006 Honda Element, which you like better than your old Accord.

Did our destroying your car make you better off? You made more money, but you are clearly not better off. You had to make more money to replace the car you had. But you might argue, "I like the new Element better than my old Accord, so I must be better off." No. You could have sold the old Accord and purchased the new Element, leaving you exactly as you are now plus with the money you made from the sale. The fact that you did not do this voluntarily indicates that we did not make you happier.

When it comes to international trade, nations often want to export more than they import under the perception that they make more "profits" that way. For an individual business exporting more than one imports means the business is turning a profit. However, this analogy does not extend to nations. Suppose that the United States exports cars to Belize in exchange for bananas. Exports and imports are measured in currency, so let's just use dollars. Normally exports will equal imports. For example, we will export $5 million worth of cars to Belize in return for $5 million worth of banana imports from Belize.

But country leaders often want exports to exceed imports, as if that is the key to becoming richer. Suppose the United States pursues this strategy and exports $5 million of cars in exchange for only $3 million of bananas. Exports exceed imports by $2 million, but what really changed for Americans? We sold more than we bought, but all that really changed was that Americans consume fewer bananas. We have fewer toys. A desire for more exports than imports confuses dollars with toys, and it is toys that make people happy.

Nation leaders often buy into this export-driven strategy. China is a good example. We import many goods from China, and being Americans, pay for the goods in U.S. dollars. China could use those dollars to import goods from the United States but instead locks away some of those dollars to limit imports and ensure that their exports exceed imports. China as a whole does not benefit by this. By locking away this U.S. currency they deny themselves American products and their number of toys is fewer than it could be.

People often like to compare dollars across different years and places. For example, in 1960 the per person income was around $3,000 per year, and in 2005 it was $42,300. However, a dollar in 1960 purchased more toys than a dollar in 2005, so we are not now as rich as these two numbers would imply. Since 1960, the price of all goods and services has risen each year, so a dollar purchases less each year. This is called inflation. The U.S. government maintains a statistic known as the consumer price index (CPI) that measures inflation. The CPI can be used to take dollars in one year and calculate its equivalent purchasing power in another year. This method is covered in Chapter 6. Using the CPI, we discover that the per person income of $3,000 in 1960 is equivalent to an income of $20,000 in 2005. As far as ability to purchase toys, people in 2005 *are* richer than those in 1960, but only 2 times richer, not 14 times richer.

The same can be said for dollars in different locations at the same time period. One of our students was offered two jobs, one in Tulsa, Oklahoma, and one in New York City. The Tulsa job paid $40,000 per year, while the New York job paid $55,000 per year. Based on dollars alone the New York job pays more, but remember to think toys and not dollars. It is well known that toys cost more in New York. After using her economic skills gained in college she calculated that a salary of $55,000 in New York buys the same amount of toys (transportation, boarding, food, clothing, etc.) as $23,000 in Tulsa. Clearly, the Tulsa job gives her more toys.

Farmers clearly understand the difference between dollars and toys. Farm incomes fell substantially during the Great Depression, leading government to introduce a series of programs to enhance farm income. Some of these programs focused on obtaining parity prices for farm products. The period 1910–1914 was a good time for farmers. The price of agricultural commodities was higher than it had ever been, and farmers could buy more toys from selling one bushel of corn than ever before. In the 1930s farmers wanted the government to enact programs so that one bushel of corn could buy as many toys as it did during the 1910–1914 era. This is a *parity price*, a price in one time period that provides the same purchasing power as another time period. The purchasing power of a dollar had declined from 1914 to 1930, so farmers knew that the parity price calculation should take inflation into account. That is, things were not as simple as saying, "the price of corn was $1.50 in 1910–1914, so we should make sure the price is $1.50 in 1930." Instead, so that one bushel of corn could purchase the same number of toys, due to inflation, the price would have to be higher than $1.50 in 1930.

Economic Lesson 2: Beware the Law of Unintended Consequences

Often, government imposes regulations with the best of intentions, but those regulations can change people's incentives, leading to an undesirable outcome. Perhaps the best example occurred in Mexico City, where local officials attempted to curb air pollution from automobiles through regulation. The regulation required that all cars stay off the streets at least once a week. The particular day on which the car could not be driven depended on the license plate number. The intention was to decrease automobile use, which would lead to lower automobile emissions and cleaner air. Unfortunately, the opposite occurred.

Law of Unintended Consequences: When a policy creates perverse incentives, leading to an outcome the opposite of the policy's intent.

Many Mexico City residents found the regulation inconvenient and decided it was worth it to them to purchase a second car so that they could drive on any day of the week. The problem is that if you are to purchase a second car that will only be driven on one day of the week, you are likely to purchase an old, cheap car, which incidentally creates more air pollution than newer cars. As a result, because there were more old cars on the road, gasoline consumption rose, and the air became dirtier. The law seeking to reduce air pollution led to an unintended consequence: greater air pollution (Preston 1996).

Policymakers often fail to consider how regulation alters incentives. After the *Exxon Valdez* oil spill, many states created laws placing unlimited liability on tanker operations. This means that in the next oil spill, there is no limit to how much one could sue the company. The intent was to send a clear and direct signal to the oil companies that no more oil spills would be tolerated. No company wants to face unlimited lawsuits, and so oil companies are less inclined to ship oil because they face greater risks. One alternative is to pay another company to ship the oil for you, which is exactly what the Royal Dutch/Shell Group did. This company possessed 46 of the most modern and safest oil tankers, but given the threat of unlimited liability they chose to hire less reputable oil tankers. These less reputable oil tankers with less reliable ships have a higher likelihood of creating a spill (Norton 2005).

As this chapter was being written, Hurricane Katrina tore through the Louisiana area. You have seen the result on television. Hurricanes are nothing new; both the Florida and North Carolina coast have been hit multiple times. What is peculiar is that people exhibit almost no aversion to building expensive houses in areas with a high chance of natural disaster. One reason is that the government provides disaster relief and subsidized insurance for living in disaster-prone areas. Subsidized insurance is needed because private insurance companies will not offer such insurance. This is because insurance companies have learned that people are not willing to pay premiums that would allow the company to recoup their costs if the insurance is needed. Basically, people do not value their homes as much as it costs to build in the disaster-prone area, but because you the taxpayer help bail them out, they build anyway.

An economic model is a simplified version of the real world where many complexities are assumed away to concentrate on a single question.

Let us look at this issue more in depth using an *economic model*, which is a simplified version of the world. It is a fable, a thought experiment. Models are useful because we can strip away all the complexities of the world and concentrate only on the important issues at hand. Suppose there are two towns: Floodville and Highville. People have the same preferences and the only difference between the two towns is that Floodville experiences a huge flood about once every fifty years, whereas Highville never floods. For the Indifference Principle to hold, and people to be indifferent over where they live, it must cost less to live in Floodville, so let us suppose that it costs $20,000 a year less to live in Floodville.

Even though Floodville residents save $20,000 per year in living expenses, every 50 years they must pay to restore their house and property due to floods. If people are indifferent between living in the two towns, the $20,000 per year savings must just offset the reconstruction costs. Now suppose that when the flood occurs in Floodville, the government decides to provide relief by taxing the residents in Highville and giving the money to Floodville citizens. People now prefer to live in Floodville, because the cost of living is lower and they are compensated for flood damage. As people move to Floodville, they raise the price of houses and property until people are again indifferent between living in Floodville and Highville.

Now let us look at the net result of compensating flood victims. People are indifferent between living in Floodville or Highville before and after the government decides to compensate flood victims. The only thing that really changes is that society as a whole loses toys. There are more people living in Floodville, so there are more homes destroyed by floods. Now, a portion of people's incomes that were spent on toys is now spent on compensating more flood victims. Is anyone made happier from the government compensation plan? There are three types of people: (1) people who move from Highville to Floodville, (2) original Floodville residents, and (3) original Highville residents who do not move. Since the two towns are exactly alike, people are not happier after they move from Highville because house prices change to make them indifferent between which town they live. Original Floodville residents are worse off because they pay higher house prices, and original Highville residents who do not move are worse off because they pay higher taxes to compensate flood victims. Overall, society loses toys.

This is not to say that governments should not provide relief effort in the face of natural disasters. This does say that governments should avoid subsidizing people who face *predictable* natural disasters. Compensating people for predictable losses creates perverse incentives for people to make no effort avoiding these losses. In the end, the compensation schemes hurts society by lowering the number of toys it consumes, by making people less happy.

Whenever government enacts policies, economists look for the Law of Unintended Consequences. And since government enacts many policies involving agriculture, agricultural economists can spend their career writing about this law. One example is regulation of swine manure. The State of North Carolina was considering laws limiting the amount of swine manure that could be applied to crops. The idea was to reduce phosphorus runoff that pollutes rivers and streams. However, economists discovered that if this law was passed, the farmers' incentives were to substitute chemical fertilizers for the manure in a way that water pollution would actually increase (Norwood and Chvosta 2005). The reason is more complicated than the previous examples and requires sophisticated economic models that employ large amounts of math and computer computations. That is what economists do. They develop sophisticated models to determine the outcome of government policies, and is why economists must earn Ph.D.'s to do their job well. Well-intentioned policies can have adverse outcomes, in agriculture and every other industry. Determining when this occurs requires a basic understanding of economics, and is partly why you are taking this course.

Economic Lesson 3: Markets Work Well (If Prices Reflect All Costs and Benefits)

Most of our toys are obtained through markets. Markets are nothing more than a collection of buyers and sellers who negotiate transactions and prices. If you want a house, you either purchase it from a builder or from a previous owner. The same can be said for a car. Your food is purchased through restaurants and grocery stores, and many of you even purchase your education by attending private school. We rely on markets to provide us with most of what we need and want, even our most basic necessities like food, and markets operate based on the profit motive. As Adam Smith described, food producers ensure us with a reliable food supply not because they are compassionate and kind, but because they like money (money for buying toys).

The reason we let the profit motive ensure our basic needs are met is that it seems to work pretty well. Compare our system to Russia, who in the past relied almost exclusively on the government to decide what goods to produce and how much to produce, and only recently is transitioning to a capitalistic society. The average per person income in the United States is about $40,000 per year, whereas in Russia this number is only $4,000 per year (The World Bank 2005). You have heard the saying "money cannot buy happiness," but in this case there is at least a correlation. A survey was conducted across the world in which nations' citizens were asked, "Taking all things together, would you say you are very happy, quite happy, not very happy, or not happy at all." A total of 39% of U.S. citizens said they were very happy, compared to only 6% for Russia. There are many reasons for this difference, but the fact that the U.S. market system provides its citizens with more of the goods and services it needs, and more of the toys it wants compared to Russia's government-run economy, plays an important role.

All market transactions make the buyer and seller better off.

Why is it that markets serve a society so well? The answer is simple. Markets are nothing more than a collection of buyers and sellers striking deals. Buyers do not buy unless they will be better off. Sellers do not sell unless it makes them better off. We can therefore say that all market transactions make buyers and sellers better off. Every person is a buyer in one sense and a seller in another. The authors sell economic services to our university and use the proceeds to purchase goods and services. Markets allow buyers and sellers to work together in a voluntary fashion to find deals that make both better off. Simply put, markets allow people to pursue their own happiness, and according to history, they do better than governments at making people happy.

However, in some circumstances markets do not work well, and some form of government intervention is warranted. Fortunately, one can often determine when markets fail. Remember we said that all market transactions make the buyer and seller better off, but sometimes a third party is affected by the transaction. You purchase chicken from the grocery store, who ultimately purchased it from a chicken processor (e.g., Tyson Foods). As discussed in Chapter 14, poultry production produces water pollution.[5] Your purchase of chicken at the grocery store makes you better off; otherwise you would not make the purchase. The sale of the chicken made the grocery store better off; otherwise it would not sell it. Unfortunately, the users of surface waters are made worse off because the raising of the chicken leads to water pollution. All things considered, some people were made better off, some were made worse off, and we do not know if society as a whole is better off or not.

This is a case of *market failure*. A market transaction made the buyer and seller better off but harmed a third party. This harm to the third party is referred to an *externality*, specifically, a negative externality. There are several remedies for this type of market failure. One is to tax chicken producers by an amount equal to the cost of the

[5]Chicken manure is applied to land, sometimes at excessive levels, and excess nutrients from the manure are delivered to surface waters where they encourage large algae blooms.

Market failure occurs when a third party is benefited or harmed from a market transaction. If a third party is harmed, a negative externality exists, whereas if benefited, a positive externality exists.

pollution. The tax money can be used to compensate water users or to clean up the pollution, so that even if pollution exists, the water users are not made worse off. Because the water users are not affected, and market transactions make buyers and sellers better off, societal happiness can only be improved by the buying and selling of chicken.

In other situations, market transactions make the buyer and seller better off, and benefit a third party. A *positive externality* exists. There may not appear to be a problem here, but there is. The problem is that there will not be enough market transactions for that particular good. Consider research and its impact on society. A researcher contributes to human knowledge, perhaps by researching new pharmaceutical drugs. The researcher sells her services to a pharmaceutical company, again, making both the buyer and the seller better off. This research increases our knowledge of human disease treatment, which leads to more discoveries and better health care down the road. A third party (society as a whole) benefits from the research. The pharmaceutical company cannot capture all the benefits to the third party, so these benefits are not reflected in the number of researchers the company hires. The end result is too little research. Even though the pharmaceutical firm would not profit from more research, society would. The solution? Government should subsidize research. The same argument is made for education. Society, including yourself, benefits from your college education, which is why government helps fund education for many students.

The point is that markets work well so long as the price negotiated by buyers and sellers fully reflect *all* benefits and costs. This includes benefits and costs to the buyer and seller and third parties. In the case of pollution from chicken production, the tax is used to raise chicken prices to reflect the costs of water pollution. In the case of research, researcher salaries are increased by subsidies to reflect the fact that basic research benefits all of society, in addition to the firms and universities employing researchers.

Economic Lesson 4: The Only Scientific Definition of "Value" Is People's Maximum Willingness-to-Pay

A man is homeless, has no money, and has not eaten for three days. What is the value of a hot meal to this man? Almost infinite, you may say, and you are correct that obtaining this meal means more to him than your entire possessions combined mean to you. This may sound cruel, and even weird, but economists would say the value of a hot meal to this man is zero. His maximum willingness-to-pay for the meal is zero because he has no money. The value of something to an individual is defined by economists as the maximum amount the individual will pay for it.

In an old Buddist tale, a student went to his Zen teacher asking, "What is the most valuable thing in the world?" The teacher replied, "the head of a dead cat." When asked why, the teacher explained it was because no one can name its price. However, if we use the economics definition of value, the Zen teacher is wrong. The value of a dead cat's head is zero because no one would pay any money for the head. If willingness-to-pay is zero, then value is zero.

Economics is a science, and science requires observation and measurement. Whatever definition of "value" economists adopt, we must be able to measure it.

As the story above illustrates, there is a difference between value and happiness. For a long time scientists thought happiness could not be measured. Recent research suggests it can, and economists often measure happiness as part of their research. When people think value, they usually associate it with money. That is, value is thought to be a way of expressing happiness through dollars.

This "value" must be measurable for it to become part of a scientific discipline. We can only measure value by estimating people's maximum willingness-to-pay for something. For example, what is the value of an autographed Paul McCartney bass guitar to you? If you are a die-hard Beatles fan you may be tempted to say the value is infinite. Suppose the bass guitar is being auctioned off and the current bid is $1,000. If you bid $1,100, I know your maximum willingness-to-pay is more than $1,100. The bid keeps going up and is now at $3,300. You keep bidding, so I know you value it more than $3,300. Now suppose the bids reach $5,500, and you stop bidding. I now know your maximum willingness-to-pay for the bass guitar is between $3,300 and $5,500. That, from an economist's point-of-view, is how much you value the guitar.

This is not a trivial topic. Economists are often asked to value intangible goods, even the value of a life. Many of the highways you drive separate north- and south-bound traffic by a median. The wider the median, the more lives that are saved from fewer head-on collusions. The median is not free, though. To use a wider median, the government must purchase more land. How wide should the median be? The Department of Transportation determines the median width by balancing the median costs with the value of saving lives. You may be tempted to say you cannot put a value on life, but you would contradict yourself every time you drove a car and risked an accident.

Economists often measure the "value of a statistical life" for use in setting government policies. Every day, people participate in activities that increase the risk of death. They drive motorcycles, smoke, eat unhealthy food, work dangerous jobs, have unprotected sex, etc. Often we can observe these activities to measure how much people value their lives. People with risky jobs tend to get paid more than similar people with less risky jobs. If salaries are $10,000 higher in jobs with a 1/1,000 chance of death than jobs with a 1/10,000 chance of death, we know that people value decreasing their risk of death from 0.001 to 0.0001 less than or equal to $10,000. If the value was more than $10,000, people would not take the risky job. This value of a statistical life is widely used by government. The Environmental Protection Agency uses it to estimate the value of lowering carcinogens in food (which can cause cancer and death) and the Department of Transportation uses it to determine how speed limits should be set. Currently, the value of a statistical life used is somewhere between $3–$7 million, depending on the government agency (Brannon 2005).

Our nation regulates food production extensively to ensure that food is safe but is also not too expensive. The government understands the trade-off between safety and costs and goes to great lengths to balance the benefits and costs of greater food safety. What is the value of safer food? Economists define this value strictly as consumers' maximum willingness-to-pay for safer food. For example, the U.S. government is considering the establishment of a system that can track where beef originated. For example, if you consume a hamburger, the government can track where the hamburger was processed and where the cow that produced the beef was born and raised. This ensures beef was raised on farms using safe production practices (i.e., produce safer meat). This system will cost money and will raise beef prices. But is it worth it to consumers? How much more are consumers willing to pay for "traceable beef" relative to other beef? To answer this question, two agricultural economists from Utah State University conducted an experiment. Subjects from all walks of life were invited into a classroom for a free lunch, where they were given beef sandwiches made from nontraceable beef and $15 in cash. Then, the subjects were given the opportunity to bid money to upgrade to a sandwich made from traceable beef. On average people were willing to pay $0.23 to upgrade to the sandwich made from traceable beef (Dickinson and Bailey 2002). If it costs less than $0.23 to produce these sandwiches made from traceable beef, then indeed the tracking system being considered by the government may benefit society.

Other researchers use different methods to calculate consumer's maximum willingness-to-pay for safer food. In the mid-1990s the government considered enacting a policy to ensure that meat was processed under more sanitary conditions, which would increase meat production costs by $1.3 to $2.1 million. To determine if the law was in society's interest, they had to measure the value of safer meat to consumers. This was accomplished by estimating the medical expenses saved from reduced illness. If households saved $200 each year in medical expenses due to safer food, then they are willing to pay up to but no more than $200 for safer food each year. Using medical savings calculations, and some other calculations using the value of a statistical life discussed earlier, it was discovered the value of the safer food was $8.5 to $43 million. It was clear, consumers were willing to pay more for the safer food than it costs to produce the safer food, and the government did indeed implement this policy (Crutchfield et al. 1997).

Now back to the homeless man with no money to purchase food. Indeed, the *private value* of a meal to him is zero because he has no money. Yet, there are altruistic people willing to give up their own money to keep this man alive. The *social value* of a meal given to this man is not zero, because society has money and will pay money for him to eat.

Economic Lesson 5: People's Actions Are Largely Driven by Opportunity Costs

We say that people perform an action whenever the benefit is greater than the cost. More specifically, people perform an action when the benefit is greater than the *opportunity cost*. When we think of "costs," we usually think of how much money is paid for something. This, however, is not enough to explain people's behavior. The real cost you pay for any activity is the value of your next best alternative.

Opportunity Cost: The value of the next best alternative.

Consider three examples; the first two are fictional. First, you run a bar and make $100,000 in revenues each year. Your annual accounting costs are $50,000.

Accounting costs refer to wages paid to bartenders, cost of beer and liquor, rent on the bar, energy bill, taxes, and so on. To run the bar, you gave up your job as a teacher with a $40,000 salary. If you did not run the bar, your next best alternative would be to return to teaching at your $40,000 salary. Assume that only money matters and that you are just as happy running the bar as you are teaching if the money made from both are equal. The benefits of running the bar are $100,000, but what are the costs? Your accounting costs are $50,000, so your accounting profits are $100,000 − $50,000 = $50,000. However, to run the bar you forego the salary you could have earned as a teacher. The value of the next best alternative (your benefit if you switched to your next best alternative) is the $50,000 you would save each year in accounting costs plus your $40,000 teaching salary. Your opportunity costs are $50,000 + $40,000 = $90,000. Thus, your economic profits from running the bar are revenues minus opportunity costs, which equal $100,000 − $90,000 = $10,000. Note that if your teacher salary increased to $60,000, your economic profits of the bar would equal negative $10,000, and you would be better off teaching.

Economic Profits = Revenues − Opportunity Costs

Now for another example. Someone gives you a free ticket to see The Rolling Stones in concert. You like The Rolling Stones and value the ticket at about $20, meaning you would pay no more than $20 for the ticket. However, the concert is being held at the same time as a Dave Matthews Band concert. You like the Dave Matthews Band much more than The Rolling Stones and are willing to pay up to $150 for a ticket (but no more). Suppose that if you go to the Dave Matthews Band concert you would pay $100 for a ticket—so you extract $50 ($150 − $100) of benefits. Of all the things you could do this night, you would rather go to one of the two concerts.

How much does it cost you to go to The Rolling Stones concert? You might want to say nothing, because the ticket was free, but you would be wrong. By seeing The Rolling Stones, you gave up the opportunity to see the Dave Matthews Band, which yields $50 of benefits. Thus, the *opportunity cost*, the value of the next best alternative, is $50. The benefit is only $20, so you are better off rocking out to the Dave Matthews Band.

Using pesticides drastically increases crop yields and lowers the cost of food; however, it carries the risk of causing cancer among some segments of the population. Some estimates suggest that using pesticides in food production causes about 20 deaths per year in the United States. Economists have estimated that it would cost U.S. consumers and agricultural producers at least $20 billion per year to completely phase out pesticides. This amounts to about $1 billion spent per life saved. Should pesticides be phased out? Someone who argues that a life is priceless might be tempted to say yes. But, if we spent $20 billion on saving 20 lives by eliminating pesticides, this means $20 billion less to spend on other things. How else might that $20 billion be spent? What is the opportunity cost of this money? Some estimates suggest that a government regulation that mandated tests for radon could save about 15,000 lives in the United States for less than $20 billion. So, we could spend $20 billion by eliminating pesticides and save 20 lives, or we could spend about $20 billion to regulate radon testing and save 15,000 lives. The opportunity cost of funding food safety regulations should be considered when passing those regulations (Lomborg 2001; Loomis 1997).

Finally, consider a real example that has great bearing on a timely policy issue: obesity. There is no denying that Americans are getting fatter. The obesity rate climbed from 14% in 1970 to 28% in 2003. This is an American phenomenon. The obesity rate in other rich countries is much lower: 19% in the United Kingdom, 12% in Belgium, and 3% in Japan. Obesity leads to many health problems. It causes people to die earlier and increases health costs to the individual and the government. Obesity, like smoking, is unhealthy.

Whether the U.S. government should take measures to curb obesity is debatable. If people voluntary eat in ways that cause obesity, they would not do so unless they are happier that way. Plus, there appears little evidence for market failure. If people eat more, what third party is affected? You may say that taxpayers are affected since they subsidize health care of the obese. There is some validity to that argument, but it is still not clear whether actions taken by the government to reduce obesity would make society happier.

However, if government does decide to curb obesity, what should it do? It depends on the cause of obesity. Four explanations have been offered: (1) Americans are wealthier than in the past and eat more in response; (2) food prices have fallen, inducing Americans to eat more; (3) Americans eat more food away from home than they used to, and restaurant food is less healthy; and (4) the opportunity cost of snacking has fallen due to new food processing technologies. Only the fourth explanation can really explain the rise in obesity documented in the last 30 years. The first three explanations have contributed some to obesity, but not nearly as much as the fourth.

Figure 1.10 shows the change in calorie consumption each day for U.S. males. Total calories consumed each day has jumped from 2080 to 2347 from the 1970s to the 1990s. The figure clearly shows that most of the increase is due to more snacking. We just can't put down that candy bar. Advances in food technologies have greatly reduced the opportunity cost of snacking. One example is the discovery of better preservatives, which allow food manufacturers to produce snacks with a longer shelf life. This means that snack machines can be stocked without worry that the snacks will spoil. Snack machines are now more profitable, the snacks can be produced at lower cost, and so there are more snack machines. Similarly, the preservatives make it easier and cheaper for convenience stores to provide a greater variety of snacks. One can stock up on many snacks, such as candy bars and cinnamon rolls, and keep the snacks in one's desk because the snacks never spoil. All this greatly reduces the

	1977–1978	1994–1996	Percent of Total Change[a]
Breakfast	384	420	13%
Lunch	517	567	19%
Dinner	918	859	−22%
Snacks	261	501	90%
Total	2080	2347	100%

[a] Shows the percent of increase in total calorie consumption due to each factor.

FIGURE 1.10 Change in Calories per Day Consumed for U.S. Males.

Source: Cutler, Gleaser, and Shapiro.

time involved in snacking. Due to food processing regulations abroad, most of these technologies have been adopted only in the United States (Cutler, Gleaser, and Shapiro 2004).

Before these food technologies, one would have to go to a snack bar or a grocery store looking for fresh doughnuts or fruits. These snacks would spoil quickly, meaning more trips and time involved in getting snacks. In short, the opportunity cost involved with snacking includes the money *and time* involved in obtaining and consuming snacks. Over the years, the time cost fell dramatically, leading to a lower opportunity cost of snacking and a rise in calories consumed for snacking. Although not the only contributor, snacking is the major contributor to obesity. Without an understanding of opportunity cost one could never arrive at this conclusion.

Let us briefly return to the issue of government regulation and obesity. Would society be better off if government sought measures to curb obesity, like a tax on snacks? Consider this fact. Technological developments have reduced the time involved in food and snack consumption by 20 minutes per person per day. This has increased the average person's weight by 20 lbs. However, these 20 lbs. can be shed by exercising 15 minutes per day. This provides individuals—on average—with five extra minutes per day, if they opt not to increase their weight. Of course, people are free to use these extra 20 minutes however they please. And if they are happier watching reruns of *Happy Days* than exercising, who is government to make them do otherwise?

Economic Lesson 6: The Time Value of Money

If we offered you the choice of $100 today or $101 in one year, chances are you would take the $100 today. Yet, if we offered you the choice of $100 today or $1,000 in one year, you would most certainly take the $1,000 in one year. The value of money depends on when that money will be received. Certainly, you would agree that a dollar today is worth more than a dollar tomorrow. The reason is that money has an opportunity cost. Consider again the choice of $100 today or $101 in one year. If you forego the money today, electing instead to take the $101 in one year, you are really giving up money. The reason is that you could have taken the $100 today, invested it in a safe interest-earning instrument, and have more than $101 next year. At the very least, you can purchase a CD (certificate of deposit) at a bank or open a savings account that pays $i = 3\%$ interest. At the end of one year, you will still have your $100 but will also have earned $\$100 \times i = \$100 \times 0.03 = \$3$ in interest. This can be written as $\$100(1 + i) = \$100(1 + 0.03) = \$103$. Thus, $100 today is worth more than $101 next year, but $100 today is worth the same as $103 next year, at the interest rate of 3%.

The term *interest rates* is usually associated with loans and investments. Yet individuals may view the trade-off between money today and money tomorrow differently from banks, so we prefer to use the term *discount rate*. A discount rate is the rate of return on money that makes one indifferent between money today and money tomorrow. It is best interpreted as the opportunity cost of money. Suppose that I give you the choice of $1,000 today or $1,050 next year and you choose $1,000 today. Then I give you the choice of $1,000 today or $1,100 next year, and you say you do not care—you are indifferent. Finally, given the choice of $1,000 today and $1,101 next year, you prefer the money next year. The discount rate (denoted r) that makes you indifferent between money today and next year is then $\$1{,}000(1 + r) = \$1{,}100$. Solving for r, we get $r = 1100/1000 - 1 = 0.1$ or 10%. The rate at which you discount money in the future is 10%.

The discount rate of 10% is interpreted as follows. So long as you can use money today to purchase a financial instrument that yields 10% or more in interest, you will forego the money today, purchasing the instrument, and earning interest. Think back to the choice of $1,000 today or $1,100 tomorrow. The $1,000 today is the *present value*, because it is money today. The $1,100 is the *future value*, because it is the money paid out later. The discount rate is just a number that converts present values to future values. Using the formula above, we see that $(\textit{Present Value})(1 + r) = \textit{Future Value}$, or $\textit{Present Value} = (1 + r)^{-1}(\textit{Future Value})$.Using this formula, the present value of $1,100 today, at a discount rate of $r = 10\%$, is $\textit{Present Value} = (1 + 0.1)^{-1}(\$1{,}100) = \$1{,}100/1.1 = \$1{,}000$.

This formula is extremely important in economics and finance, where we are often concerned with the value of future money. Managers often must consider investments that require monetary outlays today but do not provide profits until future time periods. To determine if the costs incurred today are worth the future profits, we must use the discount rate to convert future profits to a present value. Consider a more complicated example. Suppose you have an investment that pays 10% interest for three years. This interest is compound interest, meaning the interest you earn in period 1 can be used to earn greater interest in period 2. You place $1,000 in this investment, and after one year receive $\$1{,}000(1 + 0.1) = \$1{,}100$. You then invest this $1,100 back in the investment, earning $\$1{,}100(1 + 0.1) = \$1{,}210$. Notice this can be written as $\$1{,}100(1 + 0.1) = \$1{,}000(1 + 0.1)(1 + 0.1) = \$1{,}000(1 + 0.1)^2 = \$1{,}210$. Finally, investing this $1,210 back into the investment for the third year, you earn a total of $\$1{,}210(1 + 0.1) = \$1{,}331$. As before, this can be written as $\$1{,}000(1 + 0.1)(1 + 0.1)(1 + 0.1) = \$1{,}000(1 + 0.1)^3 = \$1{,}331$. If your discount rate is truly 10%, then you should be indifferent between $1,000 today and $1,331 in three years. This yields the important formula $(\textit{Present Value})(1 + r)^T = \textit{Value at time } T$, or $\textit{Present Value} = (1 + r)^{-T}(\textit{Value at time } T)$.

To see the importance of this formula, suppose you are a manager and are considering upgrading your vegetable processing plant. This upgrade will cost you $100,000 in expenses this year. Yet, it will provide you with $115,000 in extra profits in two years. Which is worth more, the $100,000 this year or $115,000 in two years? Assume that our discount rate is 10%, meaning if we did not spend the money on the upgrade, we could invest it in our next best alternative and earn 10%. Using our present value formula, we see that the present value of $115,000 in two years is $(1 + 0.1)^{-2}(115{,}000) = \$95{,}041$. The cost of the investment is $100,000 and its benefits in present value terms are $95,041. This is a bad investment, and you would be better off using the money in its next best alternative and earning 10%.

Let us now consider a more complicated investment problem. We can upgrade a vegetable processing plant, incurring costs of $150,000 today. The upgrade will yield

	Cost of Investment	Present Value of Investment Benefit
The investment costs $150,000 today, but yields $70,000 in profits each year for the next three years.		
Current Year	$150,000	$0
Year 1	$0	$(1 + 0.1)^{-1}$ (70,000) = $63,636
Year 2	$0	$(1 + 0.1)^{-2}$ (70,000) = $57,851
Year 3	$0	$(1 + 0.1)^{-3}$ (70,000) = $52,592
Sum	**$150,000**	**$174,080**

FIGURE 1.11 Discounting Example (Discount Rate = 10%).

$70,000 in extra profits for the next three years. Which is worth more, $150,000 today or $70,000 in extra money each year for the next three years? To determine this we must "discount" the $70,000 in years 1, 2, and 3. That is, we must convert them to their present value equivalents, as shown in Figure 1.11. Money in the future is worth less than money today, which is why the present value of extra $70,000 in profits in year 3 is less than that of years 2 and 1. All things considered, it is clear that the upgrade should be made. Accounting for the opportunity cost of money, the expenditure of $150,000 today yields more profits than investing in the next best alternative.

The concept of discounting is used far beyond just business management. Government even uses it to evaluate policies. Global warming is an issue many countries are taking steps to prevent. Preventing global warming requires significant costs today and the benefits will not be realized until much later. Just like money is worth more today than in the future, benefits realized today are worth more than benefits in the future. Economists have developed sophisticated computer models for analyzing the costs and benefits of fighting global warming, and they will be sure to discount future benefits of preventing global warming so that they can be compared with prevention costs that would be paid today.

SUMMARY

These are just a few lessons economics has to offer. Many more exist, especially on the macroeconomic side of economics. The six lessons in this chapter are intended only as a foundation on which we will build more complex and useful tools later. As you might notice, this chapter places more emphasis on thinking through problems than laying out a collection of facts. Contrary to this, a biology or chemistry textbook is filled with mostly facts. This is a major distinction between economics and other sciences. Economics is a social science, but it is also a method of thinking that helps us analyze social issues. One fluent in economics can easily apply their skills in addressing social issues like racial discrimination, legal issues like the impacts of gender discrimination laws, environmental issues like the impact of global warming on society, and health issues like taxes on food intended to curb obesity. Now that we have discussed economics in general, we will move to more specific topics. The next two chapters employ the economic way of thinking to better understand prices, and after that, agricultural prices specifically.

CROSSWORD PUZZLE

For two or more words, leave an empty box between words.

Across

3. Crop prices should continually rise between harvests to account for _______ costs.
4. These type of economists study the income gap between whites and minorities in the United States.
5. Suppose people are indifferent between $1,000 today or $1,100 in one year. Then $1,000 is the _______ _______ of $1,100 in one year.
7. The field of economic study dealing with large economies, especially on topics such as economic growth and inflation.
12. A(n) _______ exists when a third party is benefited or harmed by a market transaction.
13. These type of economists fill the interface between economics and psychology.
15. This textbook covers economic topics generally described as _______ economics.
16. The Law of _______ _______ states that government policies often create perverse incentives, leading to an outcome the opposite of the policy's intent.

Down

1. This law states that if transaction costs are zero, the price of identical goods should be identical across all regions.
2. The value of the next best alternative.
6. An _______ economist deals with issues like pollution and global warming.
8. The branch of economics that studies small economies, individual markets, business behavior, and individual behavior.
9. A person's maximum willingness-to-pay for something is referred to as what in economics?
10. The three I's of economic theory are incentives, interactions, and _______.
11. The act of profiting from price differences across regions or time periods.
14. Only the owner of a _______ resource can avoid the Indifference Principle.

STUDY QUESTIONS

1. Why did the buffalo almost go extinct but cattle have never come close to extinction?
2. Name four branches of economics and the topics they cover.
3. A corporation invents a new type of fertilizer that, if the new fertilizer costs the same as existing fertilizers, would make the farmer $15 more in profits per acre. The corporation owns a patent on this new fertilizer. What do you predict the price of this new fertilizer would be?
4. Wheat is grown in eastern and western Kansas. The price of wheat in eastern Kansas is $3.25 and the cost of transporting wheat across the state is $0.15. What do we know about the price of wheat in western Kansas? Answer using a number or range of numbers.
5. Soybeans are harvested in November. The price of soybeans on March 1 is $5.00 and the expected price on July 1 is $5.50 per bushel. What do we know about the monthly per bushel cost of storing soybeans?
6. The graph in Figure 1.12 shows the behavior of crop prices between harvests. How would this graph change if storage costs rose?
7. Sorghum is harvested in November. Part of the cost of storing sorghum is the interest one could have made by investing the money paid for storage cost. For example, if one pays $2,000 in storage costs and the bank offers a 5% interest rate

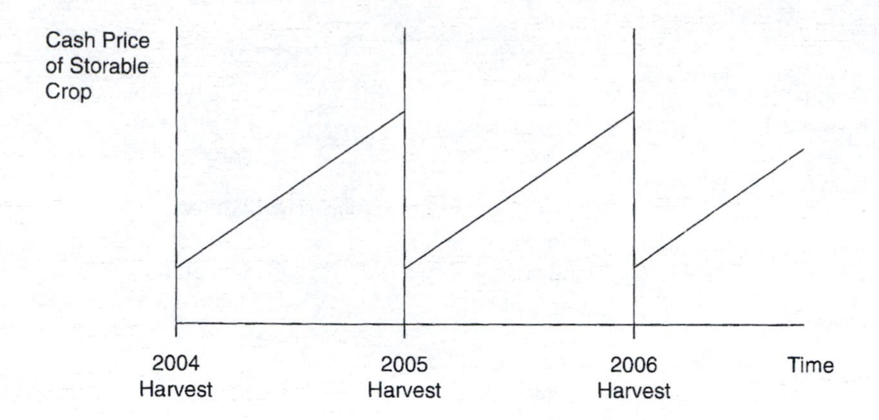

FIGURE 1.12

on certificate of deposits, by paying the storage cost one forgoes the $2,000 × 0.05 = $100 that one could have earned. If interest rates fall, what should happen to the price difference for sorghum in April and May?

8. In 1960 the price of corn was $1.00, whereas the corn price was $2.12 in 2005. Are corn farmers receiving a higher real price in 2005?
9. Suppose the government is considering a ban on a specific pesticide. The ban would raise production costs and thus food prices, harming the economy by $20 million. However, the ban would save lives due to fewer carcinogens in food. Explain how the government would use the value of a statistical life to determine if the ban benefits or harms society.
10. A farmer can make accounting profits of $80,000 per year planting corn, $60,000 per year planting canola, or $40,000 per year planting sorghum. Assume only one crop will be planted. What is the opportunity cost of planting corn? What are economic profits of planting corn?
11. A meatpacking plant is considering upgrading its processing facilities. This will cost $575,000 today, but will increase profits each year by $125,000 for five years. Use the concept of present value to determine if this upgrade is profitable. Assume an 8% discount rate.

CHAPTER TWO

Basic Price Analysis: Supply and Demand

Movie Lines

In the movie *Harold and Kumar Go to White Castle,* two young men named Harold and Kumar go to a college campus to purchase marijuana. They find a hippie-looking guy who looks like he might sell pot, and indeed he does. The following are the movie lines about their attempt to negotiate a price. (*Harold and Kumar Go to White Castle* 2004).

Hippie Student: [handing over the marijuana to Harold and Kumar] Here, that's 80 bucks.

Kumar: 80 bucks?

Hippie Student: Yeah, 80 bucks.

Kumar: Yo, this is worth 40 tops bro!

Hippie Student: Bro? I'm not your bro, bro. OK? And that's 80 bucks. . . .

Kumar: What kind of hippie are you?

Hippie Student: What kind of hippie am I? Man, I'm a business hippie, I understand the concept of supply and demand!

INTRODUCTION

Few people consider themselves certified economists, but almost everyone has heard the phrase "supply and demand." As the conversation between Kumar and the Hippie Student illustrates, the concept of supply and demand is fundamental to understanding why prices are what they are and why prices change. Ultimately, we are interested in developing a general framework to understand price formation, but first we must understand a simple market model where price is determined only by supply and demand. Supply and demand are not the only determinants of price, as we show toward the end of this chapter, but they are the basic building blocks. This chapter has three main objectives:

1. to present the basics of supply and demand
2. to discuss price formation and price changes in the model of perfect competition
3. to present a general theory of prices

In Chapter 1, one of our economic lessons was that much of the economic behavior we observe is driven by opportunity costs and consumers' willingness-to-pay. When negotiating prices, buyers want a low price and sellers want a high price. Buyers have a limit to what price they will pay, and sellers have a limit to what price they will accept. All prices between these two limits are acceptable to both, but they will each try to negotiate a price in their favor. Understanding what drives the minimum price sellers will accept is our first step in understanding prices.

OPPORTUNITY COSTS OF PRODUCTION: THE SUPPLY CURVE

Recall that opportunity cost is the value of the next best alternative. Suppose a farmer could make $50,000 per year raising corn, $40,000 per year raising soybeans, or $30,000 working at a local factory. The opportunity cost of producing corn is then the $40,000 the farmer gives up from not engaging in her next best alternative of soybean production. You are probably used to thinking of accounting costs, costs where actual money changes hands. These costs matter, but they are not the only cost that matters. The farmer also gives up her time to produce corn, and that time could be spent raising soybeans or earning wages at the factory. Second, the money paid for inputs like corn seed and pesticide costs could have been invested and earned interest. By using the money to buy inputs, the farmer foregoes that interest.

People casually throw around the word *profits*, but profits take on a specific meaning to economists. Most people define profits as revenues minus accounting costs, but economists view *economic profits* as revenues minus opportunity costs. So long as economic profits are positive, one is making more money than they could in their next best alternative. In our farming example above, economic profits are the $50,000 from raising corn minus the $40,000 opportunity cost (the value of the next best alternative, soybean production, is $40,000). Economic profits are then $10,000. Economic profits are positive, indicating the farmer is better off planting corn than the next best alternative. If economic profits are negative, they are better off going to their next best alternative, like switching from corn to some other crop, or switching from farming to teaching. The difference between accounting and economic profits is this: Accounting profits tell you whether you are making money, but economic profits tell you whether you are making the most money you can.

Economic Profits = Revenues − Opportunity Costs

Suppose that the opportunity cost of corn production is $2.25 per bushel. As long as the farmer can receive a price of $2.25 per bushel or more, she can make economic profits from raising corn and is better off producing corn than doing anything else. If the farmer does not expect to receive a price greater than $2.25, she will simply turn to her next best alternative. The point is that the producer will not produce and sell corn at a price less than $2.25 per bushel. The opportunity cost of production sets a minimum price for the product.[1]

The term *production* refers to the transformation of inputs to outputs. Beef production entails using grass, hay, grain, water, pasture, fences, antibiotics, labor, machinery, cows, and so on, to produce beef. There are all types of costs—total costs, average costs, and so on. Perhaps the most important cost is *marginal opportunity cost*: the additional opportunity cost of producing one more unit. We will drop the "opportunity" part of the phrase, but keep in mind that the word *cost* in this chapter always refers to opportunity costs. The marginal cost of corn is the extra cost of producing one more bushel of corn; the marginal cost of beef is the extra cost of producing one more pound of beef. You can think about marginal costs in terms of jogging. Suppose that you jog in one-mile intervals. Marginal cost is analogous to the time it takes to run an additional mile. During your first mile you are full of energy and complete the mile rather fast. During the second mile you have less energy and so it takes you longer—marginal cost rises. For every mile run, you become more tired and it takes you longer to run each additional mile. The marginal cost of running a mile increases the more miles you run.

Marginal Opportunity Cost = Opportunity Cost of Producing One More Unit

Economists make one important assumption about the marginal opportunity cost of production: that (just like our jogging example) marginal opportunity cost is increasing in the number of units produced. The reason is that some inputs to production are always limited. In farming, the number of arable acres is limited. Fertilizer is made from natural gas, and the amount of natural gas available is limited. Production of any good requires workers, and the number of potential employees is limited. To increase output, one must use more of these inputs. Some inputs can remain fixed, but at least one input must rise for production to increase. For example, a farmer can produce more soybeans using the same amount of land, as long as she uses more irrigation or fertilizer. Two things happen when industry output rises and the availability of some inputs is limited. First, when firms try to increase their use of inputs that are limited, they bid up the price of those inputs, which raises production costs. If farmers collectively increase production by using more pesticides, they will bid up the price of pesticides and in the process increase the cost of production. Marginal cost rises when industry production rises.

Second, the more one uses an input, the less productive that input becomes. Like the jogger, inputs in a sense become "tired." The first ten pounds of fertilizer applied to crops increases yield more than the second ten pounds. The more days you feed hogs the larger they will grow; but just like you grew faster when you were younger, hogs grow slower as they grow older. Thus, it becomes more and more expensive to grow heavier hogs the longer you feed them, and there will even become a point when they become adults and stop growing. Feed becomes a less productive input the more feed that is used, and thus the cost of putting on an extra pound for each hog

[1]That is, opportunity costs of production usually set a lower bound on price. Sometimes a firm will sell below cost to drive a competitor out of business, but after a while it must raise the price above costs or it will go bankrupt.

Marginal Product: The additional output realized from a one unit increase in input use.

Diminishing Marginal Product: Concept describing the fact that the more input one uses, the smaller the marginal product.

becomes higher. The change in output realized from increasing input use by one is referred to the *marginal product* of an input. Feeding a hog one extra pound of corn each day may increase daily weight gain by 0.5 lbs. But increasing daily feed by another pound (on top of the previous one pound increase) may only increase daily weight gain by 0.25 lbs. This is referred to as the *diminishing marginal product*. As input use rises, the contribution of each additional input to production falls.

You can relate to this better than you may think. Consider studying for a test, where you "produce" points on a test and your input is hours studied. The first hour has a huge impact on your grade, probably producing around 60 points on your test. Thus, the marginal product of the first hour is 60 points. Your second hour of studying adds less to your grade, let us say 20 points. Marginal product falls from 60 to 20 points. You keep studying and are now in your sixth consecutive hour studying. Studying one more hour now pays off very little in terms of extra points, and may even detract from your score if it keeps you from getting enough sleep. The more hours you have already studied, the more hours you must study to obtain an additional 10 points on a test. *Hours studied* is an input, and it experiences a diminishing marginal product. Notice that as the marginal product falls, marginal cost rises. If each hour studied produces less additional points, the cost of attaining more points rises the more one studies. No matter what we are producing, the marginal cost tends to rise with the number of units produced, as illustrated in Figure 2.1.

We can take all producers' marginal costs, aggregate them, and construct a marginal cost curve for the industry as a whole as shown in Figure 2.1. If we are talking about barley production, the curve tells us the marginal opportunity cost for all barley producers. Implicit in the curve is an assumption that low-cost firms produce first. As the industry increases its production level, the low-cost firms hit capacity and cannot produce any more. The additional production must now come from high-cost firms. This is like saying barley is first grown on fertile land with plentiful

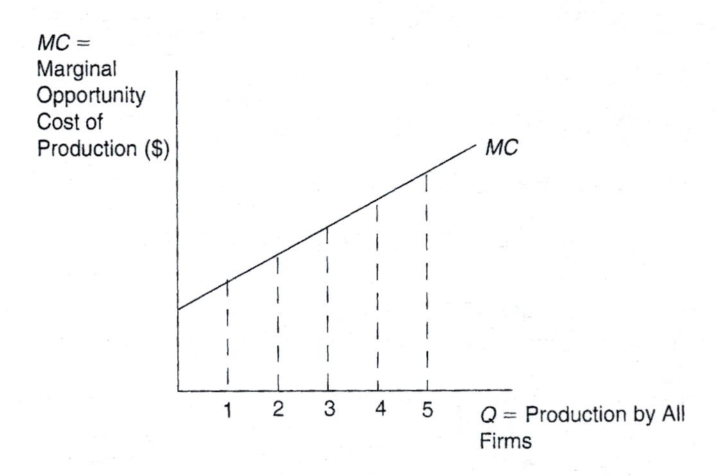

FIGURE 2.1 The Marginal Opportunity Cost of Production Increases with the Number of Units Produced.

rainfall, and if prices are high enough, it will also be produced on less fertile land that receives less rainfall.

As production of a good increases, the marginal opportunity cost of production increases as well.

Sometimes a firm has no control over the price it receives. For example, ask any wheat farmer whether she can control wheat prices to any degree and her answer will be an emphatic "no." We call these firms "price takers." This story is depicted in Figure 2.2 where the price is a constant horizontal line. At this price, how much will the firm produce? If firms produce the first unit and sell it, they will receive a price higher than the marginal cost of production (the cost of increasing production from zero to one). That is, they make money on the first unit. Anytime firms can sell for a price greater than their cost of production, they earn more money.

Notice that the marginal cost of producing the second unit is also lower than the price, so firms will want to produce the second unit as well. The marginal cost of the third unit is just equal to the price, so we will say they produce the third unit also. Once the industry has produced three units, the marginal cost of producing another unit is equal to the price. Firms would lose money on the fourth unit (marginal opportunity cost is greater than the price), so they cease producing at three units. Firms produce where the price equals the marginal opportunity cost of production. This is why we call the marginal cost curve a *supply curve*. The supply curve tells us exactly how much firms will produce at a given price.

In some cases we are interested in the well-being of an industry. Figure 2.2 shows a convenient method of estimating industry welfare. If a firm sells a good for $1.00 more than its marginal opportunity cost, it increases its economic profits by $1.00. This is no more than saying if you produce a good at a cost of $0.50 and sell it for $1.50, you make $1.00 in profits. And if you produce a good for $0.75 and sell it for $1.00, you make $0.25 in profits. Thus, the difference between price and marginal

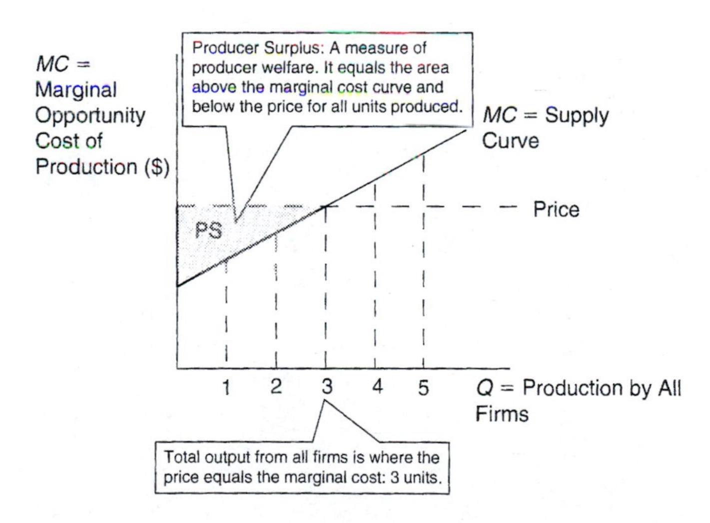

FIGURE 2.2 If Firms Are Price Takers, They Will Produce Where the Price Equals the Marginal Opportunity Cost of Production.

opportunity cost is a measure of economic profits for that unit sold—something we call producer surplus. *Producer surplus* then equals price minus marginal costs for all units sold. In Figure 2.2, this is the area below price and above marginal cost for all units produced. Economists often want to calculate producer surplus to determine how an industry is benefited or harmed by factors such as taxes, drought, hurricanes, technological developments, international trade, and so on.

There is a direct relationship between producer surplus and economic profits, but to see this relationship, we must learn one more thing about costs. Costs can be grouped into two types: *fixed* and *variable* costs. Fixed costs are those that are the same no matter how many units are produced. If a whoopie-cushion manufacturer takes out a one-year lease on a building for $3,000 per month, it must pay $3,000 in rent regardless of whether it produces zero or a million whoopie-cushions. The rent is a fixed cost. However, the rubber used to make whoopie-cushions is a variable cost because each additional unit made requires more rubber. Other inputs may also vary with the number of whoopie-cushions made—including labor, electricity, and so on. Marginal cost—the cost of producing one additional unit—only includes variable costs. Consider the cost of studying for a test. You have already purchased your textbook, so it is a fixed cost that stays the same regardless of how many hours you study. Time is a variable cost. To study more, you consume more time. Thus, to spend one more hour studying does not require another textbook, but it does require more of your time. Marginal cost includes only variable costs.

Therefore, producer surplus equals total revenues minus total *variable* opportunity costs. This is profits not including fixed costs. Because fixed costs do not vary with the number of units produced, a one dollar increase in producer surplus implies a one dollar increase in industry profits. A thousand dollar decrease in producer surplus implies a thousand dollar decrease in industry profits. Changes in industry profits can be measured directly from changes in producer surplus.

Shifts in the supply curve are caused by

(1) Price of related outputs
(2) Price of inputs
(3) Technology

Supply Curve Shifts Figure 2.2 shows a single supply curve. The curve tells us how many more units will be produced as the price increases and how many fewer units will be produced as the price decreases. These are referred to as changes *along* the supply curve. However, any factor that changes the marginal opportunity cost of production changes the position of the supply curve causing a *change in supply* or a *shift in the supply curve*. Any factor that tends to increase the marginal opportunity cost of product will shift the supply curve upward and cause a *decrease in supply*. Conversely, any factor that tends to decrease the marginal opportunity cost of product will shift the supply curve downward and cause an *increase in supply*. There are many factors that can change the marginal opportunity cost of production and they can generally be categorized as follows: price of related outputs, price of inputs, and technology.

Consider a soybean farmer. What will affect the farmer's marginal opportunity cost of production? One obvious answer is the price of soybean seed—the price of an input. If the price of seed increases, the marginal cost increases, and there will be a decrease in supply. What if technology changes the seed itself? About 10 years ago, scientists developed and commercialized genetically modified seed, which lowered production costs. The new technology thus caused an increase in supply. Marginal opportunity costs of production are also influenced by the prices of other products the farmer can produce. Consider an increase in the price of cotton. It might be difficult to initially see why a change in the cotton price affects production costs for soybeans, but recall it is the marginal *opportunity* cost that is relevant. If cotton is the soybean farmer's next best production alternative, then increases in cotton prices will indeed affect the marginal opportunity cost of producing soybeans—it will cause a decrease in supply of soybeans. As the cotton price rises, some farmers will cease growing soybeans and begin growing cotton, causing the quantity supplied of soybeans to fall. This is illustrated by the left diagram in Figure 2.3. Cotton prices rise, causing the soybean supply curve to shift leftward. At any given price, farmers now produce fewer soybeans. Conversely, genetically modified soybean seed allows soybeans to be produced at lower cost. As shown in the right diagram, supply increases and the supply curve shifts to the right. At any price, farmers will now produce more soybeans.

It might be tempting to think that increases in supply are good for producers, but this is not always true. Consider again whether farmers were made better off by the introduction of genetically modified seed. As we just argued, the new technology likely caused an increase in the supply curve by shifting the supply curve downward, because the marginal opportunity cost was lower at every quantity. This means producers, as a whole, are now producing more soybeans. But the number of soybean consumers hasn't changed. These new soybeans have to go somewhere, and to get consumers to eat more soybeans, consumers must be incited to do so by a lower price. That is, to entice consumers to increase their purchases of soybeans (so that the beans are not sitting around on the farm), farmers, as a whole, must lower their

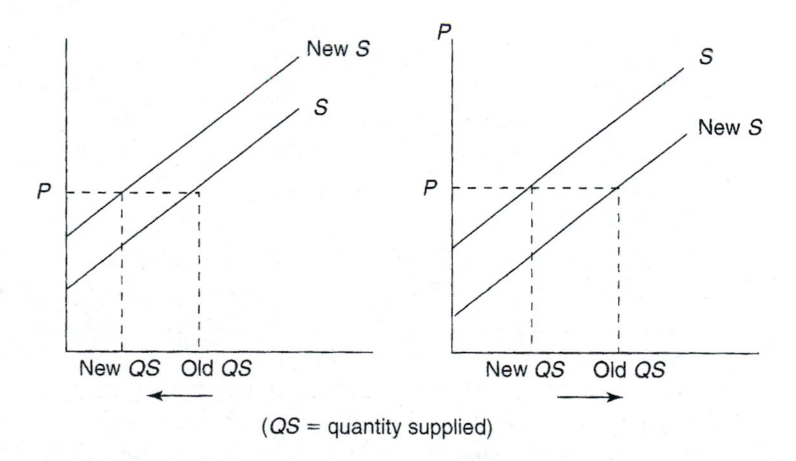

FIGURE 2.3 Shifts in Supply.

price. Although the cost of producing soybeans falls, so does the soybean price. Thus, it is unclear whether soybean producers are better off from the new technology. What this discussion does illustrate, however, is that we now need to know something about consumer demand before we can begin to talk seriously how prices are formed and how they change.

CONSUMER VALUE: THE DEMAND CURVE

If you attend the University of Colorado or Colorado State University, the most important football game is the first game during the year when the two teams play one another. Attending this game is a must for all hard-core football fans at these schools. Yes, students are poor, often living off ramen noodles for days, but many are still willing to pay a high price for a ticket to this game. Chris Winn is one of these hard-core football fans at Colorado State, and he is willing to pay up to $200 for a ticket. That is, Chris will pay $100 for a ticket, $150 for a ticket, even $199 for a ticket, but Chris will not pay $201 for a ticket. At a price of $200, Chris is indifferent between attending the game or not. Thus, Chris "values" the ticket at $200. Remember from Chapter 1 this maximum willingness-to-pay is what economists refer to as value.

In Chapter 1 we discussed the *Exxon Valdez* oil spill and the tremendous environmental damage it caused. If you ask people, "what is the value of protecting the environment from oil spills?" you will hear some people say that the value is infinite, that you cannot put a price on the environment. Economists argued the contrary; they showed that the country as a whole was not willing to pay more than around $5 billion to clean up the oil spill. At a cleanup cost of $5 billion, society was indifferent between cleaning the oil spill or not. Five billion dollars, then, is the value society places on preventing oil spills such as the *Exxon Valdez* spill.

Our examples so far regard purchases of a single unit of a good: one Colorado–Colorado State football ticket, one oil spill cleanup. In most settings, multiple units are sold. On game day thousands of football tickets are sold. Every day thousands of Coca-Colas are purchased. Every day you consume multiple meals of many different food types. In these cases, we will find it useful to discuss the *marginal consumer value* of a good—the value to consumers of one more unit of the good.

Think back to our friend Chris Winn. Attending the Colorado–Colorado State game is vital, and he will pay up to $200 for his ticket. Chris also has a girlfriend named Amanda, and Chris would like for her to attend the game as well. How much is Chris willing to pay for the second ticket? Chances are, the value of the second ticket to him is less than $200. It is vital that Chris attends but a simple pleasure if Amanda attends with him. In fact, economists have found this true for almost all goods: The marginal consumer value declines as a greater number of units are consumed. This is referred to as diminishing marginal value of consumption.

Diminishing Marginal Value of Consumption: The more units that are consumed, the less consumers will pay for additional units.

Consider a few more examples. It is early morning and you have a hankering for Krispy Kreme doughnuts. The first doughnut is heaven, absolutely heaven. The second one is pretty good too, but you are not thanking the doughnut gods like you did for the first one. The third one is still good, but you feel yourself getting full and

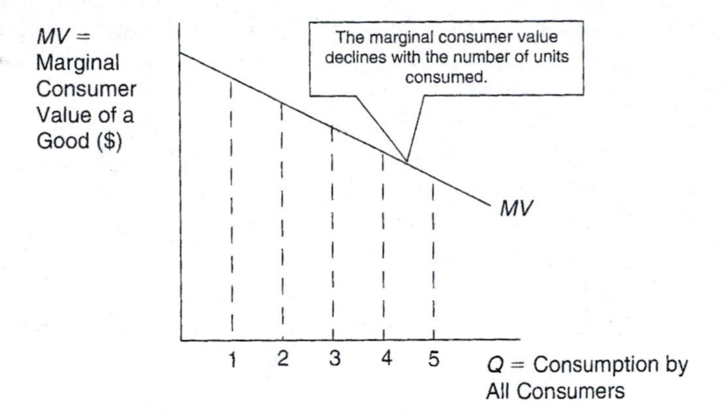

FIGURE 2.4 Marginal Consumer Value.

slightly sleepy. After finishing the third doughnut, you carefully consider whether you want a fourth. Being the compulsive doughnut eater you are, you purchase the fourth doughnut. Halfway through the fourth doughnut you can eat no more. Now someone would have to pay you to eat more, and the marginal consumer value of a doughnut (the value of one more doughnut) becomes negative.

This same phenomenon holds for groups of people as well, and when we aggregate everyone's marginal value, what we have is a market demand curve. That is, the market demand curve is not for one consumer but all consumers. In the past only computer geeks had computers, but today it seems everyone is checking their e-mail, scanning the Internet for term papers to download, and posting pictures from their latest keg party on www.facebook.com. Despite the widespread use of computers, I think we will all agree that computer geeks value their computer more than the average person. The computer geeks are the first to purchase computers, and the value they place on the computers is high. Then comes those who like computers but not as much as the geeks. Finally, the elderly are the last to purchase computers and value computers the least. Notice what happens to the marginal consumer value when consumption increases. Each additional unit is consumed by someone who values it less. Once again we find the diminishing marginal value of consumption. Economists have even gone so far as to test this proposition in rats and pigeons, and they too possess a declining marginal consumer value of consumption.[2] Some people may even purchase multiple computers. They value the first highly, and even though the second is useful, its value is lower than the first. A household may find it imperative to have one computer in the house but a mere convenience to have two. In this case, one person's value may be located at a high point and a low point on the demand

[2]See Landsburg 1995. Economists have subjected rats and pigeons to experiments where they must perform an activity to receive food, like pressing a lever. The harder that activity becomes, like making the lever harder to press, the less food the animal will consume.

curve. Remember, the demand curve reflects the value to all consumers for all units purchased.

If you value something more than it costs, you should purchase it, and you should keep purchasing more units so long as the value is greater than the price. Many times consumers have no control over the price (just like you cannot negotiate Wal-Mart prices with Wal-Mart management; you simply pay what is listed on the price tag). We call these consumers "price takers." When consumers are price takers, the marginal value curve tells us exactly how many units consumers will purchase. Using the curve in Figure 2.5, consumers keep purchasing another unit as long as the marginal value is greater than or equal to the price, and end up purchasing four units. This is why we sometimes call the marginal value curve the *demand curve*; it tells us how many units consumers will demand at any price.

In many cases we will want to know the "happiness" consumers receive from their purchases, and how much "happiness" they lose if price rises. Being economists, we want to measure this "happiness" in dollars. Think back to the computer example. A computer geek named Poindexter is willing to spend up to $10,000 for a computer, but computers today do not cost that much. Instead, he gets a low price of $1,000. Poindexter's value of the computer is $10,000, but he only pays $1,000 so he extracts $9,000 of value from the transaction. This is referred to as consumer surplus, and the larger the consumer surplus, the happier consumers are with their purchases.

Consumer surplus for a single purchase equals the marginal value minus the price. Most other people value computers less than Poindexter. Your father, for example, may value a computer at only $2,000, so if he purchases a $1,000 computer he only extracts (marginal value minus price) $1,000 of consumer surplus. More often than not we are concerned with the welfare of all consumers, and so we sum up the consumer surplus

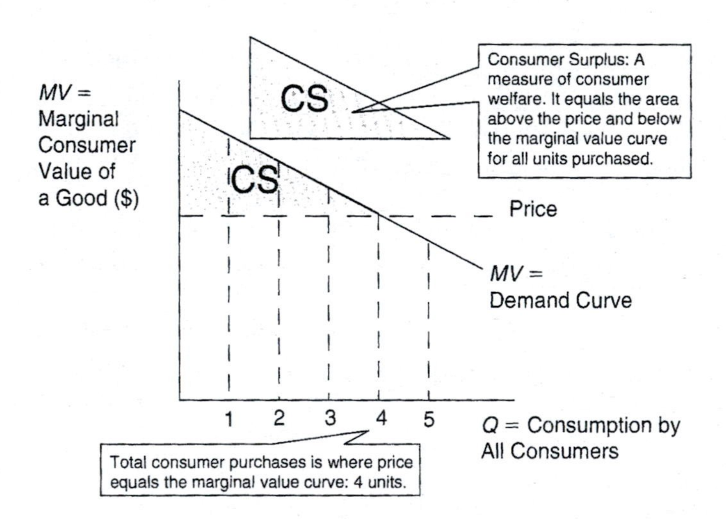

FIGURE 2.5 Consumers Purchase a Number of Units Where the Price Equals the Marginal Value Curve.

from each transaction, as shown in Figure 2.5. The area above price and below the marginal value curve for all quantities purchased equals consumer surplus.

In summary, a demand curve is simply a collection of all people's willingness-to-pay for a good in a particular area over a particular period of time. Knowing what the demand curve looks like can help you answer the following queries: If I tell you the price set for a good, can you tell me how many units will be sold? If I tell you the number of units produced and sold, can you tell me the person's willingness-to-pay for the last unit? The demand curve provides a simple mapping of marginal values to quantities consumed. The demand curve thus shows how many units will be bought at a particular price; when prices increase or decrease, we refer to this as a *movement along the demand curve*. Prices go up and people buy less; prices go down and people buy more. In addition to these movements along the demand curve, we are also interested in identifying factors that shift the demand curve, because a shift in demand will cause a change in price—as we shall soon see.

Demand Curve Shifts When we talk about factors that shift a demand curve, we are essentially talking about factors that change a person's willingness-to-pay. Think about your purchasing decisions and what causes you to change what you're willing to pay for the first unit of a good, the second unit, and so on. Suppose I offered to sell you an ice-cold Dr. Pepper. What would determine the most you are willing to pay for the Dr. Pepper? You might first think to yourself, what is the price of other soft drinks in the vending machine? Clearly, the price of other similar goods matters when deciding your willingness-to-pay. If the price in the vending machine goes up, you'll likely be willing to pay a bit more for my Dr. Pepper as well. You might also start digging around to see how much money was in your pocket, while also thinking about all the other things you wouldn't be able to buy if you spent your pocket change on the Dr. Pepper. Thus, income has an influence on your willingness-to-pay. If a $100 bill magically appeared in your wallet, you might be willing to increase your willingness-to-pay for the Dr. Pepper. You might also consider the future. If you live in Texas or Oklahoma, you can find a Dr. Pepper in just about any vending machine, restaurant, or convenience store, but if you live in other places of the United States, Dr. Pepper is harder to find. You might also think about how much money you might earn next week from your job waiting tables. This is to say that your expectations about the future will affect how much you are willing to pay now.

There are also a myriad of other factors that will influence how much you're willing to pay for the Dr. Pepper that can best be categorized as "tastes and preferences." For examples: How thirsty are you right now? Do you drink caffeinated and non-diet colas? Are you a Coca-Cola aficionado? If you change your answer to any of these questions, you will likely change your willingness-to-pay and in so doing, the demand curve for Dr. Pepper will shift. Recognizing that the market demand curve is simply a collection of all people's willingness-to-pay at a particular place at a particular time, it is clear that the market demand curve can change not only by changing any single person's willingness-to-pay but also by changing who and how many people are in the

Shifts in the demand curve are caused by
(1) Price changes of related goods (substitutes and complements)
(2) Changes in income
(3) Changes in population
(4) Changes in tastes
(5) Changes in expectations

market. Simply adding more people to a market will shift a market demand curve because there are simply more people willing to pay for the Dr. Pepper.

Economists generally say that the five main factors shifting the demand curve for any particular product are the price of related goods, income, population, tastes, and expectations. Because demand, like supply, is a fundamental concept that drives our understanding of prices, economists have defined a variety of terms to describe various shifts in demand, types of demand relationships, and types of goods based on consumers' willingness-to-pay.

First, consider terms used to describe demand curve shifts. We have said previously that anything that changes the amount consumers are willing to pay for the good will shift the demand curve. If something tends to increase consumers' willingness-to-pay, then we call this an *increase in demand* or a *rightward shift in demand*. An increase in demand also occurs if consumers, as a whole, are willing to purchase more at the same price. Back to our Dr. Pepper example, a upward shift in demand would occur if the prices of other sodas in the vending machine increased or if you had more change in your pocket. If something tends to decrease consumers' willingness-to-pay, then we call this a *decrease in demand* or a *leftward shift in demand*. If many people decided to go on a diet, willingness-to-pay for Dr. Pepper would likely fall and thus we would witness a decrease in demand.

We have previously mentioned several factors that can shift a demand curve. For two of these factors—price of related goods and income—economists have come up with terms to describe goods based on their demand relationships. First, consider the price of related goods. We have argued that most people would increase their willingness-to-pay for Dr. Pepper if the price of Coca-Cola increased in the vending machine. This means that Dr. Pepper and Coca-Cola are what we call *substitute goods*. In general, two goods are substitutes if people increase the amount they are

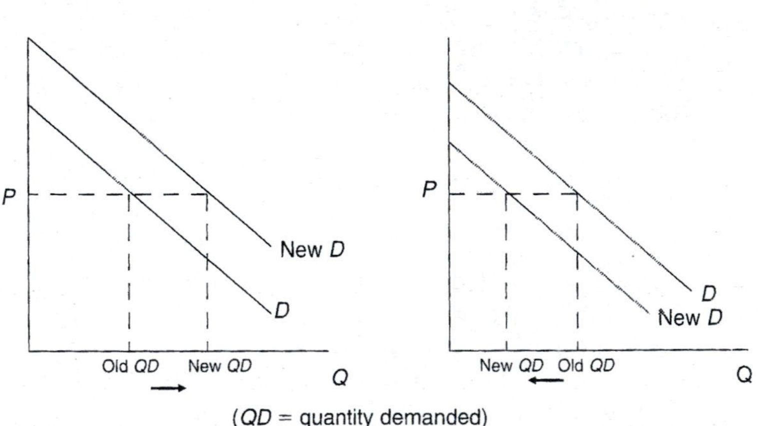

FIGURE 2.6 Shifts in Demand.

willing to pay (or buy) for one good when the price of the other good increases. But, what would happen to your willingness-to-pay for Dr. Pepper if the price of Snickers increased? If you tend to buy Dr. Pepper and Snickers at the same time for a midday snack, increasing the price of Snickers might actually reduce your willingness-to-pay for Dr. Pepper. After all, if you can't have your favorite soda, your Snickers bar just isn't going to taste as good. Thus, we say that Dr. Pepper and Snickers bars are *complement goods*. In general, two goods are complements if people decrease the amount they are willing to pay (or buy) for one good when the price of the other good increases. There are even goods that can be considered *perfect complements*, where one good is necessary to use the other. For example, your willingness-to-pay for an aluminum can is probably $0 unless it contains Dr. Pepper liquid; similarly, your willingness to pay for Dr. Pepper liquid is probably $0 unless you have something to put it in. So, aluminum cans and Dr. Pepper liquid are perfect complements.

Now, consider how income affects demand for a good. For many goods, like Dr. Pepper, we would expect that when people have more income, they will increase their willingness-to-pay for the good. These are what we refer to as *normal goods*, where demand rises as income rises and where demand falls as income falls. We can further identify a particular type of normal good called a *luxury good*. A luxury good is one where when income increases by a certain amount, demand for the good rises by a more than proportional amount. For example, if you got a 1% pay increase at work, you might increase your consumption of Dr. Pepper, but not a great deal—maybe only 0.5%. However, your 1% pay raise might increase your consumption of double espressos at Starbucks by 1.5%. If so, we would say that double espressos are a luxury good. But not all goods can be considered normal goods. When the authors were in college, we often drank Dr. Thunder, a colorfully named, but less tasty knock-off of Dr. Pepper sold at Wal-Mart and Sam's Club. Dr. Thunder didn't taste better than Dr. Pepper, but it was certainly cheaper. When we got real jobs and started earning a higher income, we put aside Dr. Thunder for Dr. Pepper. Goods like Dr. Thunder are called *inferior goods*, meaning demand for the good falls as income rises, and vice versa. Interestingly, college education is often viewed as an inferior good. In hard times when there are few jobs and people's income falls, they tend to go back to school, hoping that education will improve their situation.

We might be tempted to think that increases in demand always make consumers better off. However, consider the effects of a pay raise on the demand for Dr. Pepper. Assuming Dr. Pepper is a normal good, people will be willing to pay more for Dr. Pepper when they have more money, and thus, demand will increase, making the demand curve shift rightward. When consumer demand rises, there are now more people wanting Dr. Pepper at the same price. Because Dr. Pepper doesn't grow on trees, Dr. Pepper bottlers only have a fixed amount. At the old price of Dr. Pepper, however, consumers want to purchase more than firms have. Thus, to prevent a mob scene, Dr. Pepper bottlers will likely increase the price of Dr. Pepper. Thus, when consumer demand rises, quantity consumed rises, but so do prices. Whether consumers are better or worse off depends on exactly how price changes as a result of the intersection of supply *and* demand. This is the topic of the next section.

A PERFECTLY COMPETITIVE MARKET

Prices arise from the interaction of buyers and sellers. The preceding sections were meant to give a detailed picture of the buyers and sellers—the supply and demand curves—that can help us determine what price levels will be and how prices will change, at least under certain conditions. Here, we consider a particular condition where there are many buyers and sellers, where no one buyer or seller dominates the market. Every person buys or sells an identical product, all have identical information, and there are few obstacles to becoming a buyer or seller of the good. There are no markets that match this description exactly, but there are many markets that resemble this description. We call those markets *perfectly competitive markets*.

The market for corn resembles perfect competition. A huge number of farms produce corn in virtually every corner of the United States. There are many buyers of corn as well. Corn differs across farms according to how the plant was raised and the amount of rainfall it receives, but for the most part corn is corn. Moreover, other firms can enter and exit the corn market with relative ease. Farms not currently raising corn can easily begin raising corn. There are no laws against it, although the farmer must purchase new equipment that can be costly. Farms can cease raising corn anytime they want. The same goes for the entry and exit of buyers. At any time, if you think it will make you money, you can become a buyer of corn. What is there to stop you? At any time, a corn buyer can leave the corn business to pursue some other opportunity. Finally, there are no "secrets" to corn. You cannot just begin producing Coca-Cola because you do not know the secret recipe, but information on how to raise corn and what to do with corn if you purchase it is no secret.

Consider the market for gasoline (not the market for oil, but gasoline sales at gas stations). In any reasonable sized town, there are as many gasoline buyers as there are cars and more than a few gas stations. Some gas stations are obviously nicer than others and more conveniently located. This gives them a competitive advantage and allows them to charge higher prices. There is a limit to how much more they can charge though, because consumers can easily go to another gas station if the price is set too high. Also, if the price of gasoline starts to rise because there are too few gas stations, someone will soon build a new gas station to make profits.

Perfect Competition exists when

(1) There are many buyers and sellers, each with roughly the same market share.
(2) All sellers produce identical goods.
(3) Information on how to produce and use the good is freely available.
(4) One can become a buyer or seller with relative ease.

In perfect competition, prices are the same for buyers and sellers.

This direct competition between firms results in very similar prices across gas stations. All stations are selling gas for roughly the same price, and all buyers are paying roughly the same price. To the extent that these gas stations are identical, prices will be identical, and in perfect competition we assume identical firms. Remember the Indifference Principle: Drivers should be indifferent between where they purchase gas. Similar prices across gas stations ensure the Indifference Principle holds. When Hurricane Katrina caused gas prices to rise, they rose by roughly the same amount for all gas stations. That is, there are many buyers, many sellers, and the negotiating power of all buyers and sellers are roughly equal. This type of setting resembles a market in perfect competition, and suggests an important result of a perfectly competitive market: Prices are roughly the same for all buyers and sellers.

Producers and Consumers Together When there are many buyers and sellers and all have roughly equal negotiating power, buyers and sellers tend to take price as given. They are price takers. Corn producers look at the market price and expect nothing more or nothing less. When you buy corn from the grocery store, you pay the same price as all the other shoppers. Buyers observe the "going price" and make their purchasing decisions based on that price. Sellers also make their production decisions based on that market price. As a gasoline consumer you simply look at the posted prices and determine how much gasoline to buy. A gasoline seller looks at the price of her competitors and sets a price roughly equal to those prices (because that is the maximum amount she can charge, and if she lowers her prices, her competitors will simply lower their prices as well). However, gasoline prices can change, and when they do, they change quickly.

Let us now return to our marginal cost and marginal value curves, which we also call our supply and demand curves. See Figure 2.7 where market supply and demand curves are drawn in the same figure. If we assume perfect competition, there can be only one price that all buyers and sellers pay. Not surprisingly, this price must adhere to the Indifference Principle and must make buyers and sellers indifferent between bidding the price up or down. This price is where supply and demand cross and is referred to the *equilibrium price*.

In Figure 2.7, the equilibrium price is $4 and the equilibrium quantity is 300. The reason we refer to it as an equilibrium is that it is like the thermostat on a heater. Price may never actually equal the price, but if it is above the equilibrium, it will decrease, and if it is below the equilibrium, it will increase. The equilibrium price is the average around which prices fluctuate. At any given time, the best guess of the real price is the equilibrium price, just like if the thermostat in a room is set to 70 degrees, the best guess of the room's temperature is 70 degrees.

To see this, let us tell a story of how prices in perfect competition converge to their equilibrium. The equilibrium price in Figure 2.7 is $4, but what if the real price was only $3? The supply curve tells us that firms would produce about

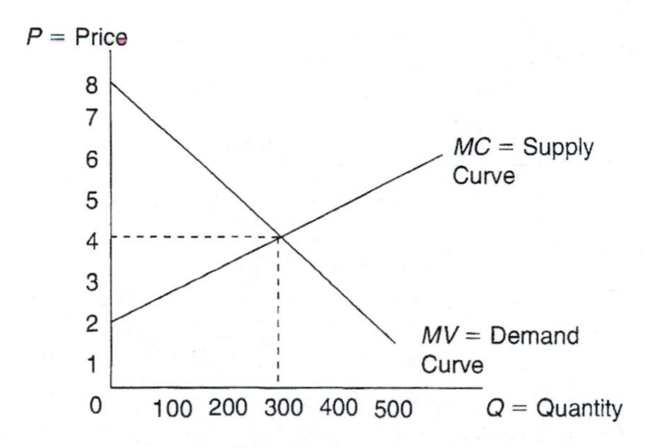

FIGURE 2.7 The Price and Quantity in Perfect Competition Is Where Supply and Demand Cross.

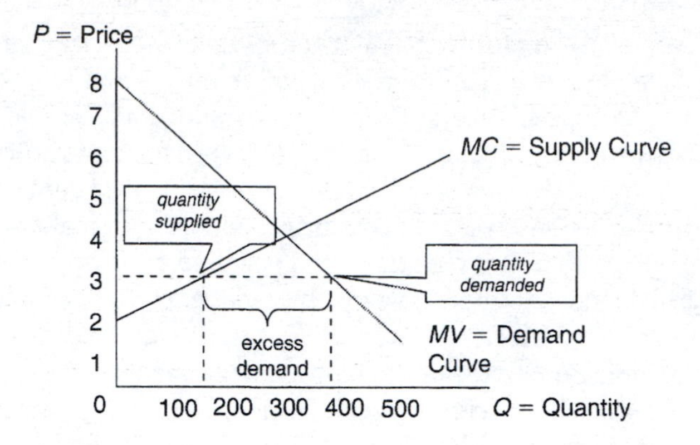

FIGURE 2.8 A Price Lower Than the Equilibrium Price Leads to an Excess Demand.

150 units (see Figure 2.8). The demand curve tells us that at a price of $3 consumers will want to purchase around 400 units. This leads to *excess demand*. Consumers want more than firms are willing to produce, so to entice firms to produce more, consumers bid up the price of the good. Conversely, if the price is above the equilibrium price firms want to produce more than consumers are willing to buy. We have an *excess supply*, and to get rid of the excess supply firms must lower their price. Although prices may fluctuate, they fluctuate around the equilibrium price given by the point where supply and demand cross. Notice at the equilibrium price of $4 quantity demanded equals quantity supplied. This is an *equilibrium price*. Buyers and sellers are in complete agreement over how much should be produced. This is an *equilibrium quantity*, and in this case the equilibrium quantity is 300.

Societal Welfare Under Perfect Competition Now that we have determined the price and quantity that would result in perfect competition, it is prudent to ask how well these competitive markets serve society. Are competitive markets "good" or is

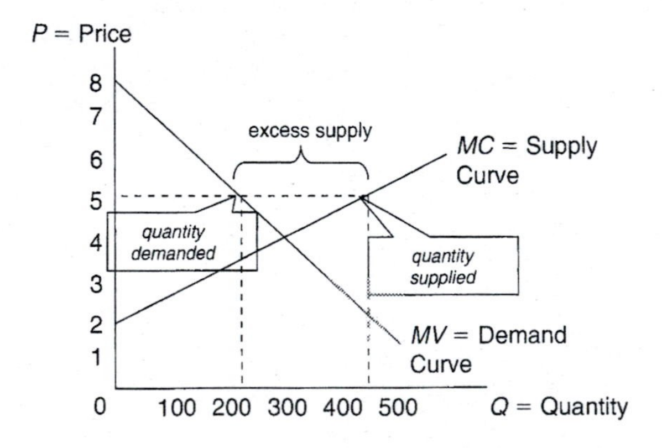

FIGURE 2.9 A Price Higher Than the Equilibrium Price Leads to an Excess Supply.

there something better? The answer is important. If competitive markets serve us well, all we need to do is let producers and consumers compete on their own terms, meaning society does not have to "do" anything except leave buyers and sellers free to strike their own deals. Figure 2.10 shows the equilibrium price and quantity in a perfectly competitive market. How well off are consumers and producers in this market? The happiness consumers get from their purchases is measured by consumer surplus: the area above price and below demand for all quantities purchased, the upper triangle in the figure. The welfare of producers is measured by producer surplus: the area above supply and below price for all quantities purchased, the lower triangle. Total societal welfare from the market is then the sum of producer and consumer surplus, what we call *total surplus*.

Notice that to calculate total surplus we can disregard the distinction between consumer and producer surplus. For any given quantity, so long as the marginal value curve is above the marginal cost curve, consumers value the unit more than it costs to produce, and the world is made a happier place by its production and consumption. For example, suppose consumers are willing to pay $12 for an organic steak and that steak only costs $8 to produce. The particular price struck will be between $12 and $8, but for purposes of calculating societal welfare, the particular price does not matter. If the price is $10, consumer, producer, and total surplus is $2, $2, and $4, respectively. If the price is $9, consumer, producer, and total surplus is $3, $1, and $4, respectively. Total surplus is the same regardless of the price and equals marginal value minus marginal cost for that unit.

Perfect competition maximizes total surplus because the equilibrium quantity is where the marginal value and marginal cost curves cross. If less than the equilibrium quantity is produced, society foregoes consumption of some items it values more than it costs to produce, and society is made worse off. If more than the equilibrium quantity is produced, then society produces goods it values less than it costs to produce, which is like burning money. Maximum happiness (if happiness is measured by

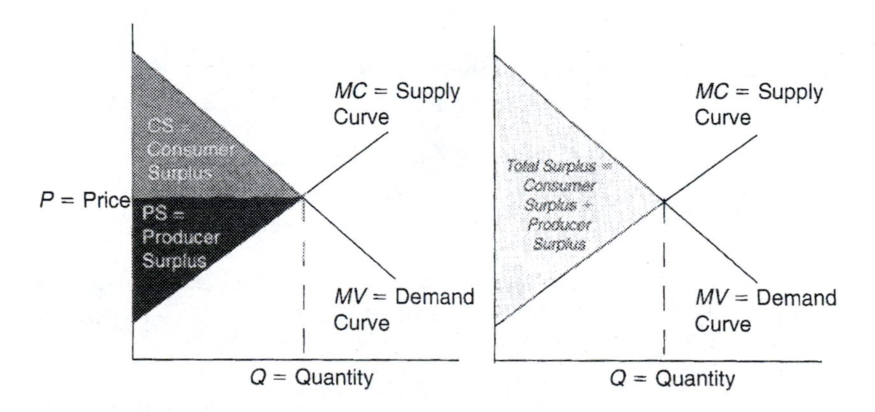

FIGURE 2.10 Consumer, Producer, and Total Surplus Under Perfect Competition.

total surplus) is then achieved at the quantity where supply and demand cross, which happens to be the exact quantity produced by perfect competition.

An economic model is a thought experiment where many complexities of human behaviors are ignored in order that one may concentrate on a few important relationships. A model may be expressed in words, as mathematical equations, or in graphs.

The Mathematics of Supply and Demand Perfect competition is an *economic model*—a model of how equilibrium prices and quantities are determined. An economic model is a fictitious economic story, where many of the complexities of the world are ignored and only a few important aspects are considered. It is like a moral fable in that it has a lesson, but not every lesson you need in life. Supply and demand tell a story of how prices change and how prices are formed. Before engineers construct a new bridge, they build a model of the bridge to see how well it works. The bridge model cannot tell you everything about how the bridge will perform in real life, but it can reveal some things. Economists use economic models for the same reason: to gain insights into real-world economic phenomena. The "stories" may be represented by words or graphs, as in Figure 2.7, or by equations. We now consider the perfectly competitive model as described by mathematical equations.

Economists are sometimes able to develop mathematical equations for supply and demand curves. A subsequent chapter illustrates how to obtain such equations, but for now, just assume we already have the supply and demand equations and wish to calculate the equilibrium price and quantity. Suppose we had the following supply and demand curves.

$$\text{Demand, } MV\text{: } P = 120 - 8(Q)$$
$$\text{Supply, } MC\text{: } P = 20 + 2(Q)$$
$$P = \text{Price; } Q = \text{Quantity}$$

One could simply graph the two curves and see that the equilibrium price is 40 and the equilibrium quantity is 10, but it will be helpful to calculate the numbers mathematically as well. The steps to solving for the equilibrium price and quantity are as follows.

Step 1: Both the supply and demand curves should correspond to the same price at the equilibrium quantity. Set the supply and demand equations equal to one another and solve for the equilibrium quantity.

$$120 - 8(Q) = 20 + 2(Q) \Rightarrow 120 - 20 = 2(Q) + 8(Q)$$
$$\Rightarrow 100 = 10(Q) \Rightarrow Q = 100/10 = 10$$
$$\textbf{Equilibrium Quantity} = 10$$

Step 2: Plug the equilibrium quantity into the supply or demand equation to calculate the equilibrium price. The equilibrium price will be the same for both equations.

$$\text{Demand: } P = 120 - 8(Q) = 120 - 8(10) = 40$$
$$\text{Supply: } P = 20 + 2(Q) = 20 + 2(10) = 40$$
$$\textbf{Equilibrium Price} = 40$$

Price and Quantity Changes The equilibrium price and quantity arise from the interaction of the supply and demand curves. Thus, any time the supply or the demand curve shifts, there will be a corresponding change in the equilibrium price and quantity. Figure 2.11 shows how the market equilibrium is altered when the supply or demand curve shifts. For example, when demand increases, the equilibrium price and quantity rises; when supply increases, quantity rises but price falls.

In fact, there is no reason supply and demand cannot shift at the same time. For example, many European countries have banned the regular feeding of antibiotics to hogs in an effort to prevent bacterial resistance to antibiotics. This will increase hog production costs because hogs grow slower without the antibiotics. This results in a decrease in supply or a leftward shift in the supply curve. Were nothing else to happen, such a supply shift would result in a higher equilibrium price and lower equilibrium quantity.

However, banning antibiotics may also increase consumer demand as many people prefer hogs raised in a more "natural environment." Consumers may also want to avoid eating food that they perceive hurts a third party (antibiotic use in swine production may lead to antibiotic-resistant bacteria, which makes treating human sickness more difficult and expensive). Also, hogs raised without antibiotics will be slaughtered at an earlier age, producing more tender meat, which most consumers prefer. This results in an increase in demand or a rightward shift in the demand curve. In sum, the antibiotic ban raises production costs, decreasing the supply of pork, but also increases consumer demand for pork. How will the price and quantity of pork respond to such a ban?

Both the supply decrease and demand increase serve to increase price, so price will definitely rise. The change in quantity is less clear. The supply decrease serves to lower quantity, while the demand increase serves to raise quantity. The

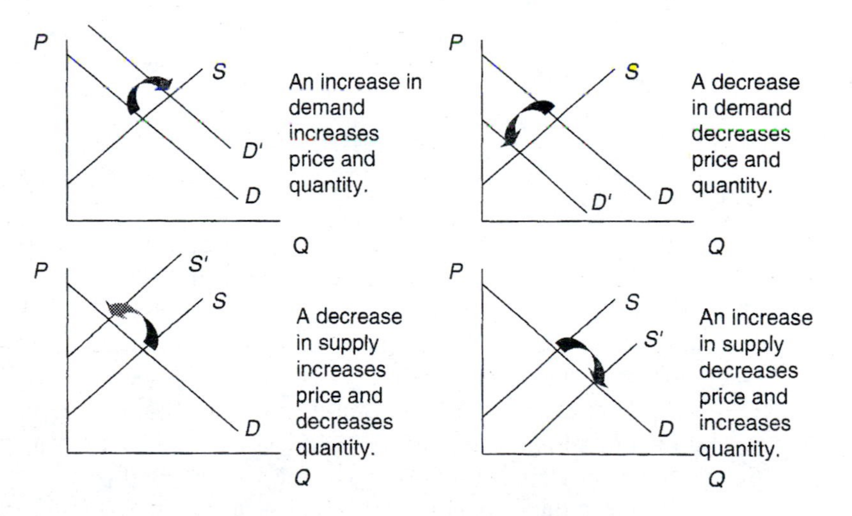

FIGURE 2.11 Equilibrium Price and Quantity Changes.

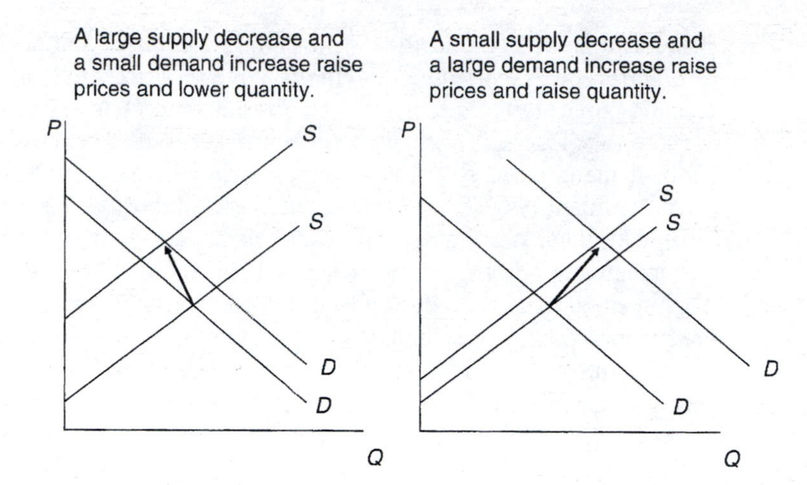

FIGURE 2.12 Simultaneous Shifts in Supply and Demand.

overall change in quantity depends on the relative magnitude of the supply and demand shifts and on the slopes of the supply and demand curves, as shown in Figure 2.12.

Here is another example. Suppose the rising popularity of high protein diets reduces the demand for wheat (because consumers purchase fewer products made from wheat, like bread), but that simultaneously the supply of wheat increases due to advances in technology, such as Round-Up Ready wheat, which makes wheat cheaper to produce. The demand decrease serves to lower prices and quantities, and the supply increase lowers price and increases quantities. In the end, all we know is that price will fall. Again, the change in quantity is unknown.

A GENERAL THEORY OF PRICES

In this last section, we discussed how prices arise in a very particular kind of market—a perfectly competitive market. Unfortunately, the real world is often more complicated. Economics is a social science involving human behavior, and humans are complicated. Prices are the result of negotiations between people, and the price formation process is anything but simple. Even though we cannot predict prices as well as physicists predict the movements of planets, economists have developed a theory of price that is logical, comprehensive, and consistent with most economic data collected. The general theory of prices is broad, broad enough that it almost cannot be wrong. Within this general theory are many "little theories" of price formation, something economists call market models. This chapter considered one of these "little theories"—the perfectly competitive market in which price was determined solely by the intersection of supply and demand. The perfectly competitive market model works well at predicting prices in some economic settings and poorly in others. Fortunately, there are many other models that cover virtually every imaginable setting. In this section, we simply want to step away from the perfectly competitive model for a few moments and speak more generally about how prices are determined. In general, prices are formed by four forces: the opportunity cost of production, consumer value, negotiating power, and psychological and social considerations.

A nursery grows hibiscus plants for sale directly to consumers. In June when the plants are typically sold, the price is about $10.00 each. What factors led to this price? One obvious factor is that some consumers must be willing to pay $10 or more. If they were not, the nursery would have to lower the price to sell the plants. Consumer value influences price. A second factor has to do with the nursery's cost. It must cost the nursery less than $10.00 to produce each plant. Otherwise the nursery would either charge a higher price or would cease growing hibiscus plants. Costs of production impact price. If the nursery has many competitors, it must charge a price closer to its cost or it risks losing customers to its rivals. Here, consumers possess more negotiating power. On the contrary, if there is only one nursery selling hibiscus plants, they can charge consumers their maximum willingness-to-pay. Here, the nursery possesses more negotiating power. The relative negotiating power of buyers and sellers impacts price. Finally, the price must pass a social test. The nursery can offer a senior citizen discount because that is socially acceptable. But it cannot offer a discount to people of a certain race—that certainly is not socially acceptable. Also, the firm may actually charge $9.99 rather than $10.00. For some reason, consumers are just more likely to purchase at $9.99. Psychological and social considerations impact price.

Prices are determined by

(1) Opportunity costs of production
(2) Consumer value
(3) Negotiating power
(4) Psychological and social considerations

This story describes the four factors comprising the general theory of price. Price is determined by (1) costs of production, (2) consumer value, (3) negotiating power, and (4) social and psychological considerations. This general theory is described below and you will find we do not exaggerate by calling it a general theory—it is so general it cannot be wrong. Due to its generality, the theory does not make specific predictions of price. It only provides a broad range in which price may fall. Earlier in the chapter we talked about the costs of production and consumer value by developing the supply and demand curves. Here we briefly discuss negotiating power and social and psychological considerations.

Negotiating Power The perfect competition model described in this chapter assumes that no single buyer or seller has any real negotiating power over the price. The market determines the price, no one buyer or seller. Many labor markets resemble perfect competition. The going salary for an agricultural reporter may be $33,000. Employers must pay this price, because, if they don't, reporters will simply work for other employers that will pay this price. The reporter herself will not ask for any more than this amount, because she knows the employer could easily replace her with someone who will work for this amount.

Not all markets are like this though. In the market for college football players there is only one real employer: the NCAA. Yes, players can choose which schools to attend, but the NCAA enacts strict rules on how the players must be compensated, so in reality NCAA is the single employer of college football players. This gives the NCAA great negotiating power—and boy do they use this power. Although a premium football player raises about $500,000 in revenues for his school each year (Brown 1993), have you ever heard of a college football player receiving anything close to this amount?

In the market for college football players, the buyer has most of the negotiating power, but in other markets it is the seller who can influence price the most. Consider Will Ferrell, who sells Will Ferrell acting services. No one can act quite like Will Ferrell, so he doesn't really have any competitors. No one else could pull off the role of Ricky Bobby in *Talladega Nights: The Legend of Ricky Bobby*. So, when it comes time to negotiate a price for Will's services, Will has most of the negotiating power and therefore negotiates a high price. This is why Will Ferrell is richer than you, and most actors for that matter.

In Chapter 3 we take on the issue of negotiating power directly. Starting with supply and demand curves, we modify the perfect competition model to reflect conditions when buyers have all the negotiating power, when sellers have all the negotiating power, and everything in between.

Psychological and Social Considerations Economics is a *social science*. At the center of economics are humans. We can never ignore human psychology or social considerations. Let us illustrate with an example. Day-care centers have a common problem of parents showing up late to pick up their kids. If people respond to incentives, then why not fine parents who show up late? Indeed, an experiment was conducted in Israel where parents were fined $3 if they arrived ten minutes late. The financial incentive backfired, and the number of late pickups more than doubled after the fine was enacted.

Does this experiment suggest people do not respond to incentives, or that they respond to incentives in an irrational manner? No, it suggests that people respond to moral and social incentives as well as financial. Chances are you would feel guilty picking up your child late, knowing workers in day-care centers have lives too. Guilt is an incentive—a social incentive. A $3 dollar fine is a low fine, which sends a signal to parents that showing up late is not that big of a deal, reducing the guilt associated with being late. Even when the fine was abolished, parents still arrived late at a greater rate than before the fine, because the low fine signaled parents should not feel too guilty about being late. Never underestimate the power of social pressure (Levitt and Dubner 2005).

We can offer no good reason why so many goods sell for $0.99, $1.99, or $199. The difference between these prices and $1.00, $2.00, and $200 shouldn't matter, but the ubiquity of prices that end in "99" suggests it does. The only explanation we have is that our brains are simply wired this way. Television infomercials always throw in something free. They may sell a skillet-radio combination for $19.99, but throw in a free set of headphones. In reality, you are buying the skillet-radio and the headphones for $19.99; there is really nothing free. But that is not how the sale is framed, and psychologists have learned that how a situation is framed plays an important role in human decisions. Otherwise, why would infomercials do this?

Christian Brothers and E & J are two competing brandies. After years of having a lower market share, E & J started gaining market share, despite the fact that both were the same price, both were readily available, and neither had an advertising advantage. In blind taste tests, people preferred both brandies equally, and market research showed people preferred the name "Christian Brothers" to "E & J." So what was E & J's secret? Their bottle. In taste tests when people saw each product's bottles, they preferred E & J. When Christian Brothers was served to people out of E & J bottles, people preferred the Christian Brothers in the E & J bottle. After redesigning their bottle to look like the E & J bottle, Christian Brothers regained their market share. This is no aberration; people never accepted margarine until it was colored yellow artificially to look like butter. Consider another interesting study. A group of neurologists put people under an MRI machine and had them taste samples of Coke and Pepsi. When people participated in blind taste tests, brain scans revealed similar neural activity when people tasted Coke and Pepsi. But when people were told they were tasting Coke, the scientists witnessed much more brain activity and a stronger preference for the Coke brand (McClure et al. 2004). The value consumers place on goods go beyond the good itself to how the good is packaged and how the good looks (Gladwell 2005).

It makes sense that consumers would like to negotiate lower prices and producers higher prices. Lower prices increase consumer surplus, and higher prices increase producer surplus. However, people hold a grudge when they feel they are treated unfairly, and businesses may forego short-term profits to ensure consumers' patronage. Ideas of fairness often enter the price formation process. Suppose there is a large snowstorm, much larger than anyone anticipated. Households will now pay a higher price for snow shovels. This gives sellers of snow shovels a rare opportunity to raise prices and extract greater profits. Will they raise prices? Perhaps, but perhaps not. When people were asked in a survey whether raising the prices is "fair" in situations like this, 82% of respondents said it was unfair. A business behaving in what consumers feel is an unfair manner is unlikely to maintain consumer patronage. To avoid this, the manager may keep the price unchanged despite the opportunity to price gouge. In this case, perceptions of fairness impact the price of snow shovels (Kahneman, Knetsch, and Thaler 2001).

Think about Thanksgiving, when almost every American household buys a turkey. The value of turkeys rises, and just like the snow shovel story, this gives sellers an opportunity to raise prices. But generally they do not. Turkey prices are about the same as the rest of the year and often seem lower. Several reasons are possible. One is that firms produce more turkeys in anticipation of the holidays, and if there are plentiful turkey supplies, consumers need not bid the price up. The second is that grocery stores want to lure consumers in with reasonable turkey prices to make money off other sales—like beer to help calm the nerves when obnoxious family members drop by. Another reason has to do with fairness. If consumers see grocery stores jacking up prices during the holidays, they will not just shrug this off as the result of economic laws. They will be disgruntled, they will complain, and they will not speak kindly of the store. No store wants this, as long as there are other competitors in town.

The value individuals place on a good can even be dependent on the cost of the good. When asked how much they would pay for a beer, individuals stated a higher price when the beer came from a fancy hotel than from a mom-and-pop store. Presumably, hotels sell beer at higher prices than mom-and-pop stores, and this information influences the value of the beer. The amount people will pay to clean up hazardous waste also depends on the cost of the cleanup. Even though we like to think the value of a good depends solely on the happiness one receives from its consumption, regardless of the cost, this does not always hold (Baron and Maxwell 1996). The point of this section is something you probably already knew. We are humans, and we live in a society with other humans. Psychological and social considerations enter into many of our decisions, including our purchasing, selling, and price negotiation decisions. This should not be ignored when discussing how prices are formed and why prices change.

SUMMARY

The price formation process is simple from one point of view but terribly complex from another. The opportunity costs of production, consumer value, negotiating power, and social and psychological considerations are the four major factors determining price. The problem is that they do not tell us exactly what the price will be. It is like asking a meteorologist tomorrow's weather forecast and she tells you it depends on the temperature, precipitation, wind speed, and barometric pressure. The forecast is correct but not very useful. The exact price in any given setting depends on the particular setting. If buyers and sellers have roughly equal negotiating power, the model of perfect competition provides a specific price prediction, just like a meteorologist provides a specific weather forecast detailing the probability of rain, the expected temperature, and expected wind speed. Like the accuracy of a weather forecast depends on the quality of the forecaster, the accuracy of a market model depends on the extent to which it properly describes the market setting.

CROSSWORD PUZZLE

For answers with more than one word, leave a blank box between each word.

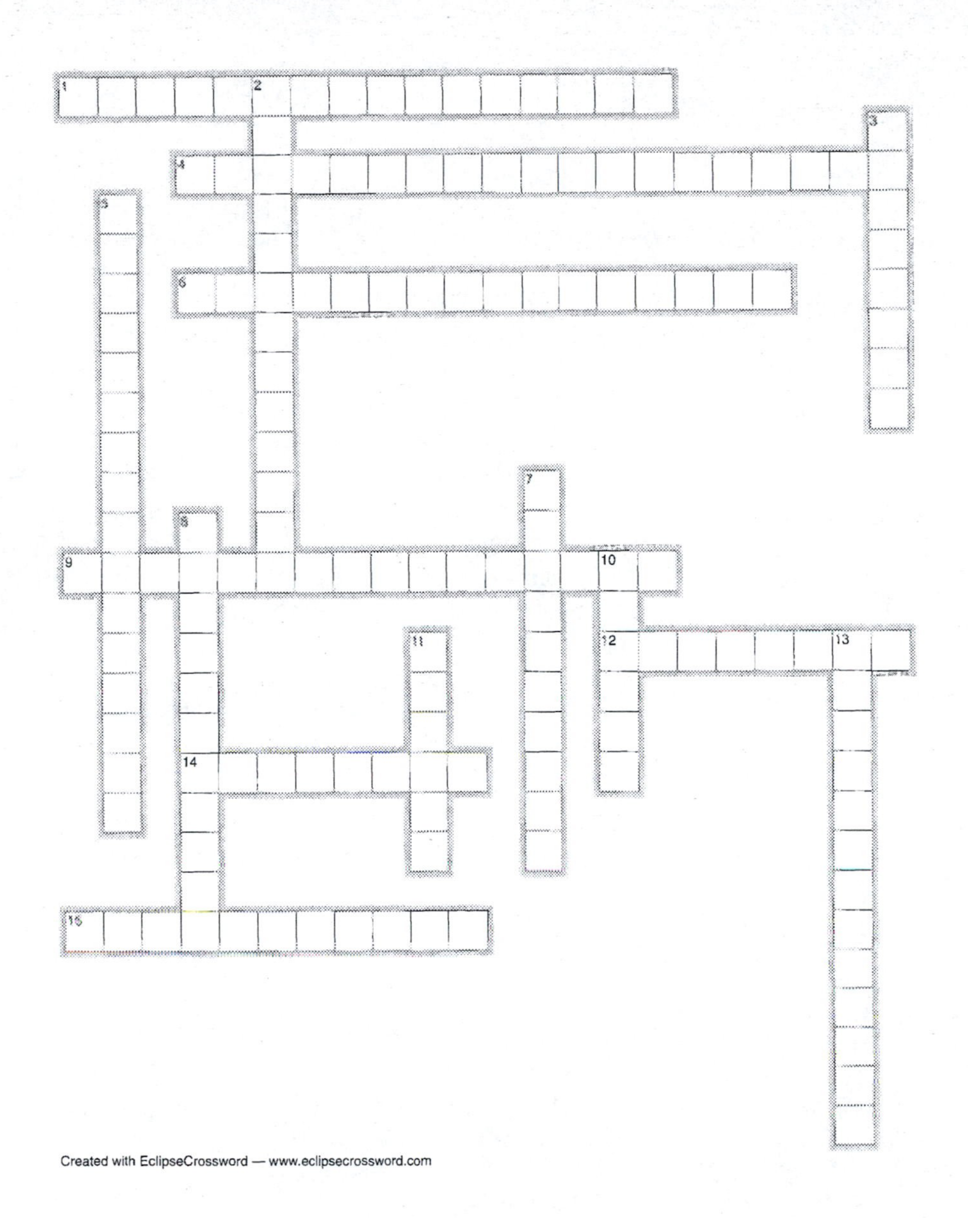

Across

1. Revenues minus opportunity costs.
4. The _______ _______ model assumes many buyers and sellers, each with roughly the same market power.
6. This term refers to the value consumers extract from the opportunity to purchase a good at a particular market price.
9. The value of the next best alternative.
12. Profits not including fixed costs are measured by the area below price and above marginal cost for all quantities produced, which is referred to as _______ surplus.
14. Consumers' willingness-to-pay for one additional unit of a good is referred to as _______ value.
15. If a price increase of one good causes the demand to increase for another good, those two goods are _______.

Down

2. The additional cost incurred from increasing output by one unit.
3. The demand for a(n) _______ good rises as incomes fall.
5. The change in output realized from increasing input use by one unit.
7. The three main supply curve shifters are the price of related outputs, the price of inputs, and _______.
8. If a price decrease of one good causes demand for another good to increase, those two goods must be _______.
10. The _______ curve indicates the quantity firms will supply at each possible market price.
11. The _______ curve indicates the quantity consumers will purchase at each possible market price.
13. A price lower than the equilibrium price leads to a(n) _______ _______.

STUDY QUESTIONS

1. Ricky Bobby walks onto a used car lot interested in purchasing an El Camino. After finding one he likes, he determines the value he places on the car at $5,000. The used car dealership initially purchased the El Camino for $3,000 and believes, if not sold to Ricky, the car could be sold soon for $4,000 but no more. What is the opportunity cost of selling the El Camino to Ricky Bobby? Will the two likely strike a deal? What do you know about the price that will be struck (give a price or range of prices)?
2. Ashley owns an agricultural advertising agency, which is currently making zero economic profits. Does this imply that Ashley is not making any money? Does this imply that unless Ashley shuts down business she will go bankrupt? The answer is no. Explain why.
3. *Fill in the blanks*. Economists assume that the marginal opportunity cost of production is _______ in quantity produced and the marginal consumer value is _______ in quantity consumed.
4. Producer surplus measures (all costs refer to opportunity costs) (*Circle all that apply*.)
   a. total revenues minus variable costs
   b. total revenues minus fixed costs
   c. total revenues minus total costs
   d. total revenues
5. A consumer's marginal value of the first and second hamburger is $5 and $3, respectively. If both hamburgers are purchased at a price of $2, what is the consumer surplus?
6. If a price is above the equilibrium price, we say there is (*Circle all that apply*.)
   a. an excess supply
   b. an excess demand
   c. neither an excess supply nor an excess demand
7. If the price of a substitute good falls, the demand for a product (*circle one*) INCREASES / DECREASES.

8. If the price of a complement good falls, the demand for a product (*circle one*) INCREASES / DECREASES.

9. If a good is a normal good and income rises, the demand for the product (*circle one*) INCREASES / DECREASES.

10. If a good is an inferior good and income rises, the demand for the product (*circle one*) INCREASES / DECREASES.

11. An increase in the marginal cost of production for a good (*circle one*) INCREASES / DECREASES / DOES NOT CHANGE the supply of that good.

12. In a supply and demand diagram, if supply decreases (the supply curve shifts to the left),
    a. the equilibrium price rises
    b. the equilibrium price falls
    c. the change in the equilibrium price is ambiguous

13. In a supply and demand diagram, if demand increases (the demand curve shifts to the right),
    a. the equilibrium price rises
    b. the equilibrium price falls
    c. the change in the equilibrium price is ambiguous

14. In a supply and demand diagram, if supply decreases (the supply curve shifts to the left),
    a. the equilibrium quantity rises
    b. the equilibrium quantity falls
    c. the change in the equilibrium quantity is ambiguous

15. In a supply and demand diagram, if demand increases (the demand curve shifts to the right),
    a. the equilibrium quantity rises
    b. the equilibrium quantity falls
    c. the change in the equilibrium quantity is ambiguous

16. In a supply and demand diagram, if demand increases and supply decreases,
    a. the equilibrium quantity rises
    b. the equilibrium quantity falls
    c. the change in the equilibrium quantity is ambiguous

17. In a supply and demand diagram, if demand decreases and supply decreases,
    a. the equilibrium quantity rises
    b. the equilibrium quantity falls
    c. the change in the equilibrium quantity is ambiguous

18. In a supply and demand diagram, if demand increases and supply decreases,
    a. the equilibrium price rises
    b. the equilibrium price falls
    c. the change in the equilibrium price is ambiguous

19. In a supply and demand diagram, if demand decreases and supply decreases,
    a. the equilibrium price rises
    b. the equilibrium price falls
    c. the change in the equilibrium price is ambiguous
20. In a supply and demand diagram, if demand increases and supply increases,
    a. the equilibrium price rises
    b. the equilibrium price falls
    c. the change in the equilibrium price is ambiguous
21. In a supply and demand diagram, if demand decreases and supply increases,
    a. the equilibrium price rises
    b. the equilibrium price falls
    c. the change in the equilibrium price is ambiguous
22. In the graph in Figure 2.13, indicate the equilibrium price and quantity, shade in consumer surplus labeling it "*CS*," and shade in producer surplus labeling it "*PS*."

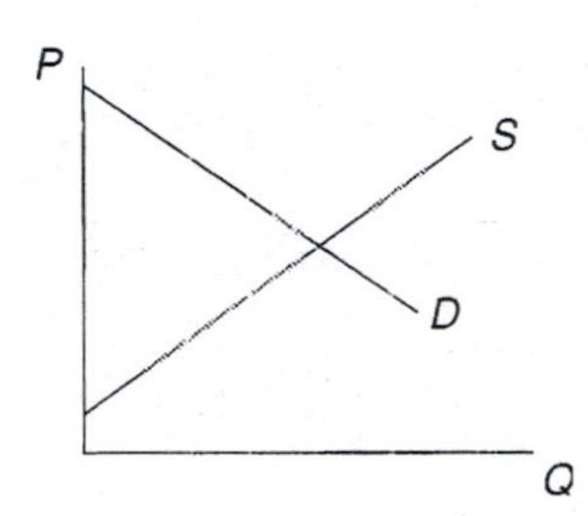

FIGURE 2.13

23. The graph in Figure 2.14 shows a price below the equilibrium price. Indicate the quantity supplied and quantity demanded at that price, and indicate the excess demand.

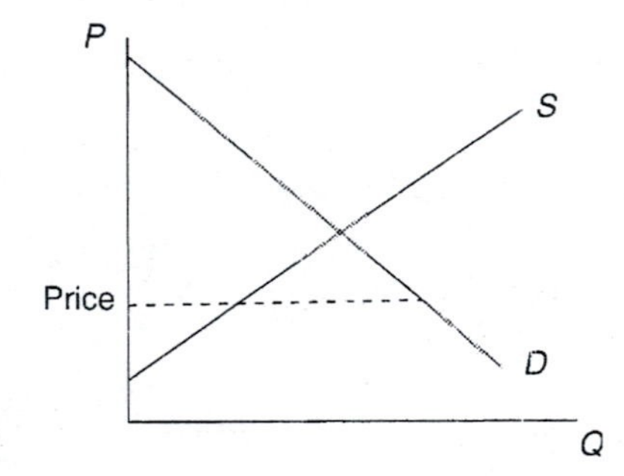

FIGURE 2.14

For the following question, use the following marginal cost and marginal value formulas.

$$\text{Marginal Cost/Supply Curve: } P = 150 + 10(Q)$$
$$\text{Marginal Value/Demand Curve: } P = 800 - 15(Q)$$

24. Calculate the price and quantity using the formulas above, assuming perfect competition.

CHAPTER THREE

Advanced Price Analysis: Mastering Supply and Demand

In the News

Americans are fatter than ever. During recent decades, the number of overweight adult Americans increased from 47% to 64%, and the number of obese adults increased from 15% to 31% (CDC 2004). Being overweight or obese is associated with numerous health problems, such as heart disease, stroke, and cancer, just to name a few. Estimates suggest that costs attributable to both overweight and obesity were as high as $92.6 billion in 2002 dollars (CDC 2002). Because roughly half of these costs were paid by Medicaid and Medicare, obesity represents a serious problem not just to people who are overweight themselves but to all Americans who must pick up the tab through higher taxes.

The increasing rate of obesity has led some health experts to propose a "fat tax" or "Twinkie tax," whereby the government would tax unhealthy foods, such as soft drinks, chips, or other junk food. Indeed, some 18 states are currently using different forms of high-calorie taxes. The idea is simple. A "fat tax" will increase the price of unhealthy food and because people do not like paying higher prices, consumers will switch to the now relatively cheaper and more healthful alternatives. Will it work? Despite the logical appeal of the "fat tax," there are many economists who think not. The reasoning is related to the steepness of the demand curve for food—or the elasticity of demand—and other demand shifters. Consider the following remarks made by economists who have studied the issue:

> "The Law of Demand states that a price increase will result in a reduction in the quantity of the good consumed. However, it is not necessarily the case that weight will also decline when ready substitutes are available." (Schroeter 2005)

> "We find own-price elasticities of demand [for dairy products] are relatively inelastic. . . . A fat tax may be an effective means to raise revenue, but will not result in a significant reduction in fat consumption." (Chouinard et al. 2005)

> "Some health activists and health researchers have argued for a tax on snack foods. . . . Price elasticities are critical information for forecasting tax impacts. . . .

> It is obvious that a small tax on salty snacks would have very small dietary impacts. Even a larger tax would not appreciably affect overall dietary quality of the average consumer." (Kuchler, Tegene, and Harris 2005)

INTRODUCTION

As illustrated in the previous chapter, supply and demand, coupled with the assumption of perfectly competitive markets, provide a framework for determining how prices are formed and how they change. The previous chapter provides the background for knowing whether prices will generally move up or down, but we need to know more if we want to say *how much* prices will increase or decrease when, for example, a fat tax is placed on Twinkies. This chapter has four main objectives:

1. to describe elasticities of supply and demand
2. to differentiate between supply and demand in the short run and long run
3. to use elasticities in an equilibrium displacement model to forecast price and quantity changes
4. to use demand functions to obtain demand elasticities and demand curves

ELASTICITIES OF DEMAND AND SUPPLY

In the previous chapter, we listed factors that cause supply and demand curves to change due to changes in other factors like the price of related goods and production costs. Often, we want to say more. We want to say exactly how much a demand curve shifts or how much a price falls. For example, the beer industry is so concerned with how beer demand is affected by the price of soda and distilled spirits that they have "file cabinets full of interesting and useful" supply and demand models (Tremblay and Tremblay 2005, foreword). A common method of forecasting market changes is to use supply and demand elasticities.

Elasticities measure sensitivity. Consider two variables: X and Y. If X causes changes in Y, the elasticity of Y with respect to X measures how much Y changes when X changes. It measures the sensitivity of Y to changes in X in percentage terms. For example, suppose X is the number of times your roommate asks you to turn down the volume on your stereo, and Y is the volume of your stereo. If you're like most college students, Y is not very sensitive to X. The elasticity of your stereo's volume to the roommate's requests is low. But, what if instead of X being the number of times your roommate asks you to turn down the volume, X is instead the number of times the coolest kid in school—the proverbial Fonzie—asks you to turn down the volume on your stereo. Now, Y is likely to be very sensitive to X. When Fonzie asks you to turn down the volume, you do it: The elasticity of your stereo's volume to Fonzie's requests is high.

There is a mathematical formula for measuring elasticity. As before, let X be a variable that causes changes in Y. Suppose we observe a percent change in X, and denote it as $\%\Delta X$ where the Δ means "change." We then observe the percent change in Y in response to the change in X, and denote it as $\%\Delta Y$. The elasticity of Y with respect to X is then calculated as

$$\text{Elasticity of } Y \text{ with Respect to } X = E_{Y,X} = \frac{\%\Delta Y}{\%\Delta X}.$$

The higher the elasticity, the more Y changes in response to changes in X—the more sensitive Y is to changes in X. Think about it. If a small percent change in X causes a large percent change in Y, the elasticity formula above will have a high value, indicating Y is very sensitive to changes in the value of X. Elasticities are useful because they have no units, they reflect only ratios. This is useful as can be seen in the above example; what units should be used to measure stereo volume? As long as we use elasticities, we can measure stereo volume using any scale. Elasticities are also useful because they are regularly used by market analysts to forecast price and quantity changes in markets. If we rearrange the elasticity equation, we obtain

$$\%\Delta Y = E_{Y,X}(\%\Delta X).$$

Thus, if economists estimate $E_{Y,X} = 0.5$, and we expect X to change by 10%, then we can predict that Y will change by $0.5 \times 10 = 5\%$.

Own-Price Elasticities of Supply and Demand

Shifts in supply and demand cause changes in prices and quantities. However, the exact change in price or quantity depends on the slope—or steepness—of both curves. Notice that the slope of a demand curve tells us how sensitive consumers are to price changes. Economists measure the sensitivity of consumers to price changes using the *own-price elasticity of demand,* which is calculated as the percent change in quantity demanded divided by the percent change in price. If the slope of the demand curve is steep, then consumers are not very sensitive to price changes, because a large price change will only modify consumer purchases by a small amount. This low sensitivity to price is referred to as an *inelastic demand.* Technically, a demand curve is considered inelastic if the own-price elasticity of demand is greater than −1 (or less than 1 in absolute value). This would mean that a 1% price change would cause a less than 1% change in the quantity demanded. Because the quantity changes by a proportionally smaller amount than the price changes, we can say that the quantity demanded is not very sensitive to price changes or that the demand is inelastic. Conversely, if the demand curve looks flat (has a small slope), a change in price will cause a large swing in quantity demanded. A demand curve with high sensitivity to price changes is called an *elastic demand.* Technically speaking, a demand curve is considered elastic if the own-price elasticity of demand is less than −1 (or greater than 1 in absolute value). This would mean that a 1% price change would cause a more than 1% change in quantity demanded. Because the

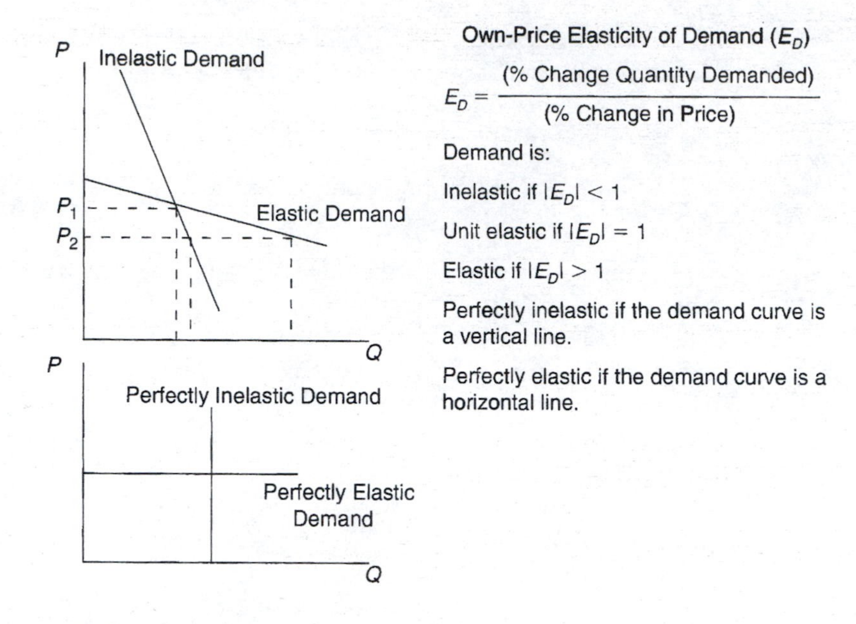

FIGURE 3.1 Elasticity of Demand.

quantity changes by a proportionally larger amount than the price changes, we can say that the quantity demanded is sensitive to price changes or that demand is elastic.

Figure 3.1 shows inelastic and elastic demand curves. For an identical price decrease from P_1 to P_2, the quantity purchased increases only a little for the inelastic demand but increases much for the elastic demand. Whether the demand for a good is inelastic or elastic (or somewhere between) depends on the availability of substitutes. There are no substitutes for food; we absolutely need it to live. Next to water, food is the most important element of life. Even if food prices double, we would still purchase a lot of food (assuming a lot of food is available for purchase), because we need it to live. Food has no substitutes, and so food demand is inelastic. Look at the formula for the elasticity of demand: (percent change in quantity demanded)/(percent change in price). Suppose food prices rise 10%, so we would put the number 10 in the denominator. Further, suppose that this rise in food prices of 10% caused consumer purchases to fall by only 1%. We would then put −1 in the numerator. The elasticity is then calculated as $-1/10 = -0.1$. The elasticity of demand is always negative and in this case is less than one in absolute value, indicating a low sensitivity to price. Food demand is inelastic.

Now consider a particular type of food: Hungry Man TV dinners. Hungry Man dinners have many substitutes. Anything you can eat besides Hungry Man dinners is a substitute. If the price of Hungry Man TV dinners increases, consumers will quickly switch to a substitute, especially other TV dinners. In this case, because there are many substitutes, consumers are very sensitive to price changes, and we say the demand is elastic. Sometimes demand can be *perfectly elastic*, where demand is a horizontal line. If you are a barley farmer who can only sell at the going market price, and since you are such a small part of the market you can sell as much as you want at that price, the demand for your barley is perfectly elastic (a horizontal line at the market price). Diabetics absolutely need insulin to live and need the same amount of insulin each day regardless of the price. In this case, insulin consumption is the same regardless of price (assuming people have the money to pay any price), making its demand curve a vertical line and its elasticity perfectly inelastic.

Demand elasticities are important for a number of reasons, one of them being that they tell us whether a change in supply will have a larger impact on price or quantity. In Figure 3.2 there are two supply and demand diagrams. The supply curves are identical in each diagram and the supply curves shift leftward by the same amount. The diagrams differ in that the left diagram shows an elastic demand and the right shows an inelastic demand. When the demand is elastic, the supply curve shift decreases the quantity by a lot, but increases the price by only a small amount. The opposite effect occurs when the demand is inelastic: The decrease in supply greatly raises the price but has only a modest impact on quantity.

Why would this matter to anyone? Suppose a government wishes to raise tax revenues by taxing producers for each unit they sell. This increases the cost of production to the firm, leading to a decrease in supply and an upward shift of the supply curve. The amount of revenue raised by the tax equals the tax amount times the number of units sold. If the demand is elastic, as in the left diagram of Figure 3.2, the tax will dramatically decrease the quantity sold, and little revenue will be raised. Conversely, if demand is inelastic, the tax will have only a small impact on the quantity, and the government can be assured of high tax revenues. Think back to our introductory example on the "fat tax." We have already argued that demand for food, in general, is inelastic. Thus, what would be the anticipated effects of a

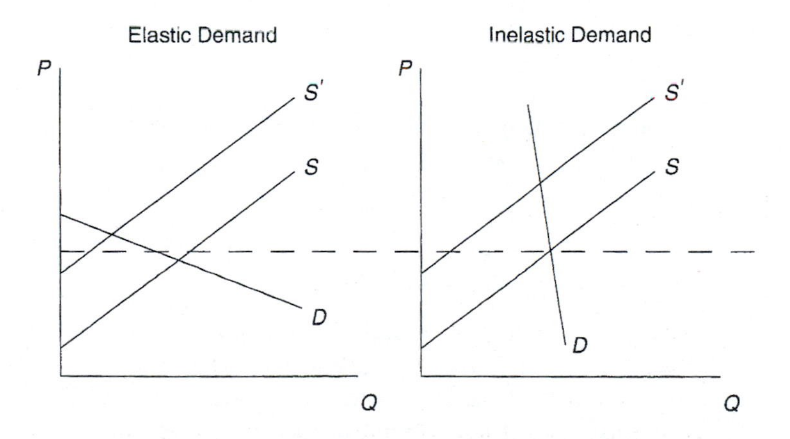

FIGURE 3.2 Impact of Supply Shifts When Demand Is Inelastic and Elastic.

tax on food? Figure 3.2 suggests that we would expect food consumption to fall, but not by very much, but it also suggests the government will collect high tax revenues.

To further illustrate, consider taxes placed on alcoholic beverages. The elasticity of demand for beer, wine, and distilled spirits is about −0.3, −1.0, and −1.5, respectively (Chaloupka, Grossman, and Saffer 2002). Beer has an inelastic demand, wine has unit elastic demand, and the demand for distilled spirits is elastic. If taxes are meant to deter alcoholic consumption, they will work well for distilled spirits, because consumers are sensitive to price changes, but not for beer. If the taxes are to raise revenue, taxing beer will be effective, but not so for wine. When he was president, Franklin Roosevelt and his administration made a particularly large blunder in failing to understand demand elasticities. In 1933, under Roosevelt's direction, the United States Department of Agriculture paid cotton farmers $100 million to plow under 10 million acres of farmland. The idea was to shift the supply curve leftward, which if demand was inelastic, would lead to a large rise in prices for cotton farmers. However, the demand for cotton was not inelastic. Wool, silk, vegetable, and synthetic fibers were perfectly good substitutes for cotton, making the demand for cotton elastic (Powell 2003). Consequently, the supply decrease did not raise prices as much as the old Democrat hoped. President Roosevelt's plan to help cotton farmers backfired because he mistook elastic demand for an inelastic demand.

We can also calculate the own-price elasticity of supply (hereafter, elasticity of supply). Take the percent change in quantity supplied due to a price change and divide it by the percent change in price. If price rises by 5% and firms respond by increasing production by 10%, the elasticity of supply is 10/5 = 2, which would be an elastic supply. The elasticity of supply is always positive because the supply curve slopes upward: firms sell more when price increases. The supply elasticity measures how sensitive firms are to price changes. If firms are sensitive to price changes, we say supply is elastic. In this case, firms will increase production by a lot when the price rises, and decrease production greatly if the price falls. Technically, if the own-price elasticity of supply is less than 1, we say that the supply is inelastic. This occurs because increasing the price by 1% causes a less than 1% increase in production. An inelastic supply describes a market where firms are not sensitive to price changes, and increase or decrease production only a little when the price changes. When the supply elasticity is greater than 1, it is considered elastic. Firms in an industry with an elastic supply change production decisions by a relatively large amount when the price changes. If the price rises X%, the industry ramps up production by more than X%.

Figure 3.3 shows supply curves with different elasticities. When the price rises from P_1 to P_2, markets with an elastic supply respond by increasing production much more than markets with an inelastic supply. Whether supply is elastic or inelastic largely depends on whether there are other goods the firm can produce. Cattle producers in areas with fertile land, deep topsoil, and good rainfall could easily raise row-crops instead. An example would be Illinois cattle producers. Contrast this with cattle producers in the foothills of eastern Kansas where there is

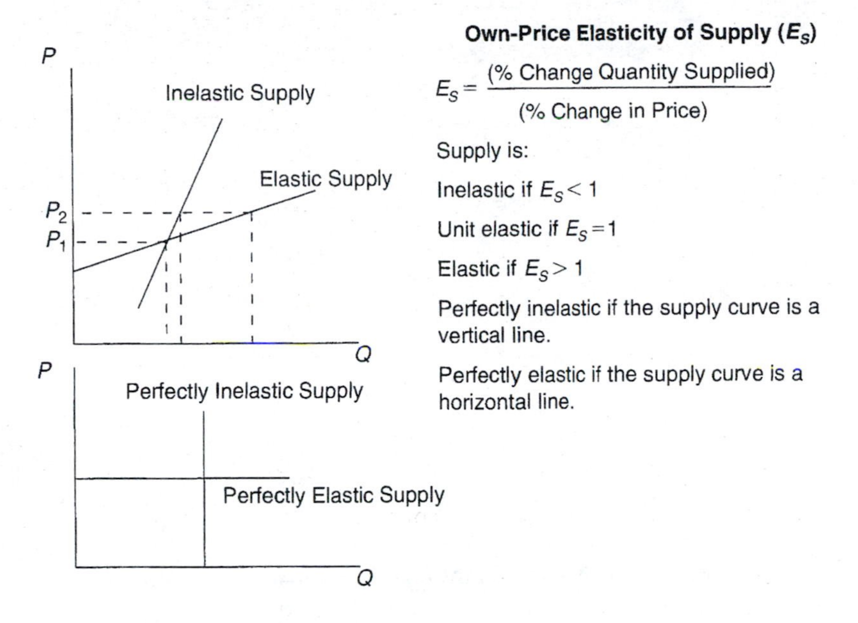

FIGURE 3.3 Elasticity of Supply.

only a thin layer of soil and everything else is rock. You cannot plow this land, and only grass grows well. Illinois cattle producers will have an elastic cattle supply. If the price of cattle falls, they can simply switch to row-crops, and will therefore reduce the quantity of cattle they supply dramatically. Kansas cattle producers have few alternatives, so they must simply take the fall in cattle prices in stride. Although they may reduce the quantity of cattle they produce, it will not be as large as the Illinois cattle producers' reduction. The supply of cattle from Kansas is inelastic, but the supply from Illinois is elastic.

In agriculture there are cases of a perfectly inelastic supply. Wheat is planted in the fall and harvested in June–July. After July, no new wheat is produced until the next harvest (notice we are ignoring the possibility of imports, which is small compared to the U.S. harvest). This fixed amount of wheat will not change with prices—it has already been produced and harvested. Regardless of whether wheat prices rise or fall throughout the year, no more wheat can be produced until the next harvest. In this case, the supply curve is a vertical line, and supply is perfectly inelastic. The market for slaughter cattle also displays something close to a perfectly inelastic supply. It takes roughly two years from the time cattle are bred until their offspring are made into beef. Even though cattle producers respond to higher prices by breeding more cattle, it takes another two years before those additional cattle are slaughtered. Thus, in any one month, the number of cattle ready to slaughter is fixed. The cattle supply today was determined two years ago. It cannot be changed for another two years, and so supply is not very responsive to price. Supply is not perfectly inelastic, however, because existing cattle can be raised to be heavier or lighter, affecting total pounds produced.

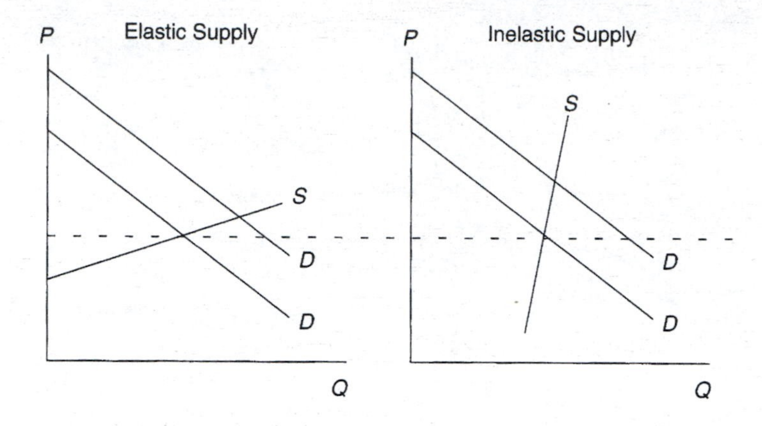

FIGURE 3.4 Impact of Demand Shifts When Supply Is Inelastic and Elastic.

Supply elasticities are useful for determining how changes in demand will impact prices and quantities. As you can see in Figure 3.4, an identical demand shift has a larger impact on prices when supply is inelastic and a larger impact on quantity when supply is elastic. The difference is important. Consider the beef marketing checkoff, an institution that taxes cattle producers and uses the revenues to fund beef advertisements like the "Beef: It's What's for Dinner" campaign. The goal of the checkoff is to increase cattle producer profits by increasing the demand for beef. Some studies have found that the advertisements have only a modest impact on prices. Should we therefore conclude that since the advertisements do not raise prices much, they do not increase profits much? No. If the elasticity of supply for beef is elastic, a rise in demand will have little impact on prices but will have a large impact on quantity. Thus, profits could significantly rise despite the small price increase. Though the price rise is small, it is spread over many more cattle. That is, people are not making much more off each animal they sell, but they are selling a lot more cattle (Davis 2005).

Hopefully you see the importance of elasticities. Elasticities do not sound like a fun concept, but understanding demand elasticities helps to forecast the impact of taxes, and understanding supply elasticities helps to determine whether advertising is beneficial to an industry. Elasticities are also used for forecasting changes in supply and demand. Notice that if we rearrange the own-price elasticity formulas, they can be written as:

$$\%\text{ Change in Quantity Supplied} = (E_S)(\%\text{ Change in Price})$$
$$\%\text{ Change in Quantity Demanded} = (E_D)(\%\text{ Change in Price})$$

By simply plugging in values for the own-price elasticities of supply and demand and changes in price, we can forecast how supply and demand will change. This will become important later in the chapter when we develop equilibrium displacement models.

Supply and Demand in the Short Run and Long Run

Until now we have said little about the time frame surrounding our supply and demand curves. The quantity could be units produced and consumed per day, per week, or over a 10-year time period. The time frame matters. For simplicity, let us look at only two time frames: the *short run* and the *long run*. In the short run some inputs cannot be varied by the firm and the number of firms in the industry is fixed. In the long run, all inputs can be varied and firms are free to enter and exit the industry.

Short Run: Period of time in which at least one input for the firm is held fixed and the number of firms is fixed.

Long Run: Period of time after which all inputs can be varied by the firm and firms may freely enter and exit the industry.

The difference between the short and long run is demonstrated nicely by considering the market for pecans. It takes around 10 years between the time pecan trees are planted and the time they start producing pecans. The short run for pecan production is then about 10 years and the long run is 10 or more years. If pecan producers wish to increase pecan production in response to higher prices, there is little they can do immediately. They can harvest more carefully, making sure every pecan is retrieved, and even increase fertilizer applications or irrigation, but the trees that are available will only produce so much. The number of trees is fixed, making it is difficult to increase production. In the short run (1–10 years) supply of pecans is inelastic. Price increases elicit only small increases in production, because there is little the farmer can do to increase production. However, given 10 years or more, farmers can plant new trees. When the price increases and farmers are given time to plant new trees, they can increase production by a greater amount. The supply curve is therefore more elastic in the long run than the short run.

Think about a beef processing firm that wants to increase the number of cattle slaughtered each week. Suppose the firm currently has a single processing plant. The firm can increase production either by paying its workers overtime or by constructing a new plant. Overtime pay can be expensive, but if a new plant is built, new workers can be hired at regular pay. On a per pound basis, it cheaper to produce more beef by opening a new plant than by having everyone work overtime. However, it takes time to build a new plant, let us say two years. The input "number of plants" is fixed in the short run but is variable in the long run. The short run is therefore two years and the long run is more than two years. When beef prices rise, beef processors can increase production more by building a new plant rather than having workers at its existing plant work overtime, making the supply curve more elastic in the long run than the short run.

Demand is also more elastic in the long run than the short run. At the time this chapter was being written gasoline prices were around $3.00, which was significantly higher than in previous years. Yet, most people consumed roughly the same amount of gas. The demand for gas, in the short run, is quite inelastic. However, if gasoline prices remain high, people will find other ways to cope. People will switch from driving SUVs to hybrids, they will begin carpooling, and even make fewer commitments that require driving. In the short run, consumers have few options to deal with rising prices, but in the long run they have more options and will therefore be more sensitive to price changes. This is just another way of saying demand is more elastic in the long run than in the short run.

Because supply and demand can be so different in the short and long run, sometimes we refer to short-run and long-run supply and demand curves. The long-run supply curve can take a particularly interesting shape. One of the authors grew up in South Carolina, where farmers frequently said that it costs at least $2.25 per bushel to grow corn. Suppose that regardless of where you grow corn in the United States, the minimum average cost is indeed $2.25 per bushel including opportunity costs. This means that whenever the price is greater than $2.25, economic profits can be made from raising corn. We would then expect that farmers would plant more corn whenever the price is greater than $2.25 per bushel. Some new farmers would even enter the picture to realize corn profits. This increases the supply of corn, depressing corn prices. If the price was less than $2.25, farmers would plant less corn and some people would leave corn farming for more profitable activities. This decreases the supply of corn, raising corn prices.

If this sounds like the Indifference Principle, you are thinking correctly (we won't insult you by calling you economists, but congratulations nevertheless!). On average, people (or, better said, those with the ability to raise corn) should be indifferent between raising corn or earning money some other way. If people prefer raising corn, more people will raise corn, increasing the corn supply and lowering corn prices. If people prefer something other than raising corn as their employment, fewer people will raise corn, decreasing the corn supply and raising prices.

Suppose that the Indifference Principle holds and the corn price exactly equals $2.25. Then, suppose demand increases, raising prices to $2.50. Farmers will respond to this high price by planting more corn and increasing the quantity of corn supplied. The corn supply will keep rising over time, and corn prices will keep falling, until the price again equals $2.25. Regardless of the demand curve, the price will be $2.25. In the long run, when firms are given the opportunity to enter and leave the corn market, firms will produce however much consumers want at $2.25. This leads to a perfectly elastic supply curve—a horizontal supply curve at $2.25, as shown in Figure 3.5. And notice the supply curve is horizontal at $2.25, the *minimum average cost of corn production*.

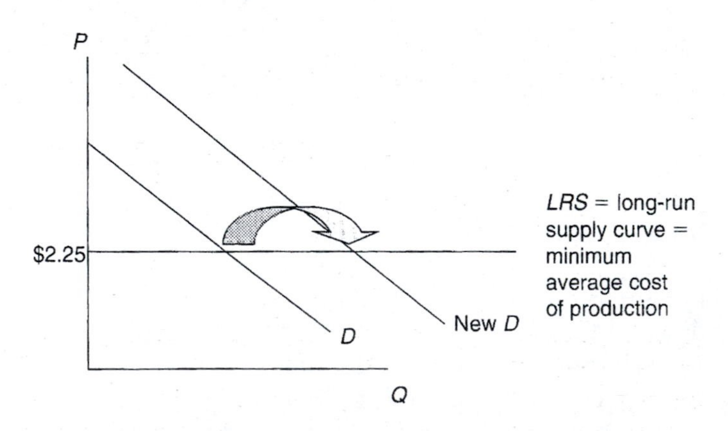

FIGURE 3.5 The Long-Run Supply Curve Can Be Perfectly Elastic.

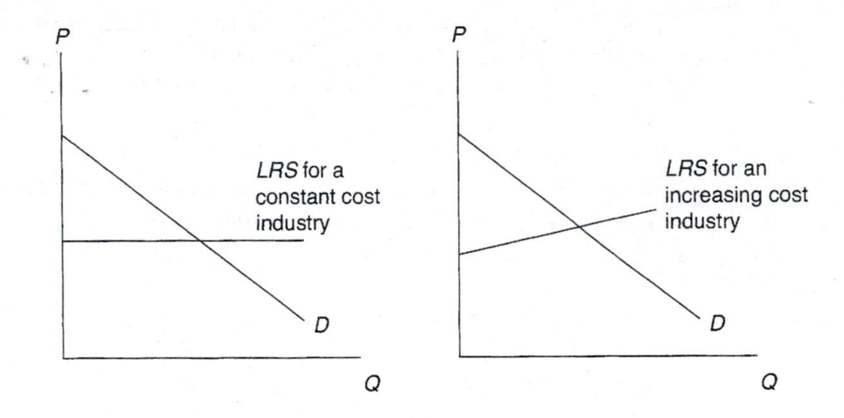

FIGURE 3.6 The Long-Run Supply Curve Equals the Minimum Average Cost of Production for Firms.

Constant Cost Industry: An industry where input prices are fixed regardless of how much the industry produces.

Increasing Cost Industry: An industry where input prices rise as the industry increases production.

There is one important assumption leading to this long-run horizontal supply curve, that the minimum average cost of $2.25 is the same regardless of the amount of corn produced. This is true only for *constant cost industries*, where input prices remain the same no matter how much of the good is produced. In a constant cost industry, the price of fertilizer and corn seed is the same if one bushel is produced or a trillion. Perhaps a more realistic industry is an *increasing cost industry*, where input prices increase with the amount of good produced. Keeping with the corn production example, if more corn is produced, more fertilizer is needed. This increases the demand for fertilizer, which should increase fertilizer price. When fertilizer prices rise, the minimum average cost of corn production must rise to something above $2.25. In this case, the long-run supply curve still equals the minimum production cost, but that minimum production cost is increasing with industry production. This leads again to a long-run supply curve that is upward sloping but relatively elastic.

A decreasing cost industry can exist as well. Sometimes the production technology results in lower costs at higher production levels. Consider the brewing industry, where the per beer cost of production changes with the firms' production level. At low levels of production, firms employ small factories that are usually labor intensive. Yet if they can sell to a larger market, they can build bigger brewing factories, using machinery that reduces the per cost of beer production. Decreasing cost industries can exist, but are considered rare and are therefore given little attention in this chapter.

Other Demand Elasticities

Managers in the beer industry are interested in the own-price elasticity of demand for beer so they can forecast how beer consumption will change if they lower or raise beer prices. In fact, the beer industry regularly estimates supply and demand curves for beer. They even claim to have "file cabinets full of interesting and useful" supply and demand models (Tremblay and Tremblay 2005, foreword). However, these beer industry experts use more than just own-price elasticities; there are many other useful elasticities. For example, managers in the beer industry are also interested in how beer consumption changes in response to the prices of distilled spirits and soda, two substitutes for beer. The cross-price elasticity of beer demand with respect to distilled spirits and soda provides such information. Another useful elasticity is the income elasticity of demand, which shows how beer demand responds to changes in consumer income.

The formulas for these elasticities are given in Figure 3.7. Suppose that we are calculating the cross-price elasticity of demand for beer with respect to whiskey. After observing whiskey prices rising 5%, when the price of beer stays the same, beer consumption rises 1%. Using the elasticity formula, the cross-price elasticity is $E_{\text{beer,whiskey}} = 1\%/5\% = 0.2$. The positive sign makes sense. Whiskey and beer are substitutes. If the price of whiskey rises, some consumers will switch to drinking more beer and less whiskey, causing beer demand to rise. In particular, the formula says that for every 1% increase in whiskey prices, *quantity demanded* of beer rises 0.2%. Saying quantity demand of beer will rise by 0.2% does *not* mean the market quantity of beer will rise by 0.2%, because market quantity depends on both supply and demand. It simply means that if the price of beer remained the same, consumption would rise by 0.2%. In reality, an increase in whiskey prices will increase the

Cross-Price Elasticity of Demand for Good *i* with Respect to the Price of Good *j* ($E_{i,j}$)

$$E_{i,j} = \frac{(\text{\% Change Quantity Demanded of Good } i)}{(\text{\% Change in Price of Good } j)}$$

- (% Change Quantity Demanded of Good i) = ($E_{i,j}$)(% Change in Price of Good j)
- If $E_{i,j} > 0$, goods i and j are substitutes; if $E_{i,j} < 0$, they are complements.

Income Elasticity of Demand for Good *i* with Respect to Consumer Income ($E_{i,\text{Income}}$)

$$E_{i,\text{Income}} = \frac{(\text{\% Change Quantity Demanded of Good } i)}{(\text{\% Change in Income})}$$

- (% Change Quantity Demanded of Good i) = ($E_{i,\text{Income}}$)(% Change in Income)
- If $E_{i,\text{Income}} > 0$, good i is a normal good; if $E_{i,\text{Income}} < 0$, good i is an inferior good.

FIGURE 3.7 Cross-Price and Income Elasticity of Demand.

demand for beer, raising the price of beer, making consumption increase less than 0.2%. Other goods are complements for beer. If you are anything like the authors, you prefer to drink beer while eating buffalo wings. The two go together nicely. If the price of buffalo wings rises, beer consumption will fall, resulting in a negative cross-price elasticity.

The concept of cross-price elasticities can also help us think about the effects of a "fat tax." The idea is not to tax every food, but only certain kinds of "unhealthy" food. However, almost every food, healthy and unhealthy, has substitutes. For example, suppose the government placed a "fat tax" on donuts but not bagels. The tax would increase the price of donuts, leading to a fall in the quantity of donuts consumed. So far, so good. However, if donuts and bagels are substitutes, an increase in the price of donuts will lead to an increase in the consumption of bagels. Whether this particular fat tax actually reduces weight will depend on the own-price elasticity of demand for donuts *and* on the cross-price elasticity of demand for bagels with respect to donuts (in addition to how many calories are in a donut and a bagel).

In addition to cross-price elasticities, it is also useful to know about income elasticities. Suppose that we observe beer consumption rise 5% when incomes rise by 20% and assume beer prices remain unchanged. Using the formula in Figure 3.7, the income elasticity is 5%/20% = 0.25. The positive elasticity indicates beer is a *normal good*. When incomes rise, beer demand rises. Specifically, a 1% rise in income increases quantity demanded of beer by 0.25%, meaning at the same beer price consumers would purchase 0.25% more. If we had calculated an income elasticity for cheap, low-quality beer (instead of all beer) like Old English 800 Malt Liquor, the income elasticity would probably be negative, indicating an inferior good. As incomes rise, people switch to higher-quality beer and the demand for malt liquor falls.

The values of cross-price and income elasticities will differ depending on whether the time frame is the long run or short run. Specifically, the elasticity values are larger in magnitude in the long run. Consider the cross-price elasticity of demand for beer with respect to whiskey. People develop habits in the food and drinks they consume. If you are used to drinking whiskey, it takes a period of time before you can become used to being a beer drinker. When whiskey prices rise, whiskey demand falls and beer consumption rises. Yet, over time as whiskey prices remain high, more and more whiskey drinkers switch to beer. The demand increase for beer is greater when consumers are given time to adjust, and the magnitude of the cross-price elasticity increases.

Several elasticities of demand for beer are given in Figure 3.8. These are real elasticities estimated by the economists Tremblay and Tremblay in their book *The U.S. Brewing Industry*. These are elasticities calculated by collecting market data on prices and consumer purchases and using statistical methods discussed later in Chapter 7. The own-price elasticities show that a 1% increase in the price of beer decreases quantity demanded of beer by 0.298% in the short run and 0.745% in the long run. Even though demand is relatively inelastic in the short run, once consumers have time to adjust to price changes, demand becomes more elastic. The cross-price elasticities with respect to

Elasticity Type	Short-Run Elasticity Value	Long-Run Elasticity Value
Own-Price Elasticity of Demand for Beer	−0.298	−0.745
Cross-Price Elasticity of Demand for Beer with Respect to the Price of Soda	0.191	0.478
Cross-Price Elasticity of Demand for Beer with Respect to the Price of Whiskey	0.015	0.038
Cross-Price Elasticity of Demand for Beer with Respect to Income	0.085	0.213

FIGURE 3.8 Elasticities of Demand for Beer.
Source: Tremblay and Tremblay (2005).

soda and whiskey are both positive, indicating they are substitutes for consumption. The higher elasticity for soda suggests that soda is more of a substitute for beer than whiskey. If you are a manager in the beer industry, this means that soda is a more threatening competitor to beer than distilled spirits. Finally, the income elasticity of demand is positive. Beer is a normal good. When incomes rise, people drink more beer.

EQUILIBRIUM DISPLACEMENT MODELS

Elasticities are useful. In the previous examples, we found that elasticities can tell us whether a tax on a specific good will raise large or small revenues, and can tell managers which goods are the greatest competitors. Elasticities are also useful for forecasting changes in prices and quantities resulting from supply and demand curve shifts. One example is where the supply of pork decreases due to tougher regulations on manure treatment and disposal. Another example is where the demand for pork increases due to an increase in the price of beef. To forecast price and quantity changes, you do not need to know the exact supply and demand curves; all you need are elasticity estimates, which are more readily available. By incorporating elasticities into something called an *equilibrium displacement model*, one can calculate price and quantity changes due to a number of outside events.

Conceptually, equilibrium displacement models operate by grouping all the variables affecting supply and demand into exogenous or endogenous variables. Given a supply and demand model consisting of a supply and demand curve, price and quantity are determined by the intersection of those curves. Because price and quantity are determined *by* or *within* the model, they are referred to as *endogenous variables*.

Endogenous Variable: One whose value is determined inside an economic model.

Exogenous Variable: One whose value is determined outside an economic model.

There are a number of factors that shift the supply and demand curves. Input prices shift the supply curve and consumer income shifts the demand curve. In the supply and demand model, the values of these supply and demand shifters are taken as given. They are viewed as fixed numbers, whose values are determined by forces outside of the supply and demand model. Because the values of these variables are determined outside of the model, they are referred to as *exogenous variables*.

Below we discuss a supply and demand model of the pork market. The pork supply curve is determined by input prices and technology, which are exogenous variables. The pork demand curve is determined by consumer income, population, the price of related goods, and tastes and preferences, which are also exogenous variables. The values of these exogenous variables determine the shape of the pork supply and demand curves. The pork price and quantity are then determined by the intersection of the two curves—a fact that lets us know the price and quantity endogenous variables. Equilibrium displacement models are used to assess how endogenous variables respond to changes in exogenous variables. An exogenous shock is assumed, which may be a rise in the price of a related good or an increase in production costs. This shock shifts either the supply or demand curve, or both. After shifting the curves, the market must find a new equilibrium where the new curves cross. We can formalize these concepts and write the consumer demand curve as

$$\%\Delta QD = E_D(\%\Delta P) + S_D.$$

Recall that $\%\Delta QD$ represents the percent change in quantity demanded, $\%\Delta P$ is the percent change in price, and E_D is the own-price elasticity of demand. Here we have introduced a new variable, S_D, which represents any exogenous demand shift. For example, if a change in any exogenous variable causes quantity demanded to change by 7%, $S_D = 7\%$. Later we will provide more concrete examples of S_D.

To determine the price and quantity change in equilibrium, we must also specify the supply curve, which can be written as

$$\%\Delta QS = E_S(\%\Delta P) + S_S.$$

Recall that $\%\Delta QS$ represents the percent change in quantity supplied, $\%\Delta P$ is the percent change in price, and E_S is the own-price elasticity of demand. The new variable, S_S, shows the percent change in quantity supplied due to a change in the value of an exogenous variable. The variables S_D and S_S are commonly referred to as demand and supply "shocks" in the sense that they refer to shifts in the demand and/or supply curves that lead the market to a new equilibrium price and quantity.

In equilibrium, the percentage change in quantity demanded must equal the percentage change in quantity supplied. Thus, we can set our supply and demand curves equal and solve for the endogenous variable, $\%\Delta P$, as a function of the exogenous shocks.

$$\begin{aligned}
\%\Delta QS &= \%\Delta QD \\
E_S(\%\Delta P) + S_S &= E_D(\%\Delta P) + S_D \\
E_S(\%\Delta P) - E_D(\%\Delta P) &= S_D - S_S \\
[E_S - E_D](\%\Delta P) &= S_D - S_S \\
(\%\Delta P) &= [S_D - S_S]/[E_S - E_D]
\end{aligned}$$

Notice the intuition behind the final equation. The denominator is always positive, because the supply elasticity E_S is positive and the demand elasticity E_D is negative. If S_D is positive, the value of an exogenous variable changed in such a way to increase demand—the demand curve shifts rightward. For example, incomes may rise, increasing the demand for the good. This should cause an increase in price. Indeed, if $S_D > 0$ and $S_S = 0$, the equation suggests an increase in price. Now suppose $S_D = 0$ and $S_S > 0$; perhaps a new technology decreases the cost of production, increasing supply and shifting the supply curve rightward. An increase in supply lowers price, and the equation above also shows that the price should fall.

Thus, once one knows the values of the shocks S_S and S_D, one can easily calculate the percent change in price. Once the percent change in price is known, it can be plugged into either the supply or demand equation $\%\Delta QS = E_S(\%\Delta P) + S_S$ or $\%\Delta QD = E_D(\%\Delta P) + S_D$ to calculate the percent change in quantity. Hopefully the examples provided below will help. But before we proceed to the examples, let us first outline the general steps in solving an equilibrium displacement model.

Step 1: Determine the value of the supply and demand shocks S_S and S_D. Either S_S or S_D may equal zero, but not both (otherwise, the equilibrium is unchanged and $\%\Delta Q = \%\Delta P = 0$).
Step 2: Specify the change in quantity supplied as $\%\Delta QS = E_S(\%\Delta P) + S_S$.
Step 3: Specify the change in quantity demanded as $\%\Delta QD = E_D(\%\Delta P) + S_D$.
Step 4: Set $\%\Delta QS = \%\Delta QD$ and solve for the $\%\Delta P$.
Step 5: Plug the calculate value for $\%\Delta P$ in to the $\%\Delta QS$ or $\%\Delta QD$ equations to calculate the percent change in quantity. The percent change in quantity should be the same for both, otherwise your calculations are wrong.

Example: Impacts of Manure Regulations in the Pork Market

Pork producers are facing tougher regulations on how they manage and treat hog manure. State and federal regulations are making it more expensive to handle manure, and consequently, it is more expensive to produce pork. How will such regulations affect the hog market? In particular, how will regulations affect the price and quantity of pork? Given an estimate of the additional cost regulations imposed on pork producers, such questions can be easily answered using elasticities and equilibrium displacement models. If pork production costs rise due to tougher regulations, the supply curve shifts leftward, increasing price and decreasing quantity. The pork market moves from one equilibrium to another, which is referred to as "equilibrium displacement." Our job is to determine the price and quantity change that would occur due to this displacement.

Pork prices are usually stated in dollars per cwt (dollars per 100 pounds). Suppose that the manure regulations increase the per cwt hog production costs by an amount equal to 2% of the current price. For example, suppose the current hog price is \$50/cwt and hog production costs rise by \$1/cwt. Paying \$1/cwt more to produce a hog is just like receiving \$1/cwt less in price, which is \$1/\$50 = 0.02 or 2% of the current price. Thus, the increase in production costs of \$1/cwt is just like a 2% reduction in price from the hog producers' point of view. *If* prices fell 2%, hog producers would reduce their quantity supplied by $\%\Delta QS = E_S(-2\%)$. Thus, the regulations are akin to a 2% price decrease, making $S_S = E_S(2\%)$. A change in manure regulations has no impact on hog demand, so $S_D = 0$.

The impact of this cost increase alone decreases the supply of pork. Figure 3.9 shows the long-run supply elasticity of pork is 2.15, meaning for every 1% decrease in pork prices, quantity supplied falls by $(2.15)(-1\%) = -2.15\%$. In this case, regulation costs are identical to a 2% decrease in pork prices, so pork supply changes by $(2.15)(-2\%) = -4.3\%$. Quantity of pork supplied falls by 4.3%, assuming the pork price doesn't change. This is the exogenous shock—now for the endogenous adjustment. In case those words confuse you, the exogenous shock is the decrease in supply, and the endogenous adjustment is the increase in price to make supply and demand equal once again. The decrease in supply will increase the pork price through the market's endogenous adjustment, which will partially offset the higher production costs. From above, we can write our supply equation as

$$\%\Delta QS = E_S(\%\Delta P) + S_S + = 2.15(\%\Delta P) + 2.15(-2\%).$$

Recall the term E_S stands for elasticity of supply and in this case it is equal to 2.15. In this supply curve, there is one equation and two unknowns: $\%\Delta P$ and $\%\Delta QS$. However, we can solve for these two unknowns by bringing in one additional equation—the demand equation. In equilibrium, quantity supplied must equal quantity demanded. When pork prices rise, consumers decrease their quantity demanded of pork. We write the demand curve as $\%\Delta QD = E_D(\%\Delta P)$. Here there is no demand shift, so $S_D = 0$. As Figure 3.9 shows, the elasticity of demand for pork is -1.96, so $E_D = -1.96$. When the supply curve shifts upward due to tougher manure management regulations, price rises. When the market reaches its new equilibrium, at this higher price, the percent change in quantity demanded must equal the percent change in quantity supplied. Otherwise the market would not be in equilibrium. This implies that we should set our two equations together to solve for the percent change in price.

Elasticity of Pork Supply	2.15
Own-Price Elasticity of Pork Demand	−1.96
Cross-Price Elasticity of Pork Demand with Respect to Beef Prices	0.60
Elasticity of Beef Supply	*0.40*
Own-Price Elasticity of Beef Demand	*−0.90*
Cross-Price Elasticity of Beef Demand with Respect to Pork Prices	*0.26*

FIGURE 3.9 Long-Run Elasticities of Supply and Demand for Pork (and Beef).
Source: The pork elasticities are derived and calculated in Chapter 6. The beef elasticities are adjusted from the short-run elasticities in Lusk and Anderson (2004).

$$\begin{aligned}
\%\Delta QS &= \%\Delta QD \\
E_S(\%\Delta P) + S_S &= E_D(\%\Delta P) + S_D \\
2.15(\%\Delta P) + 2.15(-2) &= -1.96(\%\Delta P) + 0 \\
2.15(\%\Delta P) + 1.96(\%\Delta P) &= 2.15(2) \\
[2.15 + 1.96](\%\Delta P) &= 2.15(2) \\
(\%\Delta P) &= 2.15(2)/[2.15 + 1.96] \\
(\%\Delta P) &= 1.05\%
\end{aligned}$$

Although pork production costs increase 2% due to tougher manure regulations, pork prices rise by only 1.05%. Next, we can use the supply or demand equations to calculate the percent change in quantity.

$$\begin{aligned}
\%\Delta QS &= E_S(\%\Delta P) + E_S(-2) = 2.15(1.05) + 2.15(-2) = -2.05\% \\
\%\Delta QD &= E_D(\%\Delta P) = -1.96(1.05) = -2.05\%
\end{aligned}$$

Both quantity supplied and quantity demanded must change by the same amount or the market would not be in equilibrium. The result is that the increase in pork production costs increase pork prices by 1.05% and decrease pork production by 2.05%. This is the answer for the long run, when markets have ample time to adjust to the new regulations. To ascertain the short-run impacts of the regulations, simply replace the long-run elasticities with short-run elasticities.

Example: Impact of Higher Beef Prices

In the previous example we explored how an increase in pork production costs may impact the pork market. Suppose now that you are a pork market analyst, and beef prices are expected to rise dramatically, as they did in the early 2000s. Specifically, they are expected to rise 10%. This is a demand shock, because the shock shifts the demand curve but not the supply curve. Because beef and pork are substitutes, this price change is good for the pork market, but how good? Specifically, you are asked to calculate the percent change in pork price and quantity expected to result from a 10% rise in beef prices. Again, this is easily calculated using elasticities and equilibrium displacement models.

The first step is to determine the size of the demand shock, S_D. Recall, S_D represents the percent change in quantity demanded of pork resulting from the demand shock. The cross-price elasticity of pork demand with respect to beef is 0.6, as shown in Figure 3.9. This elasticity implies that a 10% rise in beef prices increases the quantity of pork demand by $(0.6)(10) = 6\%$. Thus, the value of S_D is $(0.6)(10) = 6\%$, and since the supply curve is unaffected, $S_S = 0$. Using the same steps as described above, the percent change in price due to the demand shock is calculated as

$$\begin{aligned}
\%\Delta QS &= \%\Delta QD \\
E_S(\%\Delta P) + S_S &= E_D(\%\Delta P) + S_D \\
2.15(\%\Delta P) + 0 &= -1.96(\%\Delta P) + (0.6)(10) \\
2.15(\%\Delta P) + 1.96(\%\Delta P) &= (0.6)(10) \\
[2.15 + 1.96](\%\Delta P) &= (0.6)(10) \\
(\%\Delta P) &= (0.6)(10)/[2.15 + 1.96] \\
(\%\Delta P) &= 1.46\%
\end{aligned}$$

The pork price is expected to increase 1.46% after the 10% rise in beef price. The change in pork quantity can be solved for as

$$\%\Delta QS = (2.15)(1.46) = 3.14\%$$
$$\%\Delta QD = (-1.96)(1.46) + 6 = 3.14\%$$

Thus, a 10% rise in beef prices would be expected to increase pork prices by 1.46% and pork quantities by 3.14% in the long run. If one is more interested in immediate, short-run changes, simply replace the long-run elasticities with short-run elasticities, and perform the same algebra.

Twice we mentioned that market analysts in the beer industry have "file cabinets full of interesting and useful" supply and demand models (Tremblay and Tremblay 2005, foreword). What we meant is they have file cabinets full of supply and demand elasticities and formulas like the ones above for the purposes of forecasting market changes. The beer industry is not alone. Most large industries employ economists to develop and use elasticities in equilibrium displacement models like those described above. Government agencies also use these models to perform benefit-cost analyses. In the early 2000s the State of North Carolina considered requiring swine farms to adopt costly manure management technologies, yet they wanted to make sure the regulations did not impose a heavy burden on the farms. To determine how many farms might go out of business due to the regulations, they asked agricultural economists at North Carolina State University to calculate the economic impact, and those economists used equilibrium displacement models much like the ones described above.

GENERAL EQUILIBRIUM MODELS

Clemson University football coach Tommy Bowden is playing against Texas Tech football coach Mike Leach. Coach Bowden has thus far been passing the ball, but few yards were gained. Seeing Texas Tech's current defense strategy, Coach Bowden decides a running game will be more effective. However, Bowden knows that Texas Tech will change its defensive strategy if Clemson begins running the ball. Taking this into account, Coach Bowden decides his best strategy is still to pass the ball. There are first-order effects and second-order effects. In this example, the first-order effect would be Clemson's success if they ran the ball with Texas Tech's current defensive strategy. The second-order effect is that Texas Tech will alter their defense if Clemson alters their offensive strategy. Good coaches account for these second-order effects, and so do good economists.

Refer back to our example where the government imposes stricter swine manure management regulations on pork producers. Raising the cost of pork production, the pork supply curve shifts leftward resulting in a higher equilibrium pork price and a lower equilibrium pork quantity. In reality, this is only a first-order effect because we have not accounted for how higher pork prices will affect other markets, and how changes in these other markets would come back to impact pork prices. A second-order effect may take place—no market is an island. Beef and pork are substitutes, and a rise in pork prices increases the demand for beef, raising beef prices. Higher beef prices, in turn, increase pork demand. Changes in pork prices affect beef prices and vice versa, and this interplay between the two prices continues until both prices settle at a point where supply and demand in both markets clear. This is referred to as a *general equilibrium* because it accounts for the fact that supply and demand in different markets are related to one another.

Until now we have only considered *partial equilibrium models*, where a shift in supply and/or demand in one market is not assumed to shift the supply and/or demand in other markets. Put differently, we only considered the supply and demand for a single market in isolation. *General equilibrium models*, on the other hand, assume that the supply and demand for one good is related to the supply and demand for other goods. A general equilibrium model is depicted in Figure 3.10. Due to costly government regulations, the pork supply curve shifts upward. A partial equilibrium exists where the new supply curve intersects the old demand curve (denoted by the lighter circle). A general equilibrium model goes further to consider how this higher pork price impacts beef prices, and the subsequent impact of beef prices back on pork prices. The higher pork price causes a rise in beef demand and beef prices, because pork and beef are substitutes (as shown in the right diagram in Figure 3.10). Consequently, this higher beef price increases pork demand and pork prices.

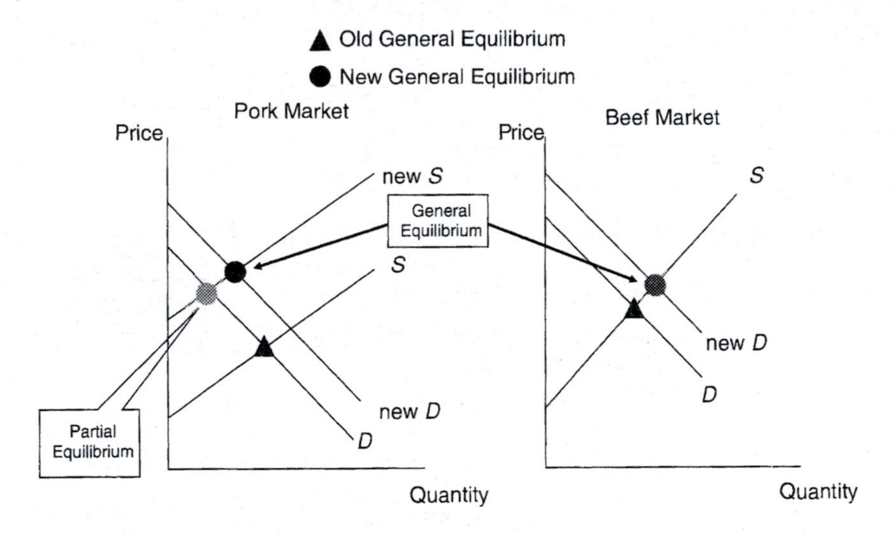

FIGURE 3.10 Impact of Higher Pork Production Costs in a General Equilibrium Model. The government imposes tighter environment regulations on pork producers, which raises the cost of pork production and shifts the pork supply curve leftward. The pork price rises. Beef and pork being substitutes, the higher pork price increases the demand for beef, shifting the beef demand curve upward and raising beef prices. The higher beef price, in turn, increase the pork demand, increasing pork prices further. The new *general equilibrium* is the pork and beef price that cause and demand to equal in both the pork and beef market simultaneously.

Thus, the increase in pork prices due to the government regulation is higher in a general equilibrium model than a partial equilibrium model. The partial equilibrium model fails to account for the fact that the government regulation will ultimately raise pork prices indirectly by raising beef prices. This demand increase partially offsets the higher pork production costs. General equilibrium models are preferred over partial equilibrium models because they are more realistic. As seen here, the use of a partial equilibrium model would overestimate the burden government regulations place on pork producers.

The equilibrium displacement model used in the previous section was a partial equilibrium model, but can easily be expanded to a general equilibrium model. The basic model setup and logic are the same; there are just more equations to solve. Let Q^{Pork} and Q^{Beef} denote the quantity of pork and beef, respectively, and P^{Pork} and P^{Beef} be the price of pork and beef, respectively. As before, $\%\Delta X$ denotes a percent change in the value of variable X. In keeping with the previous example, suppose a new regulation raises the marginal cost by an amount equivalent to 2% of the current pork price. That is, the increase in pork production costs has the same effect on profits as a 2% fall in pork prices. The pork supply curve shifts left. The percent change in quantity supplied of pork can be written as (using the elasticities from Figure 3.9).

$$\%\Delta QS^{\text{Pork}} = E_S^{\text{Pork}}(\%\Delta P) + E_S^{\text{Pork}}(-2) = 2.15(\%\Delta P) + 2.15(-2\%).$$

In the general equilibrium framework, the decrease in supply will raise pork prices and beef prices. Given that beef prices will change, shifting the demand curve, the change in quantity demanded of pork now goes by the formula

$$\begin{aligned}\%\Delta QD^{\text{Pork}} &= E_D^{\text{Pork}}(\%\Delta P^{\text{Pork}}) + E_{\text{Pork,Beef}}^{\text{Pork}}(\%\Delta P^{\text{Beef}}) \\ &= (-1.96)(\%\Delta P^{\text{Pork}}) + (0.60)(\%\Delta P^{\text{Beef}}).\end{aligned}$$

The first term $((-1.96)(\%\Delta P^{\text{Pork}}))$ indicates how the percent change in pork quantity demanded is affected by the change in pork prices. It is the own-price elasticity of demand for pork multiplied by the percent change in pork price. The second term $(0.60)(\%\Delta P^{\text{Beef}})$ is the cross-price elasticity of pork demand with respect to beef prices, multiplied by the percent change in beef prices—indicating how pork demand is altered by a change in beef prices. Remember, the increase in cost of pork production raises pork prices, which increases beef demand due to the substitutability of beef for pork, which raises beef prices and ultimately increases pork demand. All things considered, the supply and demand in the pork market are

$$\begin{aligned}\%\Delta QS^{\text{Pork}} &= 2.15(\%\Delta P^{\text{Pork}}) + 2.15(-2\%) \\ \%\Delta QD^{\text{Pork}} &= (-1.96)(\%\Delta P^{\text{Pork}}) + (0.60)(\%\Delta P^{\text{Beef}}).\end{aligned}$$

Notice that there are two equations but three unknowns. More equations are needed to solve for the equilibrium. The pork market equations contain the term $\%\Delta P^{\text{Beef}}$, but this is a number that must be calculated within the general equilibrium model. It is *endogenous* to the model. Now let us consider changes in the beef market. The beef market responds to a change in the pork price by shifting the beef demand curve, thereby changing the beef price. The beef demand curve is given by

$$\%\Delta QD^{\text{Beef}} = (-0.90)(\%\Delta P^{\text{Beef}}) + (0.26)(\%\Delta P^{\text{Pork}}).$$

The first term shows how quantity demanded moves in response to its own price, while the second articulates how the demand curve shifts in response to a change in the pork price. The beef supply curve does not change, although quantity supplied changes due to the beef price increase. We need one final equation. The beef supply curve is

$$\%\Delta QS^{\text{Beef}} = (0.40)(\%\Delta P^{\text{Beef}}).$$

Our general equilibrium now has four equations: the percent change in (1) quantity supplied of pork, (2) quantity demanded of pork, (3) quantity supplied of beef, and (4) quantity demanded of beef. All together, the equations comprising the general equilibrium displacement model are

$$\begin{aligned}
\%\Delta QS^{\text{Pork}} &= 2.15(\%\Delta P^{\text{Pork}}) + 2.15(-2\%) \\
\%\Delta QS^{\text{Beef}} &= (0.40)(\%\Delta P^{\text{Beef}}) \\
\%\Delta QD^{\text{Pork}} &= (-1.96)(\%\Delta P^{\text{Pork}}) + (0.60)(\%\Delta P^{\text{Beef}}) \\
\%\Delta QD^{\text{Beef}} &= (-0.90)(\%\Delta P^{\text{Beef}}) + (0.26)(\%\Delta P^{\text{Pork}})
\end{aligned}$$

Because there are four equations and four unknowns, there is a unique solution to this equation. The solution may be hard to calculate by hand, so most economists would use a computer program like Microsoft Excel. In fact, Excel has a Solver Add-In that makes solving such models quite easy. If you calculate equilibrium, you should obtain the following results:

$$\%\Delta Q^{\text{Pork}} = -1.98\%,\ \%\Delta Q^{\text{Beef}} = 0.09\%,\ \%\Delta P^{\text{Pork}} = 1.08\%,\ \text{and}\ \%\Delta P^{\text{Beef}} = 0.22\%.$$

General equilibrium displacement models used in practice may be larger than four equations. For example, the State of North Carolina was considering stricter swine manure regulations, but wanted to know the economic impact of raising the cost of pork production in the state. The North Carolina attorney general's office asked Dr. Michael Wohlgenant, a distinguished agricultural economist from North Carolina State University, to estimate this impact, and he did so using an equilibrium displacement similar in spirit to the one in this section. It was a larger model though, consisting of more than eight equations. Other general equilibrium displacement models contain dozens (and sometimes even hundreds) of equations and are widely used to project the impact of government policies. The expertise needed to develop and use real equilibrium displacement models requires knowledge beyond this section. However, this section provides a basic understanding of how the models are developed and used. Chances are, you are thinking this subject is not important to you because you will never be in charge of developing and using equilibrium displacement models. But you never know, just ask Megan Provost!

Megan graduated with a master's degree in Agricultural Economics from Oklahoma State University in 2003. Like many students, she found lectures on elasticities boring, and never thought she would use them in her career. Wrong! After graduating, Megan took a job with the American Farm Bureau Federation (AFBF) in Washington, D.C., as the trade economist. And what does she spend a large part of her job doing? Developing and using general equilibrium displacement models! The United States is continually striking trade agreements with individual countries, with groups of countries through The World Trade Organization. These agreements, among other things, lower the tariffs (i.e., taxes) placed on agricultural imports. Before the United States commits to any change in policy, however, they seek to estimate the economic impact of that change. This is where AFBF and Megan Provost enters. Megan regularly considers policy proposals, runs then through her equilibrium displacement model, and writes reports for trade negotiators and policymakers on the economic impact. Like the models used in this textbook, they rely extensively on elasticities.

"Sitting in my agricultural economics classes at OSU, I never thought I would debate elasticities with heads of delegations from the European Union, Japan, Canada and even the Director-General of the World Trade Organization! But today, I use elasticities on a regular basis."

—Megan Provost

FIGURE 3.11 Meet Megan Provost . . .

FROM DEMAND FUNCTIONS TO DEMAND ELASTICITIES

You might wonder where the demand elasticities in Figures 3.8 and 3.9 come from. How do economists calculate these elasticities? Economists start by first calculating a demand function, which is an equation predicting the quantity of consumer purchases based on consumer demand factors. This is easiest seen with an example. After collecting data on consumer purchases of beer, the price of beer, the price of soda, the price of whiskey, and consumer income, economists at Oregon State University developed the following equation to predict beer purchases by consumers. The equation was developed so that it "fits the data" well, meaning it generates predictions that are reasonably close to actual consumer purchases. Just like the best meteorologists are chosen based on who provides the most accurate weather forecasts, the numbers making up the equation below were chosen as those that provide the most accurate predictions of consumer purchases.

$$\text{Short-Run Demand: } \textit{Quantity Demanded of Beer} = 57243.55 - 345.31 \times (\textit{Price of Beer}) + 295.86(\textit{Price of Soda}) + 13.91(\textit{Price of Whiskey}) + 6.06(\textit{Income}) + 0.40(\textit{Quantity Demanded of Beer Last Year})$$

We can write this equation more succinctly as

$$\text{Short-Run Demand: } Q_t = 57243.55 - 345.31 \times (P_t^{Beer}) + 295.86(P_t^{Soda}) + 13.91(P_t^{Whiskey}) + 6.06(I_t) + 0.60(Q_{t-1})$$

A description of each variable is given in Figure 3.12. For historical reasons, the beer industry measures output in 31-gallon barrels. The price of beer, soda, and whiskey is given by a price index, where a higher price indicates higher real prices. Income refers to disposable income (income after taxes) in real 1982 dollars. Q_t refers to annual quantity demanded in year t, and Q_{t-1} refers to annual demand in the previous year. Consumers form consumption habits, especially with alcoholic beverages. If people drink beer today, they are more likely to drink beer in the future. If they drink more this year, they are likely to drink more next year. If beer consumption this year falls, people will consume less next year because beer has become a less important part of their consumption habits. That is why the quantity demanded of beer last year is included in the equation; it helps to predict beer consumption in the current year.

The demand equation above is a short-run demand function. Quantity demanded this year is partly dependent upon demand last year, meaning consumer habits affect demand. Whenever habits influence demand, we say it refers to the short run. For example, a fall in the price of soda decreases beer demand, but people are still accustomed to drinking beer so the demand decrease is limited. Demand falls, but demand frictions derived from consumer habits limit its fall. You can think of it like the demand curve wants to fall a lot, but habits slow down the fall. Over time the demand frictions caused by consumer habits will taper away. In the long-run, quantity demanded this year equals quantity demanded last year. This result provides a simple method for transforming the short-run demand curve into a long-run demand curve: set $Q_{t-1} = Q_t$. After setting these

Variable	Units (Per Year)	Average Value from 1953–1995
Quantity Demanded of Beer (Q_t)	Thousand 31-gallon barrels	140,650.00
Beer Price (P_t^{Beer})	Index	121.38
Soda Price (P_t^{Soda})	Index	90.80
Whiskey Price ($P_t^{Whiskey}$)	Index	151.63
Consumer Income (I_t)	Disposable Income in Billion Dollars	1,973.00

FIGURE 3.12 Beer Demand Function Variable Descriptions.

Source: Denney et al. (2004).

two variables equal, we then rearrange the demand curve so that Q_t is alone on the left-hand side.

$$Q_t = 57243.55 - 345.31 \times (P_t^{Beer}) + 295.86(P_t^{Soda}) + 13.91(P_t^{Whiskey}) + 6.06(I_t) + 0.60(Q_{t-1} = Q_t)$$

$$\rightarrow Q_t = 57243.55 - 345.31 \times (P_t^{Beer}) + 295.86(P_t^{Soda}) + 13.91(P_t^{Whiskey}) + 6.06(I_t) + 0.60(Q_t)$$

$$\rightarrow Q_t - 0.60(Q_t) = 57243.55 - 345.31 \times (P_t^{Beer}) + 295.86(P_t^{Soda}) + 13.91(P_t^{Whiskey}) + 6.06(I_t)$$

$$\rightarrow Q_t(1 - 0.60) = 57243.55 - 345.31 \times (P_t^{Beer}) + 295.86(P_t^{Soda}) + 13.91(P_t^{Whiskey}) + 6.06(I_t)$$

$$\rightarrow Q_t(0.40) = 57243.55 - 345.31 \times (P_t^{Beer}) + 295.86(P_t^{Soda}) + 13.91(P_t^{Whiskey}) + 6.06(I_t)$$

$$\rightarrow Q_t = [57243.55 - 345.31 \times (P_t^{Beer}) + 295.86(P_t^{Soda}) + 13.91(P_t^{Whiskey}) + 6.06(I_t)]/0.40$$

$$\rightarrow Q_t = 143108.88 - 863.28 \times (P_t^{Beer}) + 739.65(P_t^{Soda}) + 34.78(P_t^{Whiskey}) + 15.15(I_t)]$$

$$\text{Long-Run Demand: } Q_t = 143108.88 - 863.28(P_t^{Beer}) + 739.65(P_t^{Soda}) + 34.78(P_t^{Whiskey}) + 15.15(I_t)$$

Industry experts and economists can use this equation to help predict how beer consumption will change as the price of beer, soda, and whiskey changes. The impacts of higher or lower incomes on beer consumption can also be ascertained from the equation. In particular, the equation can be used to calculate the own-price, cross-price, and income elasticities in the short and long run shown in Figure 3.13. The general formula for an elasticity of demand with respect to any variable X is

$$\text{Elasticity of Demand with Respect to } X = \frac{\%\Delta Q}{\%\Delta X} = \frac{\left[\frac{\Delta Q}{Q}\right]}{\left[\frac{\Delta X}{X}\right]} = \left[\frac{\Delta Q}{\Delta X}\right]\left[\frac{X}{Q}\right].$$

The term Δ means change, so ΔX means "change in X." For the own-price elasticity of demand, X is the good's price, and for the income elasticity X refers to income. Using this formula the own-price elasticity of demand can be calculated from the demand function using a simple thought experiment. Pretend that the beer price increases by one dollar, making $\Delta P_t^{Beer} = 1$. According to the demand function, this increase in the beer price of 1 will decrease quantity demanded of beer by 345.31 in the short run and 863.28 in the long run. Therefore, in the elasticity equation we substitute $\Delta X = \Delta P_t^{Beer} = 1$ and $\Delta Q = -345.31$ for a short-run elasticity or -863.28 for a long-run elasticity. All that remains in the equation are values for Q and $X = P_t^{Beer}$.

Variable	Short-Run Elasticity	Long-Run Elasticity
Own-Price Elasticity	$\left(\frac{\Delta Q_t}{\Delta P_t^{Beer}}\right)\left(\frac{P_t^{Beer}}{Q_t}\right) =$ $\left(\frac{-345.31}{1}\right)\left(\frac{121.38}{140650}\right) = -0.298$	$\left(\frac{\Delta Q_t}{\Delta P_t^{Beer}}\right)\left(\frac{P_t^{Beer}}{Q_t}\right) =$ $\left(\frac{-863.28}{1}\right)\left(\frac{121.38}{140650}\right) = -0.745$
Cross-Price Elasticity with Respect to Soda	$\left(\frac{\Delta Q_t}{\Delta P_t^{Soda}}\right)\left(\frac{P_t^{Soda}}{Q_t}\right) =$ $\left(\frac{295.86}{1}\right)\left(\frac{90.80}{140650}\right) = 0.191$	$\left(\frac{\Delta Q_t}{\Delta P_t^{Soda}}\right)\left(\frac{P_t^{Soda}}{Q_t}\right) =$ $\left(\frac{739.65}{1}\right)\left(\frac{90.80}{140650}\right) = 0.478$
Cross-Price Elasticity with Respect to Whiskey	$\left(\frac{\Delta Q_t}{\Delta P_t^{Whiskey}}\right)\left(\frac{P_t^{Whiskey}}{Q_t}\right) =$ $\left(\frac{13.91}{1}\right)\left(\frac{151.63}{140650}\right) = 0.015$	$\left(\frac{\Delta Q_t}{\Delta P_t^{Whiskey}}\right)\left(\frac{P_t^{Whiskey}}{Q_t}\right) =$ $\left(\frac{34.78}{1}\right)\left(\frac{151.63}{140650}\right) = 0.038$
Income	$\left(\frac{\Delta Q_t}{\Delta I_t}\right)\left(\frac{I_t}{Q_t}\right) =$ $\left(\frac{6.06}{1}\right)\left(\frac{1973}{140650}\right) = 0.085$	$\left(\frac{\Delta Q_t}{\Delta I_t}\right)\left(\frac{I_t}{Q_t}\right) =$ $\left(\frac{15.15}{1}\right)\left(\frac{1973}{140650}\right) = 0.213$

FIGURE 3.13 Beer Demand Elasticities.

Sources: Denney et al. (2004) and Tremblay and Tremblay (2005).

We usually just substitute the average values of the variables, as shown in Figure 3.13. The calculations show a short- and long-run own-price elasticity of −0.298 and −0.745, which are the exact elasticities given in Figure 3.13. Using the same thought experiment one can easily calculate cross-price and income elasticities as illustrated in Figure 3.13.

FROM DEMAND FUNCTIONS TO DEMAND CURVES

Recall our long-run demand function for beer: Long-Run Demand: $Q_t = 143108.88 - 863.28 \times (P_t^{Beer}) + 739.65(P_t^{Soda}) + 34.78(P_t^{Whiskey}) + 15.15(I_t)$. If one substitutes values for the three prices and income, the function predicts the quantity of beer consumers will purchase. In previous chapters we were concerned with demand *curves*, not demand *functions*. Demand curves are a line with price on the y-axis and quantity on the x-axis. Demand curves predict purchases as a function of price *alone*, whereas demand functions are equations that predict consumer purchases as a function of *numerous* variables. It turns out that demand curves come straight from these demand functions. The purpose of this section is to illustrate how to derive demand curves from demand functions like the beer demand function above.

A demand curve is a diagram showing the relationship between the price of a good and the quantity demanded by consumers. Quantity is on the x-axis and price is on the y-axis. To obtain a beer demand curve we must simplify the beer demand function so that the only variables are the quantity demanded of beer (Q_t) and the price of beer (P_t^{Beer}). Numerical values must be assigned to the remaining variables P_t^{Soda}, $P_t^{Whiskey}$, and I_t. The obvious value to assign are their variable's average values as reported in Figure 3.12. After substituting the average values for P_t^{Soda}, $P_t^{Whiskey}$, and I_t, the demand function is then rearranged so that the price of beer is on the left-hand side. Once the price of beer is stated as a function of the quantity of beer, one can plot points on a diagram with quantity on the x-axis and price on the y-axis, yielding a beer demand curve.

$$Q_t = 143108.88 - 863.28 \times (P_t^{Beer}) + 739.65(P_t^{Soda} = 90.80) + 34.78(P_t^{Whiskey} = 151.63) + 15.15(I_t = 1973)$$

$$Q_t = 143108.88 - 863.28 \times (P_t^{Beer}) + 67150.22 + 5273.69 + 29890.95$$

$$Q_t = 245433.74 - 863.28 \times (P_t^{Beer})$$

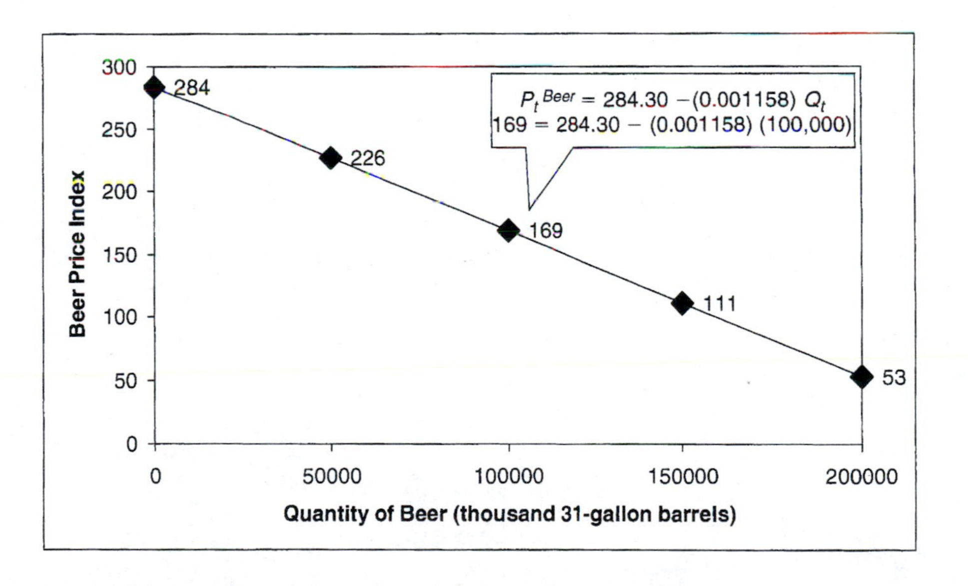

FIGURE 3.14 Long-Run Demand Curve for Beer.

$$863.28 \times (P_t^{Beer}) = 245433.74 - Q_t$$
$$(P_t^{Beer}) = [245433.74 - Q_t]/863.28$$
$$(P_t^{Beer}) = 284.30 - (0.001158)Q_t$$

$$\text{Long-Run Beer Demand Curve: } P_t^{Beer} = 284.30 - (0.001158)Q_t$$

Plugging in different values for Q_t and calculating the beer price then yields points on the beer demand curve, as shown in Figure 3.14. At a quantity of 100,000, the beer price is $169. This means that if the beer price index is 169, consumers will purchase 100,000 thousand 31-gallon barrels of beer. Consider what would happen to the demand curve if we changed the values of the soda price, whiskey price, or income. If any one of these variables increased in value, the intercept on the demand curve would rise above 284. The demand curve shifts upward, indicating a demand increase. If any of the variables decrease in value, the intercept decreases in value, shifting the demand curve downward. Although this section has focused on how demand elasticities and curves are obtained from demand functions, it is important to note that the same basic principle can be used to derive supply elasticities and curves from supply functions.

SUMMARY

Understanding why prices change is crucial to understanding agricultural markets. The supply and demand model is a useful market model for predicting price changes. Learning how to shift supply and demand curves is useful for predicting the direction of price changes, but predicting exactly how much price will change requires the use of demand and supply elasticities. Perhaps the most important lesson gleaned from this chapter is that supply and/or demand shifts curves affect producers and consumers alike. We saw an example of government placing costly regulations on pork producers. Using supply and demand analysis, along with an equilibrium displacement model, we saw that although pork producers face higher costs, these costs are partially offset by higher pork prices. Consumers pay higher pork prices, so in a sense they are helping pay for the costly regulations. The point is that you can't just say a government regulation that imposes costs on producers has no effect on consumers. In similar vein, one cannot say that taxing consumers when they buy products has no effect on producers of the product. The true cost of regulations and taxes on producers and consumers can only be calculated using the economic methods described in this chapter. One day, you may be an economic consultant whose job is to estimate the impact of a government regulation on producers in an industry. To complete this task, you must understand supply and demand elasticities and equilibrium displacement models.

CROSSWORD PUZZLE

If the answer contains more than one word, leave a blank space between each word.

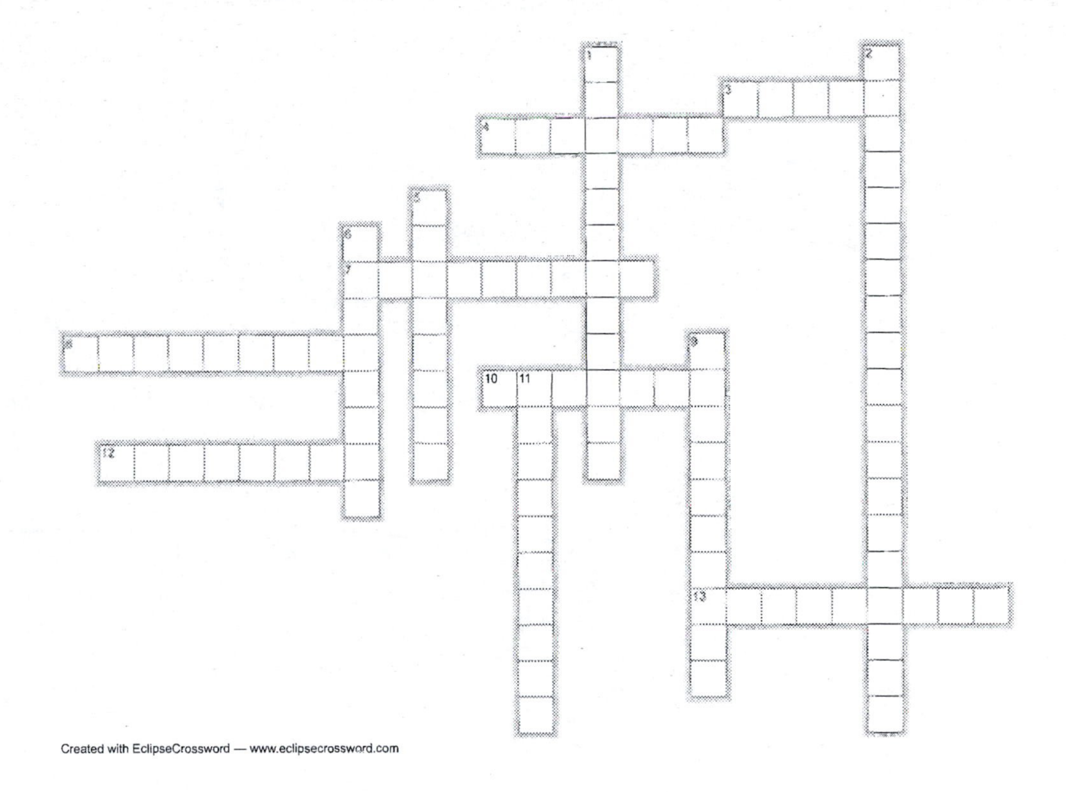

Across

3. A demand _______ shows the relationship between quantity demand and a single variable, its own price.
4. If quantity demanded is very sensitive to price changes, we say that demand is _______.
7. The _______-_______ elasticity of demand equals the percent change in quantity demanded divided by the percent change in price.
8. An _______ variable's value is determined outside a model, and its value is taken as given by a model.
10. A _______ equilibrium model calculates the equilibrium for multiple markets simultaneously.
12. A demand _______ is an equation detailing quantity demanded as a function of numerous variables, such as own-price, the price of related goods, and income.
13. If the own-price elasticity of demand is greater than 1 (or less than one in absolute value), we say the demand is _______.

Down

1. An equilibrium _______ model uses elasticities to determine how price and quantity will change in response to supply or demand shocks.
2. If the demand curve is a vertical line, we say the demand is _______ _______.
5. A period of time in which firms can vary all their inputs and consumers have fully adjusted to price changes.
6. If input prices for an industry do not change as industry production rises and falls, we say this is a _______ cost industry.
9. A measure of how sensitive one variable (Y) is to changes in another variable (X).
11. An _______ variable's value is determined by, or within, a model.

STUDY QUESTIONS

1. When the price of a good falls by 8%, consumers respond by increasing their purchases 16%. Calculate the own-price elasticity for this good and indicate whether it is inelastic, elastic, or unit elastic.
2. When the price of a good rises by 8%, producers respond by increasing their production of the good by 4%. Calculate the own-price elasticity for this good and indicate whether it is inelastic, elastic, or unit elastic.
3. The cross-price elasticity for beer and soda is positive. Are the two goods substitutes or complements?
4. Agricultural economists at the United States Department of Agriculture have estimated the following elasticities (Huang and Lin 2000). If you are a producer in the beef industry, which is a more threatening competitor for you—the poultry, pork, or fish industry? Explain why.

Cross-price elasticity of demand for beef with respect to pork	–0.0781
Cross-price elasticity of demand for beef with respect to poultry	–0.0417
Cross-price elasticity of demand for beef with respect to fish	–0.0241

5. The elasticity of demand for rice in the United States and Vietnam is –0.55 and –0.15, respectively (Cramer, Wailes, and Shangnan 1993). Think about the differences in the cultures of these two countries that might cause such disparities. Explain in a clear, logical paragraph. There is no right or wrong answer; we are just looking for a logical argument.
6. The elasticity of supply for rice in the United States and Vietnam is 0.4 and 0.15, respectively (Cramer, Wailes, and Shangnan 1993). Think about the differences in these two countries that might cause such disparities. Explain in a clear, logical paragraph. There is no right or wrong answer; we are just looking for a logical argument.

Use the elasticities in the table below to answer questions 7 and 8.

Own-Price Elasticity of Beef Demand	–0.56
Own-Price Elasticity of Beef Supply	0.15
Cross-Price Elasticity of Beef Supply with Respect to Pork	0.10

7. Due to Mad Cow Disease scares, a new government regulation forces beef producers to maintain records on all cattle they buy and sell so that contaminated beef can be traced back to the farm where the animal was raised. This is expected to increase the marginal cost of beef production by an amount equivalent to 15% of

beef prices. That is, the extra costs decrease the per cwt profit of beef production equivalent to loss in profits if the per cwt price fell 15%. Using an equilibrium displacement model, calculate how beef prices and quantities will respond to the new animal identification regulation.

8. A new pork production technology is expected to reduce the cost of producing pork, and subsequently, reduce pork prices by 25%. Using an equilibrium displacement model, estimate the impact of this new technology on beef prices and production levels.

Use the information below to answer questions 9 through 14.

The demand for pork can be described by the following demand function:

$$\begin{aligned} \textit{Quantity Demanded of Pork in million lbs} = {} & 367 - 3.12(\textit{Price of Pork in \$/cwt}) \\ & + 1.35(\textit{Price of Beef in \$/cwt}) \\ & + 1.69(\textit{Price of Poultry in \$/cwt}) \\ & + 0.12(\textit{Income in billion dollars}) \\ & + 0.52182(\textit{Quantity Demanded} \\ & \quad \textit{of Pork Last Year}) \end{aligned}$$

The average price of beef is \$226/cwt, the average price of poultry is \$78/cwt, and the average income is \$3069. The average quantity and price of pork is 1325 million lbs/year and \$172/cwt, respectively. Prices and income are in real 1982 dollars. Note that this is a short-run demand function because quantity demanded today depends on the quantity consumed in the past. This *short-run* demand function can be written more succinctly as

$$Q_t = 367 - 3.12(P_t^{Pork}) + 1.35(P_t^{Beef}) + 1.69(P_t^{Poultry}) + 0.12(I_t) + 0.52(Q_{t-1})$$

9. By setting $Q_t = Q_{t-1}$ in the short-run demand function above, prove that the long-run demand function is $Q_t = 765 - 6.50(P_t^{Pork}) + 2.81(P_t^{Beef}) + 3.52(P_t^{Poultry}) + 0.25(I_t)$
10. Calculate the short-run own-price elasticity of demand for pork, and indicate whether it is inelastic, elastic, or unit elastic.
11. Calculate the long-run cross-price elasticity of demand for pork with respect to the price of beef. Are pork and beef substitutes or complements?
12. Calculate the short-run cross-price elasticity of demand for pork with respect to the price of poultry. Are pork and poultry substitutes or complements?
13. Calculate the long-run income elasticity of demand for pork. Is pork a normal or inferior good?
14. Using your long-run pork demand function, calculate the pork demand *curve*, and graph it below with price on the y-axis and quantity on the x-axis.

CHAPTER FOUR

Advanced Price Analysis: Imperfect Competition

In the News

Hurricane Katrina brought many changes to our country. It revealed our inability to deal effectively with catastrophes. It reminded us of how differently the poor and middle class live. It also led to a spike in gas prices. For the first time, gas prices hit $3.00 a gallon. This had never happened before, and many gas pumps were not even able to display prices higher than $3.00 on the pump.

Some of this price increase was justified. Gas is made from crude oil. The cost of making gasoline includes the cost of purchasing crude oil and the cost of transforming oil into gasoline. When Hurricane Katrina hit, it disrupted the supply of crude oil. Less oil means higher oil prices and, consequently, higher gas prices. The problem is that an economist estimated the rise in oil prices increased the cost of producing gasoline by about $0.23 per gallon (Nichols 2005), yet the price of gasoline rose by about $0.45 per gallon (MSN Money 2005). Either the cost estimate was wrong, or someone was using Hurricane Katrina to rip consumers off. As a result, Congress held hearings on the rising gas prices. Consider the following remarks from this hearing.

> "Why are the oil companies making record profits, and what are they doing with them? Our job is to make sure that one, price gouging, and, two, unfair profiteering and unconscionable profiteering, do not take place and especially does not take place as a result of the hurricane."
>
> —Senator Pete V. Domenici
>
> "The distrust is enormous. I hope somebody out there is listening who controls these prices."
>
> —Senator Dianne Feinstein
>
> Senator Maria Cantwell suggested the president use his emergency powers, *"to look at this issue of price gouging and look at what level of price increase is realistic."*
>
> (Kirkpatrick 2005)

This is a serious issue. If companies are really profiteering from the hurricane, government involvement in the gasoline market may benefit consumers. But if the gasoline price increase is really due to higher gasoline production costs, government involvement will likely make things worse. Government tried regulating gas prices in the 1970s and it led to widespread gasoline shortages, with people waiting hours at the gas station to fill up with gas. To determine whether $3.00 gasoline is really "justified," one must understand how prices are formed. That is the purpose of this chapter.

INTRODUCTION

Up to this point we have only concerned ourselves with competitive markets, where no one buyer or seller has any significant control over the market price. This, of course, does not adequately describe many markets. Microsoft completely controls the price of Microsoft Windows because they have a patent on Windows, making them the single seller. The NCAA completely controls the compensation college football players receive because they are the only buyers of college football players. When there is a single buyer or seller, that one buyer or seller has complete control over the price. In other markets there are a few buyers and sellers, given them some but not complete control over price. If you are a cattle producer in west Texas, it is unlikely you can find more than two buyers for your cattle. There are only a few sellers of credit cards but many buyers, and certainly this gives those credit card companies some market power. Other companies produce a differentiated product that gives them market power. Coca-Cola, for example, has some control over its price because no one knows their secret formula.

The purpose of this chapter is to develop market models to describe situations where

1. there is only one seller
2. there is only one buyer
3. there are only a few buyers or sellers
4. firms produce differentiated products

These models are different from perfect competition because certain buyers or sellers *do* have some control over the market price. For this reason, they are referred to as models of *im*perfect competition.

THE MONOPOLY MODEL

In the perfect competition model, there are many producers and consumers and no one buyer or seller has an advantage in negotiating price. This is a realistic model for some goods like corn, but unrealistic for goods like Monsanto's Bt cotton. There is only one company who can produced Bt cotton. Monsanto has the patent and is therefore the only seller, and there are no close substitutes for

A market with many buyers but one seller of a good with no close substitutes is a *monopoly.*

Bt cotton. *When there is only one seller of a good with no close substitutes, we call this a monopoly.*

The extent to which a firm is a monopoly hinges crucially on whether there are close substitutes available. KFC (formerly, Kentucky Fried Chicken) sells a unique type of chicken with 11 herbs and spices. No one knows their secret recipe, so no one can sell chicken exactly like KFC. KFC is not a monopoly, however, because there are many firms selling similar fried chicken. Due to their competitors, KFC cannot set any price it wants. If they set their price too high, people will simply purchase fried chicken from other vendors (we highly suggest Popeye's). However, Microsoft can set a very high price for Windows because there are no substitutes (except for the computer savy who can use Linux, and the counter-culture people purchasing an Apple). Consumers either pay the high price or they do not get a Windows-like product. Microsoft Windows is a monopoly; KFC is not. The absence of close substitutes is an absolute must for a monopoly to exist. In 2005 Visa and MasterCard were sued on the basis that they exerted illegal market power, meaning they behaved like a monopoly. Mastercard's response to the claim was that they could not be a monopoly because there are many close substitutes to credit cards, namely, cash and checks (Shepherd 1997).

Producer and Consumer Behavior in Monopoly

A monopoly is the only seller of a good, so it can set whatever price it wants. The seller has all the negotiating power. The monopoly faces a trade-off though. The higher the price it charges, the less consumers will buy. Consumers simply observe the price the monopoly sets, and then purchases the quantity given by the demand curve. The monopoly therefore sets the price that maximizes its profits, and consumers must live with that price. What exactly is the profit maximizing price? The monopoly follows a simple rule: Sell an additional unit whenever the marginal revenue is greater than the marginal cost. We have already covered marginal cost, but not marginal revenue. Marginal revenue is the additional revenue from selling one more unit. So long as marginal revenue is greater than marginal cost, the monopoly can add to its profits by producing an additional unit.

Marginal revenue is illustrated with an example in Figure 4.1. If the monopoly wants to sell one unit, it sets a price of $4. This is the maximum price it can charge and still sell one unit to consumers. To sell two units it lowers the price to $3. For each additional unit sold, it must lower its price by one. Notice what happens to total revenue as the price is lowered and sales rise. Revenues rise and then fall. Marginal revenue tells us the change in total revenue for each additional unit sold. By reducing price from $4 to $3, sales increase from 1 to 2 and revenues increase from $4 to $6. Therefore, marginal revenue at one unit is $6 − $4 = $4. Notice that if we plot marginal revenue in Figure 4.1, the marginal revenue curve lies underneath the

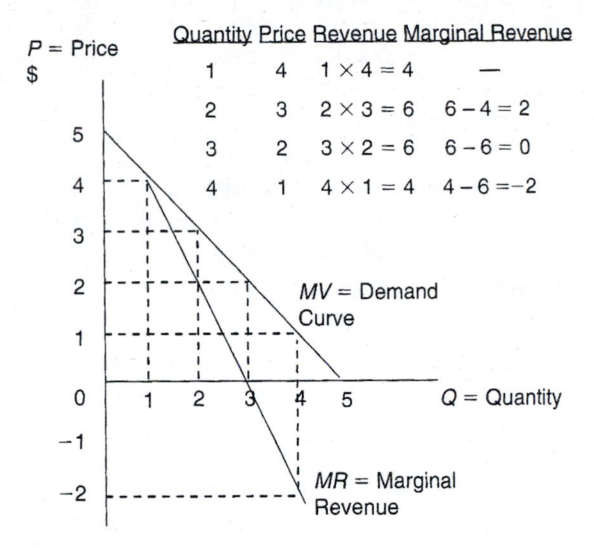

Quantity	Price	Revenue	Marginal Revenue
1	4	1 × 4 = 4	—
2	3	2 × 3 = 6	6 − 4 = 2
3	2	3 × 2 = 6	6 − 6 = 0
4	1	4 × 1 = 4	4 − 6 =−2

FIGURE 4.1 For a Monopoly, the Marginal Revenue Curve Lies Underneath the Demand Curve.

demand curve. This is an important concept: *The marginal revenue curve always lies underneath the demand curve.*

Once we know the marginal cost curve and the marginal revenue curve, we know exactly how a monopoly will behave. Consult Figure 4.2, which is exactly like the

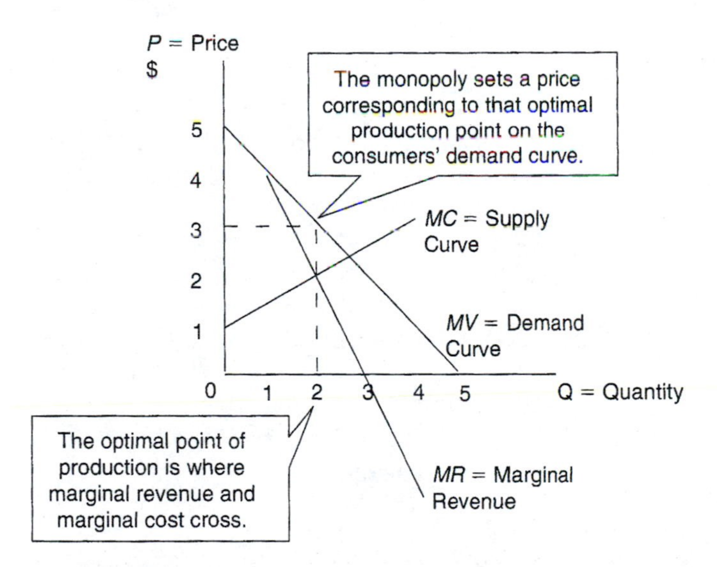

FIGURE 4.2 The Monopoly Produces a Quantity Where Marginal Revenue and Marginal Cost Cross. Here, the monopoly will produce 2 units and sell them for $3 each.

previous figure except that we include the monopoly's marginal cost curve. If the monopoly increases its production from zero to one unit, marginal revenue is \$4 and marginal cost is 1.5, so the firm makes \$4 − \$1.5 = \$2.5 on the first unit. Thus, the monopoly will sell at least 1 unit. To sell a second unit, it must lower the price to \$3. Marginal revenues of \$2 exactly equals marginal costs of \$2. The firm is indifferent between producing the second unit or not, and in this case we simply assume the monopoly will produce it. The monopoly keeps selling additional units until marginal revenue equals marginal costs. At the optimal quantity of 2, you know the monopoly will charge \$3 per unit because that is the maximum price it can set and still sell 3 units. A price higher than \$3 means consumers will buy less than 3 units, which is not the profit-maximizing quantity. A price lower than \$3 means consumers will still buy 2 units but the firm receives less money for each unit sold. The best price is then the point on the demand curve corresponding to the profit-maximizing quantity. As you can see from the graph, a monopoly always charges a higher price and sells fewer units compared to perfect competition (because price is higher and quantity is lower than where supply and demand cross).

The Mathematics of Monopoly Like the perfect competition model, if we know the formulas for the supply and demand curves, we can calculate exactly what a monopoly will charge. Let us use the same equations as in perfect competition (from Chapter 2) and assume the marginal value (demand) and marginal cost (supply) curves are as

$$\text{Demand}, MV: P = 120 - 8(Q)$$
$$\text{Supply}, MC: P = 20 + 2(Q)$$

As shown previously, the monopoly will lower its price and sell more units until marginal revenue equals marginal cost. This means if we can obtain a formula for the marginal revenue curve, we can set it equal to the marginal cost curve and solve for the profit-maximizing quantity. It turns out that if the demand curve follows a simple linear formula like $P = a - bQ$ (where a and b can be replaced with any positive number), the marginal revenue curve will equal $P = a - (2 \times b)Q$.[1] The marginal revenue and demand curves share the same intercept, but the marginal revenue curve falls twice as fast as the demand curve. Therefore, if the demand curve is $P = 120 - 8(Q)$, the marginal revenue curve is $MR = 120 - 2 \times 8(Q) = 120 - 16(Q)$. Remember, the marginal revenue curve always lies underneath the demand curve. Now that we have the marginal revenue curve, we can easily calculate the price and quantity for a monopoly using the following steps.

[1]This can be seen using calculus. *If* $P = a - bQ$, where Q is quantity sold, then total revenue is $P \times Q$ or *total revenue* $= [a - bQ]Q = aQ - bQ^2$. Marginal revenue is the change, or derivative, of total revenue with respect to Q, which is $a - 2bQ$.

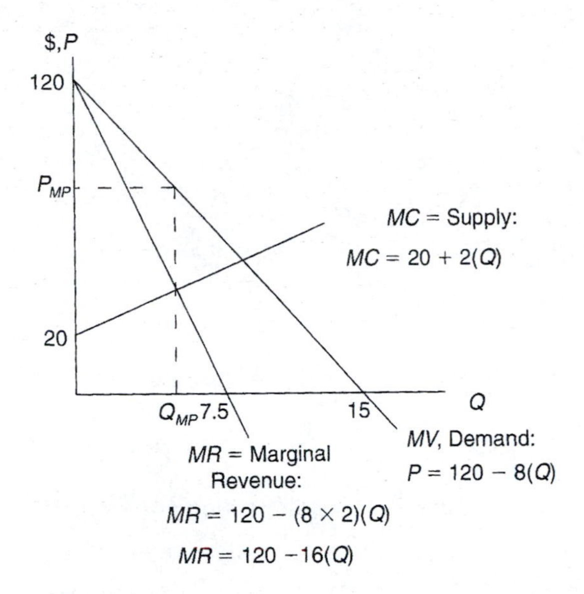

FIGURE 4.3 A Monopoly Sells a Quantity, Which Sets Marginal Revenue Equal to Marginal Cost. $Q_{MP} = 5.6$ and $P_{MP} = 75.2$.

Note: The marginal revenue curve intersects the demand curve at the *y*-axis, instead of at $Q = 1$ as in the previous two graphs, because this figure assumes one can sell fractions of a unit.

Step 1: Use the demand curve to obtain the marginal revenue curve.

$$\text{Demand: } P = 120 - 8(Q) \text{ implies}$$
$$\text{Marginal Revenue} = 120 - (8)(2)(Q) = 120 - 16(Q)$$

Step 2: Solve for the quantity that sets marginal revenue equal to marginal cost.

$$\begin{aligned} \text{Marginal Revenue} &= \text{Marginal Cost} \\ 120 - 16(Q) &= 20 + 2(Q) \\ 120 - 20 &= 16(Q) + 2(Q) \\ 100 &= 18(Q) \\ Q_{MP} &= 100/18 = 5.6 \end{aligned}$$

Step 3: Solve for the price on the demand curve corresponding to the quantity obtained for in step 2.

$$\begin{aligned} \text{Demand, } MV\text{: } P &= 120 - 8(Q = 5.6) \\ P_{MP} &= 120 - 8(5.6) = 75.2 \end{aligned}$$

Compare the price and quantity under a monopoly to that under perfect competition (calculated in Chapter 2). The price is higher and the quantity is lower. The monopoly charges a higher price and consumers respond by purchasing less. You probably already knew this, but what you did not know is that we can prove society is better off under perfect competition than under a monopoly. See Figure 4.4. Producer surplus for a monopoly is greater than under perfect competition, and consumer surplus is smaller. Perfect

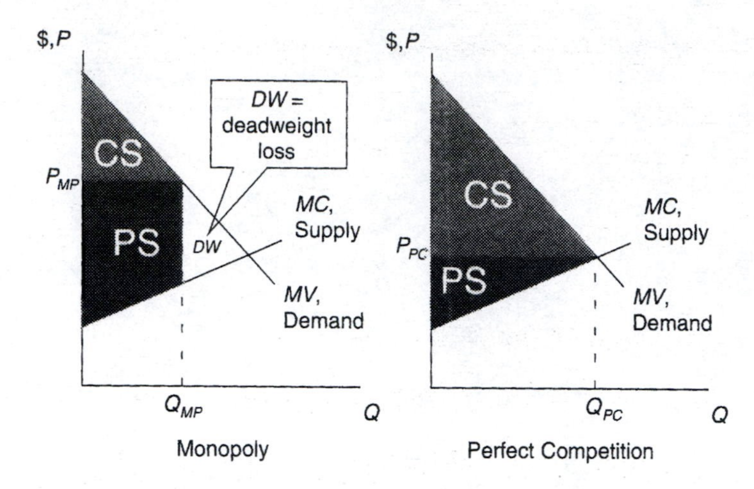

FIGURE 4.4 Comparing Perfect Competition and Monopoly.

competition generates a lower price and greater consumption than monopoly. The figure also highlights a very important difference between the two market structures.

Consumer surplus is lower under a monopoly, but it is obvious consumers are worse off when faced with higher prices. Producer surplus is higher under a monopoly, but it is obvious a monopoly should be able to negotiate higher prices and make more profits. What may not be obvious—until you learn to use these market models—is that society as a whole is worse off under a monopoly. Societal welfare can be measured as the sum of producer and consumer surplus. This sum is referred to as *total surplus* and takes into account consumer benefits and producer profits. Producers benefit from a monopoly and consumers are hurt, but it is clear from Figure 4.4 that consumers are hurt more than producers benefit.

Total Surplus = Consumer Surplus + Producer Surplus

This figure clearly shows that total surplus is lower under a monopoly than perfect competition, and therefore society as a whole would be better off under perfect competition. A monopoly creates what is called a *deadweight loss* (shown in Figure 4.4). Remember from Chapter 1 to think toys, not dollars. Society consumes less toys under a monopoly (Q_{MP} is less than Q_{PC}), and the deadweight loss is the value of this foregone consumption. This is why our government has laws to promote competition among sellers and outlaws monopolies. This is why Microsoft was taken to court by the U.S. government, and why courts heard arguments from cattlemen claiming that Tyson Foods exerts too much market power.

THE MONOPSONY MODEL

The monopoly model above assumed there are many buyers but only one seller, giving the seller all the negotiating power. Now let us turn the table and assume many sellers but only one buyer, giving the buyer all the negotiation power. *When there are many sellers but only one buyer of a good with no close substitutes, we call this a monopsony.* A good example of monopsony is college athletes. Even though college athletes have their choice of schools to attend, the National Collegiate Athletic Association (NCAA) limits the compensation they can receive. So in a sense, all college athletes work for the NCAA, and the NCAA has complete control over the prices paid to athletes (in terms of financial assistance). They use this power. Although a premium football player raises about $500,000 in revenues for his school each year, have you ever heard of a college football player receiving anything close to this amount (Brown 1993)?

A market with many sellers but only one buyer of a good with no close substitutes is a *monopsony*.

Producer and Consumer Behavior in Monopoly Like all firms, we assume the monopsony is out to maximize its welfare. The monopsony is a consumer—it is the *only* consumer—and so it seeks to maximize its consumer surplus. You have probably already guessed that the monopsony will negotiate a lower price than the perfect competition price. The monopsony calls all the shots. It determines the price it will pay, and then producers observe this price and determine how much they will sell. Announcing a lower price presents a trade-off for the monopsony. A lower price decreases the amount it pays for each unit purchased but also decreases the number of units producers are willing to sell. To maximize its profits, the monopsony must pick the one perfect price that balances this trade-off.

A monopsony follows a simple plan to maximize its profits. Think about how you study for a test. You study the first hour only if you think the benefits are greater than the opportunity cost of your time. Should you study the second hour? Only if the benefits of studying the second hour are worth the opportunity costs of the second hour. At some point, you realize the benefit from an extra hour of studying is less than the opportunity cost (which is the value of your next best alternative, which on Tuesday nights is watching *House*), and you cease studying. A monopsony follows a similar logic when deciding how many units to purchase. They purchase one more unit when the benefit of the extra unit outweighs the cost. The benefits of purchasing another unit is given by the marginal value curve. The cost of purchasing an additional unit is given by the marginal expenditure curve, shown in Figure 4.5.

Using Figure 4.5, suppose you are a monopsony who knows it will purchase at least one unit. To entice producers to sell you 1 unit, you must offer them a price of at least $1. You are now considering purchasing 2 units instead of 1. For producers to sell 2 units, they must receive a price of at least $2 (because the supply curve at 2 units equals $2). Observe how total expenditures change when moving from 1 to 2 units purchased; it increases from $1 to $4. The marginal expenditures from increasing purchases from 1 to 2 units is $4 − $1 = $3. Similarly, the marginal expenditures of the third and fourth units are $5 and $7, respectively. If we plot these points, we see that the marginal expenditure curve lies above the marginal cost curve. Figure 4.6 is the same as the previous diagram except that a demand curve is included. Using the rule: Purchase another unit whenever the marginal value is greater than or equal to the marginal expenditure, the monopsony will purchase 2 units and will pay producers $2 per unit.

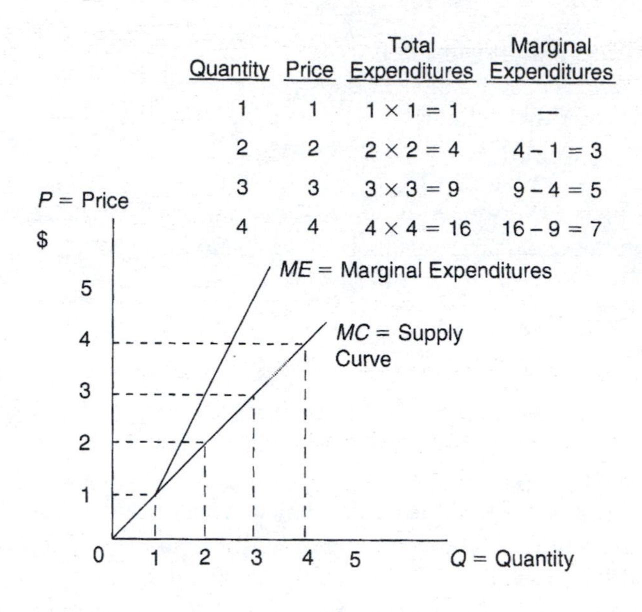

Quantity	Price	Total Expenditures	Marginal Expenditures
1	1	1 × 1 = 1	—
2	2	2 × 2 = 4	4 − 1 = 3
3	3	3 × 3 = 9	9 − 4 = 5
4	4	4 × 4 = 16	16 − 9 = 7

FIGURE 4.5 The Marginal Expenditure Curve Lies Above the Marginal Cost Curve.

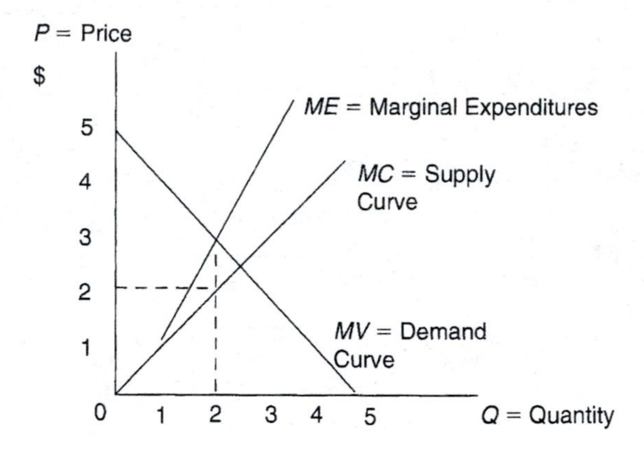

FIGURE 4.6 The Monopsony Purchases a Quantity Where Marginal Expenditure and Marginal Value Curves Cross. Here, the monopsony will purchase 2 units, paying producers $2 per unit.

It is easy to get confused between marginal cost and marginal expenditures, so let's stop for a second and distinguish the two. Marginal cost refers to the additional cost *producers incur* to produce another unit. Marginal expenditures refer to the additional cost *producers must be paid* to produce another unit. The difference has to do with producer surplus. For producers to sell more units, they must be given a price that covers their marginal costs and provides them with extra producer surplus.

The Mathematics of Monopsony

In keeping with the previous sections, we illustrate how to calculate the exact price a monopsony will set given an equation for the marginal value and marginal cost curves. As before, the marginal value and marginal cost equations are

$$\text{Demand}, MV{:}\ P = 120 - 8(Q)$$
$$\text{Supply}, MC{:}\ P = 20 + 2(Q)$$

It turns out that with a simple marginal cost curve like $P = a + b(Q)$, the marginal expenditure curve equals $P = a + 2b(Q)$.[2] The marginal expenditure curve rises twice as fast as the supply curve. The steps to calculating the monopsony price and quantity are as follows.

Step 1: Use the marginal cost formula to obtain the marginal expenditures formula.

$$\text{Supply}, MC{:}\ P = 20 + 2(Q) \text{ implies}$$
$$\text{Marginal Expenditures} = 20 + (2)(2)(Q) = 20 + 4(Q)$$

Step 2: Solve for the quantity that sets marginal expenditures equal to marginal value.

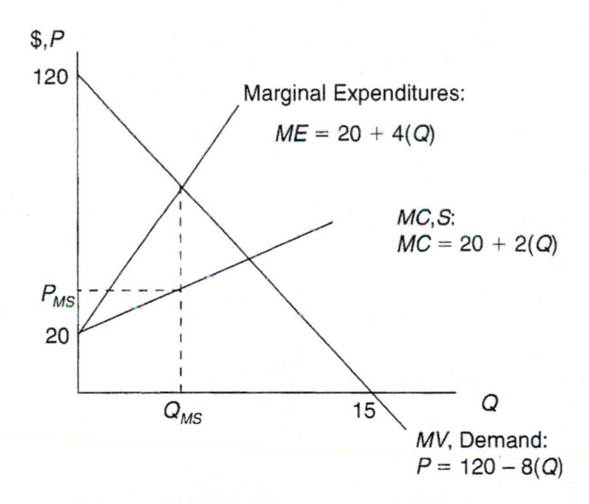

FIGURE 4.7 A Monopsony Purchases a Quantity That Sets Marginal Expenditures Equal to Marginal Value. Q_{MS} = 8.33 and P_{MS} = 36.66.

Note: The marginal expenditure curve intersects the supply curve at the y-axis, instead of at $Q = 1$ as in the previous two graphs, because this figure assumes one can sell fractions of a unit.

---

[2]This can be seen using calculus. Total expenditures equal price (given by $P = a + b(Q)$) times quantity purchased, Q, making *total expenditures* $= P \times Q = [a + bQ]Q = aQ + bQ^2$. Marginal expenditures are the change, or derivative, of total expenditures with respect to quantity, which equals $a + 2bQ$.

$$\begin{aligned}
\text{Marginal Expenditures} &= \text{Marginal Value} \\
20 + 4(Q) &= 120 - 8(Q) \\
4(Q) + 8(Q) &= 120 - 20 \\
12(Q) &= 100 \\
Q_{MS} &= 100/12 = 8.33
\end{aligned}$$

Step 3: Solve for the price on the supply curve corresponding to the quantity solved for in step 2.

$$\begin{aligned}
\text{Supply, } MC\text{: } P &= 20 + 2(Q = 8.33) \\
P_{MS} &= 20 + 2(8.33) = 36.66
\end{aligned}$$

The monopsony price is 36.68 and the monopsony quantity is 8.34. Compare this to the perfect competition price and quantity of 40 and 10, respectively, and you will see that a monopsony results in a lower price and lower quantity. If you peer behind the equations and curves, what you find is that the monopsony simply purchases less than buyers would in perfect competition. With less pressure by buyers (now there is only one buyer), sellers receive a lower price.

Before we found that a monopoly results in lower social welfare than perfect competition, so let's see if the same holds true for a monopsony. Figure 4.8 shows the outcome when perfect competition and monopsony exist. The monopsony forces the price downward from the perfect competition level, decreasing the quantity produced and decreasing producer surplus in the process. It places downward pressure on prices so that it can extract greater consumer surplus. Although it may not be obvious from the figure, consumer surplus is definitely larger under a monopsony than under perfect competition. Social welfare, defined as producer surplus plus consumer surplus, falls when a monopsony is present. By forcing prices downward, the monopsony induces producers to produce and sell fewer units than they would in perfect competition. This means society foregoes goods and services it otherwise could have enjoyed. Less toys means less happiness. This decline in consumption is referred to as deadweight loss.

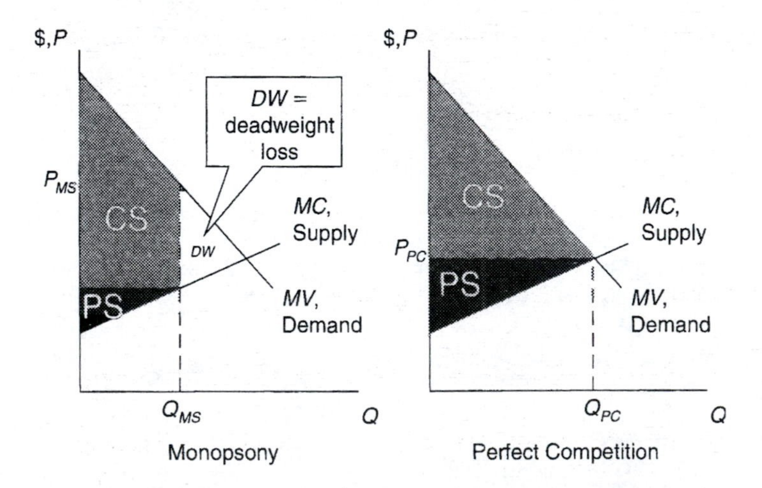

FIGURE 4.8 Comparing Perfect Competition and Monopsony.

OTHER MARKET MODELS

Of all the different ways buyers and sellers interact to strike deals, we have discussed three: perfect competition, monopoly, and monopsony. Perfect competition describes a setting where there are many buyers and sellers and all possess roughly equal negotiating power. The price resulting from perfect competition is usually referred to as the *competitive price*, and in most cases is the ideal market because it maximizes the sum of producer and consumer surplus, which we call total surplus. When buyers or sellers gain negotiating power over the other, the net result is a lower quantity bought and sold. A monopoly (single seller) restricts quantity sold to drive up prices. A monopsony (single buyer) restricts purchases to drive price down. Either way, quantity is lower than it would be under perfect competition. There are less toys, and society is the worse for it. Few markets can be exactly described as perfect competition, monopoly, or monopsony. As you have probably noticed, the market for many goods lies somewhere between perfect competition and monopoly. There are only a few sellers of credit cards: MasterCard, Visa, Discover, and American Express. Anheiser-Busch, Miller, and Coors collectively sell 90% of all beers purchased in the United States (Tremblay and Tremblay 2005). The soda market is dominated by Coca-Cola and Pepsi, and five firms sell 80% of all beef processed in the United States (McMahon 1998). These markets are best described as *oligopolies: markets where there are a few sellers of identical goods and many buyers*.

Oligopolies can exist for a number of reasons. The most common reason has to do with *economies of scale*. Budweiser makes thousands of gallons of beer at one time. One of the authors is a homebrewer, meaning he makes homemade beer. He purchases the same ingredients as Budweiser, but Budweiser uses huge factories and thousands of workers, whereas the author brews beer in five-gallon containers with no help from anyone. As a result, Budweiser can produce beer at a much lower per unit cost than the author. Sometimes, to produce goods at a low per unit cost, you need to "get big." Given there is limited demand for any one product, a few "big" firms can easily supply the market, and anyone who tries to enter the market and compete would also have spend millions of dollars in capital. Once the market is saturated by a few large firms, entering the market as a new competitor would not be profitable. The result is a few large firms supplying the entire market—an oligopoly.

The price that results from an oligopoly follows no simple formula. The reason is that oligopolies engage in strategic behavior. When Pepsi increases its advertising, so does Coca-Cola. When General Motors announced its "employee discount" special and reduced the price on all its automobiles, so did some other car manufacturers. Firms in an oligopoly have incentives to both collude and compete. First, let's talk about collusion. Sometimes, without explicit planning oligopoly firms will charge similar high prices (if they explicitly planned high prices, that would be illegal), and the price will be close to the monopoly price. This is referred to as *tacit collusion*. No actual discussion takes place, but there is an unspoken understanding that both will keep their prices high. Neither reduces their price, because they are scared the other will retaliate with a price reduction in turn, and no business wants a price war. Tacit collusion usually occurs through *price leadership*, where one firm announces a price and all other firms respond with similar prices. Tacit collusion through price leadership is difficult to prove because firms may raise prices to cover extra costs, not just to extract more profits. However, economists have noted price patterns that strongly suggest tacit collusion—including the meatpacking industry from 1890 to 1920, the cigarette and steel industries in the 1930s, and the steel and auto industries in the 1950s (Shepherd 1997). Sometimes firms explicitly, and illegally, collude to set high prices, essentially acting like a monopoly. The animal feed market is an oligopoly and is renown for price fixing. Three companies—Akzo Nobel, BASF, and UCB—control 80% of the world market for vitamin B4 used in animal feed. In 2004, the European Commission fined the companies $66.34 million for meeting in secret to agree to set high prices (*Feedstuffs* 2004).

In other cases, oligopoly firms will engage in price warfare that leads to low prices close to that of perfect competition. Consider airline tickets. There are only a few airlines, so it is an oligopoly market, but every airline struggles to avoid bankruptcy. Many do go bankrupt. Next time you visit the grocery store, notice there are only a few sellers of soup. In some stores, Campbell's Soup is the only brand available. Yet, soup is not expensive. It is hard to imagine soup getting cheaper. Campbell's Soup knows the second it starts charging high prices, grocery stores will simply buy from someone else. So they keep their prices low to ensure a high market share. Price can be high or low in oligopolies. For now, all we will say about oligopolies is that the price will be somewhere

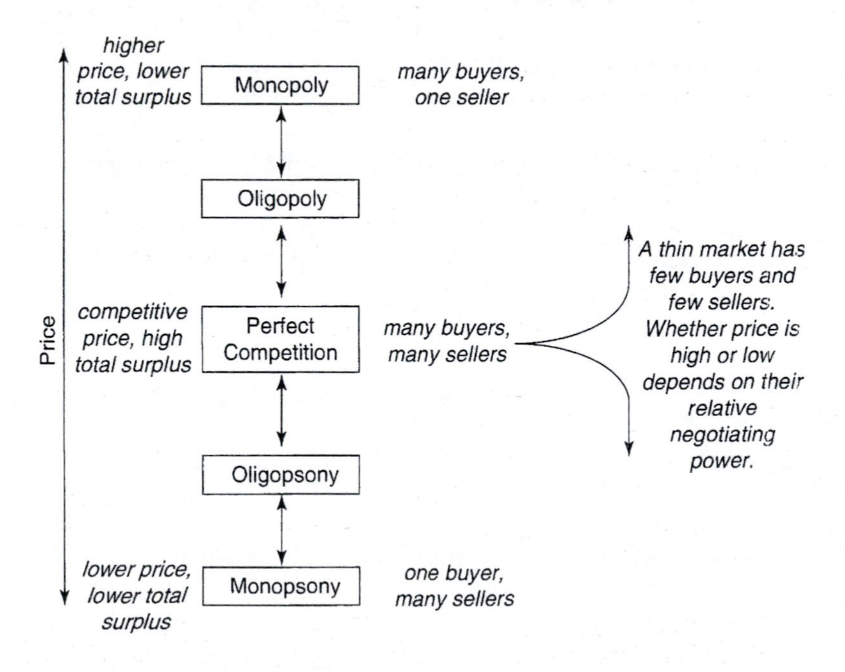

FIGURE 4.9 Market Structures.

between the monopoly and the perfect competition price. The fewer the number of sellers in the oligopoly, the closer the oligopoly price should come to the monopoly price.

An *oligopsony* exists when there are only *a few buyers of an identical product*. Oligopsonies are prevalent in agriculture. Throughout the Midwest are thousands of cattle producers, yet there are only a few firms who will purchase cattle. These cattle buyers (the meat packers) have the power to depress prices, if they colluded with one another to keep prices down. However, the buyers also compete with one another for a limited number of cattle. Elements of perfect competition and monopsony are present. The presence of only a few firms gives the oligopsony a price negotiating advantage, yet the firms are still competing with one another. In the end, whether prices are close to the perfect competition or monopsony price will depend on the strategic interactions between firms. They may collude and set high prices, they may engage in price warfare setting low prices, or something between the two. For now, all we will say about oligopsonies is that the price will be somewhere between the monopsony and the perfect competition price. The fewer the number of buyers in the oligopoly, the closer the oligopoly price should come to the monopsony price.

Finally, there is a type of market referred to as a *thin market*. Only a few buyers and sellers exist in a thin market. An example would be the market for miniature horses. Few people want them; few people sell them. The price that results depends on the negotiating power between the buyer and seller. The price could resemble the competitive price, or it could be more in favor of the buyer or seller.

MONOPOLISTIC COMPETITION

All the market structures previously discussed assume that firms sell an identical good. As you well know, there are many markets where firms produce different varieties of a good. Each variety is differentiated from each other somewhat. There is no textbook just like this textbook, but there are many others you could have used in your class instead (thank you for not using the others, by the way). The same can be said of soda, candy, even fast food. McDonalds sells hamburgers similar to Wendy's but not identical. We all know Wendy's burgers are better. These market settings are best described as *monopolistic competition*.

Monopolistic competition exists when there are many competing varieties of a good, each different in some way, but each variety being a close substitute for one another. Firms can freely introduce new varieties, decreasing the demand for all old varieties.

Since each firm produces a different good, there is no single marginal value or marginal cost curve facing the industry. There is however a single marginal value curve for each variety, and a single marginal cost curve for each variety. Each firm produces a differentiated good, which is not imitated perfectly by any other firm. However, other firms produce varieties that are close substitutes. Thus, each firm is like a "small monopoly." Because each firm has a unique variety, it can set any price it wants for that variety. It faces a downward sloping demand curve, unlike firms in perfect competition who must sell at the going market price.

Unlike a real monopoly there is free entry and exit of other firms that produce close substitutes. Each time a new firm enters with its own variety, the demand curve for all other firms falls. Remember, the demand curve shows the marginal value consumers place on the good. The greater the variety, the less each consumer is willing to pay for

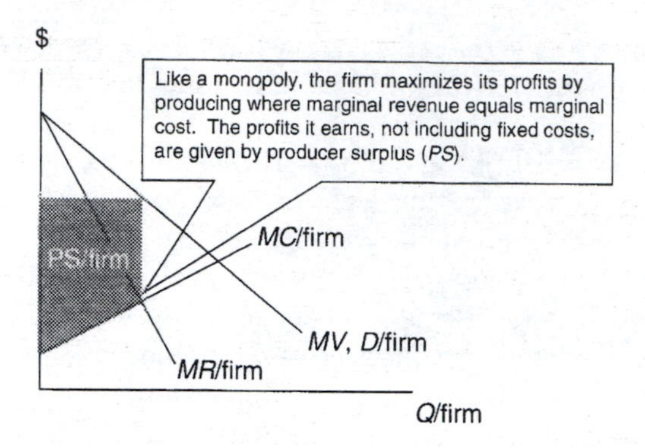

FIGURE 4.10 In Monopolistic Competition, Each Firm Produces a Differentiated Version of a General Good. This gives the firm market power, allowing them to set their own price like a monopoly.

any single variety. The less the variety, the more each consumer is willing to pay for any single variety. For example, if Dr. Pepper was no longer sold, many of its loyal customers (yours truly included) would switch to other sodas and increase the demand for other sodas. For convenience, we will assume that each variety has an equal market share. If there are only two firms, each has a 50% market share. If there are four firms, each has a 25% market share. This is referred to as *symmetric* monopolistic competition.

All firms have the same market share in a symmetric monopolistic competition model. In an asymmetric model, some firms have more market share than others.

In Figure 4.11, the left graph shows a situation with just five firms. Each firm receives producer surplus equal to the area *PS/firm*. Recall that producer surplus equals profits above fixed costs. If producer surplus per firm is greater than fixed costs, then the firms are earning money. More importantly, they are earning *economic profits*. They are making more money producing in this monopolistic competition market than their next best alternative. There is nothing in a monopolistic competition market to stop other firms from entering the market and producing their own unique variety. If firms are making money in this market, other firms will want to enter the market to make money as well.

Thus, we would expect that whenever producer surplus per firm is greater than fixed costs, new firms will enter the market with new varieties, decreasing the demand curve facing all firms. Similarly, when producer surplus per firm is less than fixed costs, firms are losing money. Their economic profits are negative, meaning they would be better off in their next best alternative, and firms will begin to leave the market. With some varieties now off the shelf, the demand curves facing the remaining firms rise. As firms enter and exit the market, they will eventually settle on an equilibrium where economic profits are zero. At zero economic profits, each firm is making just as much money as it could in their next best alternative. Producer surplus per firm equals the fixed cost for the firm. Firms are indifferent between producing or not—the Indifference Principle again!

The assumption of symmetric monopolistic competition seems unrealistic in some settings. Consider the market for soda ("coke" for Southerners or "pop" for Northerners). The market for sodas resembles monopolistic competition because no two firms produce an identical soda. Each soda is differentiated in some way, yet the

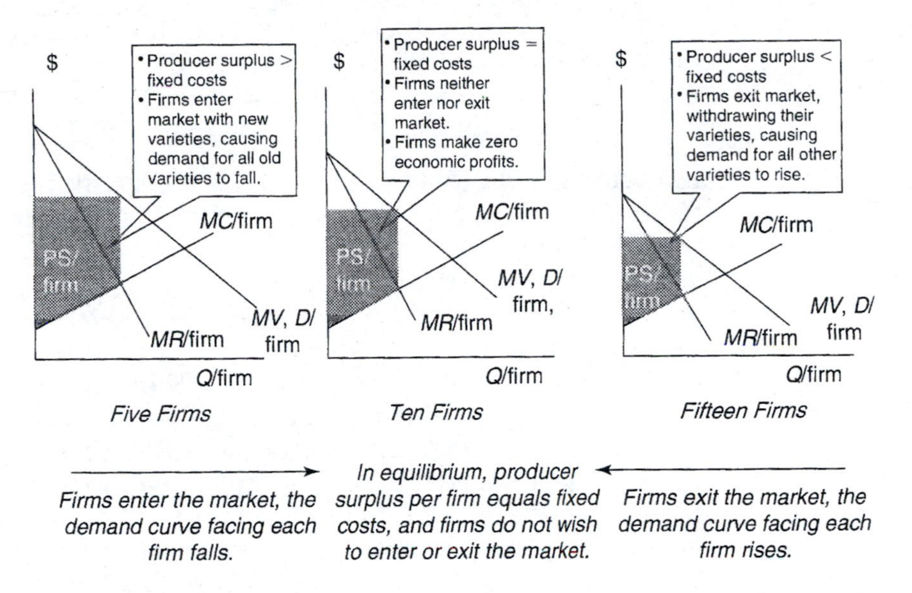

FIGURE 4.11 In Symmetric Monopolistic Competition, Firms Will Enter or Exit the Market Until Producer Surplus Equals Fixed Costs, and Economic Profits for All Firms are Zero.

sodas are strong substitutes for one another. But all sodas do not have the same market share. Coca-Cola and Pepsi possess over 70% of the market share for sodas with numerous other sodas sharing the remaining market. Clearly, there is a greater demand for Coca-Cola and Pepsi than there is for Sprite (Bhatnager 2005).

In these cases, we may want to use an *asymmetric* monopolistic competition model. The "asymmetric" part simply means that some firms will have a larger market share than others. Suppose we start in a market with "dominant" firms—those with the greatest market share. Then, other firms enter the market producing their own differentiated product. Each new firm produces a different product, but receives a smaller market share than the older firms, and consequently a smaller profit. New firms keep entering, and each new firm makes less profits than the previous firm until the producer surplus for the next firm entering just equals fixed costs. At this point, no firm will want to enter because it knows it will receive the smallest market share and its producer surplus will be less than its fixed costs (it will make negative profits). No firm will want to exit because producer surplus is equal to or greater than fixed costs. We have an equilibrium, where the last firm makes no [economic] profits, but all other firms do (Norwood 2005).

Market Models in Perspective

A word of warning: The market models described above are just the first step in understanding real markets. These models have strengths and reflect certain aspects of reality, but they also have weaknesses. Understanding these weaknesses is a must in applying economics to real-world markets. To illustrate, consider the beefpacking industry, which includes firms who purchase live-cattle, slaughter the cattle, and sell cuts of beef to wholesalers and grocery stores. Beefpackers are the buyers of live-cattle. Feedlots are the sellers of live-cattle. These are producers who take cattle weighing around 400–800 lbs and place them on a high-energy grain diet until the cattle are around 1200 lbs, after which they are ready for sale.

Beefpacking in the United States The degree of market power is often measured by the four firm concentration ratio, measured by the percent of market sales accounted for by the four largest firms. This ratio has increased from 41% in 1982 to 80% in 2004 (MacDonald and Ollinger 2000). Currently, 80% of all the beef you eat was slaughtered by one of four beefpackers. Moreover, certain beefpackers have a dominating presence in particular areas. In the Panhandle of Texas, four large beefpackers purchase 90% of cattle produced by 300 feedlots (Crespi and Sexton 2004). With 300 sellers against four buyers, it is easy to conclude the buyers have more negotiating power. Even if they do not explicitly collude, buyers for the beefpacking plants may tacitly respect each other's territory. If they know one plant tends to buy cattle from one set of feedlots, and they buy from another set, both may be reluctant to invade each other's territory and fight for more cattle in fear that it will drive up the price they pay. Consider the following quote from a feedlot manager.

> "No one else will buy my cattle. Excel buys 80% and are only in the market for about a week."
>
> (Crespi and Sexton 2004)

Furthermore, beefpackers often hire cattle buyers to procure their cattle supplies, and these buyers often represent more than one beefpacker, making the market power of cattle buyers even greater. Given these considerations, it is easy to conclude that the four large beefpacking firms employ their market power to negotiate low prices for the live-cattle they purchase. Similarly, if they dominate the beefpacking industry, they also possess power on the selling side. Only four firms sell most of the beef you eat to grocery stores, and it seems plausible they can negotiate higher prices for the beef they sell in addition to low prices for the cattle they buy. The four beefpacking firms (Tyson Foods, Excel Corporation, Swift & Company, and Smithfield Foods) are an oligopsony in the live-cattle market and an oligopoly in the beef market.

Earlier in this chapter, using our market models, we showed that society is worse off when markets move from perfect competition to oligopsony, and when it moves from perfect competition to oligopoly. You may revisit this in Figures 4.4 and 4.8. Based on these models, we would conclude that the increase in market power gained by the beefpacking plants is bad from society's perspective, and government policy should reduce the four firm concentration ratio. Yet, this conclusion may be wrong.

In Figure 4.8 where we compare perfect competition to monopsony, the demand/marginal value curves are the same regardless of the market structure. Thus, we assumed that the value of fed-cattle to the beefpacker is the same regardless of whether the market is perfect competition or monopsony. This assumption is not valid for the beefpacking industry. Studies have shown that the large increase in concentration in beefpacking was not due to the firms' desire to gain market power but to the firms' desire to reduce production costs. Economies of scale exist in the beefpacking industry. To reduce the per head cost of slaughtering and processing cattle, one must use larger slaughtering facilities. Today, 58% of all beefpacking plants slaughter more than a million head of cattle each year. In 1977 the number was only 12%. Because some firms employ larger plants with lower costs, they are able to drive out smaller plants who cannot compete due to their high costs. The end result is a few large firms slaughtering most of the cattle. This shift to fewer but larger beefpacking plants has reduced the cost of beef processing by 28%—no trivial number.

Therefore, as the beefpacking industry consolidated into a few large firms, the cost of production for these firms has fallen. On the one hand, greater market power by the beefpackers probably leads to lower cattle prices and higher beef prices, which is bad for society. On the other hand, lower beefpacking costs increase the amount of money available for purchasing live-cattle, which could increase live-cattle prices. Lower beefpacking costs also allow beefpackers to reduce consumer beef prices. The net effect on the increase in beefpacking market concentration on society is therefore unknown.

Consider one possibility shown in Figure 4.12 where we focus on live-cattle prices. Before there is industry concentration, the market is in perfect competition as shown in the left diagram. After industry concentration, beefpackers (the buyers of live-cattle) possess market power and negotiate a price below that where supply and demand cross, as shown in the right diagram. However, industry concentration occurred in pursuit of larger plants, which process live-cattle at a lower cost. Because live-cattle can be processed cheaper, they are more valuable to the beefpackers and the marginal value curve for live-cattle rises. The demand curve for live-cattle shifts upward. The demand increase in Figure 4.12 shows a scenario where the increase in demand for live-cattle just offsets the gain in market power by beefpackers, and the price of live-cattle remains unchanged. Producer surplus is unchanged as well. Consumer surplus increases though, implying that total surplus (the sum of consumer and producer surplus) rises.

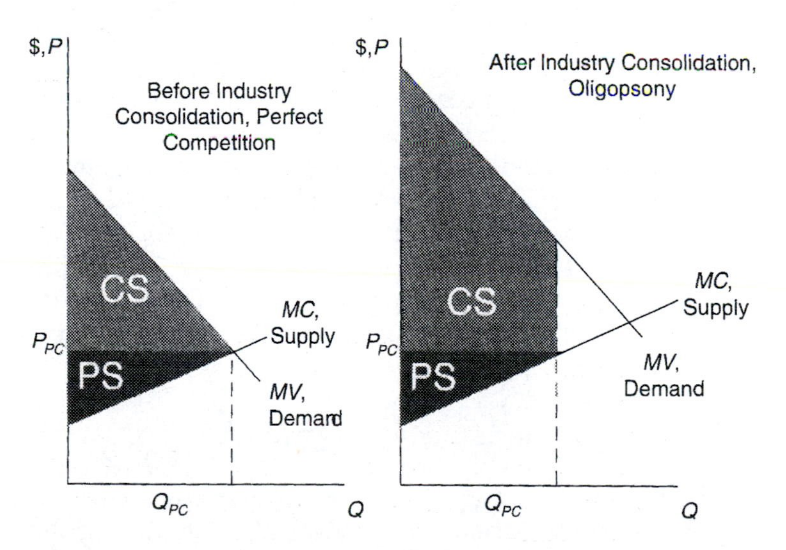

FIGURE 4.12 Comparing the Market for Live-Cattle Before and After Beefpacking Industry Consolidation.

Although the increase in the four firm concentration ratio resulted in a less competitive market, society as a whole is made better off.

It is worth noting that if we drew an increase in demand as a smaller shift, society may be made worse off and live-cattle prices may fall. The point is that in cases where the number of buyers and sellers decrease and markets become less competitive, society is not necessarily worse off. If the change in number of buyers and sellers is associated with a supply or demand shift, society may be better off.

The increase in buyer market power in the live-cattle market may or may not depress live-cattle prices and profits. The only way to tell is to look at data and see how prices and profits change as buyer market power increases. A total of 35 studies have looked at such data to see what they say. Not all studies agree, but in general the studies suggest that the increase in beef processor concentration has reduced live-cattle prices, but not by very much (Gardner 2002, 158).

Patents, Market Power, and Bt Cotton A monopoly, we said earlier, is less desirable than perfect competition. When a market moves from perfect competition to monopoly, total surplus, and hence societal welfare, falls. Monsanto currently has a monopoly on Bt cotton. Monsanto invented Bt cotton, received a patent for their invention, and therefore is the only firm who can legally sell Bt cotton seed. According to our previous analysis of monopolies, this is bad, and the government should strip away Monsanto's patent and let any firm sell Bt cotton.

Again, this is a situation where our simple economic models need refining to capture the peculiarities of this topic. If Monsanto's patent were removed, the price of Bt cotton would indeed fall and total surplus would rise. The impact of the patent removal does not stop here though. Monsanto developed Bt cotton through years and millions of dollars in research. Never would they have invested so much if they were not assured of a patent, allowing them to make profits to pay for their investment. If Monsanto's patent were stripped by the government, many (or perhaps all) firms would cease research in genetically modified crops in fear their future patents will be stripped as well. In fact, research in basically every industry would fall. Firms only engage in research if the incentives of patents and the profits they represent are assured. Less research means less product development and less technological advancement, the two major factors of today's large wealth.

Governments regularly issue and protect patents to encourage invention and discovery. These lead to new products, new markets that provide consumer and producer surplus. Eliminating patents may increase total surplus for existing products, but they ensure potential total surplus from future products will never corne to fruition.

SUMMARY

In this chapter, a number of models were introduced that run the gamut of complete buyer negotiating power to complete seller negotiating power. Some models assumed all firms sold the same product, whereas others allow product variety to differ across firms. All were models of *im*perfect competition, where some buyer or seller had greater price negotiating than another. Generally speaking, we saw that perfect competition is preferred to imperfect competition because it provides more surplus to society, but "generally speaking" is not always correct. When comparing competitive to uncompetitive markets, one must always account for other differences in the two markets besides who has market power. Buyers of live-cattle now have more power over price but they can process beef at a lower cost. Sellers of a patented product have more power over price, but this power is the reward of their discovery. A thorough analysis of any market requires more than the simple perfect and imperfect competition models; however, these models are the necessary building blocks.

CROSSWORD PUZZLE

If the answer contains more than one word, leave a blank space between each word.

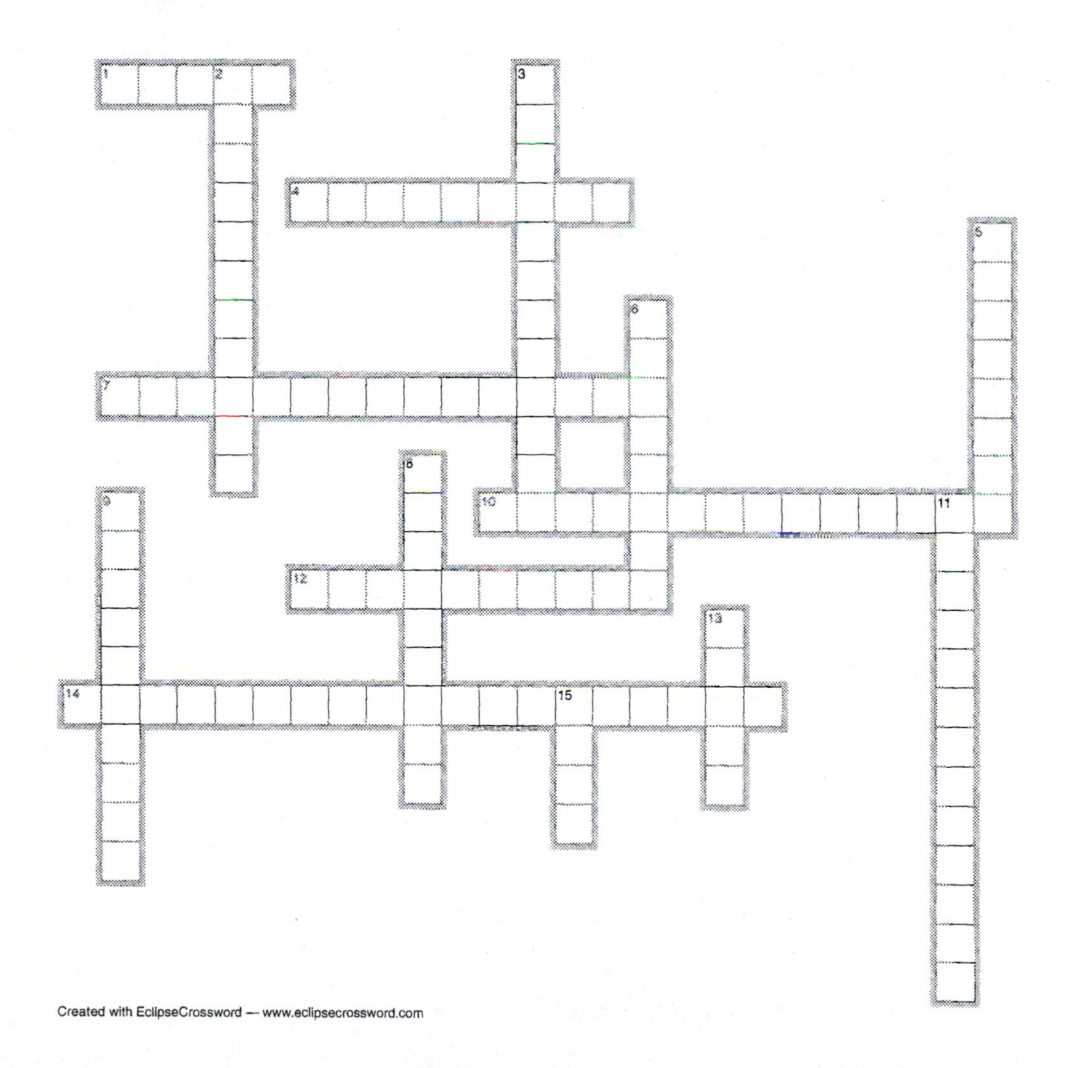

Across

1. In a monopoly, the marginal revenue curve lies (above, below) _______ the demand curve.
4. A market structure where there are many sellers but only one buyer of a good with no close substitute.
7. An unspoken, but understood agreement to collude and set high prices.
10. A fictitious economic story, where many complexities of the world are ignored and only a few important aspects are analyzed.
12. A market structure with many sellers and a few buyers.
14. A market structure where there are many buyers and sellers.

Down

2. The four major factors determining price are (1) _______ cost of production, (2) consumer value, (3) negotiating power, and (4) psychological and social considerations.
3. A form of competition where each firm produces a differentiated product and firms may freely enter and exit the industry.
5. A word referring to "one more" or "one additional unit."
6. A market structure where there are many buyers but only one seller of a good with no close substitutes.
8. A market structure with many buyers and a few sellers.
9. In _______ monopolistic competition, each firm can have a different market share.
11. If price is below the equilibrium market in perfect competition, a(n) _______ _______ is said to exist.
13. In a monopsony model, the marginal expenditure curve lies _______ the marginal cost curve.
15. A _______ market contains ony a few buyers and sellers.

STUDY QUESTIONS

1. Suppose a monopoly faces the demand curve illustrated below, and whose coordinates are shown in the following table.

   In the table below, calculate the total revenue and marginal revenue for each quantity sold. To the best of your ability, plot the marginal revenue curve in the graph in Figure 4.13.

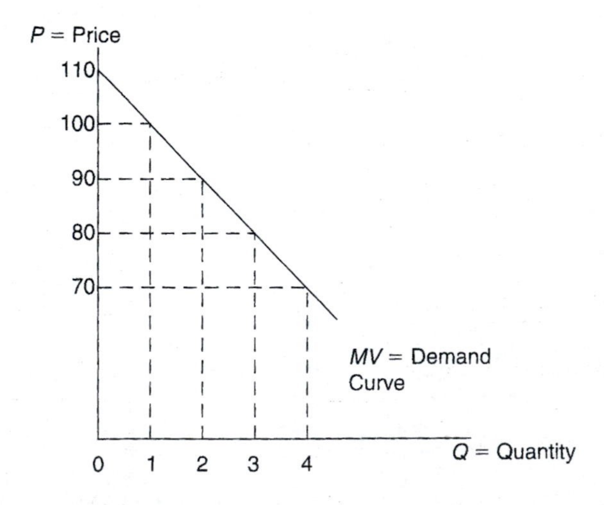

FIGURE 4.13

Quantity Demanded	Price	Total Revenues	Marginal Revenue
0	$110		
1	$100		
2	$90		
3	$80		
4	$70		

2. Suppose a monopsony faces the supply curve illustrated below, and whose coordinates are shown in the table below.

   In the table below, calculate the total expenditures and marginal expenditures for each quantity purchased. To the best of your ability, plot the marginal expenditures curve in the graph in Figure 4.14.

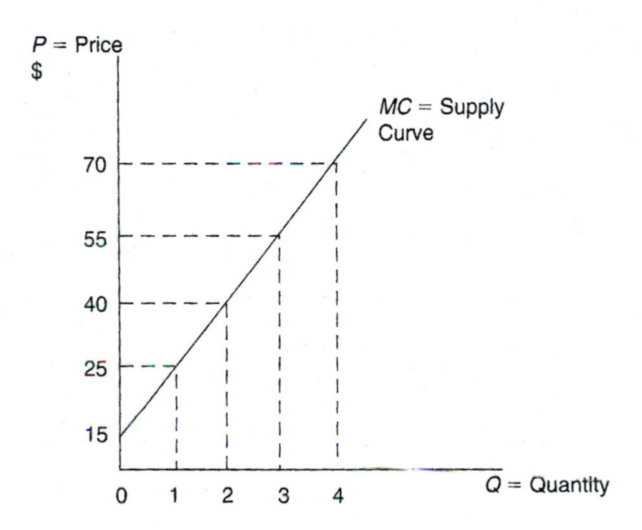

FIGURE 4.14

Quantity Supplied	Price	Total Expenditures	Marginal Expenditures
0	$15		
1	$25		
2	$40		
3	$55		
4	$70		

3. A monopoly market, compared to perfect competition, will have a (*circle one*) HIGHER / LOWER price and a (*circle one*) HIGHER / LOWER quantity.

4. A monopsony market, compared to perfect competition, will have a (*circle one*) HIGHER / LOWER price and a (*circle one*) HIGHER / LOWER quantity.
5. The price in an oligopoly should be between the price in perfect competition and the price in (*circle one*) MONOPOLY / MONOPSONY.
6. The price in an oligopsony should be between the price in perfect competition and the price in (*circle one*) MONOPOLY / MONOPSONY.

For the following three questions, use the following marginal cost and marginal value formulas.

$$\text{Marginal Cost/Supply Curve: } P = 150 + 10(Q)$$
$$\text{Marginal Value/Demand Curve: } P = 800 - 15(Q)$$

7. Calculate the price and quantity using the formulas above assuming perfect competition.
8. Calculate the price and quantity using the formulas above assuming a monopoly.
9. Calculate the price and quantity using the formulas above assuming a monopsony.
10. The number of beefpackers (those who purchase live-cattle and process them into consumable beef) has decreased dramatically over the last 30 years. The main reason is that beefpackers have learned they can process beef at less cost if they build very large processing facilities. As a result, fewer beefpackers are needed to process beef. Now there are only a few buyers of live-cattle but still thousands of sellers of live-cattle. Although this does bestow beefpackers with market power and they can use this power to reduce the price they pay for live-cattle (resembling a monopsony), explain why live-cattle prices could actually rise due to these structural changes.

CHAPTER FIVE

Agricultural Prices

"Legalize gambling, why let farmers have all the fun?"

—Anonymous quote seen on a bumper sticker, referring to the fact that farm prices are so unpredictable it makes farming seem like gambling.

INTRODUCTION

Most people are employed in jobs where their salary or hourly wage is fixed. This gives them security and allows them to plan their expenditures well in advance. The income from other professions is not so certain. Even though lawyers may make a good living, their monthly profits will vary depending on the nature of the cases they accept and their success in court. Restaurant profits depend on unpredictable changes in number of customers. Entrepreneurs must take economic uncertainty as a fact of life, especially those in the agricultural sector. The price farmers receive is rarely fixed and is difficult to predict. This also makes life difficult for those who purchase farm products. In order to develop a viable strategy for dealing with volatile farm prices, one must understand how and why agricultural prices change.

The purpose of this chapter is to extend our supply and demand model to incorporate specific features of agriculture to better understand agriculture prices. Specifically, this chapter

1. covers four determinants of agricultural price changes
2. shows how to construct time-series diagrams in the presence of seasonality, market shocks, and production lags
3. reviews the causes and nature of price cycles

Cow-calf producers are those who breed cattle, raise calves, and sell the calves after they are weaned, usually 210 days after the calf is born. It costs around $0.95 to add one pound to weaned calves (OSUa 2005). Back in 1998, cow-calf producers received prices around $0.78 per pound. They were losing money, but patience has its rewards. In 2004, prices rose to over $1.20 per pound (LMICa 2005). Breeding a cow is like gambling. The outcome is uncertain. There is a big difference between agriculture and the roulette wheel though. The number chosen by the roulette wheel is random. One number is just as likely to be chosen as the next, and there is no way to predict which number will appear.

Agricultural prices are not random; however, they can be complicated to understand. There is a big difference between random and complicated. Both roulette wheels and the weather are difficult to predict. The weather is directly caused by world events like changes in the season, and understanding those events helps to predict the weather. Nothing helps to predict the roulette wheel. Like the weather, prices can be predicted to some degree, once one understands what causes prices.

Prices are negotiated by people, and people are motivated by incentives. For a farmer or an agribusiness to be competitive, a good long-term strategy for predicting prices is paramount. By studying the incentives that dictate prices, one can develop forecasts of prices in the immediate future and in the long run. Agribusiness strategies must be designed to take advantage of and protect against price movements. Profits will go to the firms that make the best use of information, and this includes information about agricultural prices.

UNDERSTANDING AGRICULTURAL PRICES

In previous chapters we studied the perfect competition model where price is determined by the intersection of supply and demand. When people say "supply and demand," they are usually referring to markets that are close to perfect competition, or at least one where no one buyer or seller has a huge impact on market price. Price in these settings is the result of intense competition among the buyers and sellers. Most agricultural markets are adequately described by this setting, so we will use supply and demand to describe the general behavior of agricultural prices.

Prices are determined by negotiations between buyers and sellers, that is, people. People are complicated, influenced by a myriad of motivations from selfish greed to admirable kindness. Because people are complicated, price formation is complicated. If we focused on the true price formation process, attempting to identify each factor influencing price exactly, this chapter would be longer than the IRS code, and more importantly, you wouldn't read it. Also, businesspeople need to communicate price information in a clear and simple manner. Businesspeople talk about "price trends" as opposed to the numerous factors leading to those trends. They talk about "market adjustments" as opposed to intricate and detailed interactions between people that give rise to those adjustments.

Changes in agricultural prices are caused by

(1) Changes in long-run supply and demand
(2) Seasonality
(3) Supply and demand (or market) shocks
(4) Market adjustments

In short, people have developed ways of talking—a market lingo—about agricultural prices that relays much information in a simple, succinct, and clear format. If you read this chapter carefully, you will be able to converse in this market lingo. Think of anything that may influence the price of any agricultural product, whether it be corn, hogs, or flowers, and you can probably group it into one of four determinates of agricultural prices: changes in long-run supply and demand, seasonality, market shocks, and market adjustments. Each factor will be considered individually at first, giving you a clear understanding of its role in agricultural price formation. At the end of the chapter they will be brought together for a holistic view of agricultural prices.

Changes in Long-Run Supply and Demand

Successful agribusinesses develop their strategies with a long-run view. Livestock producers build expensive production facilities based not only on expected prices next year, but on the expected price of livestock for the next 10 years. The same goes for livestock processing plants. You do not spend millions of dollars building a processing plant if you think meat prices will plummet over the next 15 years. Between 2003 and 2005 beef prices rose to historically high levels. Cattle production was more profitable than ever, but should cattle producers expand their herd to produce more cattle? Should others enter the cattle business to take advantage of these high prices? It depends on the reason for these high prices. Part of the reason for the high prices was the ban on beef imports from other countries due to the BSE scare, and part was due to the popularity of high protein diets.

The BSE scare will eventually become history, and high protein diets may be a fad. If so, these high prices were a short-term phenomena. The long-term outlook, at least over the next 20 years, may not be as promising. Over many years, prices are determined by long-run supply and demand. Recall that the long-run supply curve tells us how much firms will produce at a price, given many years to adjust production levels. The long-run demand curve tells us how much consumers will purchase at a price, given they have many years to adjust consumption levels. Also recall that supply and demand are relatively elastic in the long run. The more time firms and consumers are given to adjust to a price change, the more sensitive they will be to that price change.

Over many years, price changes are due to movements of the long-run supply and demand curves. When long-run supply increases, the long-run supply curve shifts downward and the price begins a decline. The price fall may not be fast though, and it

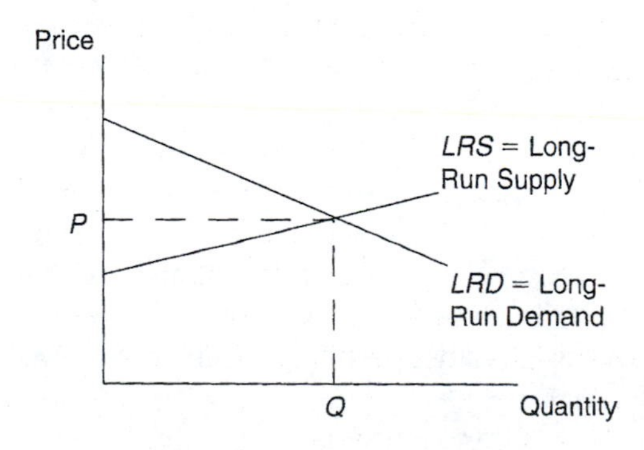

FIGURE 5.1 Over Many Years, Agricultural Prices Are Determined by the Intersection of Long-Run Supply and Demand.

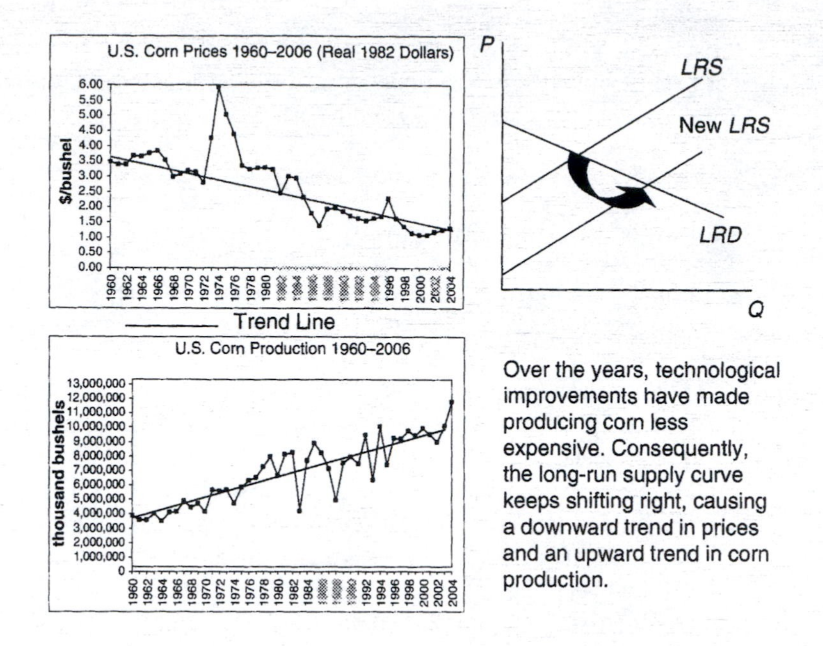

FIGURE 5.2 Corn Through the Years.

may take years to reach the new long-run equilibrium. Thus, we will not see a rapid price decline but rather a steady decline over time. For example, consider the corn market over the last century. See Figure 5.2 showing the price and quantity of corn over time.[1] Price has trended down and quantity has trended up. There is only one thing that causes a simultaneous increase in quantity and decrease in price—increasing supply (i.e., a rightward shift in the long-run supply curve). The increase in supply is due to constant technological improvements in corn production, which makes it less expensive to produce corn each year, shifting the supply curve to the right. This is not to say that the demand for corn has not changed, only that the major force in the corn market over the past 100 years is an increasing supply, resulting from technological advancements.

The fed-cattle market provides another example of how long-run supply and demand affect prices. Fed-cattle refer to cattle that are raised for beef, have been fed to maturity, and are about two years old. At this point, the cattle are ready for slaughter and processing into retail beef. Fed-cattle are also known as live-cattle. See Figure 5.3, which shows fed-cattle price and fed-cattle supplies. Supplies are measured by the quantity of steers (castrated male cows). Because the number of male and female calves born is roughly equal, the number of fed-cattle steers is a good measure of the total number of fed-cattle.[2] From the early 1970s to 1994, both the price and quantity of fed-cattle trended downward. Think back to supply and

---

[1]Source for Figures 5.2 and 5.3 is the National Agricultural Statistics Service (NASS).
[2]Uncastrated males comprise only a very, very small portion of fed-cattle.

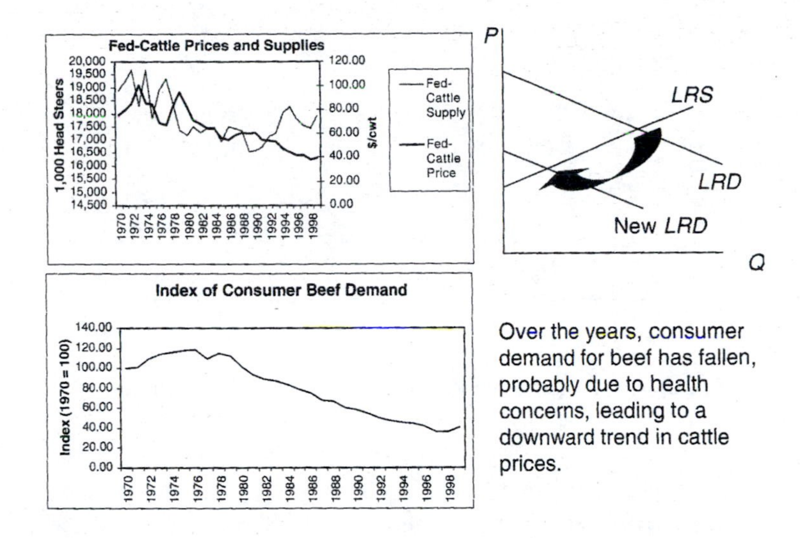

FIGURE 5.3 Fed-Cattle Market Through the Years.

demand: What would cause both price and quantity to fall? Decreasing demand (i.e., a demand curve shifting to the left) is the answer.

Indeed, the bottom graph of Figure 5.3 shows an index for beef demand provided by Montana State University (Marsh 2003). If the index is rising, beef demand is rising, meaning the beef demand curve is shifting upward (to the right). If the index is falling, beef demand is falling, meaning the demand curve is shifting downward (to the left). As the bottom graph shows, consumer demand for beef fell throughout the 70s, 80s, and most of the 90s. Lower demand for beef by the consumer is translated down to the farmer, and the farmer receives a lower price for fed-cattle. Although no one knows for certain why demand fell during this period, most think it was due to health concerns and greater competition from poultry.

In the corn market, the long-run supply curve was increasing, causing prices to trend downward over time. Demand was falling in the fed-cattle market, also causing prices to trend down. When the intersection of the long-run supply and demand curves change, that is, when the long-run equilibrium changes, we see prices moving from their old to their new equilibrium along a path. Prices in the long run do not jump, they trend. Earlier, we said beef demand was rising in the early 2000s due to the popularity of high protein diets. It was suggested that this was a fad, but it may not be. If high protein diets increase in popularity, beef demand will continue to rise, pulling fed-cattle prices up with it. Instead of seeing cattle prices trend down, they will trend upward. Often, we like to illustrate these trends using a time-series diagram on which time (e.g., years) is on the x-axis and the price is plotted on the y-axis. If we are concerned with long-run price trends, then the time-series diagram will display either an upward trend, a downward trend, or no trend. Figure 5.4 illustrates these three scenarios. The left diagram shows an upward trending long-run equilibrium, meaning each year we expect prices to be higher than before. Each year, urban sprawl reaches out to formerly rural areas, and

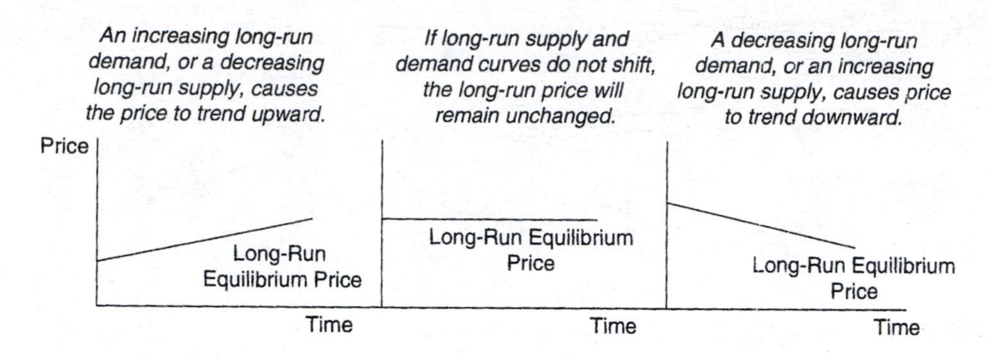

FIGURE 5.4 Time-Series Diagram of Long-Run Prices.

every year farmland is turned into shopping malls, parking lots, and trailer parks. Each year, farmland becomes more scarce. Holding all else constant, we would expect the receding rural-urban interface to lead to higher farmland prices. The price of farmland would trend upward, as the left graph shows.

The center graph describes a situation where long-run supply and demand are not changing. In this case we would expect the average price over many years to remain constant, and so the center time-series diagram shows a horizontal line. The right graph portrays the corn and cattle markets discussed previously. Increasing long-run supply pushed corn prices down, and a decreasing long-run demand forced cattle prices down. Each year, the average price one expects is lower than the previous year, and the time-series diagram shows a downward trend.

One final word about the long-run equilibrium price. Remember that the equilibrium price is like a thermostat. The temperature never exactly equals the thermostat setting, but it's always heading in that direction. The same goes for long-run equilibrium prices. No agricultural price ever equals its long-run equilibrium. But if price is greater than the long-run price, it will eventually fall. If it is lower than the long-run price, it will eventually rise. No matter how hard you hit a softball, it will eventually fall to the ground, and no matter what happens in a market, prices will trend toward their long-run equilibrium value.

Seasonality

Agricultural production ultimately depends on sunlight, and the sun shines brighter during some seasons than others. The impact of seasonality is most obviously seen in crop production. Our primary crops (corn, soybeans, and wheat) produce seed only once a year. This means we must harvest once a year and store the grain for continual consumption until the next harvest. Of course, we import some grain from countries south of the equator who harvest at different times, but these imports only partially dampen the effect of seasonality. In Chapter 1 we discussed how price should continually rise in the months between harvests to provide incentives for people to store the grain. Although we only harvest once each year, we

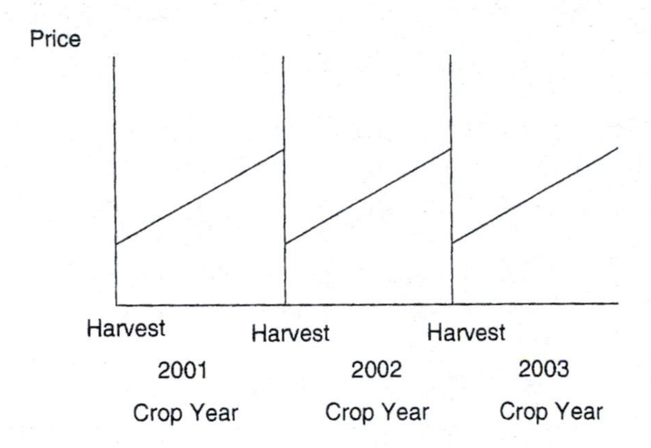

Figure 5.5 The Price of a Crop Should Continually Rise Between Harvests.

consume year-round, so it is important that some grain is always in storage between harvests. The Indifference Principle states that people should be indifferent between storing corn and not storing corn, so the price difference between months must equal storage cost between months. Thus, price must rise each month between harvests to compensate those who incur storage costs, as illustrated in Figure 5.5.

Corn is harvested between September and November, so one would expect corn prices to continually rise between December and August. However, corn prices do not behave in this manner, as illustrated in Figure 5.6. Prices rises until May, but then steadily decline until the next harvest. People are compensated for storing grain the first few months, but not the last few months before harvest. Some people purchase grain in April to store for use in August, knowing that they could simply purchase the grain cheaper in August and not pay storage costs. They choose to store expensive grain instead of purchasing cheap grain later. Why, then, would anyone want to

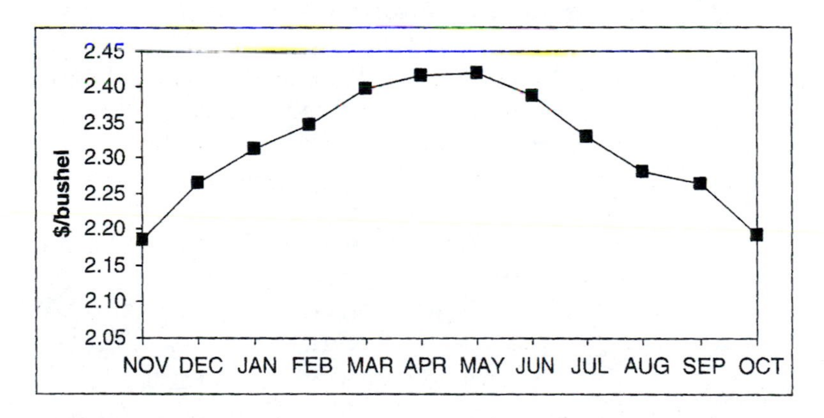

FIGURE 5.6 U. S. Corn Prices Between Harvests (Corn Is Harvested Around November).

Source: LMICb (2005).

Note: Prices reported for each month are the average monthly price between 1990 and 2005.

store grain when they know they can purchase it cheaper later? Does the Indifference Principle not hold here, or are we failing to consider something important? Unfortunately, agricultural economists have not found a clear answer about why prices begin falling well in advance of the next harvest.[3]

What follows is a "stylized explanation"[4] for the drop in corn prices after May. First, Figure 5.5 assumes that each year's harvest produces the same amount of corn. However, corn production each year varies according to the weather. Corn is harvested September–November, but once August rolls around, corn has begun tasseling[5] and the people have a very good idea of the next harvest's size. If a drought occurs and the market believes the next harvest will produce very little corn, people will be reluctant to sell the corn supplies they have in storage. Less corn will be available on the market, and corn prices will rise.

Many food manufacturers will want to protect themselves against tight corn supplies and the corresponding high commodity prices. Thus, they are willing to pay higher prices for corn earlier in the season as insurance against low supplies. For example, Coca-Cola uses corn syrup as a sweetener and absolutely must have corn available at all times to keep their plants running efficiently. Coca-Cola and other food processors may make most of their purchases from November to April, securing the supplies of corn they will need for the remainder of the year and possibly longer. When making their corn purchases in February, March, and April, they know they could wait until May and June for cheaper corn. After all, they have seen Figure 5.6. But come June it could turn out that projections for the next harvest are low, increasing corn prices, forcing them to pay high corn prices in June, and having to scramble for their corn needs. They pay more for corn early in the season as insurance against the unlikely event that corn prices will skyrocket later. This is just like you buying car insurance against the unlikely event that you get into a wreck.

This gives rise to the upside-down U-shape of corn prices in Figure 5.6. Corn is simply more valuable early in the crop year. It is more convenient to secure one's corn supplies early than have to scramble around in search of corn should supplies run low. This has been termed *convenience yield*. College students are very familiar with convenience yield, at least in the bar. When the bartender yells "last call," it is students' final chance to purchase beer. Everyone rushes to the bar, fighting the crowd for their beer, and many go away empty-handed. The smart students will purchase a couple of beers in anticipation of the last call. This reduces the risk of not getting a beer and is more convenient because you do not have to fight everyone else at the bar. In fact, some would pay more to have a little extra beer on hand to avoid the last call rush—that is convenience yield. Who thought buying corn and buying beer would have so much in common?

The seasonal behavior of most crops follow this pattern. Prices rise in the months following harvest, but then begin falling the months before the next harvest. See

---

[3]One might suspect that the United States begins importing corn from other countries around May, but imports as a percentage of U.S. production is small.

[4]A stylized explanation is an explanation that is popular but not grounded in fact.

[5]The "tassel" is the small flowers produced by the corn plant containing pollen. Experts can measure the tassel and predict with good accuracy the number and size of ears the corn will produce.

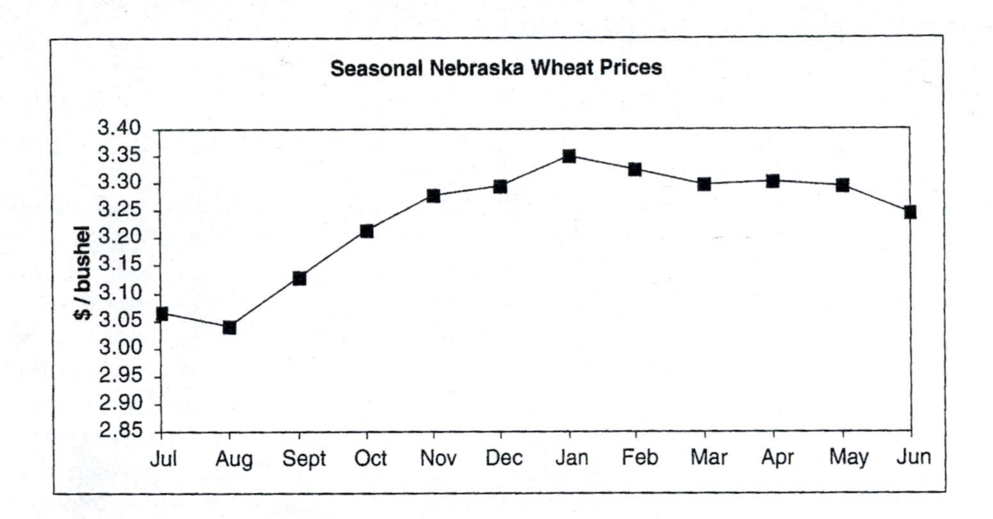

FIGURE 5.7 Nebraska Wheat Prices Between Harvests (Wheat Is Harvested Around July).

Source: NASS (2005).

Note: Prices reported for each month are the average monthly price between 1990 and 2005.

Figure 5.7 showing Nebraska wheat prices. After wheat is harvested in July, the price rises to account for storage costs, but halfway through the crop year the price start to decline. The decline is not as pronounced as the corn market, but it is present. Now, let us return to our time-series diagram and illustrate the seasonal behavior of grain prices. See Figure 5.8. At the top is a situation where long-run supply and demand are not changing, so the long-run equilibrium price is a horizontal line. Notice how the price changes between harvests. At harvest, the price is at its lowest point, and between harvests the price hits its highest point.

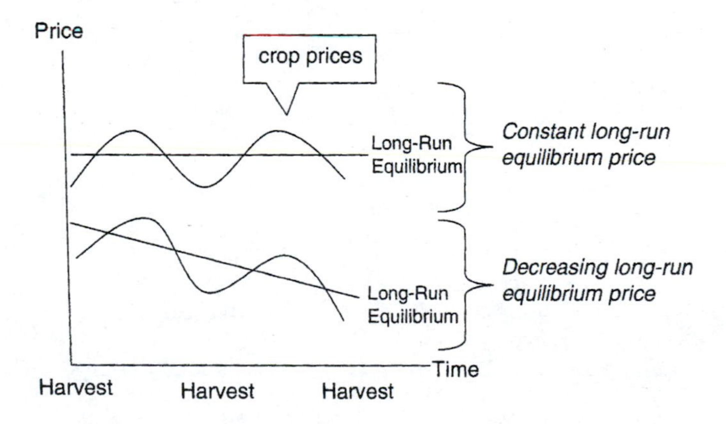

FIGURE 5.8 Time-Series Diagram of Crop Prices with Long-Run Equilibrium and Seasonal Variation.

Recall in the last section we said that the long-run price of corn is decreasing because the long-run supply curve is continually shifting to the right. This means the long-run equilibrium price is trending downward, as shown at the bottom of Figure 5.8. Corn prices still exhibit their seasonal cycle. Within a year, the price is lowest at harvest and highest between harvests. But in any given month, the price is lower than it was last year due to the decreasing long-run equilibrium price.

Livestock prices also exhibit seasonality, especially cattle prices. Driving through the countryside of eastern Oklahoma, Kansas, and Texas (cattle country as some people would say) in February, you are sure to notice newborn calves dotting the landscape. Most calves are born around February, and for good reason. Calves and their mothers need the most food 3 to 8 months after the calf is born. Grass is most plentiful in the summer months, so if calves are born in February, grass is most plentiful when the calves and their mothers need it the most. One could breed cows to give birth in the fall, but would have to supplement the lack of grass with expensive feed and hay, which increases the cost of production. Calves are less expensive to raise if born in February.

Once calves are weaned, they are referred to as stocker-calves.[6] Because most calves are born in February and are weaned at 7 months of age, the supply of stocker-calves is greatest around September–October. As a result, stocker-calf prices are lowest in these two months, as shown in Figure 5.9. Of course, not all calves are born in February. Appealing to our old friend the Indifference Principle, cattle producers should be indifferent between calving in February, July, or September. Periodically, you will run into someone who calves in August, weaning the calves and selling them as stocker-calves in March (at 7 months of age). A glance at Figure 5.9 reveals why: Feeder-cattle prices are highest in March. Although it costs more to raise calves born in August, you get a higher price. The higher price should just equal the additional

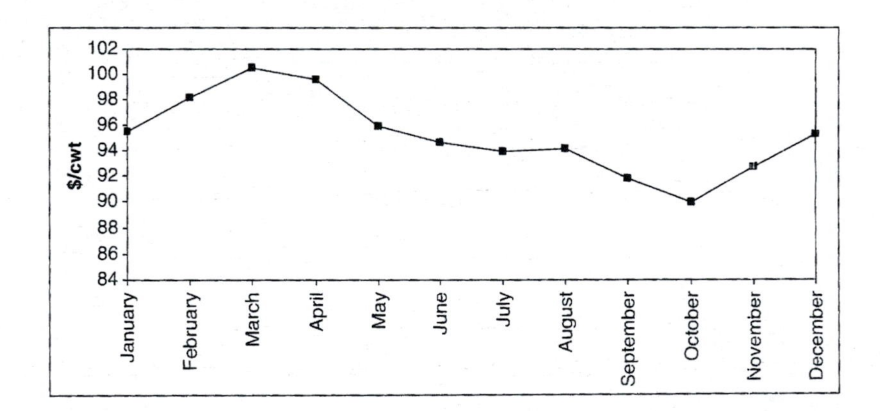

FIGURE 5.9 Stocker-Calf Prices by Month.

Source: LMICa (2005).

Note: Prices are the average Oklahoma feeder-cattle prices 400–500 lbs, 1992–2001.

---

[6]The term *stocker-calves* means they are ready to eat only grass, hay, and grain (no more mother's milk) until they are ready for slaughter.

cost, making one indifferent between calving in February or August. This explains the seasonal variation in feeder-cattle prices. Prices are highest in March because it costs more to raise calves born in August (7 months prior). Similarly, prices are lowest in October, because it is cheapest to raise calves born in February (7 months prior).

In this example, the Indifference Principle does not hold for all cow-calf producers, but it holds for enough of them to ensure a steady supply of beef. Producers in the northern region of the United States would never calve in the fall because it is simply too cold. These producers *are not* indifferent about when to calve. The force of the Indifference Principle is felt in southern regions with mild winters though. Producers in this region have the option of when to calve, and some elect to calve in the fall to receive higher prices. Also, it should be noted that there are other sectors of the beef industry that help to ensure a steady supply of cattle. Most calves are born in February or soon after. Once these calves become stocker-calves, they are typically placed on wheat pasture. They may remain on wheat pasture for a long or short period of time. As they grow in the pasture, they become referred to as "stockers" rather than "stocker-calves." There is no fixed timeline in which stockers remain on pasture. They can be kept just a few months and then sold to a feedlot where they will receive a high protein diet and will grow fast. Or, they can be kept on pasture many months.

As you might suspect, the decision of how long to keep stockers on pasture depends on expectations about future prices. If producers expect a price premium six months from now, they will keep stockers on pasture longer to reap higher profits. If prices are expected to fall in six months, they will sell their stockers soon. Thus, the price of stocker-calves plays an integral role in smoothing out beef supplies. Most calves are born in the spring, and if all cattle producers followed the same strategy, all cattle would be turned into beef during the same month. This would give us lots of beef in some months but little beef in others. Instead, prices move so that some calves remain stockers only a short period and others a long period, depending on expectations about beef demand. It is this facilitating role of price, whether it be the price for stocker-calves or stockers, that ensures you can always find fresh beef at the supermarket, despite the fact that most calves are born in February–March.

The discussion thus far has considered only seasonality in supply, but clearly demand may vary across seasons. Turkey demand is highest in November and December and beef demand peaks in the summer due to the popularity of backyard barbeques. Moreover, seasonality in demand enters into producers' production decisions, leading to seasonal changes in supply not due to the weather. Turkey demand is highest in November, but not turkey prices. The reason is that turkey producers increase the supply of turkey in November to ensure adequate supplies are available.

Market (Supply and Demand) Shocks

Some aspects of agricultural prices are predictable, like downward trends and seasonal variation. Other aspects are not predictable and appear somewhat random. Figure 5.2 plots real corn prices across time and shows an extraordinary period of high prices in 1974–1976. These high prices were not expected. If they were, many more people would have planted more corn in anticipation, driving prices back down to normal levels. The high corn prices of the mid-1970s were caused by a large wheat failure in the USSR. With little wheat of its own to feed its people, the USSR purchased large amounts of U.S. wheat, corn, and soybeans, driving up the demand for U.S. grains and U.S. grain prices. Up to one-sixth of the U.S. wheat crop was exported to the USSR (Gardner 2002). This demand increase was temporary. Crop production in the USSR soon rose to its normal level, no longer requiring them to purchase U.S. grain. This unexpected, temporary surge in demand is referred to as a *demand shock*. It represents a temporary deviation from the corn market's long-run equilibrium. Price jumped up quickly, but over a few years settled back to its long-run equilibrium price.

Market Shocks

- Positive Demand Shock: unexpected, temporary increase in demand
- Negative Demand Shock: unexpected, temporary decrease in demand
- Positive Supply Shock: unexpected, temporary increase in supply
- Negative Supply Shock: unexpected, temporary decrease in supply

The previous example refers to a *positive* demand shock, one that increases demand. Hurricane Katrina not only affected Louisiana farmers, but corn and soybean farmers far up the Mississippi River. Much of the grain grown in states like Illinois and Iowa are shipped in barges down the Mississippi to be exported out of Louisiana's ports. Hurricane Katrina shut down these ports, decreasing the number of buyers for Iowa corn, and consequently the demand for Iowa corn. Iowa corn farmers faced a 15% cut in prices, but the cut was temporary (Jome 2005). The ports were rebuilt and exports soon resumed. In this case, the price temporarily fell below its long-run equilibrium due to a *negative* demand shock. A negative demand shock is a temporary, unexpected decrease in demand.

There are supply shocks as well. The grain market in the United States differs from the market in the former USSR. Russian and U.S. grain are produced far enough apart that supply and demand curves differ for each. If you were a Russian farmer in 1974, the USSR crop failure was a supply shock. Wheat crop yields were dismal across the Soviet Union and weather was the cause. The Russian supply of wheat fell, leading to higher wheat prices. This led to an increase in demand for U.S. wheat by Russians. Low yields decreased the supply of USSR wheat, driving up wheat prices. But as we said earlier, this was a temporary event and the supply curve eventually returned to normal, and so did wheat prices. The supply shock in the USSR wheat market caused a demand shock in the U.S. wheat market. These unexpected events that cause a temporary decrease in supply are called *negative supply shocks*. Conversely, if surprisingly good weather temporary increases the wheat supply, we call this a *positive supply shock*.

To make sure we understand the impact of market shocks on prices, let us illustrate these shocks in a time-series diagram. Consider the top diagram in Figure 5.10. We have a constant long-run equilibrium price, meaning neither long-run supply nor demand is shifting. The market temporarily deviates from this price due to a market shock. Price is driven up either by a positive demand shock or negative supply shock. This could be an increase in demand for U.S. grain by USSR or a decrease in supply of USSR grain due to bad weather. The bottom diagram is slightly more complicated. Here, we assume an increasing long-run equilibrium price. Either the long-run demand is increasing or the long-run supply is decreasing. Whatever the cause, each year we expect the price to be higher than last year. However, a market shock occurs that temporarily drives price down. If this was the beef market, the culprit might be a

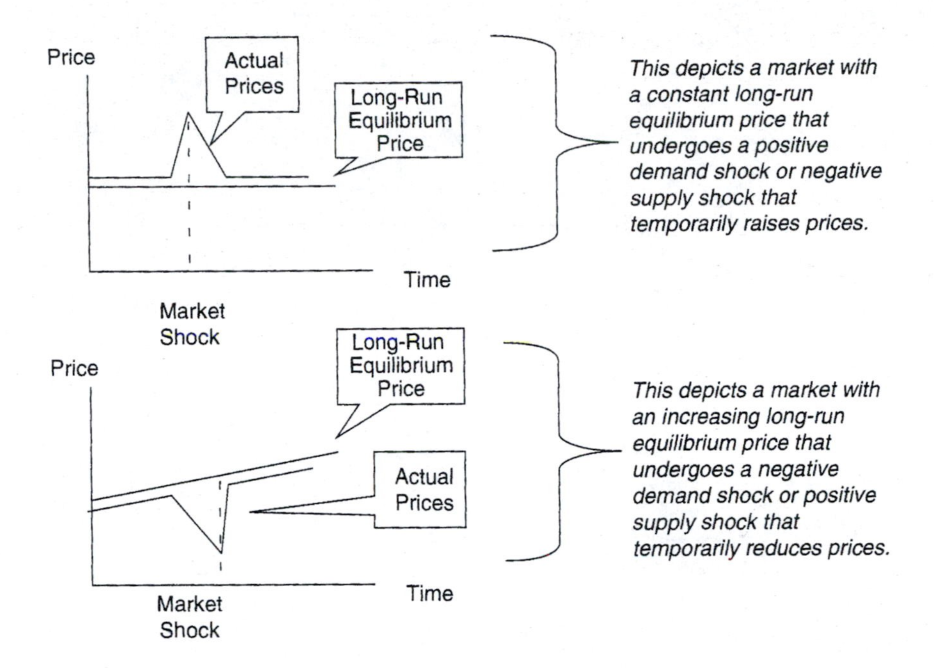

FIGURE 5.10 Time-Series Diagram of Market Shocks.

decrease in demand due to health scares about BSE, or an unexpected increase in supply of beef supplies due to extraordinarily good weather.

Market Adjustments

The previous section made it appear that with any temporary price change, prices will quickly jump right back to their long-run equilibrium price. This is not the case in most agricultural markets. In reality, markets have to rediscover their long-run equilibrium, and that search can be long and wild. The reason markets must adjust to shocks is that agricultural production experiences production lags: time lags between the time production decisions are made and the output is produced. It takes about two years between the time a cow is bred and her offspring is ready for slaughter. This production lag of two years has important implications. It means the beef we consume today is largely based on farmers' decisions and market prices two years prior. The production lag for hogs is shorter, around one year, and chickens a little more than a month.

Crops experience a production lag of about one year. A farmer decides how much winter wheat to plant in September, and you harvest in July, resulting in a production lag of 10 months. This production lag causes a delay in signals sent from consumers to producers. Basically, producers have to make their production decisions based on the price they *expect* to receive, and their expectations are not always correct. In this section, we will develop a popular model known as the *Cobweb Model*. As an example, we will use the beef market where the production lag is about two years. To illustrate the model, we will tell a story, a story that begins in the year 2007 in Figure 5.11.

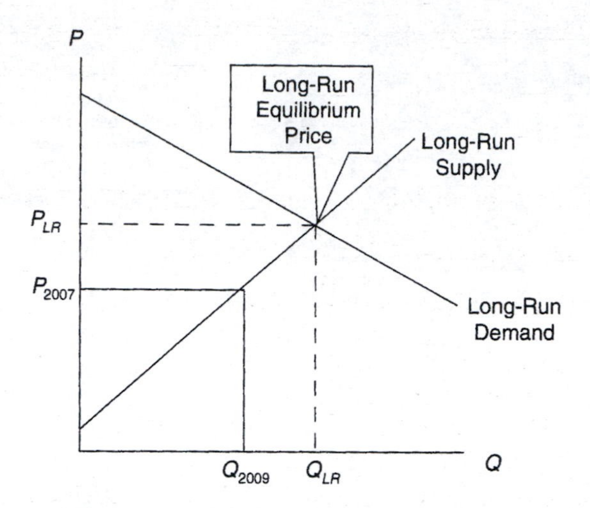

FIGURE 5.11 The Cobweb Model in 2007.

In 2007, a market shock occurs that sends prices below their long-run equilibrium price to P_{2007}. Perhaps demand temporarily fell due to a health scare. Cow-calf producers must decide how many animals to breed this year, and seeing the low price, produce less than the long-run equilibrium quantity. They breed a number of cattle to produce Q_{2009}. The cows will be bred in 2007 and will give birth, and those calves must be fed and raised until they are ready for slaughter two years after breeding. In 2009, the Q_{2009} amount of cattle are processed into beef and presented to the consumer for purchase. The demand shock has since passed. Compared to the long-run equilibrium quantity, Q_{2009} is a small amount of beef, and consumers bid the price of

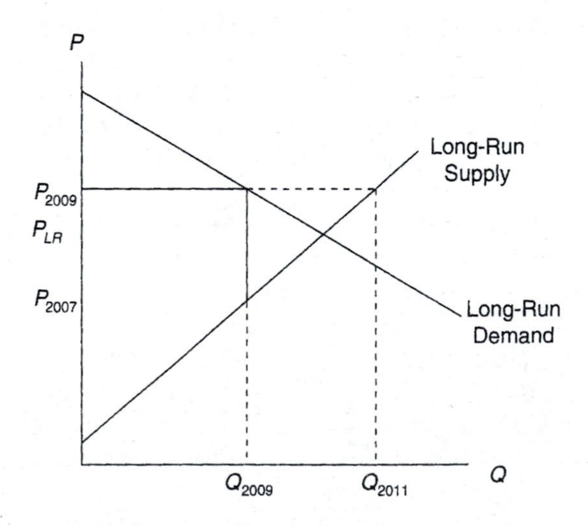

FIGURE 5.12 The Cobweb Model in 2009.

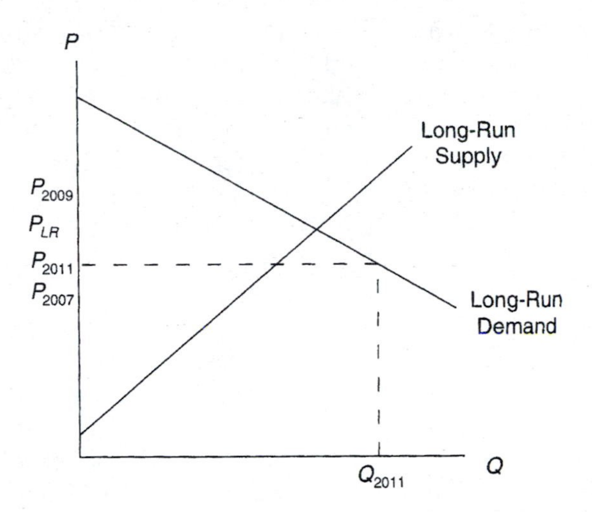

FIGURE 5.13 The Cobweb Model in 2011.

beef up to P_{2009} (see Figure 5.12). Once again, in 2009, cattlemen must decide how many cattle to breed. Seeing the high price of P_{2009}, they decide to increase production to Q_{2011}.

Notice the quantity Q_{2011} is greater than the long-run equilibrium price, so to entice consumers to purchase all units, the price must fall below the long-run equilibrium price. The price inducing consumers to purchase all Q_{2011} units is given by P_{2011}. In Figure 5.13 the story continues. Cattle producers plan their production for 2013 based on the price they see in 2011, and on and on. The price keeps going above, below, above, and below the long-run equilibrium price, but each two years coming closer to the long-run equilibrium price. If you trace this weaving pattern of price,

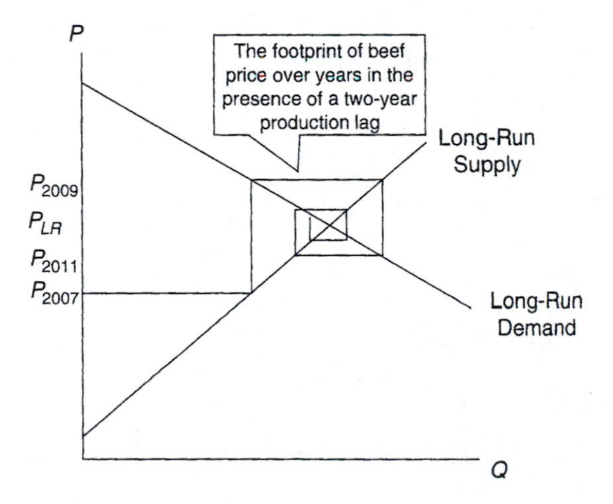

FIGURE 5.14 The Cobweb Model.

you get a cobweb-like figure as shown in Figure 5.14. It takes the price many years to settle back to its long-run equilibrium. This is what we call a *market adjustment* and is due to the production lag that slows producer and consumer feedback. If there was no production lag, producers could quickly adjust their production decisions and reach the long-run equilibrium in a matter of weeks. Instead, they must wait two years to see how their production decisions impact price.

Assumptions of Cobweb Model

(1) Production lag
(2) Producers make production decisions for the future based on current prices

At this point, we should make clear the assumptions of the Cobweb Model that led to these market adjustments. These assumptions are (1) a production lag and (2) producers plan their future production levels based only on current prices. The first assumption is a biological truth; the second is open to debate. Do producers make all their future production decisions based solely on the current price? Suppose you were a hog producer who saw market prices in the range of \$60–\$70/cwt for 20 years. Then, out of nowhere, prices jump to \$100/cwt. Would you assume that prices in the future will always be \$100/cwt, or would you think that this high price is just an aberration, and that in the future prices will fall back to normal levels? Are farmers naïve or rational? It is unlikely that farmers base *all* their production decisions solely off the *current* price, but we do have reason to believe they give the current price much weight. In this case, the main results of the Cobweb Model remain, and production lags will lead to the sort of market adjustments described here.

Recall how we compared the long-run equilibrium to a thermostat. When the temperature is greater than the thermostat setting, the air conditioner will kick in and bring the temperature down. The room varies in temperature, but does not deviate far from the thermostat setting because the air conditioner is continually measuring the room's temperature and adjusting accordingly. Now, imagine a thermostat that could only read the temperature every hour. The room's temperature would swing wildly. The room temperature would go from hot to the set temperature, hot to the set temperature, back and forth again. Just like a delay in a thermostat's reading increases the volatility of the room temperature, a production lag increases the volatility of market prices.

Now let us incorporate these market adjustments into a time-series diagram. In Figure 5.15 we show a market that begins with price at its long-run equilibrium.

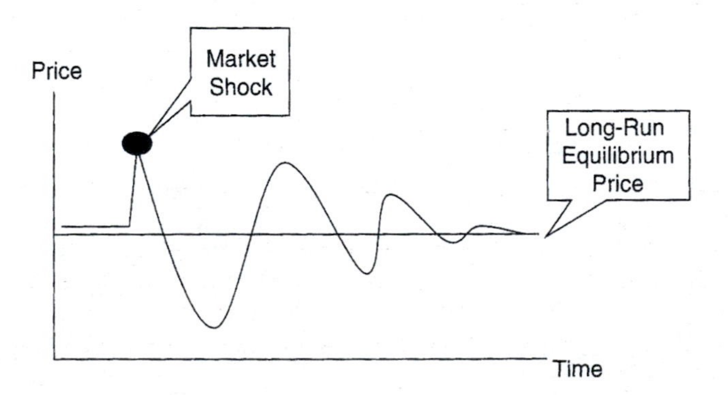

FIGURE 5.15 Time-Series Diagram with Market Shock and Market Adjustments.

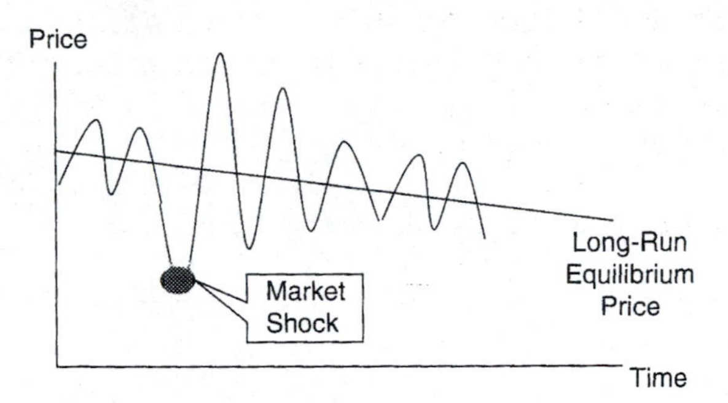

FIGURE 5.16 Time-Series Diagram with Market Shock, Seasonality, Market Adjustments, and a Declining Long-Run Equilibrium Price.

Then, due to some market shock, the price jumps up. But this shock is temporary and the market must now find its way back to its long-run equilibrium. At first, prices will adjust too much, undershooting the long-run equilibrium. Then it will overshoot. Then it will undershoot again, though not as much, and it will overshoot again, though not as much. Eventually, the market adjustments will taper off and price will equal its long-run equilibrium.

Ready for a really complicated scenario? Suppose you have a market with a decreasing long-run equilibrium price, seasonality, a market shock that temporarily lowers prices, and a production lag that leads to market adjustments. This scenario is shown in Figure 5.16. Before the market shock, prices moved up and down in regular intervals due to seasonality. A market shock then hits, such as a large increase in supply, forcing prices to a temporary low. As the market adjusts to the shock trying to rediscover its long-run equilibrium, it overshoots and then undershoots the long-run price. Prices surge up then plummet down. Over time, it surges up and down, with the surge intensity decaying, and prices come closer to the long-run equilibrium. The effects of the shock have dissipated, the market has fully adjusted, and prices are only affected by the declining long-run equilibrium price and seasonality.

Price Cycles

Agricultural prices, and especially livestock prices, are known for exhibiting price cycles beyond that explained by seasonality. Prices go up, then prices go down, and those ups and downs are partially predictable. As a rough rule, the hog cycle lasts about four years, meaning every four years the price will hit a recent high and every four years the price will hit a recent low (Sterns and Petry 1996). The cattle cycle is about 10 years—again, these are very rough estimates (Lawrence 2001). What causes these cycles? Our previous discussion of agricultural prices gives us some tools to understand these price cycles.

The previous section described the interplay between market shocks and market adjustments as follows. A shock occurs, then the price over- and undershoots its long-run equilibrium value as it rediscovers the long-run equilibrium. The magnitude of this over- and undershooting dampens with time, and the price settles back to its long-run value. These shocks are not infrequent, but occur all the time in agriculture. Markets are continually adjusting to the shocks, and the market never really finds its long-run equilibrium price. Market prices, adjusting to a barrage of constant shocks, continually over- and undershoots its long-run price.

What follows is a story illustrating the causes of livestock price cycles, accompanied by the illustration in Figure 5.17. This could be used for the pork or beef industry, but we will say this is pork. We need a starting point, so let's assume prices are lower than their long-run equilibrium, but prices are rising. As the price rises, hog producers respond by producing more hogs. But to produce more hogs, they need more females and males to breed. Although producers want to ramp up production, they temporarily reduce the number of animals sold to provide more breeding stock. We call this the *expansion phase* because producers are building up their breeding stock to produce more hogs in the future.

Notice that because fewer hogs are available for slaughter (more are being reserved for breeding), this drives up the price even more. Prices keep rising, and hog herds keep expanding. Eventually the price is greater than the long-run equilibrium price, and pork floods the grocery stores. Prices must fall to entice consumers to purchase this extra pork. As the price falls, hog producers respond by decreasing their planned production levels. Planning to produce fewer hogs in the future, they sell more of their breeding stock. This is the *liquidation* or *contraction stage*. As this breeding stock hits the market, increasing hog supplies, prices are depressed even further. Eventually, the breeding stock falls to a new low, there is little pork on the

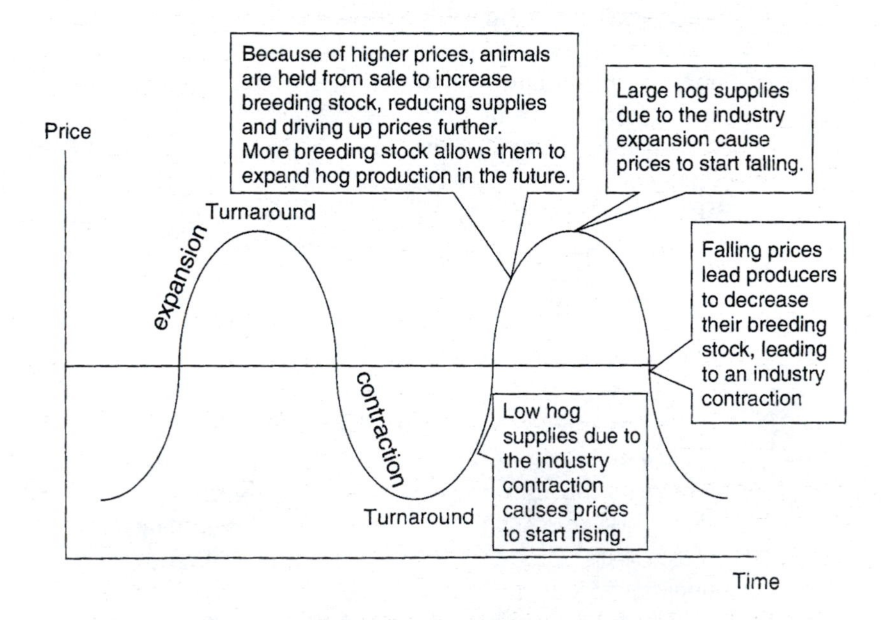

FIGURE 5.17 Illustration of Livestock Price Cycles.

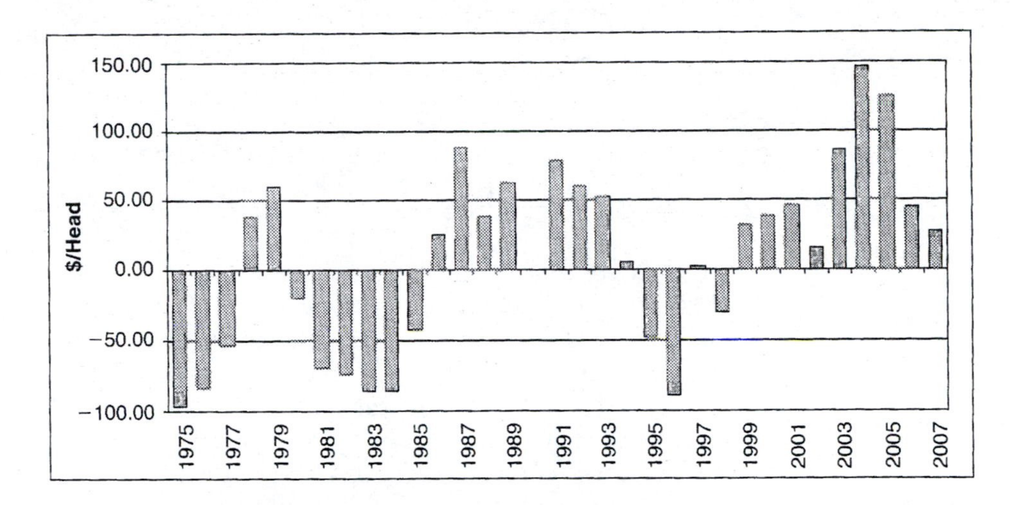

FIGURE 5.18 Profits from Cow-Calf Production.
Source: *Feedstuffs* (2005).

market, and consumers bid up the price of hogs again. The price begins its ascent, and the cycle repeats itself.

Consider now how these livestock cycles affect profits. Figure 5.18 shows profits from cattle production across years. It is evident that to remain in cattle production, one must be willing to accept losses in some years to realize gains in other years. In some years producers have lots of cattle to sell but get a low price. In other years the price is high but they have few cattle for sale. In the best years the price is moderately high and producers have a fair amount to sell, resulting in the highest profits.

It is important to understand terminology like that used to describe livestock prices above. Many cow-calf producers subscribe to newspapers like *Feedstuffs* for market information. On November 14, 2005, *Feedstuffs* ran a special article projecting how the cattle market will look in 2006. Consider the following excerpts from this article.

> "Ranchers have shifted the beef cow herd from contraction to expansion, concluding the lengthiest liquidation in history—nine years—which explains the feeder supply's tightness and points to cyclically lower prices."

> "Key indicators support that the expansion is on, he said. The number of heifers held back for the beef cow herd was up 5.1%."

From 1996 to 2005, beef producers were selling most of their heifers, leaving few females for breeding. This means fewer calves were born and processed into beef, driving up prices. In response to these prices, *Feedstuffs* indicates that the market has reached its turnaround and an expansion is underway. When they say "the number of heifers was held back," this means heifers formerly sent to slaughter are now being saved for breeding to produce more calves in the future. Hopefully, this instills an understanding of the importance of knowing the causes and terminology of livestock cycles. Without this knowledge, reading everyday market reports would be a cumbersome task.

SUMMARY

The previous chapters developed an understanding of prices in general, whether for agricultural and nonagricultural products. This chapter extends those concepts to understanding agricultural prices. The general upward trend, downward trend, or absence of any trend in price depends on the long-run supply and demand conditions. Prices rarely equal their long-run trend. They temporarily deviate and return. One cause of these deviations is seasonality. All agricultural production relies on the perpetual burning of the sun. The sun shines brighter during some months than others, making agricultural prices dependent upon seasons as well.

Another cause is market shocks. Demand temporarily rises and falls based on temporary changes in preferences like fads or economic recessions. Agriculture is dependent on the weather, and in some years rainfall is plentiful and in other years it is not. Plant and animal diseases can temporarily disrupt meat, vegetable, and grain supplies. These market shocks are temporary, unpredictable, and a perpetual force in agriculture. Market adjustments make it difficult for agricultural markets to respond to shocks. Agricultural markets do not just experience a shock and then recover quickly. Knock down a man with glasses and he will not recover quickly; he spends time fumbling around for his glasses. In a similar vein, markets do not recover their equilibrium prices quickly after shocks. Like a man searching the grass somewhat blindly for glasses, markets spend a considerable amount of time rediscovering their equilibrium.

And finally there are price cycles in livestock. Livestock markets are complex and experience too many shocks to be stable over many years. Like a cat chasing its tail, livestock producers chase prices, seeking to maximize their profits given where prices are going, only to have their decisions alter the course of prices. These concepts and terminology used here are not mere academic exercises. If you plan to work in an agribusiness that involves the buying or selling of agricultural commodities, it is a must that you understand price cycles, market adjustments, seasonality, shocks, and long-run trends. Not only must you understand the concepts, but you must use the terminology contained in this chapter. It is how agribusiness people talk about agricultural prices. Agricultural lingo is as important to agriculture as the Spanish language is to the country of Spain. Well, that's a bit of an exaggeration, but it is important!

CROSSWORD PUZZLE

If the answer contains more than one word, leave a blank space between each word.

Across

2. In response to rising prices, livestock producers withhold animals for slaughter to increase their breeding stock, increasing prices even further. This is the _______ phase of the livestock price cycle.
4. The _______-_______ is the period between the expansion and contraction phase of the livestock price cycle when price is at its highest point.
5. As prices fall, livestock producers sell more of their breeding stock, depressing prices even further. This is the _______ phase of the livestock price cycle.
7. A market _______ is an unexpected but temporary event affecting agricultural prices.
8. A _______ _______ is the length of time between when production decisions are made and the final product is ready for consumption.

Down

1. Changes in _______-_______ supply and demand cause prices to trend upward or downward over time.
3. Corn prices are lowest at harvest and highest between harvests due to _______.
5. Regular ups and downs in livestock prices are referred to as livestock price _______.
6. Due to production lags, there are market _______ in response to market shocks.

STUDY QUESTIONS

1. The diagram below illustrates a decrease in the long-run supply of a commodity. To the right of the diagram in Figure 5.19 is a blank time-series diagram. Illustrate the long-run equilibrium price trend associated with a decreasing long-run supply curve.

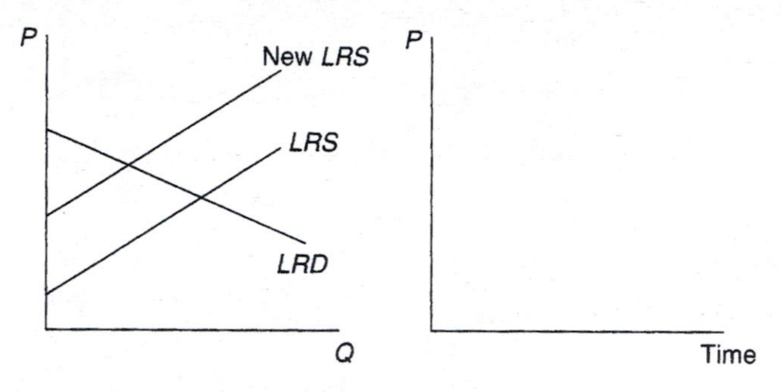

FIGURE 5.19

2. Consider the market for corn. Assume that the long-run equilibrium price of corn is decreasing over time. Illustrate the seasonality and long-run trend of corn prices in the time-series diagram in Figure 5.20. Assume that a convenience yield for corn exists. Be sure to clearly indicate which line is the long-run equilibrium price and which is the actual price at any point in time.

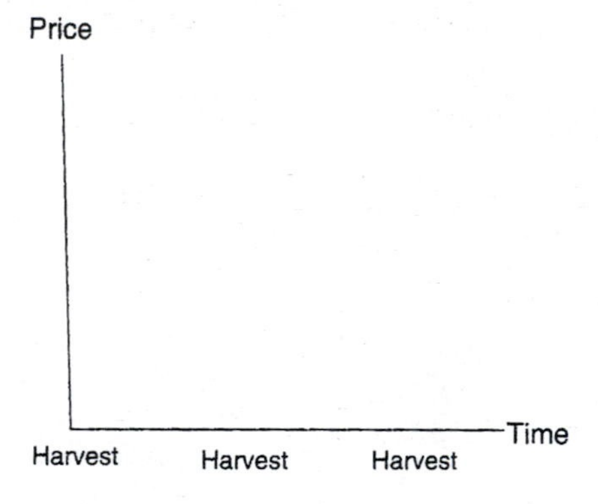

FIGURE 5.20

3. Consider the price of stocker-calves. Suppose the long-run equilibrium price is rising over time. Illustrate the behavior of stocker-calf prices in the time-series diagram in Figure 5.21, taking into account seasonality and the long-run trend. Be sure to clearly indicate which line is the long-run equilibrium price and which is the actual price at any point in time.

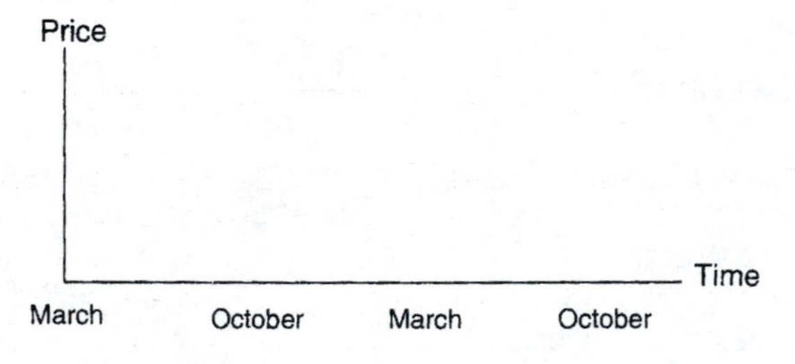

FIGURE 5.21

4. Suppose a new forage (e.g., grass) was developed that grew well in the winter. This would decrease the cost of feeding cows in the winter, and subsequently, would narrow the cost gap between calves born in February and calves born in August. Explain how this would affect the seasonal behavior of stocker-calf prices.
5. Consider the market for barley, where the long-run equilibrium price is increasing. Suppose a negative supply shock occurs due to plant disease, reducing the barley harvest and causing a temporary spike in price. Illustrate this in the time-series diagram in Figure 5.22, taking into account the production lag in barley. Be sure to clearly indicate which line is the long-run equilibrium price and which is the actual price at any point in time.

FIGURE 5.22

Consider the supply and demand for pork shown in Figure 5.23. Assume that the production lag for pork is exactly one year. A market shock occurs in 2005, temporarily decreasing prices from its long-run equilibrium of 4 to 2.5. Use the Cobweb Model to answer the questions below.

6. How much pork will be produced in 2006? ______________
   What will be the pork price in 2006? ______________
   How much pork will be produced in 2007? ______________

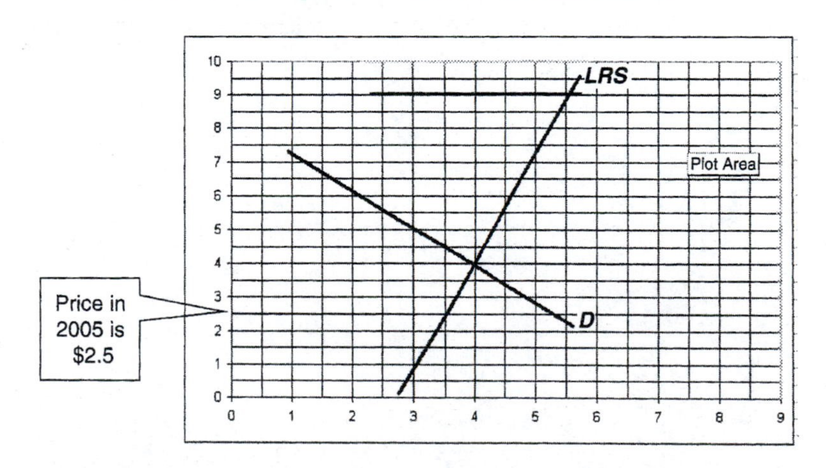

FIGURE 5.23

7. Following from question 6, show how the market adjustment process affects prices from the time of the initial demand shock to the time price settles to its long-run equilibrium of $4.00 in the time-series diagram in Figure 5.24.

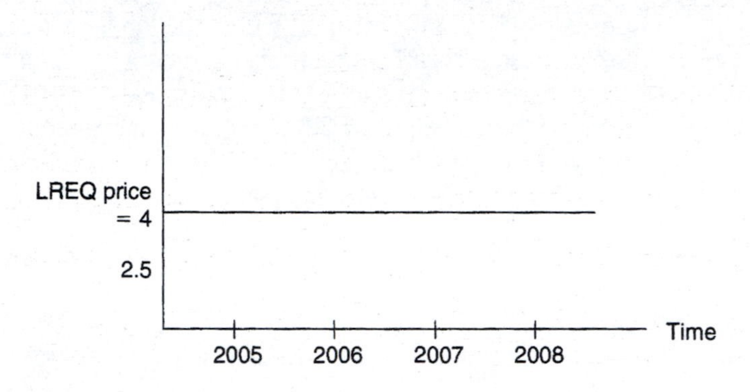

FIGURE 5.24

CHAPTER SIX

The Food Marketing Channel

What my mother believed about cooking is that if you worked hard and prospered, someone else would do it for you.

—*Nora Ephron,* movie director and writer

The times they are a changing.

—*Bob Dylan*

INTRODUCTION

The food industry has undergone significant changes the past 50 years. As women began entering the labor force in large numbers during and after World War II, wives had less time to devote to food preparation and cooking. Yet, the extra money they brought to the family by working could be used to pay others for prepared food. In 1963, 76% of all food consumption occurred at home. In 2002, only 59% of food was consumed in the home (Economic Research Service 2005b). Moreover, the type of food consumed at home has altered significantly. In 1965, the average person spent 44 minutes per day in food preparation and 21 minutes in food cleanup. These numbers dropped to 27 minutes for food preparation and only 4 minutes for food cleanup in 1996 (Culter, Gleaser, and Shapiro 2004). Part of the reason less time is spent in the kitchen is due to the fact that more time is spent in restaurants. Another reason is that food processors have developed new food items that involve less time in preparation, cooking, and cleanup. In a matter of minutes, one can microwave and cook an entire meal. In short, food processors have assumed a larger role in food preparation, cooking, and cleaning. They do this because consumers are willing to pay for this service.

Now more than ever, one must understand the entire food marketing channel, which is the progression of food from the farm product to the consumer item. That is the purpose of this chapter. Specifically, we aim to

1. provide an overview of the food marketing channel and the role of each sector in the food marketing channel
2. discuss how firms at various points in the food marketing channel coordinate their activities
3. build a supply and demand model able to accommodate multiple stages of the food marketing channel.

Marketing Bill or Marketing Margin: The cost of transforming farm production into a consumer good.

The money spent on taking a farm product (e.g., a live-cow or a bushel of raw corn), processing it into a consumer product, and marketing the product is referred to as the *marketing bill* or the *marketing margin*. For every dollar consumers spend on food, a portion goes to the farmer and the remainder is spent on food processing and marketing—the marketing bill. Figure 6.1 shows the breakdown of the food dollar between farm value and the marketing bill, and as we said earlier, the marketing bill now dominates the food dollar. In 1954, the marketing bill was about the same as the value of farm output. This means retail food prices were only about twice the prices farmers received. Now, the marketing bill is six times larger than the value of farm output, and farmers receive only one-sixth of each dollar paid by the consumer.

Of all consumer food expenditures, a portion is captured by the farmer in terms of her product sales. As Figure 6.2 shows, this percentage has declined. In 1950 the farmer received a little over $0.40 for each consumer dollar spent on food, whereas today they receive only $0.20. Does this mean the farmer is worse off today than 50 years ago? Is it unfair that farmers now only receive 20% of food expenditures when they received so much more before? The answer to this question is: not

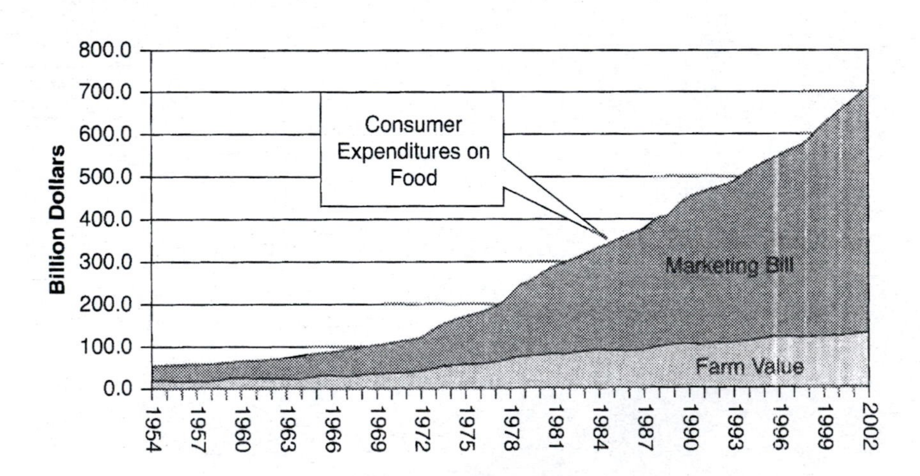

FIGURE 6.1 The Marketing Bill and Value of Farm Production Over Time.
Source: Economic Research Service (2005b).

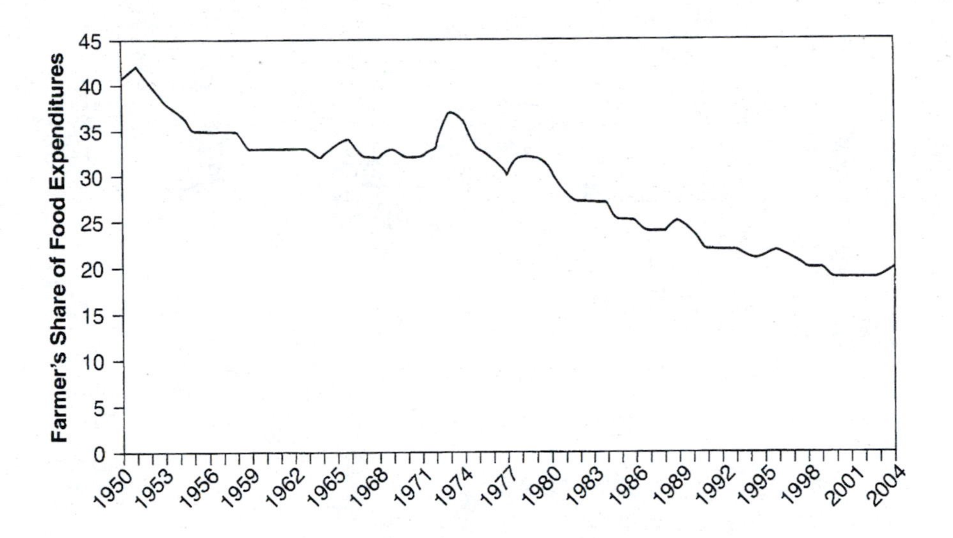

FIGURE 6.2 Farmer's Share of Food Expenditures Over Time.
Source: Data used to construct figure were obtained from the Economic Research Service website.

necessarily. Food purchased today has undergone greater processing than before, which naturally raises processing costs and food expenditures. Today you may purchase prepared ground beef burritos, precooked roasts, prepared hamburger patties, and so on. This extra processing costs money, so consumers must pay a higher price for greater processed food. Consumers are eating just as much or more food while paying more for food processing services, so food expenditures have risen over time. Although farmers receive less of *each* dollar spent, there are more *total* dollars spent, so farmers could actually be receiving greater revenues even if their share of food expenditures falls. Indeed, Figure 6.1 shows that the value of farm production has risen over time. Because food has undergone greater processing over time, it is only natural for processing costs to capture a larger portion of the food dollar and farm production a smaller portion.

To visualize where all the money spent on food goes, see the food dollar in Figure 6.3. For every dollar spent on food, $0.21 goes to the farmer. Most ($0.39) goes to labor at the food processing, wholesale, and retail levels. This includes the people killing the cow, packaging the meat, washing the vegetables, the accountants keeping track of all the sales, the secretaries answering the phones, the managers, and even the George Castanzas of the world, who get paid for doing very little. When you spend a dollar on food, $0.04 goes to advertising, $0.085 goes to the cost of packaging, and only $0.035 goes to profits. These "profits" refer to profits made by the owners of the food processors, wholesalers, and retailers. It is important to consider profits as a part of costs. Investors will only invest money in capital if it provides them a return. Food processing requires expensive capital, and those profits are just part of the capital cost.

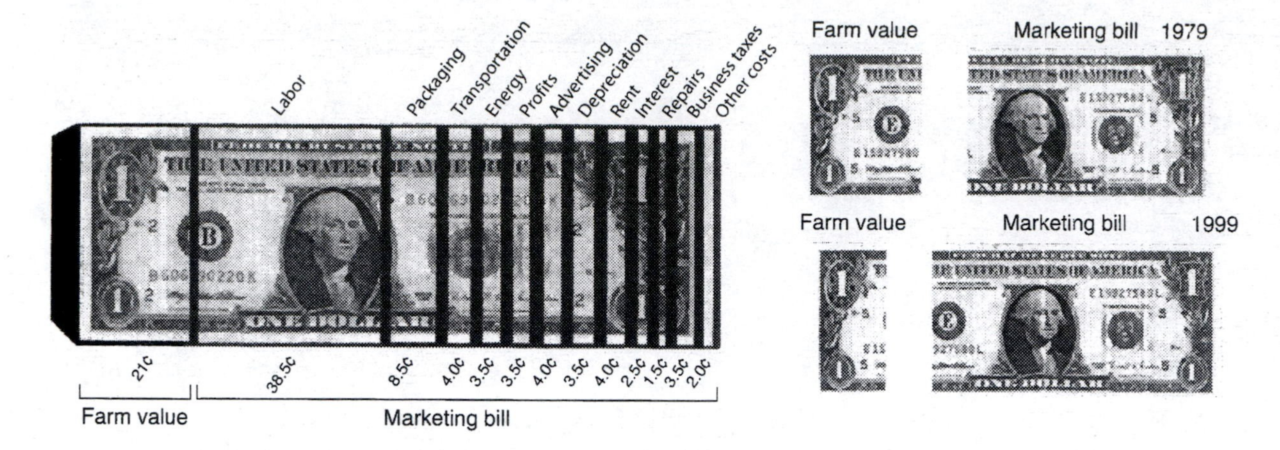

FIGURE 6.3 Breakdown of the Food Dollar.
Source: Economic Research Service (2005c) and Elitzak (1999).

To understand food markets one must be able to visualize the entire food marketing channel. What happens at the farm level impacts food on the grocery store shelf. Changes in consumer behavior are transferred all the way to the farm (and those who sell inputs to the farm). Every business who touches food as it moves from the farm to the consumer's mouth is part of the food marketing channel. Each part of the channel plays a role in increasing the value of food. Farm production, food processing, and marketing activities all create food value. Economists often refer to value as *utility*, and identify four distinct forms of utility: form, time, place, and possession utility. Form utility is the easiest form to understand. Very few people go out to buy a live-cow for its beef, because the beef is not in the *form* we desire. Consumers desire cattle to be slaughtered, processed, packaged, sanitized, and treated so that it has a desirable taste. The most obvious reason that beef prices are higher than the prices farmers receive for cattle is that consumers prefer cattle to undergo expensive processing to transform it into a desirable form. Consumers do not buy raw wheat; they buy flour, so they must pay the cost of transforming wheat into flour. Form utility refers to the act of transforming an agricultural product into an acceptable consumer food.

Production is the creation of utility, or value or happiness. Utility can be grouped into four distinct types.

(1) Form Utility
(2) Time Utility
(3) Place Utility
(4) Possession Utility

Wheat is only harvested once a year, yet consumers desire to consume wheat products the entire year. This means consumers must pay someone to store wheat between harvests and transform it into a food product when the time is right. Consumers want more turkey and cranberries at Thanksgiving and more corned beef on Saint Patrick's Day. We think the demand for beer may rise on Saint Patrick's Day as well (this sounds like a good class field trip!). The activity of delivering food products at the time consumers desire yields *time utility*. Except for at farmer's markets, very few people purchase directly from the farmer. But even when they do, the farmers bring the food to a convenient location. In most cases, food changes hands many times before being sold at the grocery store. Consumers want to purchase their food at a convenient location. The activity of delivering food to a convenient location for purchase yields *place utility*. Finally, consumers wish to take possession of food in particular ways. Consumers want a convenient method for taking possession of the good. Many consumers wish to pay with credit cards, so businesses that allow credit cards increase *possession utility*. Anyone who has experienced frustration at not being able to use their Discover Card for cab rides or credit cards at farmer's markets understands possession utility. Stores that concentrate on customer service provide possession utility as well. Unconditional money-back guarantees and helpful salespeople aid in you taking possession of a product, providing more value than described by the form of the good, the place of the store, and the time it is provided. Each time a company provides one of these four types of utility they add to the marketing bill.

THE FOOD MARKETING CHANNEL

The food marketing channel describes all the activities contributing towards food production. From the fertilizer producer, to the farmer, to the food processor, and to the grocery store manager, all are important and necessary elements. We call it a "channel" to emphasize the necessity of each firm involved in food production and to reflect the interdependence of these firms. What happens at the consumer level has ramifications at the farm and farm input sector, and vice versa. A simple model of the food marketing channel is provided in Figure 6.4.

Adding Form Utility: Farm Inputs and Farm Production

We begin with farm input providers. John Deere produces tractors, Monsanto produces pesticides, and Mosaic produces fertilizers, all of which are necessary ingredients for farm production. Livestock producers must purchase feed (corn, soybean

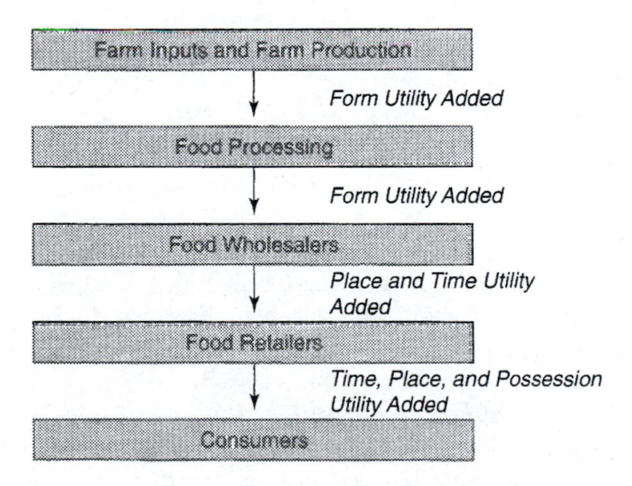

FIGURE 6.4 The Food Marketing Channel.

meal, hay), medicine, and buildings. Banks who provide farmers with loans and companies selling insurance are input providers as well. The farmer takes these inputs and provides form utility. They take inputs consumers do not want, like the nitrogen in fertilizer or the chemicals in pesticides, and turn them into something consumers do want, like cotton. Consumers usually prefer more form utility than that provided at the farm, so the farmer sells her product to a food processor, where it is transformed into a more desirable product.

Farms are falling in number but growing in size. In 1920, there were 6.5 million farms and farmers comprised 30% of the country's population. By 1991, the number of farms fell to 4.6 million and farmers currently comprise less than 2% of the population. Yet, food production has continually risen despite the fall in number of farms, and for two reasons. First, although farm numbers have fallen, the average farm size has risen. In 1910 the average farm size was 140 acres compared to 500 acres today. Second, since 1940 farm productivity has increased 2% each year, meaning each year farmers could produce 2% more farm output using the same amount of inputs (Gardner 2002).

Adding Form Utility: Food Processing and Manufacturing

On the livestock side, meat processors slaughter and process the carcass to provide cuts of meat consumers like to purchase. Some go further and process the meat into other consumer products, like Slim Jims and McRib sandwiches. On the crops side, processors take raw commodities and produce a food item, like taking canola and producing canola oil. Food processing adds form utility. In large food sectors like meat and dairy, the number of processors has been falling while the average processor size has been growing. Most of this is due to mergers and acquisitions. The four-firm concentration ratio (the share of total output by the four largest firms) in the hog industry rose from 34% in 1977 to 50% in 1996 (MacDonald and Ollinger 2000); in the chicken industry this ratio rose from 23% in 1967 to 41% in 1992 (Ollinger, MacDonald, and Madison 2005); and in the beef industry rose from 41% in 1982 to 80% in 2002 (MacDonald and Ollinger 2005).

Previously we showed that labor is the largest cost component of food production, even more than the cost of farm production. Labor is more important at the processing level than any other part of the food marketing channel, so it is important to see which food-processing sectors hire the most employees. Figure 6.5 shows the number of employees in each food-processing sector. Meat by far employs the most people, with preserved fruits and vegetables, bakery products, and beverages coming in second, third, and fourth. The figure also shows the change in number of firms for each sector. Some sectors have experienced a decline in the number of firms, meaning a few firms are dominating a larger part of the market. However, across all food types the number of processing plants has grown, mainly due to an increase in the number of small specialty food processors captured in the miscellaneous food category. For example, over 500 new small salsa makers entered the food-processing

Processing Sector	Number of Employees in 2000	Change in Number of Processing Firms, 1992–1997
Meat	505,000	−78
Preserved Fruits and Vegetables	223,000	65
Bakery Products	202,000	232
Beverages	185,000	179
Miscellaneous Foods	172,000	800
Dairy	145,000	−190
Grain Mill Products	123,000	−87
Sugar and Confections	89,000	130
Fats and Oils	29,000	−21

FIGURE 6.5 Number of Employees and Firms in Select Food-Processing Sectors.

Source: Harris et al. (2003).

industry during the 1990s (Harris et al. 2003). There is still room for small businesses in the food industry, but only in select food items.

Farmers often see their prices remaining stagnant while food prices rise. This has led many farmers to conclude that they are the victims of market power. That is, as food processors consolidate, they gain market power and take the portion of the food dollar that is rightfully the farmers (as some say). This is especially true in the cattle markets, where only a few packers now buy virtually all the cattle. A closer look suggests that food processors are not making off with huge profits at the farmer's expense. The average rate-of-return on stocks for food processing companies was around 23% in the late 1990s (Harris et al. 2003). This seems high, until you consider the fact that the average rate-of-return to all stocks during this period was 26.3%. Thus, food processors' profits were less than average across all industries. Economic profits (profits subtracting out opportunity costs) were essentially negative or close to zero, exactly what one would expect from a competitive industry with little to no market power.

Adding Time and Place Utility: Food Wholesalers

Once food is completely processed and ready for consumption, it is delivered to retail outlets, often via food wholesalers. Wholesalers are the link between retailers and food processors. You and I are the retailer's customer, but the retailer is the wholesaler's customer. Your local grocery store wants to make sure you may purchase your food at a convenient place and time. Thus, they locate at a convenient location and stock their shelves with the goods you desire at the time of year you desire. They cannot perform this task unless the wholesaler will sell to them at this location and at the appropriate time. The retailer and wholesaler must work together to ensure that consumers can purchase their food at the right time and place. Thus, both wholesalers and retailers provide time and place utility to food. There are three types of wholesalers: merchants, MSBOs, and brokers. Merchants are those who buy and resell food. They purchase directly from food processors and resell the food to retailers. MSBOs refer to "manufacturers' sales-branches and offices" but is best explained with an example. A food processor purchases wheat and transforms it into flour. Once the flour is made at the processing facility, it could be sold from the facility directly to the retailer but is instead shipped to an MSBO. The MSBO is owned by the same company that made the flour but specializes in marketing the flour to retail outlets. The only difference between an MSBO and a merchant is that the MSBO is owned by the same company that processed the food. A broker is simply paid a commission to procure food for a retailer. For example, a grocery store may hire a broker to pick up and deliver produce from a vegetable processor and are compensated for their time and transportation costs. Brokers work on commission, never assuming ownership of the food item. Of all the wholesalers, 56% are merchants, 25% are MSBOs, and brokers comprise the remaining 19% (Harris et al. 2003).

Wherever you live, chances are that a Wal-Mart Supercenter has moved to your town in the past 15 years. Wal-Mart has the ability to offer lower prices on consumer goods, including food. Part of Wal-Mart's success is due to its distribution system. Wal-Mart operates its own distribution centers; it acts as a wholesaler and retailer. There is a trend in the food industry for retailers to own their own distributing centers. The retailer assumes the duties of the wholesaler. These are referred to as *self-distributing retailers*, and include all the big grocery store chains: Wal-Mart, Kroger, Albertsons, and Safeway. Even though the wholesaler is cut out of the picture, the wholesaler's job must still be done by someone. You can cut out the middleman, but not the middleman's job. Self-distributing retailers are aggressive adopters of technologies that deliver food to grocery stores in a more efficient manner (often called supply chain management technologies). More food items are moved through warehouses of self-distributing retailers per hour, giving it lower distribution costs. Wal-Mart's distribution network is considered especially efficient, making its prices especially low.

However, this does not imply that self-distributing food retailers will overtake wholesaler-supplied food retailers. Although they have higher distribution costs, wholesaler-supplied retailers tend to have less labor costs because they are less likely to be unionized and place less emphasis on hiring and training skilled employees. Due to its smaller emphasis on efficient distribution networks, wholesaler-supplied retailers can focus more on adapting to the consumers' needs. Of all the food distribution centers in the United States, 34% are considered self-distributing retailers, 38% are third-party wholesalers, and the remaining 28% are processor-delivered retailers (King 2003). In this last group, the food processor delivers and stocks the shelves themselves and often uses scan-based trading where retailers are not billed for the good until it is sold to the consumer.

Adding Time, Place, and Possession Utility: Food Retailers

Food retailers are those that sell food directly to consumers. Of all the sales from food retailers, 70% come from supermarkets, 15% from small grocery stores, 10% from convenience stores, and 5% from specialty stores like seafood markets. Of utmost importance is a convenient location. Consumers demand place utility in their food. Time utility is important as well. Stores that do not provide enough turkeys at Thanksgiving and hot dogs on the Fourth of July will lose their business to competitors that do. Finally, retailers provide possession utility by allowing consumers to purchase food conveniently. This includes allowing consumers to use credit cards for their purchase, speedy checkout lines, nice shopping carts, and Starbuck's coffee while they shop.

The trend in food retail outlets is to combine food with nonfood items in the same store for consumer convenience. Figure 6.6 shows the number of supermarket types in 1990 and in 2000. A "supermarket" is a grocery store having at least $2 million or more in food sales, measured in 1980 dollars. The conventional supermarket contains all major food items and a limited general merchandise selection and may provide deli and bakery services. A superstore is larger, and general merchandise accounts for at least 10% of sales. As Figure 6.6 shows, the number of conventional supermarkets is falling, whereas the number of superstores is growing. Many stores now combine superstores that contain a pharmacy and are referred to as combination food and drug stores. Warehouse stores are also growing in number, have a limited product variety and fewer services, but tend to sell in bulk. Superwarehouses are warehouses that also provide a service deli, as well as meat, seafood, and bakery departments. Finally, there are hypermarkets, which are the largest supermarkets with at least 150,000 square feet of service area and a large selection of food and merchandise items (Harris et al. 2003).

Supermarket Type	Number of Stores in 1990	Number of Stores in 2000
Conventional	13,200	9,900
Superstore	5,800	7,900
Warehouse	3,400	2,400
Combination Food and Drug	1,600	3,700
Superwarehouse	300	500
Hypermarket	100	200

FIGURE 6.6 Number of Supermarkets in 1990 and 2000.
Source: Harris et al. (2003).

Another trend in the food retail industry is towards fewer firms. The share of sales by the eight largest food retailers increased from 30% to 41% between 1997 and 2000. Most firms are growing in size by merging with or acquiring other competitors, except for Wal-Mart who has grown by introducing new stores. In just the past few years Wal-Mart has jumped to the head of the pack and is the largest seller of food with a market share of 25% of all sales. Kroger trails at second with 12.5% market share, and the third, fourth, and fifth largest food retailers are Albertsons, Safeway Inc., and Costco Wholesale Group (*Feedstuffs* 2005a).

Putting the Food Marketing Channel Back Together

Now that we have dissected the components of the food marketing channel, let us combine them again and evaluate the relative importance of each sector. Each sector adds value to food by creating form, time, place, and/or possession utility, but by different amounts. Figure 6.7 shows the value added to the food and fiber system by different sectors. You may find it surprising that of the total value of food and fiber, farming creates only 7% of that value. The inputs to farming comprise a larger 34%, processing adds 17% of all value, and wholesaling and retailing is the most valuable sectors, with 42% of total value.

Yet, each component of the food marketing channel depends on the others. Grocery stores cannot sell grits if farmers do not plant corn, and corn farmers cannot sell their corn unless there is someone transforming the corn into grits. In one sense, the sectors must work together and coordinate with one another to ensure an effective food system. In another sense, they must compete with one another. Each

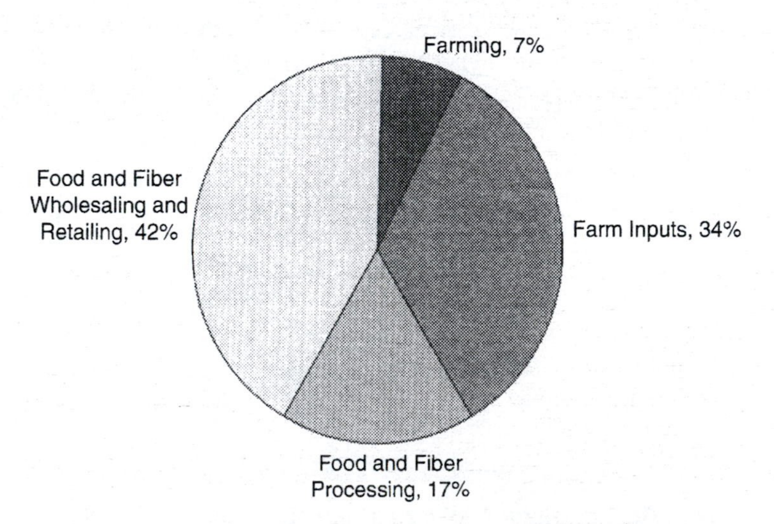

FIGURE 6.7 Value Added by the Food and Fiber System in 2000.
Source: Harris et al. (2003).

sector competes for a portion of the food dollar, and the portion of that dollar they obtain depends partially on their negotiating power. In the next section, we explore the issue of vertical coordination, with special emphasis in studying how different sectors in the food marketing channel coordinate with one another across different food items.

VERTICAL COORDINATION IN AGRICULTURE

Jackson enters Kentucky Fried Chicken with an appetite for fried chicken. Jackson ate here yesterday, and the fried chicken breast he tried was delicious, bringing him back again. However, today the chicken breast was substantially smaller and overcooked. Jackson complained to the manager and received an additional chicken breast for free. The manager, named Harrison, laments at giving away free chicken. Harrison's boss has been complaining about profits lately, and giving away free chicken will not help. The problem is that the chicken breasts Harrison receives are inconsistent. Some breasts are much bigger than others. This makes it difficult to provide his customers with a consistent product and provide even frying to a batch of chicken breasts. Harrison has complained to his chicken provider, but the provider claims there is little he can do because the chickens the farmer grows vary in size. For chicken breasts to be a consistent size, the chickens themselves must be of a consistent size. If consumers purchased chickens directly from the farmer and did the frying themselves, we might not have this problem because the consumer could ask the farmer to produce more consistent chickens directly. This is not the case though. The chicken changes hands many times on its way up the food marketing channel before being purchased by Jackson. At each point in the channel, firms' actions are dictated by price incentives. As price goes up, more is produced and vice versa. And if firms receive higher profits from producing a consistent size of chicken, they will. Unfortunately, market prices can be imperfect signals, imperfect forms of communication, and firms along the food marketing channel seek alternatives to coordinate their activities.

Vertical Coordination: The act of coordinating activities between sectors along the food marketing channel.

This describes a vertical coordination problem along the food marketing channel. The longer the food marketing channel, the more difficult it becomes to relay information from the consumer to the farmer. There was a game we used to play in elementary school. All the students would sit in a circle. The game began by one person whispering a comment to the student on their left. That student then passed the comment to the person on their left, and so on. Eventually, it got back to the student who made the original student, and the comment had always changed to something different. It changed from something like "I like recess" to "see that pretty dress." Even though the game was meant to illustrate the dangers of gossip, it is also a metaphor for the difficulty of passing along information in the food marketing channel.

Consider another vertical coordination problem. Hogs can exhibit carcass defects that impose costs on the pork processor. Pork processors have a harder time processing hogs that are too heavy. It is more expensive for them to process carcasses that have bruises, and some hogs have genes that cause them to exhibit PSE meat problems, which stands for pale, soft, and watery. Quality problems cost packers around $10.08 for each head of hog processed (Martinez 1999). However, 81% of these costs are controlled by the farmer. This means that to reduce these carcass problems the pork processor must convince the farmer to employ different hog production strategies.

Carcass bruises can be avoided by properly administering vaccinations and antibiotics, as well as humane treatment of the hogs. Overweight hogs can be avoided through proper management, and PSE problems can be completely eradicated by improving genetics. All of these costs the farmer money, but the evidence suggests that the increase in the carcass value outweighs these costs, so it is in the pork industry's interest to address these problems. The processor can more than compensate the farmer for the extra costs, and both earn more money in the process. The problem is that the farmer incurs the cost while the processor receives the benefit. If the processor can find a way to compensate the farmer, both can make more money. The processor and farmer must coordinate for them to reap higher profits. But how, exactly, should they coordinate?

Vertical Coordination Through Average Pricing

There are numerous ways for firms to coordinate their activities vertically along the food marketing channel. One is through markets and the reliance on prices as information signals. Prices are often efficient signals of information. High beef prices signal that consumers want more beef than is currently produced, and producers respond to high beef prices by producing more beef. It is as if consumers kindly asked cattle producers to increase production, but instead the conversation was held by negotiating prices. Suppose that the cost of improving hog carcasses with more intensive management practices is $2.00 per head to the farmer but improves the value of the carcass to the processor by $3.00 per head. The processor should place a premium between $2.00 and $3.00 per head on such hogs. For example, if processors pay $2.50 more for such hogs, the farmer and the processor both benefit by $0.50 per hog—everyone wins. This is a price signal, and producers will respond by improving their hog management practices, which increases profits to hog producers and pork processors.

The problem is that, in the past, most animals were not sold on an individual basis. They were more likely to be sold in groups. Pork processors would purchase a herd of pigs, perhaps several hundred, and pay a price for each animal based on the perceived average quality of the group. This is referred to as *average pricing*. The price would most likely be determined in some sort of auction where the buyer and seller do not directly interact. This single price is then a function of numerous factors. It depends on the perceived carcass quality of each hog, as well as supply and demand conditions. If producers receive a higher price, it is difficult for them to determine exactly why they received a higher price. It could be that hog supplies have fallen, pork demand has risen, or that the herd seems to be of high quality. It could even be the order in which the hogs are sold in the auction. It is well known that prices fall as auctions near their end. The price "signal" is really a mix of signals, and if the producer receives a higher price because the processor thinks the carcasses will be of high quality, he may not know it. Other producers will not know it either. Although they see another producer receiving a higher price, they are not sure if it is because they are the proper weight, if hogs seem to be handled more humanely and are less likely to contain carcass bruises, or if the particular breed of hog produces a higher-quality carcass.

Another problem is that the animal is purchased before slaughter, yet it is not until after slaughter that the processor discovers the quality of the meat. There are many quality characteristics the food processor simply cannot see in the live animal, which makes it impossible to pay more for a live animal with a superior carcass. And if producers do not get paid more for producing better carcasses, they will not take costly actions that lead to a better carcass. The point is that average pricing of live animals, as opposed to assigning a unique price to each animal, reduces the information contained in price signals and makes it impossible to send certain signals. There are several ways around this information problem. One is to simply price each animal carcass according to its quality. Instead of paying for the animal while it is alive, the grower and processor can agree on a formula for the animal's price based on its carcass. This process is used by the cattle and pork industries and is referred to as grid pricing. Another solution is for the processor to write contracts with hog producers specifying the type of hog to be raised and how it should be raised. Such contracts are used extensively in the hog and poultry industry. Finally, one of the easiest ways to improve coordination between the farm and processing sector is to place them under the same ownership and management. This is referred to as vertical integration and is popular in the egg and turkey industry.

Vertical Coordination Through Grid Pricing

Cattle markets have historically relied almost exclusively on average pricing of live animals, leading to the coordination problems previously discussed. After years of watching beef demand decline due to consumer dissatisfaction, the industry decided to improve beef quality through the use of grid pricing, where each carcass is assigned a unique price based on its quality characteristics. The term *grid pricing* is used because the premiums and discounts assigned to each carcass are based on a table or grid, reflecting the value processors place on different carcass traits.

An example of a grid is shown in Figure 6.8. Each carcass starts at a base price of $108.59 per cwt. This base price will differ as supply and demand conditions vary. The quality grade adjustment assigns carcasses with higher-quality meat price premiums. Quality is determined by visually inspecting the meat and measuring its marbling (intramuscular fat content). The more marbling, the fattier the beef and the better the beef tastes. The best beef is designated as Prime. Most prime steaks are sold to restaurants. Choice steaks are the baseline (given neither a premium nor a discount)

Base price	$108.59
Quality grade adjustment	
Prime	$5.17
Choice	—
Select	–$12.33
Standard	–$21.00
Yield grade adjustment	
1.0–2.0	$1.67
2.0–3.0	$0.75
3.0–3.5	–$0.17
3.5–4.0	–$0.33
4.0–5.0	–$16.83
>5.0	–$21.83
Carcass weight adjustment	
400–500 lbs.	–$20.67
500–550 lbs.	–$17.33
950–1,000 lbs.	–$18.67
>1,000 lbs.	–$24.50

FIGURE 6.8 Example of Cattle Pricing Grid (Prices Are in $/cwt of Carcass).
Source: Lusk et al. (2003). Reprint permission made available by Blackwell Publishing.

and are generally sold in restaurants and supermarkets. Select and Standard are poorer-quality beef and receive discounts. Generally, if you see a steak in a supermarket that is not labeled Choice, it is Select Grade or ungraded beef. Even though some consumers may prefer Select beef due to its lower fat content, the discount assigned to Select beef indicates that most consumers prefer taste to leanness. Some carcasses produce more retail cuts of meat because they contain less fat. The yield grade adjustment assigns a higher price to carcasses yielding more beef. Finally, there is a weight adjustment. Beef processing plants are designed to process carcasses around 600–900 lbs. Excessively small or large carcasses are more difficult and expensive to handle and therefore receive a discount.

In just a short period of time grid pricing has become a standard pricing practice. In 1996, only 16% of cattle were sold under grid pricing, but that number has risen to 62% in just 10 years. This has introduced a remarkable improvement in vertical coordination between the consumers and producers of beef. Suppose that consumers' tastes change so that they now prefer leaner beef, even if leaner beef is less tasty. Consumers begin purchasing more Select beef in the grocery store. The grocery stores tell their wholesalers they need more Select beef, and the wholesalers deliver the same message to beef processors. To induce cattlemen to produce cattle with leaner beef, they lower the discount assigned to Select beef. In Figure 6.8, Select cattle receive a $12.33 per cwt discount. Now, that discount may fall to $6.00. Select beef is cheaper to produce, so cattlemen respond by producing more Select beef. The price system works to facilitate vertical coordination because prices are assigned based on the supply and demand of individual carcass traits.

Vertical Coordination Through Production and Marketing Contracts

Spot Market: A market where the price is negotiated and the exchange takes place immediately.

Contracts are a popular tool in the business world, and agriculture is no exception. In the early 1900s, poultry and egg farmers and processors coordinated through spot markets. *Spot markets* refer to markets where commodities are sold "on the spot." That is, prices are negotiated at the same time ownership is transferred. If a poultry processor wanted to process 5,000 chickens this week, it must go to auctions or negotiate prices with individual farmers for 5,000 chickens that week. Over time, broiler meat (meat from young chickens, which is the chicken you consume the most) processors sought greater control over chicken genetics and how the chickens were raised to respond to consumers' call for more uniform and consistent meat. Processors were also interested in providing a more stable supply of chickens than could be achieved from spot markets. This was achieved through *production contracts*.

To understand how production contracts work, consider the story of a chicken grower named Buster. Buster signs a legal contract with a chicken processor like Tyson Foods. Tyson Foods owns the chickens; Tyson does all the breeding and hatching of the chicks. The chicks are then delivered to Buster's farm. Under the contract, Buster is required to purchase and maintain a poultry production facility, like the one shown in Chapter 14. Tyson owns and delivers the chicken feed to Buster's farm and even provides veterinary services to Buster. Buster's job is to take the chicks and feed given to him and raise the chickens until they are ready for slaughter, at which time the chickens are called broilers. Tyson Foods picks up the chickens and pays Buster a fixed fee based on the pounds of broilers produced. The fee may also include a bonus if his production efficiency was better than other growers.

Production Contract: An arrangement where the food processor supplies the animals, feed, and other inputs, and the farmer supplies the farm facilities and is responsible for animal growth. The farmer is paid a flat fee or a fee based on animal performance.

It may seem as if Buster is little more than an employee of Tyson Foods. Indeed, that is the point of production contracts. The processor gains greater control over the production process, including the supply of broilers, but also assumes all market risk. It allows the processor to take an active role in improving meat quality and smoothing broiler supplies. Regardless of whether broiler prices rise or fall, the fee the grower receives remains (roughly) the same (of course, the contract specifications in the long run can be modified). The processor assumes all the risk but receives all the benefits when market conditions are favorable. Today, virtually all broiler production takes place in the presence of production contracts.

Marketing Contract: A contract specifying the amount, price, and type of good to be exchanged in advance.

The dairy, fruits and vegetables, sugar beet, and cotton industries rely instead on *marketing contracts*, where the farmer and processor agree on the quantity and price of farm product to be exchanged months in advance. The contracts may also contain a specification of quality and may use something similar to grid pricing to assign discounts to producers failing to meet those specifications. As of 2001, 57% of fruit, 94% of sugar beet, and 17% of livestock production was conducted under marketing contracts (MacDonald et al. 2004).

The pork industry has pursued a mix of marketing and production contracts. Roughly 61% of all hogs in 2001 were produced under one of these contracts. Of all the hogs produced in the United States, 7% are produced under marketing contracts and 53% under production contracts (MacDonald et al. 2004). The rest are sold on spot markets. In areas with a history of hog production like Iowa, marketing contracts are more popular, where the predetermined price is based off a formula. The formula might be sophisticated like the cattle pricing grid, or as simple as an agreement to pay whatever the spot price is at the time of exchange. The hog industry has rapidly expanded in areas like North Carolina, Texas, Missouri, and Oklahoma. In these areas, virtually all hogs are grown under production contracts.

Vertical Coordination Through Vertical Integration

Vertical Integration: The process by which two or more steps of the production process are under the same ownership.

The turkey industry has also pursued production contracts to facilitate vertical coordination. Approximately 56% of all turkey production is coordinated through production contracts. Of the remainder, 32% of turkey production is under *vertical integration*. Vertical integration occurs when a firm in one sector of the marketing channel assumes ownership of firms in another sector of the channel. This owner is referred to commonly as the integrator. An extreme example is Braums, a fast-food chain and ice cream store located in the Oklahoma area. Braums owns the cows that produce the milk, the processing facilities that turn the milk into ice cream, and the retail outlet that sells the ice cream. Every part of the food marketing channel is owned by Braums. In many food industries, vertical integration occurs by a food processor taking ownership of the farm. This is referred to as *downstream integration*, where one firm begins producing inputs, which they previously purchased in a market. *Upstream integration* can also occur, where firms begin performing the function of a firm that previously purchased their production. The most common form of upstream integration is farmer cooperatives, which will be discussed shortly.

In the case of turkeys, 32% of all turkeys are produced in a setting where the food processor owns the turkey growing facilities and all employees are wage or salary laborers. With production contracts, our grower named Buster was "like" an employee of the integrator. With vertical integration, Buster is an employee. Vertical integration is more common in the egg industry, accounting for 60% of all eggs (Martinez 2002).

A word of caution: Not everyone uses the exact same terminology when describing vertical integration. Some people include the use of production contracts as vertical integration and refer to the processing facility issuing the contracts as the integrator. Although this textbook defines vertical integration as the case where two or more steps of the production process are under the *ownership* of one firm, others defined it as the two or more steps being *controlled* by one firm. Under production contracts, the food processor exerts so much control over the farm production and owns so much of the inputs used at the farm level (e.g., the animals, feed, etc.) that some people refer to production contracts as vertical integration.

Reasons for Contracting and Integrating

As we have seen, poultry and egg processors have aggressively pursued vertical coordination through integrating (taking ownership of the farm) and production contracts. In reality, there is little difference between vertical integration and production contracts. In both cases the processor owns the animal, provides the feed, and dictates how the animal should be raised. Thus, we refer to both as *vertical control*. Vertical control refers to the extent to which one sector of the food marketing channel (in livestock industries we are usually referring to meat processors) controls activities in other sectors. In contrast to poultry, there is little vertical control by beef processors. Beef processors rarely own the animals before slaughter and have very little control over how the animal is bred, fed, or raised. At this point, it is useful to discuss the incentives of food processors to vertically integrate or secure their supplies through production contracts and vertical integration, and why these incentives are more pronounced for poultry and pork than beef.

Incentives for Vertical Control: Reduce Transaction Costs. One can think of "transaction costs" as simply the cost of negotiating the buying and selling of production. Transaction costs are like a broad definition of price. For example, one may travel 100 miles to purchase a ton of hogs at $40/cwt. Even though the "price" is $40/cwt, the transaction costs is the total cost of the purchase, which includes the price, fuel costs, the opportunity cost of time involved in traveling, and so on. Sometimes transaction costs are assumed to be a separate part of price. For instance, the price of hogs purchased may be $40/cwt, and the transaction costs include all additional costs of procuring the hogs. However, in this chapter we assume the "price" of a good is included in the total transaction cost of obtaining the good.

The beginning of this section told a story about Jackson the chicken customer and Harrison the Kentucky Fried Chicken manager. The problem in the story was that farmers produced chickens of various size, leading to chicken breasts of various sizes, which makes it harder to cook all chicken breasts to the same temperature. One way to alleviate this problem is for the meat processor to segregate chicken breasts of different weights and only sell Harrison breasts of a uniform size. This is one way to produce uniform chicken breasts, but if consumers want their chicken breasts to always be the same size, the poultry processor must find some way to market the remaining breasts that do not fit consumers' tastes. For example, the extra large and small breasts may be exported to Russia. By forcing processors to deal with irregular-sized chickens, it increases their processing costs, which ultimately raises the price of chicken. A better alternative is to convince chicken producers to grow more uniform chickens. This gives consumers the product they prefer without forcing processors to deal with the irregular chicken breasts. As a result, the transaction cost of acquiring the right size of chicken breasts is reduced, and the food marketing channel becomes more efficient.

Earlier we discussed how market prices can be fuzzy signals of information, especially when animals are priced the same according to the average quality of the group.

Producers selling on spot markets receive higher prices one month and lower prices other months and are never sure exactly why their price changed. This makes it difficult for processors to signal information through prices. If a processor wants to minimize carcass problems by using only a specific type of genetics, it is far easier for them to make growers use those genetics by owning and providing them with the animals themselves. This is why pork and poultry processors have been so aggressive in providing the grower with animals whose genes were chosen by the processor. The *transaction costs* of procuring animals with the desired genetics is often less when relying on production contracts than spot markets, causing firms to favor production contracts.

The quality of an animal depends on many characteristics, some of which are not easily measured in the live animal or even in the carcass. A processor purchasing live hogs on the spot market cannot give the grower a premium for producing a lean meat carcass, because one needs to process the hog to determine if the meat is lean. This makes it hard for processors to sell lean pork if they purchase solely from spot markets. However, if the processor owns the hogs, they can purposely breed only boars and sows that produce lean meat. Moreover, certain carcass traits that produce good tasting meat cannot be identified until the animal is fully processed and consumed. That is, some meat traits can only be detected by the consumers' mouth. In these instances, the only way to know if the animal has those traits is to know the animal's genes.

Processors are also concerned with the type of feed administered to the animal. We have all seen from mad cow disease what dangers can arise from improper feeding. The taste of meat also depends heavily on the type of grain the animal was fed. The best way for the processor to determine an animal's diet is to either own the farm or write production contracts that require the grower to use the feed provided by the processor. Vertical control is used by processors to ensure product quality by giving them greater control over what animals are bred and how the animals are raised. Thus, contracts ensure clear signals and clear measurements of how the animal was raised, and consequently, how the animal will taste. The transaction costs of achieving the desired consumer product can be lower under vertical control (contracts and integration) than when relying on spot markets, so the use of spot markets falls and vertical control rises.

Incentives for Vertical Control: Risk-Sharing. Hog producers who obtain inputs and market their hogs on spot markets face greater risk than their counterparts who enter production contracts. These "independent" hog producers must buy their own corn, and corn prices can be quite volatile. Their counterparts receive corn free from their contractor. Revenues for independent hog producers depend on the hog spot market price, which also varies greatly. Conversely, hog producers under production contracts receive the same compensation regardless of hog market prices. Thus, the use of production contracts eliminates much of the risk for individual hog producers. Where does the risk go? The food processor absorbs the risk. This means that the processor absorbs the losses when prices are unfavorable for pork production, but reaps the benefits in favorable times.

This risk-sharing can be win-win for the hog producer and processor. Processors tend to be large firms, with greater ability to withstand unfavorable market conditions. These processors also tend to be more diversified, where losses in one industry (like the hog industry) are often offset by gains in another industry (like corporate stocks or crops). Independent hog farmers, on the other hand, often receive a majority of their income from hog production. For these farmers, a steady farm income means a steady total income. If the main purpose of production contracts was risk-sharing, we would expect the farmer to receive less compensation under less risky production contracts than they do as independent farmers where they face greater risk (making them indifferent between the two business arrangements). Yet, this does not seem to be the case, indicating that risk-sharing is not the only reason for production contracts.

Incentives for Vertical Control: Efficiency Gains. Private enterprise often conjures the image of a "may the best man win" setting. In a capitalistic society, less efficient firms tend to be conquered by their more efficient counterparts. We have seen Wal-Mart come to dominate the food retail industry, largely by selling at lower prices than their competitors. Achieving efficiency gains through a well-organized distribution system and their tough negotiations with input suppliers, Wal-Mart can place food on their shelves at lower prices than their competitors. Thus, we see less efficient grocery stores being replaced by Wal-Mart—the best man won, and the best man was Wal-Mart.

Similar stories can be witnessed in every sector of agriculture. Small inefficient farms go out of business, whose land is purchased by larger more efficient farms. Small, inefficient meat-processing facilities have been replaced with much larger processing facilities. Within any one sector of the food marketing channel, inefficient firms get replaced by more efficient firms. Why are some firms more efficient than others? A lot has to do with management. Better managers lead to more efficient firms, and as they replace less efficient firms, their management style and strategy is applied to a greater percentage of food production.

There is no reason why more efficient firms, with their superior management, should not replace less efficient firms at different sectors of the food marketing channel as well. Consider a story of the Douglass family, who are poultry growers in North Carolina. Father Douglass was an independent poultry producer, owning his own facilities, owning the chickens, and owning the grain fed to the chickens. This job included many activities. One included purchasing and determining the exact diet to administer to the chickens. Another pertained to selecting which hens and roosters should be bred. A third concerned how the chickens should be raised, and the fourth task was marketing the chickens. Marketing included finding a buyer for the chickens and negotiating a price. Father Douglass had so many tasks to perform that he could not be as efficient in any one task as someone else who performed only one of those tasks.

Father Douglass has since retired, but his son has taken over the chicken farm. Things have changed on the farm. His son raises chickens under a poultry contract with Tyson Foods. Tyson delivers the chickens and the chicken feed, and the son is in charge of maintaining the building and raising the chickens. The son has far less duties, allowing him to concentrate on raising healthy, fast-growing chickens. With so much more time devoted to the chickens, the son is a better chicken grower than his father. In fact, the son spends so much time specializing in raising chickens, he is one of the most efficient chicken growers Tyson can find. Of course, the son knows nothing about purchasing and blending chicken feed, breeding chickens, or marketing chickens. The son does not need to know these things; Tyson Foods does.

Now let us peer into the workings of Tyson Foods. Tyson will have one unit devoted solely to the procurement and blending of grains for chicken feed. Being specialists in chicken feed, they are good at what they do. They produce feed efficiently. Having more time to concentrate on chicken feed than father Douglass, they can procure better feed at a lower cost than father Douglass ever could. Tyson also has a team dedicated to chicken breeding, and another team specializing in marketing the chicken meat. The former studied animal science in college, and the latter studied agricultural economics. Both units are experts in their field. Although they are not any smarter than father Douglass, they accumulated human capital in different particular areas and can breed better chickens and market chickens better than father Douglass ever could.

Father Douglass never conducted his own research in poultry production. Never did he consider using rare chicken genes to produce a better tasting meat product, and never did he experiment with alternative buildings for raising chickens. The reason is that research is expensive, and if the research is only used to improve one farm, the research will not pay for itself. Large food processors who own or contract with many farms have more incentives to experiment; the research results of a single experiment can be applied to hundreds of farms. Indeed, large food processors like Smithfield Foods and Tyson Foods spend millions in research, and as we will see shortly, this research has led to significant efficiency gains.

Recall that vertical control is the act of one firm exerting greater control over the production practices in another sector of the food marketing channel. Typically, the firm exhibiting greater control does so because it believes its production strategies are more efficient. Tyson Foods took control over poultry genetics and feed procurement because it thought itself to be more efficient at this than the Douglass family. The Douglass family was more efficient at poultry production though, so Tyson left most of the hog production practices to the Douglass son. By both parties concentrating in the area they are more efficient, they can produce food at lower cost. Indeed, if production contracts are a more efficient form of vertical coordination than spot markets, firms using production contracts will gain market share over those using spot markets. In the end, the best man will win, and in agriculture, often the "best man" represents the firm using production contracts. This is not just speculation. Economists have proven that livestock growers under production contracts are more efficient that those not under production contracts (Key and McBride 2003).

In other settings, one firm believes its management strategy is so efficient that it will pursue vertical integration, where it possesses greater control over how a product is produced. It is said that the owner of Braums once believed he could produce food cartons at less cost than his supplier, so he built his own food carton production facilities and found he was correct. Braums now makes its own food cartons. By pursuing vertical integration in the food carton sector, it lowered the cost of fast-food production.

Why Not Beef? We just presented logical reasons why a poultry or pork processor would want to obtain vertical control over farm production. Most pork and poultry is under vertical control, either through vertical integration or production contracts. But if vertical control has so many advantages, why is it virtually absent in the beef market? Below are four general reasons: long biological cycle, multiple stages of production, disperse geographic concentration, and reliance on land.[1]

Poultry and pork producers have made great strides in improving meat quality by investing in better genetics. This is less profitable in the beef industry because the biological cycle is much longer. The period of time between breeding and slaughter is five months for poultry and one year for pork, but two years for beef. Improvements in beef genetics take much longer before they are realized at the retail level. Thus, investments in beef genetics are less profitable than poultry and pork genetics. Second, there are more stages of production in cattle production. Poultry production has two basic stages: hatching and growing. Farrowing and growing are the basic two stages of pork production. Beef, on the other hand, contains three stages: cow-calf, stocker, and feeding. Often, ownership of the animal changes hands at each stage This means that ascertaining vertical control requires control over more stages of production and is more complicated. More complicated generally implies more costly.

Third, cattle tend to move over a large geographic area during the three stages of production. Cow-calf production takes place where grazing is most available, which is the southern plains, southeastern, and mountain western states. Stocker production takes place mainly in Oklahoma, Texas, and Kansas, where winter wheat is grown and the climate is relatively warm. Finally, feeding tends to take place in dry climates, which include the three aforementioned states but also Colorado and Nebraska. Cattle are healthier in dry climates and therefore grow faster. Contrast this with pork and poultry, where all stages of production can be easily conducted in the same county. Clearly, the less area one needs to cover to exert vertical control, the easier it is to establish vertical control. Finally, neither poultry nor pork production require a particular type of land. Both are raised in buildings that can be located most anywhere and do not consume much land, which reduces the cost of land acquisition. Cattle, on the other hand, require large amounts of grazing land at the cow-calf and stocker stage, and this land is held by thousands of different landowners. To exert vertical integration in the cattle industry would require huge purchases of land and negotiations with numerous different landowners, making vertical control less desirable from the processors' point of view.

---

[1]Much of the discussion that follows is taken from Ward F-552.

Successes of Vertical Control

Vertical control over both the farm production and food-processing level has two main goals: to reduce production costs and improve product quality. The evidence suggests both goals have been met. The real winners are meat consumers, who now have a more consistent and higher-quality meat product at lower prices. The evidence for this is best illustrated by two graphs compiled by economist Steve Martinez at the Economic Research Service. Figure 6.9 shows one of these graphs. On the *x*-axis is the extent of vertical control by a meat industry, measured as the percentage of output produced under vertical integration or contracts. On the *y*-axis is the percentage of change in price over time. Clearly, the greater vertical control in an industry, the slower prices rise. This is due to the fact that vertical control lowers production costs. As an example, economists have studied the effect of vertical integration in the egg industry. During the period 1960 to 1984, vertical integration increased from 26% of total egg production to 89%, and the efficiency gains led to lower egg prices. From the period 1973 to 1983, egg prices dropped 8.2 cents per dozen. And if vertical integration had not been increasing during this period, egg prices would only have dropped 3.4 cents per dozen (Kinnucan and Nelson 1993).

The broiler industry's aggressive pursuit of vertical control has led to an improved product that has increased broiler demand. See Figure 6.10, which plots the quantity demanded of broilers over various years. The quantity refers to quantity demanded by consumers at each year, given the prices that occurred in each year. The black line is actual demand. The grey line is a simulated demand and is the forecasted demand if the broiler industry had not undergone greater vertical control. Clearly, demand is greater in the presence of vertical control, and this greater demand is due to improvements in meat quality.

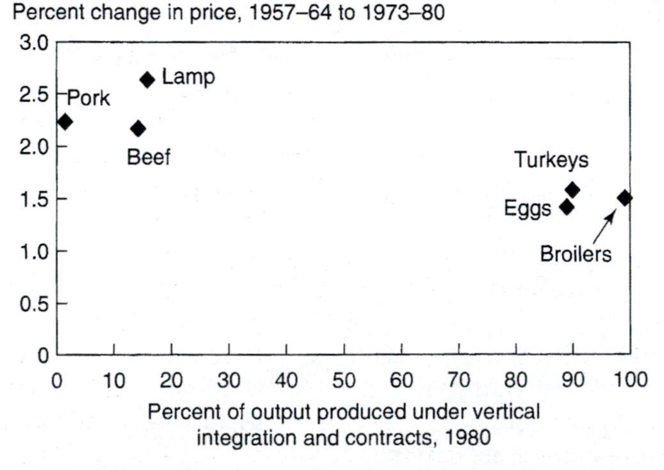

FIGURE 6.9 Impact of Vertical Control on Meat Prices.
Source: Martinez (1999).

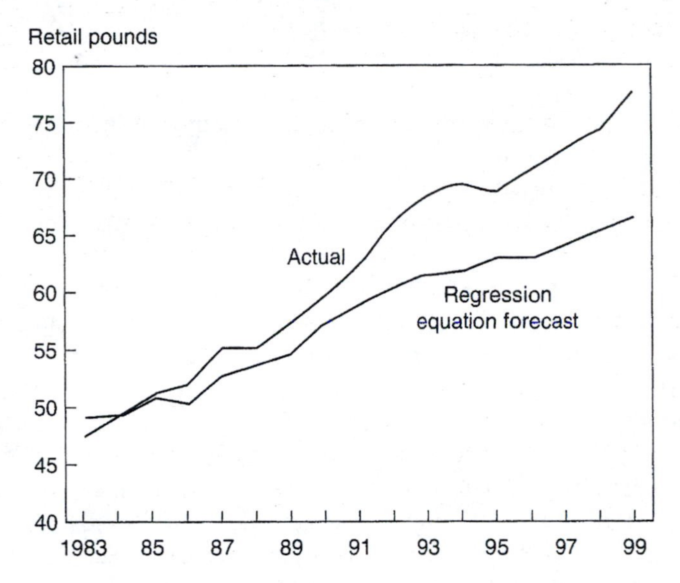

FIGURE 6.10 Actual Versus Simulated Broiler Demand.
Source: Martinez (1999).

Contracts, the Farmer, and Market Power

Many have claimed that processors gain market power when they exert vertical control through marketing or production contracts. If you want to become a hog grower in North Carolina, you will want to grow under a production contract. You could be an independent hog producer, but there are very few markets for independent producers to buy and sell hogs in North Carolina. As far as the authors know, there is only one independent hog producer of substantial size in the entire state of North Carolina. And if you want a production contract, there are only two processors in the state who enter these contracts: Smithfield Foods and Premium Standard Farms. There are hundreds of hog producers, yet only two processors. And at the time this was written, the two firms were considering merging. Clearly, this bestows the two processors with some market power in the region. The poultry market in North Carolina (and many other states) has similar characteristics. Thus, these markets are best described as an oligopsony: many sellers but only a few buyers.

In Chapter 4 we saw that an oligopsony results in a lower price for the seller compared to perfect competition. This might imply that hog and poultry producers in North Carolina would be better off if government encouraged competition by regulating the presence of the food processors. Yet, Chapter 4 also cautioned against drawing such quick conclusions. As we have seen, the use of contracts by processors to extend vertical control results in lower production costs and higher quality. This means there are more profits in the industry and therefore more profits than can be shared with the hog and poultry producer. Of course, this does not imply that the greater profits will be shared with the farmer, but the possibility that it can suggests we should not be too quick in condemning the processor's market power. If you revisit Figure 4.12 from Chapter 4, it will remind you that a movement from perfect competition to oligopsony can increase producer surplus if it increases demand.

From a conceptual point of view, vertical control through vertical integration and production contracts may or may not lead to greater market power. In these cases, we must collect data and see what it reveals. A researcher at Rutgers University collected data on market power for 38 food industries, along with data on the extent of vertical integration within those industries. Using regression techniques discussed in the next chapter, the researcher sought to determine whether industries with greater vertical integration also tend to display greater market power. No such correlation was found. If industries pursue vertical integration to attain market power, the data suggest they are unsuccessful (Bhuyan 2005).

A far more contentious issue involves the use of marketing contracts in the live-cattle industry.[2] Cattle buyers and sellers frequently enter contracts in which sellers agree to sell a certain number of cattle at a later predetermined date. The price at that date may be a fixed price, but it is usually tied to market price at that later date. The cattle market is an oligopsony: many cattle sellers but only a few buyers. Cattle under such contracts are referred to as *captive supplies*, meaning they have already been sold, but the actual exchange will take place later. From 1999–2001, the four largest live-cattle buyers procured 32% to 43% of their cattle through captive supplies.

It is clear why cattle buyers and sellers would want to enter such contracts. Beef processors operate huge processing facilities that can process thousands of cattle in a single day. These facilities process cattle at low cost, as long as they operate close to capacity. However, their costs skyrocket if operated under capacity (slaughtering few animals relative to the plant's capacity). Profits hinge crucially on the ability of cattle buyers to procure enough cattle to operate efficiently. Obtaining these cattle ahead of time through captive supplies greatly reduces the risk of operating under capacity. Sellers have incentives to enter these marketing contracts as well because it allows them to plan their sales in advance.

So far, captive supplies seem like a win-win situation, but many cattle producers would disagree. The problem lies in the fact that many of these marketing contracts tie the price of the cattle under contract to the spot price (the price of cattle not under contract sold in the market) of cattle at that later date. The cattle buyers who acquire cattle through captive supplies also acquire cattle through the spot market. Because there are few buyers, the buyers can procure much of their supplies through these marketing contracts and then do their best to negotiate low prices in the spot market later, achieving low prices for all their cattle. Indeed, economists have discovered that the use of these contracts can (but not necessarily will) bestow

---

[2]Live-cattle are cattle that are ready to be sold to the beef processor and slaughtered. They are also referred to as fed-cattle.

the buyer with greater market power and allow them to depress cattle prices (Xia and Sexton 2004). To determine if captive supplies really depress cattle prices, economists analyzed data on cattle trades and concluded an increase in captive supplies does lead to a decrease in cattle prices, although the decrease is small (Ward et al. 1996).

The captive supply issue is far from settled and has led to a lawsuit of Tyson Foods by a group of cattlemen, with no clear verdict at the time this chapter was written. Pressure from some cattlemen even led to proposed legislation, which would make it illegal for beef processors to own or control livestock 14 days prior to slaughter. This was called the Johnson Amendment, and although it received much attention, it never passed. Even if captive supplies bestow buyers with the ability to depress cash prices, the impact is small, and eliminating captive supplies impedes beef processor's ability to ensure that their plants operate near capacity. If the plants cannot operate near capacity, their costs rise, leaving them less money to pay for cattle. It is entirely possible that the Johnson Amendment would lead to lower cattle prices even if it eliminated captive supplies.

Many states still harbor fear of production contracts and vertical integration or any ownership of the live animal by the meat processor. Iowa currently has a ban on packer ownership of livestock. Meat-processing plants cannot own the animals they slaughter, unless it reaches an agreement with the Iowa attorney general. Two such processors have reached an agreement: Cargill and Smithfield Foods. Smithfield Foods is allowed to own animals it processes, as long as it purchases at least 25% of its hogs through the spot market (Arnot and Gauldin 2006).

COOPERATIVES

As the farmer's share of the food dollar has declined over years, some have postulated that the reason is a loss in farmers' negotiating power with food processors. As we have seen, many agricultural markets are easily described as a large number of farmers and relatively few buyers of the farm product. Thus, it is easy for farmers to see their share of the food dollar decline and conclude that it is being taken by food processors, wholesalers, and retailers.

However, we have also seen there are other plausible reasons why the farmer's share of the food dollar has declined. The most obvious reason is higher food-processing costs as consumers demand more processed and convenient food. Still, this farmer perception has led to legislation aimed at giving farmers greater bargaining power. Specifically, farmers are allowed to "collude" in ways other businesses cannot. The Sherman Act of 1890 prevents the formation of monopolies, with some exceptions. A firm can exist as a monopoly if it derives its market power by producing at the lowest cost or providing a superior product. Following the Sherman Act was the Clayton Act of 1914, which provided more specific details on what type of activities constitute unfair business practices. The Clayton Act prohibits tying contracts, which are contracts that prevent purchasers from buying a rival's product. It also made it illegal for one person to belong to the board of directors of more than one firm. Mergers between companies that stifle market competition were made illegal. The Clayton Act made charging different consumers different prices illegal, unless those price differences reflect different costs or product quality. Laws intended to prevent the formation of monopolies, except in cases where firms are superior due to lower costs or product quality, are referred to as antitrust laws. Of those activities that antitrust laws forbade, price-fixing is one. Price-fixing occurs when individual firms under different ownership communicate and agree on what prices should be charged. Firms are not allowed to meet and agree to set identical prices. Instead, their price decisions must be determined with no communication between firms.

Now, back to the perceived problem of low negotiating power for the farmer. The government felt it desirable to allow farmers to coordinate in order to negotiate better prices. Government wanted to give farmers the right to price-fix, but this would violate the antitrust acts. Thus, new legislation was needed. The Capper-Volstead Act was passed in 1922 to protect farmers from antitrust acts. This act is commonly referred to as the "*Magna Carta* of farmer cooperation." The act allowed farmers to form a co-op, or cooperative, in which they may improve their bargaining power by communicating and coordinating their activities. It allows farmers to pool their production and set common prices—price-fixing. Farmers, in short, can behave like a monopoly. The Capper-Volstead Act was not intended to give farmers monopoly prices; its intention was to give farmers the market power to negotiate "fair" prices. In spirit with the antitrust acts, as Pasour and Rucker (2005) state, it does not allow cooperatives to accrue "undue price enhancements." That is, the law wanted to improve farmers' bargaining power but not give the co-op so much power that it started exhibiting excess market power itself. However, Pasour and Rucker (2005) also warn that enforcement of this principle is the responsibility of the USDA and not the Department of Justice. The USDA is largely a pro-agricultural producer agency and has not once charged cooperatives as guilty of receiving these undue price enhancements.

Farmers cannot just get together and start price-fixing. There are set rules on how the cooperation can be formed. To understand these rules, we first need to review the basic forms of business organizations allowed in the United States. First, there is a *proprietorship*, where one person owns and controls the firm. The business is not taxed, but the owner's income is taxed. Once the owner dies, the proprietorship ceases to exist. Next, there is the *partnership*, where two or more individuals own and control the business. As with proprietorships, the business is not taxed but the income made by the partners is taxed. Third, we have the corporation, which accounts for most of the business income generated. A corporation is legally a person, even though it is only an organization. That is, the United States views the Microsoft company as an artificial person. This artificial person is held responsible for debts and lawsuits, and the income generated by a corporation is taxed like a person. If you own stock in a corporation, you are a part owner of the organization, and income generated by the corporation is taxed twice. It is taxed first at the corporation level—because the corporation is a person—and second when income is given to you in the form of dividends. The corporation exists—in the legal world—perpetually. Stockholders may die, but the corporation does not.

Even though a corporation is legally viewed as a person, it is not a person and therefore cannot make decisions. The people owning the corporation make the decisions. Corporations are owned by the stockholders. Specifically, the holders of common stock (the other form of stock is preferred stock, which is not covered here), hereafter referred to as stockholders, run the company. Corporations are run by a chief executive officer, or CEO. The CEO is elected either by a board of directors (who are elected by the stockholders) or by vote of stockholders. A major advantage of a corporation is its status as a legal person. If sued or bankrupt, the corporation, not the stockholders, is held responsible for damages. If you own stock in the corporation Monsanto and Monsanto goes bankrupt, courts can go after assets owned by Monsanto, but they cannot take your house or your car. By investing in a corporation, you can only lose the amount you invested. Contrast this with a sole proprietorship or partnership. If you invest $100,000 in your own proprietorship and are sued, the courts can take that $100,000 plus the amount the judge or jury deems fair. If you invest $100,000 in a corporation, the most you can lose is the $100,000. Between the extremes of sole proprietorship and corporations are limited liability corporations (LLCs). LLCs are not subject to double taxation and shareholders can only lose the amount of money invested in the LLC. However, because there are few rules dictating how LLCs should be run, many investors are hesitant to invest in LLCs, limiting the amount of capital an LLC can raise.

Back to cooperatives. Cooperatives are corporations, a special type of corporation. Owners of agricultural cooperatives must be engaged in agricultural production and are referred to as members instead of stockholders. A nonagricultural cooperative can be created, and many exist. But if individual for-profit firms want to coordinate their actions through a cooperative and avoid antitrust regulation, they must be firms involved in agricultural production. Anyone who does business with a cooperative is called a *patron* rather than a customer. Cooperatives being a corporation, the owners hold stock in the cooperative. Owners of the corporation run the corporation. Though much of the business activities are delegated to the CEO to control, the CEO is ultimately responsible to the stockholders or patrons. The stockholders and patrons make decisions through votes, or by voting in a Board of Directors. In normal corporations, the owner has a number of votes equal to the number of shares owned. In most cooperatives, each member is entitled one and only one vote, no matter the number of shares she owns.

Corporations compensate stockholders through dividends. In a regular corporation the stockholder may receive an unlimited amount of dividends. Indeed, the corporation is charged with giving the stockholder the greatest amount of dividends possible. However, a cooperative is not allowed to distribute dividends that provide members with a rate or return greater than 8%. Instead of generating profits for the members, the cooperative is supposed to transfer those profits via lower input prices or higher output prices. Income generated by normal corporations are taxed once at the corporate level and again when issued to stockholders as dividends. It is here where the cooperative structure has a unique advantage over the normal corporation. Income generated by cooperatives are not taxed at the cooperative level and are taxed only when given to the members.

Recall that the intent of cooperatives is to give the farmer greater negotiating power over its prices. Many farmers perceive that they pay higher prices for their inputs and receive lower prices for their output than they should. They perceive that they pay high input prices and receive low output prices, giving their input suppliers and their customers undeserved profits. If only they could buy inputs from nonprofit organizations or sell their output to nonprofit organizations, their profits would be higher. That is exactly the point of a cooperative—to exist as a nonprofit organization serving the farmer. That is why cooperatives are not allowed to generate a rate-of-return greater than 8%, to ensure it behaves as a nonprofit entity.

The idea of a cooperative is to act as a nonprofit organization serving to enhance farmers' profits, and there are numerous ways to enhance farmers' profits. You might be familiar with the Ocean Spray brand name, a leading seller of juice drinks. Ocean Spray is a cooperative, owned by more than 650 cranberry growers and 100 grapefruit growers. Without the Ocean Spray cooperative, these fruit growers would harvest their cranberries and grapefruit and sell the produce to a for-profit food processor. This food processor would process the fruit into a drink and sell it to the consumer. Instead, the Ocean Spray cooperative assumes the role of the food processor. Farmers deliver their harvest to the cooperative, and the cooperative makes the fruit drink and sells it to the consumer. This is an example of upstream integration, where the farmer also assumes the role of processing upstream in the food marketing channel. The cooperative then takes the revenue, pays its costs, and returns the remainder to the farmer as a price paid for the fruit. The cooperative itself seeks no profits. It seeks to deliver all possible profits that can be captured at the juice-processing stage to the farmer in the form of higher farm prices.

Consider another example. Virtually all farmers purchase fertilizer, and fertilizer can be expensive. In the absence of a cooperative, the farmer would purchase the fertilizer from a for-profit fertilizer company. Presumably the company makes profits, otherwise it would not exist. The farmer views the profits as money taken from her by excessively high fertilizer prices. In response, various cooperatives have formed to sell fertilizer directly to farmers at cost. Medford, Wisconsin, is home to the Medford Cooperative, which sells seed, grain, agronomy services, and fertilizer to the farmer. Because the cooperative is not allowed to make profits, it returns those potential profits back to the farmer in the form of lower input prices. Cooperatives can also provide a service. Roughly one-third of all agricultural loans are made through the Farm Credit Cooperative, which is a farmer-owned financial cooperative institution.

At first, the idea of a cooperative sounds appealing. We have a food marketing channel, consisting of various firms in the farm input, farming, food-processing, wholesaling, retailing, and farm service industry. Each firm makes profits. Otherwise, it would not stay in business. Each of these firms either sell the farmer something or purchase something from the farmer. If the firm went from profit to nonprofit status, it would have to either lower the price at which it sells to the farmer or raise the price at which it purchases farm products. Either way, farmers' profits rise. Following this logic, it is in the farmers' interest to replace all firms in the food marketing channel with a farmer-owned cooperative. This way, all profits in the food industry will belong to the farmer and the farmer alone.

As you might suspect, although there is some legitimacy to this argument, there are also some flaws. Profits do not just fall from the sky. Profits are earned. Often, firms that earn above-normal profits do so because they are run by above-average managers or have possession of unique fixed assets, such as a unique and efficient production process. This allows superior firms to produce a superior product, the same product at less cost, or both. The idea behind a cooperative is to enter an industry in which firms are making profits, perform the same function as those firms, but divert those profits to the farmer-owners. As an example, suppose a for-profit firm in the fertilizer sales business is making profits. This firm is owned and operated by a single person, named Selah, who has a particular talent for her field. She sells bags of fertilizer for \$2.00 each. Of that \$2.00, \$1.80 goes to paying her costs and the remaining \$0.20 are her profits. Her managerial talent allows her to produce and sell fertilizer at less cost than other firms and earns her substantial profit. Farmers see Selah's profits and decide to create a cooperative to produce and sell fertilizer directly to farmers at cost.

However, just because Selah produces fertilizer at \$1.80 per bag does not mean the cooperative can also. Selah has worked for many years in her field and has developed substantial human capital. She is good at what she does, far better than any cooperative manager who could be hired. The fertilizer cooperative opens for business, but finds it cannot compete with Selah. Even doing its best, it costs the cooperative \$2.10 to produce a bag of fertilizer. Though the cooperative is a nonprofit entity, and Selah is in it for the money, Selah sells fertilizer for less. The cooperative decides the only way to compete is to hire Selah to run the cooperative. Yet, the compensation required to induce her to leave her own business and run the cooperative must equal the profits she makes in her for-profit business. Selah can run the cooperative and produce fertilizer at \$1.80 per bag, but she must be compensated at least \$0.20 per bag to run the cooperative. In the end, farmers pay the same amount (\$2.00) for fertilizer regardless of whether they purchase from a for-profit business or a nonprofit cooperative.

Again, the point is that profits rarely fall like manna from heaven. They are earned and are usually earned by people with special skills. Profits made by a firm are earned through these skills; profits are payments for these skills. If these talented people went to work for a cooperative, they would still need to be compensated for their talents. Otherwise, they would return to their for-profit business where they earn more money. The word *profit* itself can be deceiving. Profit is often just a payment to managerial expertise, and if you eliminate the profit, you often either raise costs or fail to compete in the marketplace. Just because a cooperative does not make profits does not mean it can compete or serve the farmer better than a for-profit firm.

Cooperatives can be organized in a number of ways. Many times the organizational structure makes it difficult for the cooperative to compete with for-profit firms. For these reasons, some agricultural economists believe that cooperatives generally cannot compete with for-profit firms (Garoyan 1983). According to Pasour and Rucker (2005), the major motivation for the formation of cooperatives is its tax advantages (income is only taxed once for cooperatives, but twice for the normal corporation) and not its nonprofit orientation.

Yet there are settings where even if a cooperative cannot compete with for-profit firms, cooperatives can still benefit farmers by promoting market competition. Let us return to our example of Selah, the fertilizer manufacturer. Suppose that much of Selah's $0.20 per bag profits is not due to her exceptional managerial ability, but due to the fact that she is the only seller in the area. Her market power allows her to charge high prices and reap large profits. But when the cooperative enters the market, she now faces competition and must reduce her fertilizer price in order to compete. Her profits fall and so do fertilizer prices. This is referred to as the "yardstick of competition." If farmers' profits are low due to market power on the part of input suppliers and buyers of farm products, the presence of a cooperative spurs competition, reduces market power, and can increase farm profits. Recall the reason government allows the formation of cooperatives: to provide farmers with negotiating power to fight back against market power. If farmers do face market power from their sellers and input suppliers, then cooperatives may indeed improve the farm community. Because the cooperative in this case promotes competition, it would improve societal welfare as well. But if farmers do not face market power obstacles, we believe the general consensus is that cooperatives will only benefit farmers due to their tax advantages or may not benefit them at all. It all boils down to how profits are earned. If a firm earns profits through market power alone, cooperatives may benefit farmers. But if profits are earned due to exceptional managerial ability, the ownership of a patent, or anything that makes the firm "special," it is unlikely a cooperative will benefit the farmer.

MULTI-SECTOR MODELS

The simple supply and demand model in the previous chapters can only accommodate one sector of the food marketing channel. The price and quantity refer to one particular sector. It might be the price and quantity of harvested cotton at the farm, or clothes made from the cotton sold at Old Navy. Because the purpose of this chapter is to encourage thinking about the entire food marketing channel at once, we wish to expand this simple model to incorporate multiple sectors. As an example, let us consider the beer industry. Pure beer is made from barley, hops, yeast, and water. Some other beers may add wheat or rice, but basic beer includes only these four ingredients. Beer processors or brewers like Anheiser-Busch purchase barley and hops from farmers and water from various sources and maintain yeast colonies themselves. The sugar in barley is extracted through a malting process, mixed with water, and then yeast is added to the mixture. The yeast convert the malt sugars to alcohol via fermentation. Hops are added at various points for its taste and preservative properties. The fermented beer is then bottled and marketed to wholesalers, which in turn market the beer to retailers, which in turn market and sell the beer to consumers. Of course, the beer brewers take an active role in marketing to consumers as well. For simplicity, we will ignore the wholesalers and retailers and treat the industry as if the beer brewers sell directly to consumers.

There is a supply and demand for beer. The beer processors are the suppliers and beer drinkers make up the demand. There is a supply and demand for barley as well; barley farmers make up the supply and brewers comprise the demand. This section will now invoke two assumptions that allow us to create one diagram illustrating both the supply and demand for beer and barley. The first assumption is fixed proportions technology, which means that for every unit of good produced there is a fixed proportion of input used. In our case, this implies that for every gallon of beer produced a fixed amount of barley is used to produce that beer. This is a valid assumption for the brewing industry, where a certain amount of barley is needed for good tasting beer—no more, no less. With this assumption, if we know retail output, we know farm output. Let farm output (bushels of barley) be denoted Q^F and retail output (gallons of beer) as Q^R. Under the fixed proportions technology, we can state $Q^F = \alpha Q^R$ where α is a constant number. Farm output is then just a fraction of retail output, and $Q^{F*} = Q^F/\alpha = Q^R$ is retail-equivalent farm production.

For example, the authors brew beer at home. Every beer recipe involves roughly 1.5 lbs. of barley for each gallon of beer brewed. Here, $\alpha = 1.5$. If we know that 3 lbs. of barley are used, we know that 2 gallons of beer were brewed. If 2 gallons of beer were brewed, 3 lbs. of barley must have been used. If we know retail output, we know farm output exactly. By invoking this assumption, we can graph retail and retail-equivalent farm production on the same axis, as in Figure 6.11.

The second assumption is that the marketing bill, the cost of taking barley and transforming it into beer, on a per unit basis is constant. Sometimes this assumption is valid and other times it is not, but its use greatly simplifies our supply and demand diagram. Let the per unit marketing bill be denoted *MB*. The consumer demand for

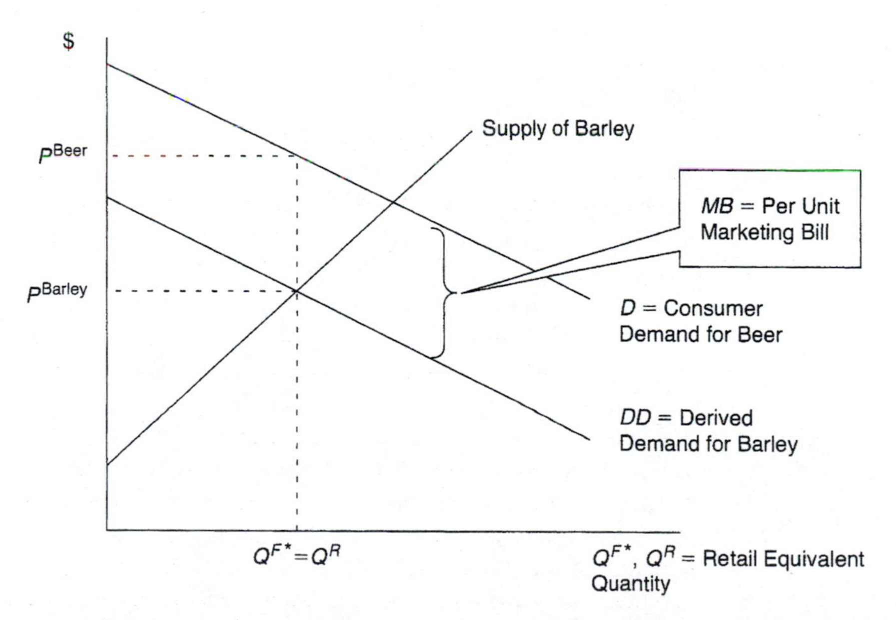

FIGURE 6.11 Consumer Demand for Beer and Derived Demand for Barley.

beer is shown in Figure 6.11. The demand for barley is derived from this demand curve, so we call it the derived demand curve. The brewers do not care for the barley per se, they only want it because they can make it into beer and sell it to consumers. If we know the consumer demand for beer and the marketing bill, we know the derived demand for barley. The key is remembering that a demand curve represents maximum willingness-to-pay. Suppose that the maximum amount consumers will pay for beer is \$1.00 per Q^R, and the marketing bill is \$0.30. If barley was free, brewers could take the free barley, make beer selling it for \$1.00, leaving them with \$1.00 − \$0.30 = \$0.70 in profits. Thus, the maximum amount brewers are willing to pay for barley is \$0.70 per Q^{F*}.

To further illustrate the point, suppose barley costs \$0.50 per Q^{F*}. Profits per beer would then be \$1.00 − \$0.30 − \$0.50 = \$0.20 per Q^R. Profits are positive, so the brewer would purchase the barley to make and sell beer. Now suppose barley cost \$0.69 per Q^{F*}; profits per beer would be \$1.00 − \$0.30 − \$0.69 = \$0.01. Profits are still positive, so the brewer would still purchase barley to make and sell beer. If barley costs rise to \$0.70, profits per beer would be zero, and the brewer would not purchase the barley. The absolute maximum the brewer will pay for barley is \$0.70 per Q^{F*}, which equals the maximum willingness to pay for beer minus the marketing bill. Thus, the derived demand curve for barley equals the consumer demand curve for beer minus the marketing bill, as shown in Figure 6.11.

Let us assume that there are enough buyers and sellers of barley to model the barley market as perfectly competitive. The supply curve for barley is drawn in the figure and is stated in retail-equivalent units. The barley market equilibrium is where the barley supply curve and barley derived demand curve cross. The barley price is P^{Barley}. Given that retail output is $Q^{F*} = Q^R$, the price that induces consumers to purchase this much beer is P^{Beer}, giving us the result that the retail price equals the input price plus the marketing bill: $P^{\text{Beer}} = P^{\text{Barley}} + MB$.

This model is useful because it allows us to analyze how changes in one sector of the food marketing channel affect the entire channel. Suppose that the marketing bill increases, perhaps because the government places higher taxes on beer brewers. The consumer demand curve stays put; consumer demand is based on people's enjoyment of beer, not taxes on brewers. The larger marketing bill causes the derived demand curve to fall lower underneath the consumer demand curve. The new barley market equilibrium is at a lower price and quantity. The lower quantity of barley implies less beer is sold, driving up the beer price and raising the retail price of beer. The illustration in Figure 6.12 depicts this story. The circles designate the old equilibrium and the triangles designate the new. The quantity of barley and beer falls; the barley price falls; the beer price rises. This should make sense to you. If it becomes more expensive to take barley and produce beer, barley is less valuable and its price falls. As beer becomes more expensive to make, less beer is brewed and the price consumers pay rises.

Now consider the effect of an increase in beer demand. An increase in consumer demand makes barley more valuable, so the derived demand for barley rises, as depicted in Figure 6.13. The consumer demand curve rises, but the marketing bill remains the same, so the derived demand curve rises just as much as the consumer

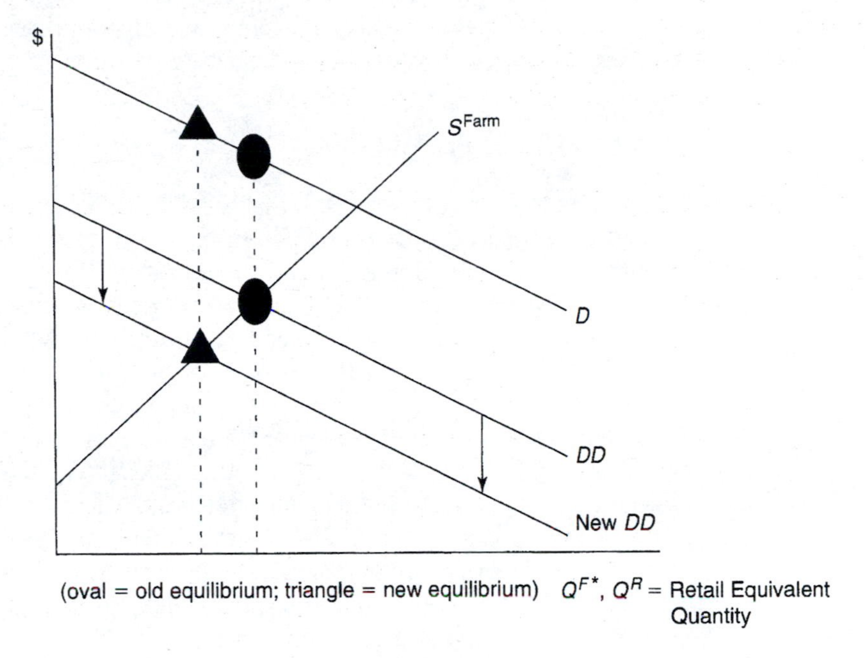

FIGURE 6.12 Increase in the Marketing Bill.

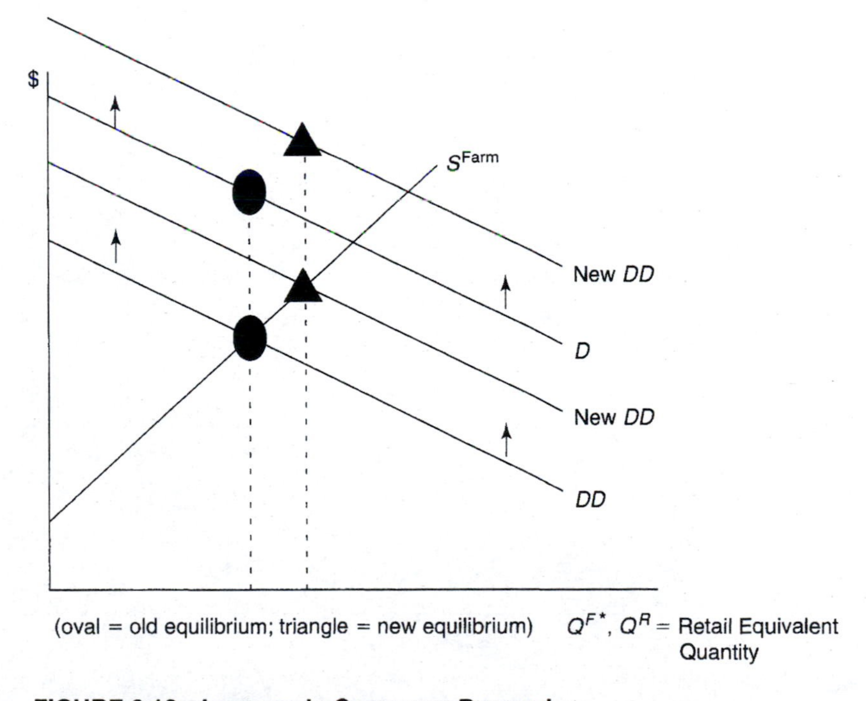

FIGURE 6.13 Increase in Consumer Demand.

demand curve. The new barley equilibrium is at a higher price and quantity, and the beer market is also at a higher price and quantity.

An Alternative Interpretation

The model above provides a key result that the difference between the farm price and the retail price equals the marketing bill. Invoking this result allows us to link retail and farm markets in a simpler diagram like that in Figure 6.14. The marketing bill creates a "wedge" between retail and farm prices. At market equilibrium, the retail-equivalent farm quantity Q^{F*} and retail quantity Q^R must equal, and the $P^{\text{Retail}} - P^{\text{Farm}} = MB$, where MB is the marketing bill. This equilibrium can be easily found as follows. First, draw a diagram with a supply curve for the farm product and a demand curve for the retail product. Create a "wedge" in the diagram—a vertical line whose height exactly equals the marketing bill MB. Then, find the unique point in the diagram where the top of the wedge touches the demand curve and the bottom touches the supply curve. This is the market equilibrium. The point on the demand curve is the retail price, the point on the supply curve is the farm price, and by construction the difference between the two prices equal the marketing bill. Moreover, at this point both Q^{F*} and Q^R equal, so we know the market is indeed in equilibrium. The quantity demanded of beer exactly equals the quantity supplied of barley, where barley quantity is expressed in retail equivalent units (i.e., as the gallons of beer the barley will produce).

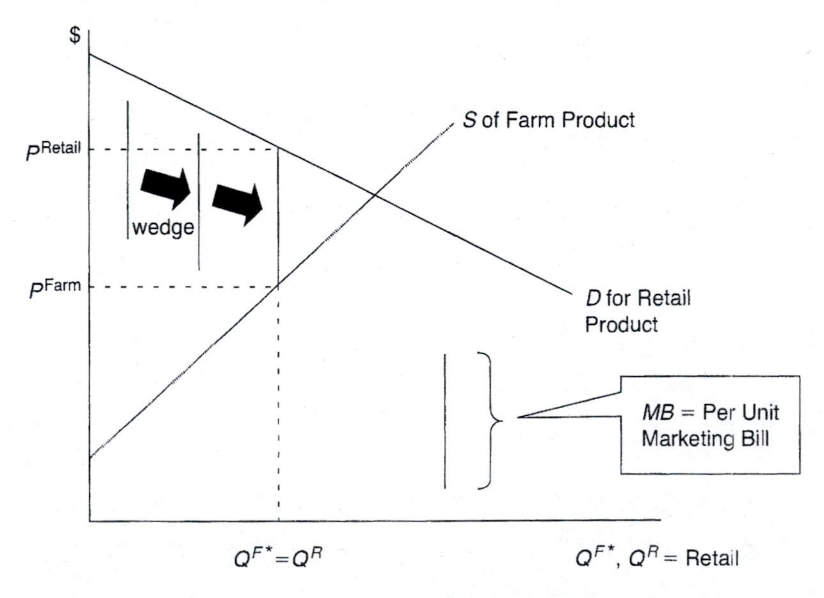

FIGURE 6.14 Multi-Sector Model Equilibrium.

Cost Transmissions Along the Food Marketing Channel

Each decade ushers in new food safety regulations, forcing processors to spend more money on sanitation, disease prevention, disease detection, and similar activities. This is not to say the processors do not undertake voluntary efforts to ensure food safety, only that regulations require them to spend more on food safety than they would voluntarily. Policymakers take food safety regulations seriously, never tightening restrictions unless they estimate the benefits of safer food to outweigh the costs. It is common for people to view regulations on food processors as affecting only the profits of those processors, but in reality, some of the costs imposed at the processing level are transmitted downward to the farmer and upward to the consumer. All sectors of the food marketing channel are linked.

To understand this link suppose a new food safety regulation increases the cost of processing food. The cost of taking the farm product and creating an edible food product is now greater, so the marketing bill now rises. As the wedge in Figure 6.15 becomes larger, the retail price will rise, the farm price will fall, and the equilibrium quantity will fall. The entire food marketing sector is impacted by the food safety regulation, as processors face higher costs, consumers pay higher prices, and farmers receive lower prices. Food processors do not pay the regulation costs alone; the costs are shared with consumers and farmers. But between consumers and farmers, who is most impacted?

The answer is that it depends on elasticities of farm supply and consumer demand. Figure 6.15 shows the consequences of a larger marketing bill on the market equilibrium. The left-hand diagram illustrates a situation where demand is more elastic than supply, and the right-hand diagram illustrates the opposite. Consider the left-hand diagram first. When the marketing bill rises, consumer prices rise only a little but farm prices fall by a much larger proportion. In this setting the farmer feels

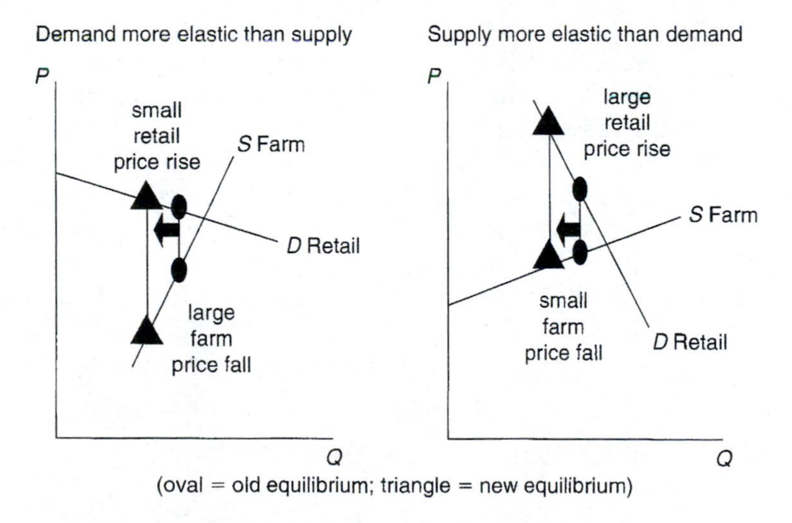

FIGURE 6.15 Impacts of Larger Marketing Bill.

greater impacts of higher food-processing costs. Yet if farm supply becomes more elastic than consumer demand (right-hand diagram), the farm price fall is modest but the consumer price rise is steep. Consumers are harmed more than farmers in this setting.

Whoever has the least elastic supply or demand pays most of the increase in marketing bill. This produces an important point. When the cost of food production rises, regardless of the specific sector who directly pays the cost, all sectors (including consumers) of the food marketing channel are affected, and those sectors with the least elasticity are affected the most. Remember from earlier chapters that elasticity is determined by the number of substitutes available. If consumers have many close substitutes for the good of interest, consumer demand will be elastic. If producers have alternative means for earning money, supply will be elastic.

SUMMARY

Take a bite out of a burrito and you are tasting the fruits of labor from hundreds of individuals. From miners of potassium deposits, pesticide manufacturers, corn farmers, wheat farmers, cattle producers, beef processors, flour millers and bakers, dairypersons, cheese manufacturers, to Taco Bell (and we have left out many people), they all played a part in that 60-cent burrito. Even though some individuals add more value than others, they are all necessary. They all add some form of utility, whether it be form, time, place, or possession utility. This chapter was concerned about painting a holistic picture of this process—the food marketing channel.

It takes a well-functioning economy for hundreds of people to coordinate their activities to produce one burrito. The United States has such an economy, consisting of a balanced blend of government and market activity. The food coordination system is even improving. The use of grid pricing, vertical coordination, and vertical integration seems to have improved our meat quality while lowering prices at the same time. Yet the food marketing system contains some contentious issues. There are many farmers selling to a relatively few number of buyers. Thus far, there is little evidence for concern, but the issue still weighs heavily on some minds.

Perhaps the most important point of this chapter is that within each sector, each person within the food marketing channel impacts the other. One should not look at changes at the farm level without considering the impacts at the processing, wholesaling, and retailing level. Similarly, changes in consumer behavior will find their way back to the farm. A cattle producer cannot assume that taxing the beef consumer will leave her unharmed. A beef consumer cannot think vice versa. Just as the quarterback relies on the offensive line, the running backs, the receivers, and the defense, each sector in the food marketing channel relies on the other sectors.

CROSSWORD PUZZLE

For answers with more than one word, leave a blank space between each word.

Created with EclipsoCrossword—www.eclipsecrossword.com

Across

3. The act of coordinating activities between sectors along the food marketing channel is referred to as _______ coordination.
6. Smithfield Foods delivers Dave live animals and feed, and pays Dave to raise the animals in production facilities owned by Dave. Dave has a _______ contract with Smithfield Foods.
8. The cost of transforming a raw farm product into a consumer item.
9. A market where the price is negotiated and the exchange takes place immediately.
10. The _______ demand curve is created by taking the consumer demand curve and lowering it on the *y*-axis by an amount equal to the marketing bill.
11. Cattle that are under contract for sale to a specific buyer before the exchange takes place are referred to as _______ supplies.
12. The _______-_______ Act of 1922 allowed the formation of cooperatives.

Down

1. A form of corporation that is owned by farmer-members, cannot earn a rate-of-return greater than 8%, and is taxed at a lower rate than normal corporations.
2. Gwenn signs a contract with Smithfield Foods to deliver 5,000 hogs in two months at a price of $65/cwt. This is an example of a _______ contract.
4. Gibson Foods is a pork processor who decides to build and operate its own farms in addition to its processing facilities. This is an example of _______ _______.
5. The four types of utility from purchasing and consuming a good are (1) form utility, (2) time utility, (3) place utility, and (4) _______.
7. In the beef industry, _______ pricing refers to the use of a table or formula to dictate how cattle will be priced based on the quality of the carcass.

STUDY QUESTIONS

1. The four types of utility discussed are form, time, place, and possession. Consider a farmer's market that is open once a week, provides organic food that some segments of food consumers prefer and cannot find elsewhere, locates five miles out of town, and only accepts cash as payment. Identify which type(s) of utility the farmer's market is specializing in providing and which type(s) it fails to provide at a level comparable to normal grocery stores.

One quality consumers seek in beef is tenderness. Other things held constant, more tender beef produces a more enjoyable eating experience. Until recently, it was difficult to determine whether a steak is tender until it is consumed. However, in the past 10 years a new machine has been developed that can test a carcass for beef tenderness, thereby allowing a processor to market "guaranteed tender beef." That is, steaks can be designated as "tender" and "nontender" before the steak is purchased. Whether cattle achieve the "tender" label is partly random, but genetics do play a role. By observing which genetics tend to produce tender beef, producers can alter their breeding decisions to increase the supply of tender beef. Assume cattle with tender beef genes can be identified with reasonable accuracy.

2. Explain how average pricing may fail to encourage the production of tender beef.
3. Explain how the grid pricing system could be used to encourage the production of tender beef.
4. Explain how an industry can utilize production contracts to encourage the production of tender beef.
5. Explain how an industry can utilize vertical integration to encourage the production of tender beef.
6. A group of cattlemen in West Texas have only two buyers of their live-cattle. The beef processors (the buyers of their live-cattle) make significant profits. The cattlemen wonder, if they purchased their own beef-processing plant as a cooperative and operated at nonprofit, whether the cooperative could purchase their cattle for a higher price than the for-profit processors. Describe under what conditions the cooperative may or may not be able to pay higher live-cattle prices.
7. Hogs are slaughtered at 250 lbs., and each hog produces around 125 lbs. of retail meat. If the hog quantity is 2,000 lbs., what is the retail-equivalent quantity (in lbs. of retail pork)?
8. One pound of raw wheat produces 0.7 lbs. of retail flour. If 120 billion lbs. of wheat is produced and made into flour, what is the retail-equivalent of this wheat production?
9. Figure 6.16 shows a consumer demand curve for flour. Suppose the cost of turning wheat into flour is $0.20 per pound of retail-equivalent wheat. That is, the

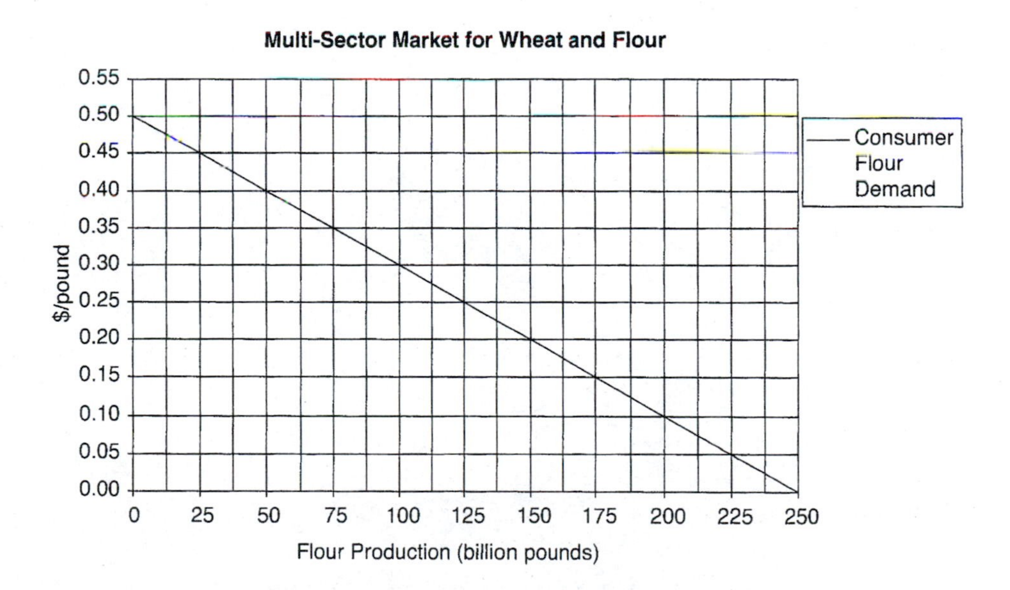

FIGURE 6.16

marketing bill for wheat made into flour is \$0.20 per pound of retail-equivalent wheat. Draw in the wheat derived demand curve and label it *DD*. Next, suppose that the wheat supply curve is given by the formula $P = \$0.05 + 0.0005(Q)$, where Q is the pounds of retail-equivalent wheat. Draw this supply curve in the diagram and label it S^{FARM}. Then, clearly label the wheat farm price as P^F, the flour retail price as P^R, and the equilibrium quantity as Q^R.

CHAPTER SEVEN

Empirical Agricultural Price Analysis

Facts are stubborn things.

—*John Adams*

INTRODUCTION

Economists are often asked to answer questions like: how will a ban on antibiotics in livestock affect pork prices, does the "Beef: It's What's for Dinner" television commercial really increase the demand for beef, and what is the best forecast of corn prices next year? Each of these questions require us to analyze data. Most often these types of questions are answered using a tool called "regression analysis." Good decision making should be grounded in facts, and the purpose of regression is to determine exactly what the facts are. The purpose of this chapter is to introduce regression analysis and illustrate how regression can be used to answer important agricultural questions. Specifically, we

1. introduce regression analysis
2. show how to convert nominal prices to real prices
3. illustrate how to estimate supply and demand curves for agricultural commodities
4. use estimated supply and demand curves to conduct policy analysis
5. show how to predict prices using time-series analysis
6. conduct hedonic price analysis

Policymakers have considered banning the frequent low doses of antibiotics administered to hogs through their feed and water. This ban would raise production costs, decreasing the pork supply by shifting the supply curve to the left. The ban might also increase consumer demand for pork, increasing pork demand and shifting the demand curve to the right. Pork prices would rise, but by how much we cannot tell unless we know the supply and demand curves and how the magnitude of each shifts.

Fortunately, we can estimate supply and demand curves, and we can estimate how much they would shift in response to an antibiotic ban in swine production.[1] Government agencies regularly compile data on livestock prices, quantities, and other items that allow us to use a technique called regression analysis to estimate the supply and demand for pork. Regression is a statistical technique used by every branch of science. This chapter will first illustrate how to conduct regression analysis. We will not cover the statistical theory behind regression or prove to you why it works. When you first learn a sport, coaches usually teach you the techniques for hitting a baseball, serving a tennis ball, or kicking a soccer ball; they do not teach you why that technique works. Similarly, we will teach you the techniques for using regression analysis, but not why it works. The "why" is left for subsequent, more advanced classes.

After using regression to estimate the supply and demand for pork, we will estimate the impact of an antibiotic ban on pork prices and quantities. Regression will also be used to determine if the "Beef: It's What's for Dinner" TV advertisements increase beef demand. This is referred to as empirical price analysis, where economic theory is combined with market data to answer important questions. The word *empirical* means based on observation and experiment, rather than theory. It is probably derived from the Greek philosopher Sextus Empiricus, a skeptic who stressed the importance of observation over belief or theory.

INTRODUCTION TO REGRESSION ANALYSIS

The idea behind regression analysis can be best illustrated using softball as an example. For three years we took students to a field and asked them to hit softballs. Underhand pitches were thrown and students were asked to try and hit each pitch as far as they could. After each hit we measured the distance from home plate to the point the ball stopped rolling. Some people hit further than others. The longest hit was 318 feet, the shortest hit was 3 feet, and the average hitting distance was 138 feet.

Why did some people hit further than others? Is there any information we can gather that will help us explain differences in hitting ability? We asked our students what they thought caused differences in hitting ability. Specifically, they were asked what kind of data could help predict hitting distance. The two most frequently noted factors were gender and experience. Our students thought that the average male can hit farther than the average female. Further, those who have many years of experience playing softball or baseball should also be able to hit further than those with less experience.

Therefore, we should be able to predict how far each student will hit based on their gender and experience. This is accomplished by constructing an equation that predicts distance based on gender and experience. For example, let us create two variables, one

[1]Note "antibiotic ban" refers to the ban on subtherapeutic antibiotic use, which is antibiotics given daily at low doses, regardless of whether the animal is sick. Antibiotics might still be given to hogs if they were sick.

measuring gender and the other measuring experience. Let *Male* be a "dummy" variable that equals one if the hitter is a male and zero otherwise. Variables that equal one or zero to indicate the presence of something are referred to as *dummy variables*. Then, let *Experience* be a variable denoting the hitter's years of experience in softball or baseball. For example, if the hitter is a male with 10 years of experience, *Male* = 1 and *Experience* = 10. Conversely, if the hitter is a female with no experience, *Male* = 0 and *Experience* = 0. A prediction equation is constructed by placing the variable being predicted on the left-hand side, and making it a linear function of the variables that help predict the dependent variable. Our equation predicting hitting distance is

$$\textit{Predicted Hitting Distance} = a_0 + a_1(\textit{Male}) + a_2(\textit{Experience})$$

where a_0, a_1, and a_2 are "coefficients" we must calculate. This equation can be written generically for any data as

$$Y = a_0 + a_1(X_1) + a_2(X_2) + \cdots + a_k(X_k).$$

In this equation, the dependent variable is denoted by Y and is being predicted using data on k explanatory variables $X_1, X_2, \ldots, X_k$. Our hitting distance equation only has two explanatory variables.

Dependent Variable: Variable being predicted by a regression equation.

Explanatory or Independent Variables: Variables used to predict a dependent variable in a regression equation.

Coefficients or Parameters: The unknown values of the regression equation calculated by a software package like Microsoft Excel.

The variable *Predicted Hitting Distance* is referred to as a *dependent variable*, because its value *depends* on the values of *Male* and *Experience*, which are referred to as *independent* or *explanatory variables*. Regression analysis is used to calculate the values of a_0, a_1, and a_2 (which we call the *coefficients* or *parameters*) that give good predictions. These parameters are easily calculated using Microsoft Excel. Excel takes the prediction equation, looks at the actual softball data, and then finds the coefficient values a_0, a_1, and a_2 that yield predictions most closely matching the data. The estimates are obtained from Excel by the following steps. First, collect and enter data for the hitting distance, gender, and experience for each individual in an Excel spreadsheet, where each row represents a different hit. Put the dependent variable in the leftmost column, and the explanatory variables in the right, adjacent columns. Each variable should enter a different column, and be sure to place the explanatory variables side by side, as illustrated in Figure 7.1. Make sure your data are entered correctly, contain only numbers, and that there is an equal number of rows with data under each column. Our data happen to begin at row 7 and end at row 319, giving us 312 observations of data.

Next, call Excel's regression routine by selecting *Tools*, then *Data Analysis*, then *Regression*.[2] It first asks you for the "*Y* Range," which are the data on your dependent variable. You give this information to Excel by placing your cursor in the textbox for "*Y* Range" and selecting the data, including the variable name. In our case, the variable name is located in cell A6, and the data are in cells A7 to A319, so we select rows A6 to A319. The "*X* Range" refers to the explanatory variables, and we give this

[2]If *Data Analysis* does not appear under the *Tools* menu, select *Tools*, *Add-Ins*, and then *Analysis Toolpack* and Excel will install it. You may need your original installation software.

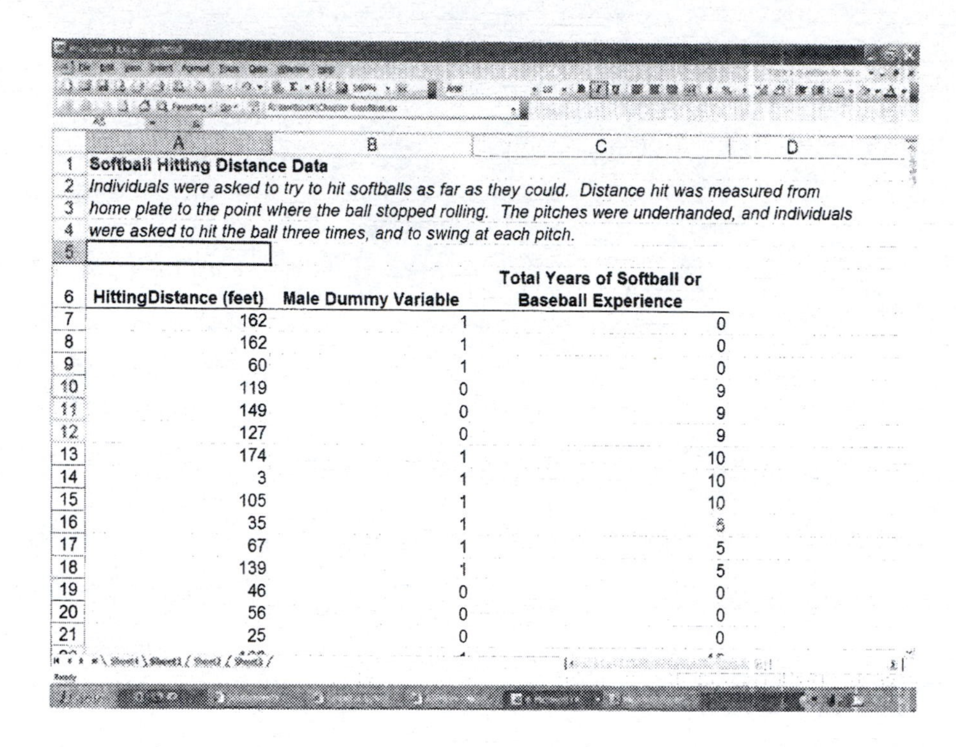

	A	B	C	D
1	**Softball Hitting Distance Data**			
2	*Individuals were asked to try to hit softballs as far as they could. Distance hit was measured from*			
3	*home plate to the point where the ball stopped rolling. The pitches were underhanded, and individuals*			
4	*were asked to hit the ball three times, and to swing at each pitch.*			
5				
6	**HittingDistance (feet)**	**Male Dummy Variable**	**Total Years of Softball or Baseball Experience**	
7	162	1	0	
8	162	1	0	
9	60	1	0	
10	119	0	9	
11	149	0	9	
12	127	0	9	
13	174	1	10	
14	3	1	10	
15	105	1	10	
16	35	1	5	
17	67	1	5	
18	139	1	5	
19	46	0	0	
20	56	0	0	
21	25	0	0	

FIGURE 7.1 Step 1 of Regression Analysis Is to Organize Your Data in Microsoft Excel.

information to Excel by selecting the data on the explanatory variables including the variable names. This selection includes cells B6 to C319. Be sure your variable name is located in the row immediately above the first data point, and make sure the name takes up no more than one row.

Finally, check *Labels* (to tell Excel your data selection includes the variable names) and click *OK*. Excel will then create a new sheet shown in Figure 7.3. As the figure shows, the values of a_0, a_1, and a_2 calculated by Excel are: $a_0 = 71.22$, $a_1 = 77.64$, and $a_2 = 2.85$. Thus, our prediction equation is

$$\textit{Predicted Hitting Distance} = 71.22 + 77.64(\textit{Male}) + 2.85(\textit{Experience})$$

This equation tells us that males hit about 77.64 feet further than females, and that each additional year of experience increases hitting distance by 2.85 feet. Excel provides two other items that are useful for regression analysis. One is the p-value. Even if there were no gender differences and females hit just as far as males, Excel would not give us a coefficient of zero for *Male*. The coefficient will always be nonzero. There is always some probability that an explanatory variable really has no impact on the dependent variable, and that probability is given by the p-value. Notice that the p-value on *Male* and *Experience* are both zero. This indicates that the probability that both variables have no impact on hitting distance is zero, which means the

Using Your Prediction Equation

A male with 6 years of experience is predicted to hit the softball: $71.22 + 77.64(\textit{Male} = 1) + 2.85(\textit{Experience} = 6) = 166$ feet.

A female with 0 years of experience is predicted to hit the softball: $71.22 + 77.64(\textit{Male} = 0) + 2.85(\textit{Experience} = 0) = 71$ feet.

A female with 20 years of experience is predicted to hit the softball: $71.22 + 77.64(\textit{Male} = 0) + 2.85(\textit{Experience} = 20) = 128$ feet.

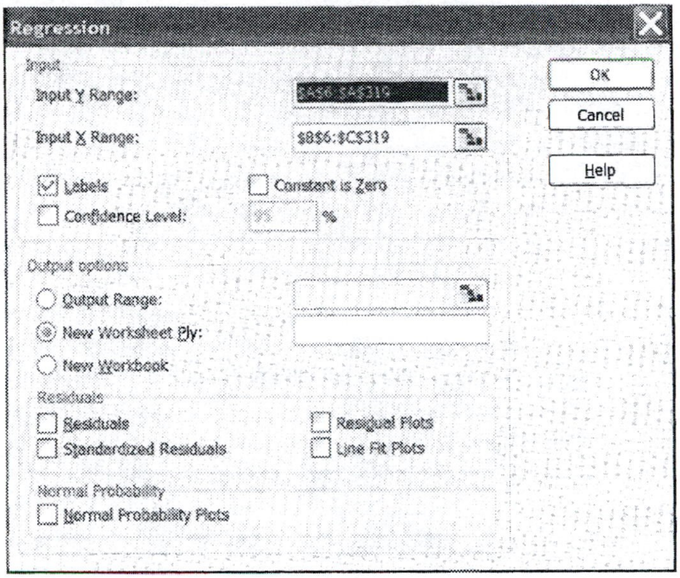

Regression

Input

Input Y Range: $A$6:$A$319

Input X Range: $B$6:$C$319

Labels

Constant is Zero

Confidence Level: 95 %

OK

Cancel

Help

Output options

Output Range:

New Worksheet Ply:

New Workbook

Residuals

Residuals

Standardized Residuals

Residual Plots

Line Fit Plots

Normal Probability

Normal Probability Plots

FIGURE 7.2 Step 2 of Regression Analysis Is to Fill in the Regression Dialog Box.

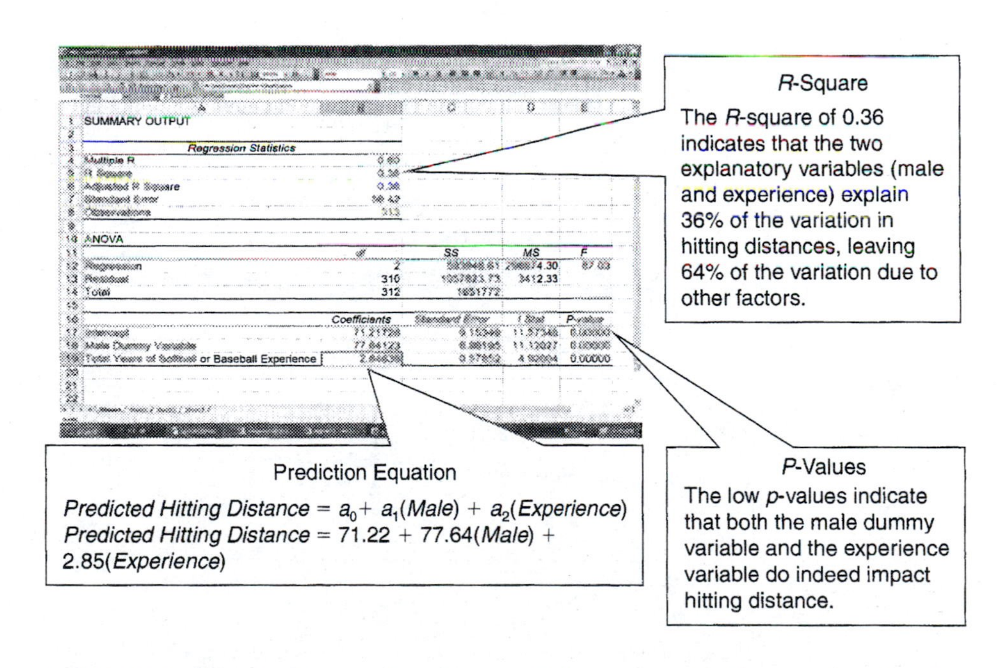

FIGURE 7.3 Step 3 of Regression Analysis Is to Interpret the Regression Results.

probability they truly impact hitting distance is 100%—both gender and experience indeed matter!

Economists do not like to conclude that one variable really impacts another unless we are very sure and the p-value for that variable is typically 5% or lower. Suppose that the p-value on *Male* equaled 0.2, meaning the probability there are no gender differences equals 20%. In this case, we would *not conclude* there are gender differences in regards to hitting distance, because we have a 20% chance of being wrong. We can live with a 5% chance of being wrong, but not 20%.

If the p-value of an explanatory variable is greater than 0.05, we conclude that variable does not impact the dependent variable.

The prediction equations given by Excel are useful, but it is important to note that they do not provide perfect predictions. The equations always predict with error, and the R-square value tells us the accuracy of the equation. In this case, our R-square is 0.36, which roughly means that the 36% of the variation in hitting distances across observations is due to differences in gender and experience. Thus, the remaining variation (64%) is due to other factors that were not included as explanatory variables. These other factors might be hand-to-eye coordination, innate strength, or simple luck.

To illustrate regression analysis, consult Figure 7.4 where actual hitting distances for the students are plotted along with the prediction equation from the regression. The dots are actual hits, while the line is the prediction for each hit. Very seldom do the dots lie on the line, suggesting that very seldom does the equation predict

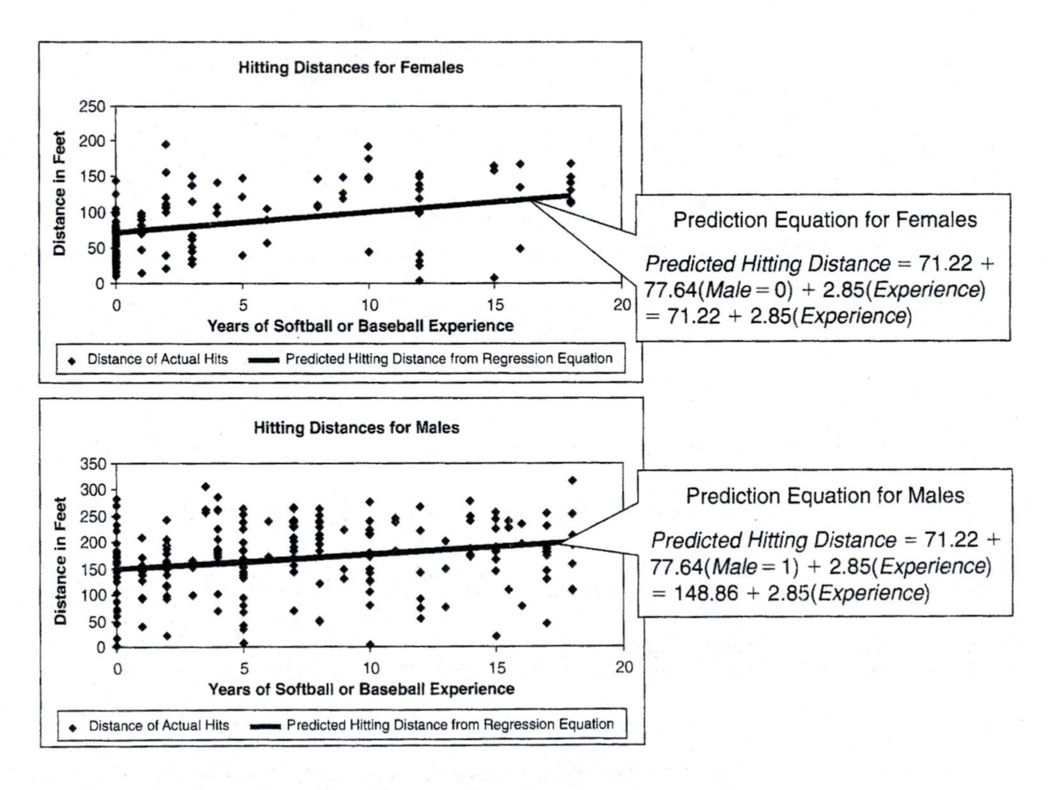

FIGURE 7.4 Predicted and Actual Hitting Distances.

perfectly. However, try to draw another line in the graph that "fits" the data better, and you cannot. The line shows the average hitting distance for each gender at each level of experience, and no other line would predict with less error.

FROM REAL TO NOMINAL PRICES

Shortly, regression analysis will be used to predict prices and identify the relationship between prices and select variables. Often, these prices change across time. For example, we may use regression analysis where price is the dependent variable (as opposed to hitting distance) and quantity is an explanatory variable (as opposed to baseball experience). However, a dollar in 1960 purchased much more than a dollar today, so we cannot compare raw (or nominal) prices across time. Nominal prices must be converted to "real" prices to compare prices across time meaningfully.

A dollar in 1960 had over six times the purchasing power compared to a dollar today. This means that if we multiply a dollar in 1960 by six, giving us \$6.00, this tells us that a dollar in 1960 is comparable to \$6.00 today. Corn was being sold for \$1.00 per bushel in 1960, which is the same as selling for \$6.00 today. Only on these terms can we compare today's dollars with dollars of the past.

Fortunately, formulas for converting dollars in different time periods to a common unit are readily available. The consumer price index (CPI) is particularly useful and freely obtained through the Internet. This is an index measuring the amount of money it takes to purchase a common basket of household goods in different time periods. The more money it takes to purchase food, homes, and cars, the less purchasing power a dollar possesses and the greater the effects of inflation. The CPI has a base year, which is usually around 1982. A base year of 1982 means only that the CPI is designed to equal 100 in 1982. If the CPI is 100 in 1982 and 103 in 1983, this means that inflation was 3% between the two years. If the CPI is 100 in 1982 and 110 in 1992, inflation over the 10-year period was 10%. Thus, the index is designed so that the percentage difference in the index between two years is exactly the inflation rate between those two years.

A simple formula exists for taking dollars across many time periods and converting them to dollars of the same year. The formula is

$$Real\ Price = (Nominal\ Price/CPI\ Base\ Year = 1982)(100)$$

The new real price is stated in the base year of the CPI index, which in this case is 1982 dollars. For example, see Figure 7.5 where nominal corn prices in 1960, 1982, and 2005 are converted to real 1982 prices. The nominal corn prices rose between 1960 and 1982, however, real prices (stated in 1982 dollars) fell. This indicates that the \$1.00 price in 1960 purchased the same amount of goods and services as \$3.27 in 1982. The nominal corn price equals the real 1982 prices because it is in the base year and the CPI equals 100 for that month. Looking at the prices we can see that selling a bushel of corn in 1960 gave the farmer more purchasing power than a bushel in 1982 or 2005. This is despite the fact that nominal prices were the lowest in 1960. In 1960, the sale of one corn bushel could purchase more goods and services than in 1982 or

Year	Nominal Price of Corn	CPI Index (Base Year = 1982)	Real Price of Corn (1982 Dollars)
1960	$1.00	30.62	($1.00/30.62)(100) = $3.27
1982	$2.54	100.00	($2.54/100.00)(100) = $2.54
2005	$2.12	201.49	($2.12/201.49)(100) = $1.05

FIGURE 7.5 Converting Nominal Corn Price to Real Corn Prices.

2005. Nominal prices rose from 1960 to 1982, but so did the price of all goods and services. After adjusting for inflation (adjusting for the fact that the price of most everything rose during this period), corn prices really fell during this time period. When comparing prices across time periods, the first and most important step is to convert all prices to real prices.

One can always change the base year. In Figure 7.5, we can easily make a new CPI whose base year is 2005. Simply divide each CPI number in the table by the CPI for 2005 (201.49), then multiply by 100. This sets the CPI in 2005 to 100, making it the base year, and adjusts the CPI in all other years accordingly. Then, using the same formula, one can calculate the real price of corn in 2005 dollars. Performing such calculations, we can see that $1.00 in 1960 is equal to (30.62/201.49)(100) = $15 today.

ESTIMATING SUPPLY AND DEMAND FOR AGRICULTURAL COMMODITIES

This book is not about softball or inflation; it is about agricultural markets. The reason regression was introduced using softball as an example is that you are probably more familiar with sports than you are with agricultural economics. You grew up playing tee ball, not supply and demand. In all likelihood, using information on gender and experience to predict hitting distance made sense to you. You know that males hit better than females—on average—and that gaining experience improves one's hitting distance. In this section we will use regression analysis to estimate supply and demand equations for pork. Just as we identified factors that influence hitting distances, we must identify factors that influence the supply and demand for pork and will therefore rely on the material covered in the previous chapters.

Prices in competitive markets are determined by supply and demand. When there is a long production lag (long time period between when production decisions are made and the final product is ready for sale), supply and demand become more complicated than two curves crossing. Recall that the production lag for pork is one year. Producers at any given time decide how many sows[3] to breed, but the sows' offspring are not ready to be slaughtered until a year later. This means that the quantity of hogs supplied today is based on production decisions made

[3]Sows are female hogs of the breeding age.

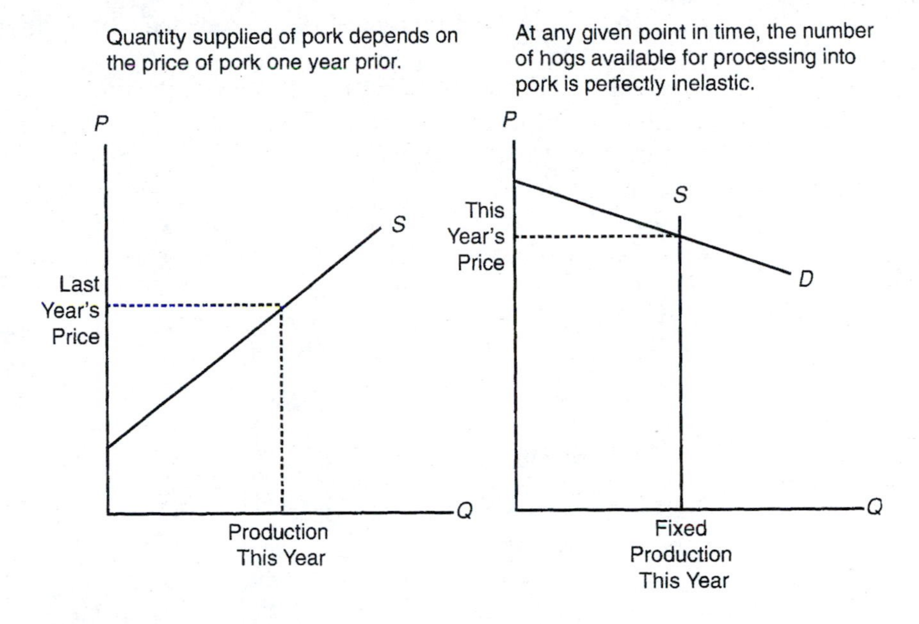

FIGURE 7.6 The Supply and Demand for Pork Given the One-Year Production Lag.

one year prior, and hence the price of pork one year prior. This relationship is illustrated in the left-hand diagram of Figure 7.6. Quantity supplied today is the point on the supply curve corresponding to last year's price. Thus, the amount produced at any given time period is caused by the price in the previous year and opportunity costs in the previous year (recall the supply curve is the marginal opportunity cost curve).

In a market with a year-long production lag

- Last year's supply curve and last year's price determines current quantity supplied.
- Current quantity supplied and the demand curve determine current prices.

After the production decision is made, the production that will result one year later is (for all practical purposes) fixed. However many sows were bred will produce a particular number of farrows,[4] and those farrows will be raised to a certain age and then slaughtered for processing into pork.[5] Regardless of how prices change during that year, the production decision was made last year, so pork production in the current year does not respond to the current price. This fixed amount of production must then be sold to consumers. See the right-hand diagram in Figure 7.6. Supply is fixed at a certain level, and the price that results is the price where this fixed supply and demand intersect. If supply is large, the resulting price is low, and if the supply is small, the price is high. Thus, the fixed quantity supplied and the demand curve determine pork prices.

[4]Farrows are baby pigs.

[5]It is true that hogs can be kept on feed for longer or shorter periods of time in response to price expectations, but the ability of producers to modify hog production within a couple of weeks or months is very limited. Hog supply in the very short run is very inelastic, inelastic enough to deem it perfectly inelastic.

In the softball prediction equation, we predicted hitting distance based on gender and experience, because gender and experience *cause* people to hit different distances. In the case of pork supply, prices and costs of production one year prior *cause* quantity supplied, so to estimate a supply curve we construct a regression model with quantity supplied as the dependent variable and explanatory variables including last year's price and last year's production costs. In the case of pork demand, price is *caused* by the intersection of the demand curve and the fixed supply. Thus, demand is estimated using a regression with price as the dependent variable and explanatory variables including the quantity supplied and factors that shift the consumer demand curve. In what follows, we will estimate the pork supply and demand equations, illustrating how they can be estimated from regularly available data, and will then bring them back together to calculate equilibrium prices and quantities.

Estimating Pork Supply

The supply curve is the marginal opportunity cost curve, so we would like to estimate a regression with quantity supplied as the dependent variable and last year's price and the opportunity cost of production as the explanatory variables. Unfortunately, we cannot observe the true opportunity cost of production like we can the pork price. We can, however, use data on variables that proxy costs of production. The major cost of pork production is feed, and hog feed consists mainly of corn, so we can replace production costs with the price of corn. Also, note that the ability of hog producers to produce additional hogs depends on how many sows they currently have. After all, mother pigs are needed to produce baby pigs. The number of sows in inventory can be easily proxied by last year's production level. If production was high last year, producers had many sows last year, and so they are likely to have many sows this year and can maintain a high production level. Current production partly depends on last year's production, so we include it as an explanatory variable as well.

$$\begin{aligned} \textit{Predicted Pork Production} &= a_0 + a_1(\textit{Last Year's Pork Price}) \\ &\quad + a_2(\textit{Last Year's Corn Price}) \\ &\quad + a_3(\textit{Last Year's Production Level}) \end{aligned}$$

Data on each of these four variables can be found at government organization websites like the Economic Research Service at the United States Department of Agriculture. Figure 7.7 shows such data organized to estimate a regression. All variables with dollar units are converted using the consumer price index with a 1982 base year, so that they represent real 1982 dollars. Pork production is measured in million pounds for each month from January 1972 to December 2003. Pork prices are in $/cwt, and the corn price is reported in $/bushel. After using the regression tool in Excel, we get the prediction equation reported in Figure 7.7. The R-square is 0.82, so the equation explains 82% of the variation in pork production. The p-values are zero for all three explanatory variables, indicating each variable indeed affects pork supply (ignore the p-value on the intercept). More importantly, the signs of each coefficient are consistent with the supply and demand model. The coefficient on pork price (2.89) is positive, indicating that as pork prices rise, producers respond by increasing production. The coefficient on corn (−96.36)

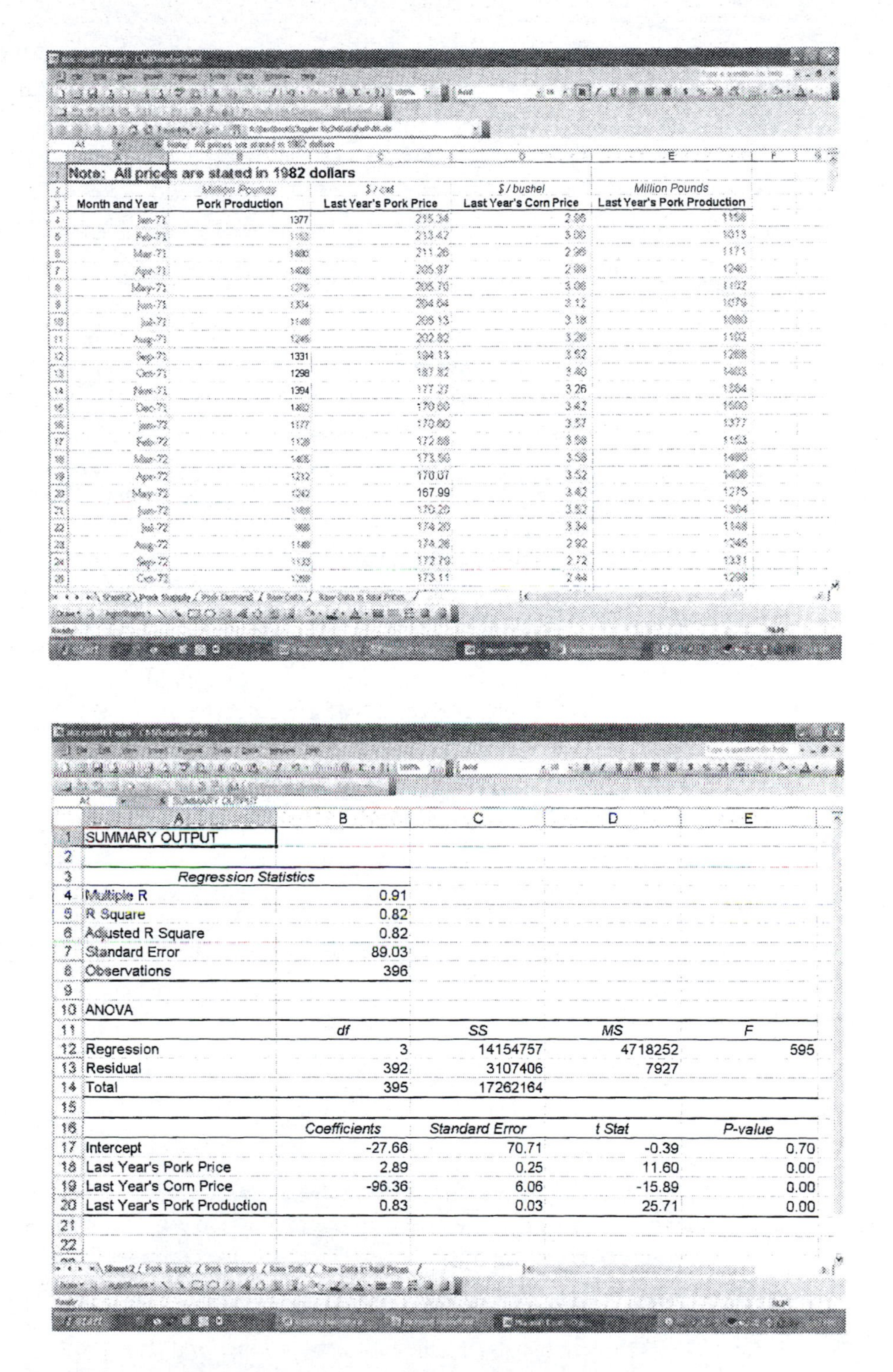

Note: All prices are stated in 1982 dollars

Month and Year	Million Pounds Pork Production	$ / cwt Last Year's Pork Price	$ / bushel Last Year's Corn Price	Million Pounds Last Year's Pork Production
Jan-71	1377	215.34	2.96	1158
Feb-71	[illegible]	213.47	3.00	1013
Mar-71	1480	211.26	2.96	1171
Apr-71	1408	205.97	2.99	1240
May-71	1276	205.70	3.08	1122
Jun-71	1304	204.64	3.12	1079
Jul-71	1148	205.15	3.18	1080
Aug-71	1246	202.82	3.26	1102
Sep-71	1331	194.13	3.52	1288
Oct-71	1298	187.82	3.40	1403
Nov-71	1394	177.27	3.26	1364
Dec-71	1402	170.60	3.42	1500
Jan-72	1177	170.80	3.57	1377
Feb-72	1128	172.88	3.58	1153
Mar-72	1406	173.56	3.58	1480
Apr-72	1212	170.07	3.52	1408
May-72	1242	167.99	3.42	1275
Jun-72	1188	170.20	3.52	1304
Jul-72	988	174.20	3.34	1148
Aug-72	1148	174.28	2.92	1246
Sep-72	1132	172.79	2.72	1331
Oct-72	1288	173.11	2.44	1298

SUMMARY OUTPUT

Regression Statistics	
Multiple R	0.91
R Square	0.82
Adjusted R Square	0.82
Standard Error	89.03
Observations	396

ANOVA

	df	SS	MS	F
Regression	3	14154757	4718252	595
Residual	392	3107406	7927	
Total	395	17262164		

	Coefficients	Standard Error	t Stat	P-value
Intercept	-27.66	70.71	-0.39	0.70
Last Year's Pork Price	2.89	0.25	11.60	0.00
Last Year's Corn Price	-96.36	6.06	-15.89	0.00
Last Year's Pork Production	0.83	0.03	25.71	0.00

FIGURE 7.7 Estimating the Pork Supply Curve.

is negative, as it should be. As corn prices rise, production falls as the supply curve shifts to the left. Finally, the coefficient on last year's production level is positive as expected. The larger last year's production level, the greater will be this year's production level.

$$\textit{Predicted Pork Production} = -27.66 + 2.89(\textit{Last Year's Pork Price}) - 96.36(\textit{Last Year's Corn Price}) + 0.83(\textit{Last Year's Production Level})$$

It will be useful to adopt a more convenient notation. Let Q_t be pork production at time t (t may stand for January 1977 or July 1999), P_t be pork prices at time t, and C_t be corn prices at time t. Time is measured, in this case, in months, so a one unit increase in t refers to a movement forward in time by one month. If Q_t is production at month t, then $Q_{t\text{-}12}$ is production one year prior, and $P_{t\text{-}12}$ is the pork price one year prior to month t. The supply equation is then more succinctly written as

$$\textit{Short-Run Supply Function: } Q_t = -27.66 + 2.89(P_{t-12}) - 95.36(C_{t-12}) + 0.83(Q_{t-12})$$

Recall that economists often differentiate between long-run and short-run supply. The equation above is a short-run supply curve because production depends on previous production levels. In the long run, pork producers have more time and options to respond to prices changes, making supply more elastic. Put differently, producers have more time to adjust to changes in the long run. The variable *Last Year's Production Level* accounts for the fact that the ability of hog producers to adjust their production levels depends on their previous production level. In the long run, producers have adequate time to adjust and will reach an equilibrium where $Q_t = Q_{t-12}$. For example, if producers wish to increase production, they may be limited by the number of sows in inventory. Each year, production rises and $Q_t > Q_{t-12}$, but eventually producers will reach their desired production level and $Q_t = Q_{t-12}$. This presents a convenient method for converting the short-run supply curve above to a long-run supply curve: We simply set $Q_{t-12} = Q_t$, and rearrange the equation so that Q_t is alone on the left-hand side.

$$\begin{aligned}
Q_t &= -27.66 + 2.89(P_{t-12}) - 96.36(C_{t-12}) + 0.83(Q_{t-12} = Q_t) \\
\rightarrow Q_t - 0.83(Q_t) &= -27.66 + 2.89(P_{t-12}) - 96.36(C_{t-12}) \\
\rightarrow 0.17(Q_t) &= -27.66 + 2.89(P_{t-12}) - 96.36(C_{t-12}) \\
\rightarrow Q_t &= [-27.66 + 2.89(P_{t-12}) - 96.36(C_{t-12})]/0.17 \\
Q_t &= -162.71 + 17.00(P_{t-12}) - 566.82(C_{t-12})
\end{aligned}$$

Lastly, note that in the long run where producers have adequate time to adjust to market changes, the effects of the production lag disappear, and we can replace all subscripts $t-12$ with t.

$$\textit{Long-Run Supply Function: } Q_t = -162.71 + 17.00(P_t) - 566.82(C_t)$$

Notice the stark difference between the long- and short-run supply curves. The long-run supply curve is more sensitive to changes in the price of corn and pork, yielding a more elastic supply curve as economic theory suggests. At this point, we hope you see the importance of understanding the supply and demand theory. Without the supply and demand model, we could not estimate both short- and long-run supply curves from the same data.

Estimating Pork Demand

Pork prices in the current time period are caused by the quantity of pork supplied and the position of the demand curve. Given a particular quantity supplied, the observed pork price tells us the position of the demand curve, so demand curves are estimated by using current pork prices as a dependent variable and quantity supplied as an explanatory variable. Other factors that shift the demand curve are also included in the prediction equation. In Chapter 2 it was shown that the prices of related goods, income, population, and tastes and expectations all shift demand. Two goods related to pork consumption are beef and chicken. These two goods are substitutes for pork. As their prices rise, consumers substitute towards pork, increasing the demand and consequently the price of pork. As with the supply curve, all prices and income (anything measured in dollars) are converted to real 1982 dollars using the Consumer Price Index.

Data on pork and poultry prices are easily obtained through the United States Department of Agriculture website. Also available from government agencies (which you can locate through a simple Google search) is the personal disposable income in the United States, which represents people's income after taxes. Also, it seems plausible that pork demand may vary by the season. To reflect seasonal changes in demand, we construct dummy variables for different quarters of the year.

Finally, it seems plausible that consumers form food-purchasing habits. If demand was high last year, it seems plausible it will be high this year because consumers are in the habit of eating pork. If demand fell last year, consumers purchase less pork out of habit, and demand should be low this year as well. Put differently, if many pork roasts were served at the last Christmas dinner, many pork roasts will likely be served this Christmas. If many pork chops were grilled last summer, people are more likely to grill pork chops this summer as well. Thus, because demand this year depends on consumption last year, we include last year's consumption as an explanatory variable. This variable will also let us differentiate between short-run and long-run demand. In the short run, consumers make adjustments to pork prices, but due to shopping habits, those adjustments are slower than in the long run when consumers have plenty of time to modify their purchasing behavior. These habits create a "friction" in demand shifts, a friction that exists in the short run but not the long run.

$$\begin{aligned} \textit{Predicted Pork Price} = {} & a_0 + a_1(\textit{Quantity of Pork Supplied}) + a_2(\textit{Beef Price}) \\ & + a_3(\textit{Poultry Price}) + a_4(\textit{Disposable Income}) \\ & + a_5(\textit{First Quarter Dummy Variable}) \\ & + a_6(\textit{Second Quarter Dummy Variable}) \\ & + a_7(\textit{Third Quarter Dummy Variable}) \\ & + a_8(\textit{Last Year's Pork Production}) \end{aligned}$$

This equation is collapsed to a more succinct equation by using abbreviations for each variable.

$$P_t = a_0 + a_1(Q_t) + a_2(B_t) + a_3(O_t) + a_4(I_t) + a_5(DQ1_t)$$
$$+ a_6(DQ2_t) + a_7(DQ3_t) + a_8(Q_{t-12})$$

Notice we do not include a dummy variable for the fourth quarter. The reason is that the fourth quarter is already represented in the intercept (a_0). In regression, if you have a set of dummy variables that describe all possible states of the world, you must exclude one of the dummy variables. In our softball regression, hitters were either males or females, but only a single dummy variable for males was included in the regression. In the pork demand equation, at any given time it is either the first, second, third, or fourth quarter, so we "drop" the fourth quarter dummy variable from the equation. Observe the regression results in Figure 7.8.

Short-Run Pork Demand Function: $P_t = 64.6979 - 0.0651(Q_t) + 0.2332(B_t)$
$$+ 1.1291(O_t) + 0.0194(I_t)$$
$$- 8.0152(DQ1_t) - 12.7642(DQ2_t)$$
$$- 11.4670(DQ3_t) + 0.0004(Q_{t-12})$$

The signs of the coefficients are as expected. The negative sign on Q_t reflects the first law of demand: A greater quantity corresponds to a lower price on the consumer demand curve. The signs on the beef and poultry prices (B_t and O_t) are negative, indicating that beef and poultry are substitutes for pork. As beef and poultry prices rise, consumers substitute away from beef and poultry and towards pork, increasing pork demand and the pork price. As income rises, the positive sign on I_t indicates consumer demand for pork rises, and pork prices rise, making pork a normal good. The negative sign on each quarterly dummy variable indicates that pork demand (and hence pork prices) are lower in the first three quarters than the fourth quarter. Finally, the positive sign on Q_{t-12} reflects habitual behavior of consumers. The greater the pork consumption in the past, the greater will be current pork and pork prices.

Next, we can consult the p-values to determine if each variable has a true, significant impact on pork demand. Recall the p-values indicate the probability that each variable has *no* impact on pork prices and pork demand. If this probability is greater than 5%, we typically conclude the explanatory variable has no significant impact on the dependent variable. Only the coefficient Q_{t-12} has a p-value above 5%. This signifies that current pork demand is not dependent upon pork demand in previous periods, and if consumers possess purchasing habits, those habits are difficult to detect in data. Thus, we say Q_{t-12} is not statistically "significant." All other variables significantly affect pork demand.

Given that Q_{t-12} is not "significant," one could remove it from the regression model and reestimate the model without it. In this case, the resulting demand equation is both the long- and short-run demand. However, to prepare you for cases where the lagged production variable is significant, it will remain in the equation. To obtain

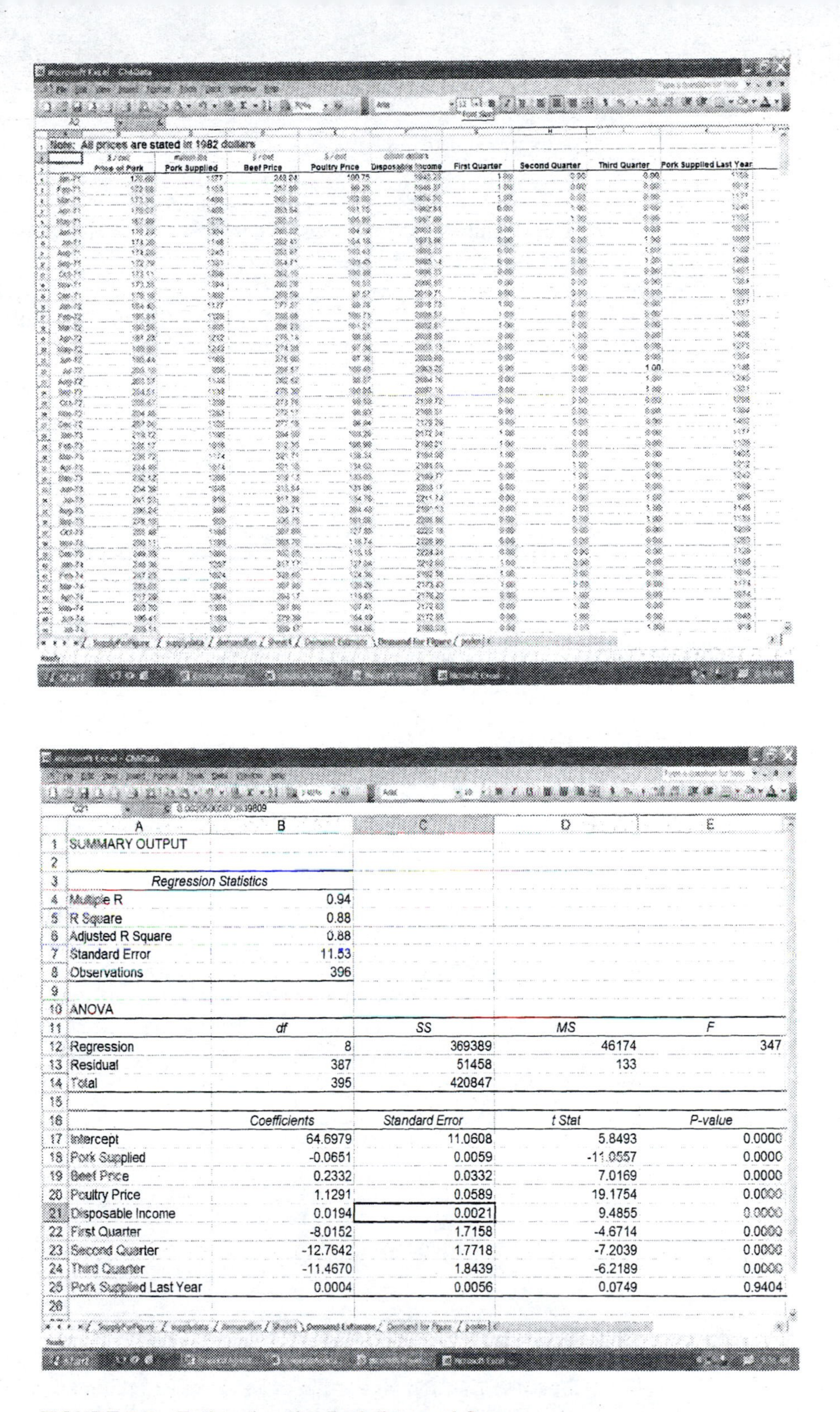

Note: All prices are stated in 1982 dollars

	Price of Pork	Pork Supplied	Beef Price	Poultry Price	Disposable Income	First Quarter	Second Quarter	Third Quarter	Pork Supplied Last Year

SUMMARY OUTPUT

Regression Statistics	
Multiple R	0.94
R Square	0.88
Adjusted R Square	0.88
Standard Error	11.53
Observations	396

ANOVA

	df	SS	MS	F
Regression	8	369389	46174	347
Residual	387	51458	133	
Total	395	420847		

	Coefficients	Standard Error	t Stat	P-value
Intercept	64.6979	11.0608	5.8493	0.0000
Pork Supplied	-0.0651	0.0059	-11.0557	0.0000
Beef Price	0.2332	0.0332	7.0169	0.0000
Poultry Price	1.1291	0.0589	19.1754	0.0000
Disposable Income	0.0194	0.0021	9.4855	0.0000
First Quarter	-8.0152	1.7158	-4.6714	0.0000
Second Quarter	-12.7642	1.7718	-7.2039	0.0000
Third Quarter	-11.4670	1.8439	-6.2189	0.0000
Pork Supplied Last Year	0.0004	0.0056	0.0749	0.9404

FIGURE 7.8 Estimating the Pork Demand Curve.

a long-run demand curve, we follow the procedures for obtaining a long-run supply curve and set $Q_t = Q_{t-12}$. When consumers are given time to adjust to price changes, the effects of shopping habits disappear, and the demand curve becomes stable and so do purchases. When Q_{t-12} is set equal Q_t, the following long-run demand curve is obtained.

$$
\begin{aligned}
P_t &= 64.6979 - 0.0651(Q_t) + 0.2332(B_t) + 1.1291(O_t) + 0.0194(I_t) \\
&\quad - 8.0152(DQ1_t) - 12.7642(DQ2_t) - 11.4670(DQ3_t) \\
&\quad + 0.0004(Q_{t-12} = Q_t) \\
\rightarrow P_t &= 64.6979 - 0.0651(Q_t) + 0.0004(Q_t) + 0.2332(B_t) + 1.1291(O_t) \\
&\quad + 0.0194(I_t) - 8.0152(DQ1_t) - 12.7642(DQ2_t) \\
&\quad - 11.4670(DQ3_t)
\end{aligned}
$$

Long-Run Pork Demand Function: $P_t = 64.6979 - 0.0647(Q_t) + 0.2332(B_t) + 1.1291(Q_t) + 0.0194(I_t) - 8.0152(DQ1_t) - 12.7642(DQ2_t) - 11.4670(DQ3_t)$

The coefficient on Q_t is a smaller negative number, and if you graph the short-run and long-run demand curves, the long-run demand curve's slope will be less steep, indicating a more elastic demand. Remember, supply and demand curves are always more elastic in the long run.

Supply and Demand Together at Last

So far, we have estimated supply and demand functions for pork. These functions can then be used to obtained supply and demand curves. Recall the long-run pork supply function: $Q_t = -162.71 + 17.00(P_t) - 566.82(C_t)$. A supply *curve* refers to a graph with price on the *y*-axis and quantity on the left-axis. To obtain this curve, we must assume some value for the price of corn. Figure 7.9 shows that the average price of corn over many years equals \$2.48, so we substitute $C_t = 2.48$ and rearrange the supply curve so that price is on the left-hand side, yielding the long-run supply curve.

Variable	Average Value
Price of Corn (\$/bushel)	\$2.48
Price of Beef (\$/cwt)	\$225
Price of Poultry (\$/cwt)	\$78
Disposable Income (billion dollars)	\$3,103

FIGURE 7.9 Average Value of Selected Variables from 1972–2003.

Note: All dollars are real 1982 dollars.

$$
\begin{aligned}
Q_t = Q_t &= -162.71 + 17.00(P_t) - 566.82(C_t = 2.48) \\
\rightarrow Q_t &= -1568.42 + 17.00(P_t) \\
\rightarrow (17.00)P_t &= 1568.42 + Q_t \\
\rightarrow P_t &= (17.00)^{-1}1568.42 + (17.00)^{-1}(Q_t) \\
\textit{Long-Run Supply Curve:}\quad P_t &= 92.26 + 0.05882(Q_t)
\end{aligned}
$$

To obtain a demand curve we employ a similar methodology. Taking the demand function, the values of B_t, O_t, and I_t are replaced with their average value, as shown in Figure 7.9. Noting that across years the average value of each quarterly dummy variable is one-fourth, $DQ1_t$, $DQ2_t$, $DQ3_t$, is replaced with 0.25. After substituting these values, we arrive at a pork demand curve.

$$
\begin{aligned}
P_t = {} & 64.6979 - 0.0647(Q_t) + 0.2332(225) + 1.1291(78) + 0.0194(3103) \\
& - 8.0152(0.25) - 12.7642(0.25) - 11.4670(0.25) \\
& \textit{Long-Run Pork Demand Curve:}\quad P_t = 257.37 - 0.0647(Q_t)
\end{aligned}
$$

Now that we have a pork supply and demand curve, we can calculate the equilibrium price and quantity, using the algebra you learned in Chapter 3.

Impacts of an Antibiotic Ban

Let us now use these supply and demand curve estimates to analyze how a ban on subtherapeutic antibiotic use would impact the pork market.[6] The ban would raise production costs. Even though the increase in costs is not known for sure and will differ across farms, the best estimate is that pork production costs will rise about 0.81% (Lusk, Norwood, and Pruitt 2006). The supply curve is the marginal cost curve, meaning we could replace the variable P_t on the left-hand side of the supply curve with the word *marginal cost*. Thus, the supply curve will shift up 0.81% if an antibiotic ban is enacted. This shift can be handled mathematically by multiplying the entire right-hand side of the supply curve equation by 1.0081.

$$
\begin{aligned}
\textit{Supply Curve Before Antibiotic Ban:}\quad P_t &= 92.26 + 0.05882(Q_t) \\
\textit{Supply Curve After Antibiotic Ban:}\quad P_t &= [92.26 + 0.05882(Q_t)](1.0081) \\
&= 93.01 + 0.0593(Q_t)
\end{aligned}
$$

The ban will also increase consumer demand for pork, because antibiotic-free pork is considered a more natural and healthy product. Exactly how much the demand curve will shift upwards is difficult to determine though. However, research has shown that consumers will pay up to 78% more for antibiotic-free pork chops than regular pork chops (Lusk, Norwood, and Pruitt 2006). Although it is unclear whether this large premium extends to all pork products, it gives us a starting point for estimating the size of the demand shift. There are a number of reasons why this might be an upper bound to willingness-to-pay. For example, people tend to behave more socially responsible when they know their actions are being recorded by a

[6]The ban considered here refers only the finishing stage of pork production.

researcher than in an anonymous shopping setting. Also, this premium exists only when consumers are given information about antibiotic use in swine production, and if they know whether a ban is in place. It may be that if an antibiotic ban is enacted, very few consumers will be knowledgeable of the ban, and the demand shift will be only slight. In response to this uncertainty, we will consider two possibilities. Scenario A assumes that pork demand increases 25% due to a ban. This takes the 78% premium and reduces it, albeit arbitrarily, to be conservative. Scenario B assumes demand does not change in response to a ban. In considering Scenario A we must shift the pork demand curve upwards by 25%, which is accomplished by multiplying the right-hand side of the long-run demand curve equation by 1.25.

$$\textit{Demand Curve Before Antibiotic Ban:}\quad P_t = 257.37 - 0.0647(Q_t)$$

$$\textit{Demand Curve After Antibiotic Ban, Scenario A:}\quad P_t = [257.37 - 0.0647(Q_t)](1.25) = 321.71 - 0.0809(Q_t)$$

$$\textit{Demand Curve After Antibiotic Ban, Scenario B:}\quad P_t = 257.37 - 0.0647(Q_t)$$

Then, using the same math as in Figure 7.10, one can calculate the equilibrium price and quantity after the antibiotic ban under Scenarios A and B. If you performed the algebra correctly, you will get the equilibrium prices and quantities shown in Figure 7.11. In Scenario A where consumers are well informed of the antibiotic issue and are knowledgeable of the ban, price and quantity increases. The demand increase far outweighs the supply decrease, leading to higher prices and more pork. For pork producers, this implies greater profits for the entire industry. Thus, if Scenario A is correct, pork producers should impose a voluntary antibiotic ban on themselves.

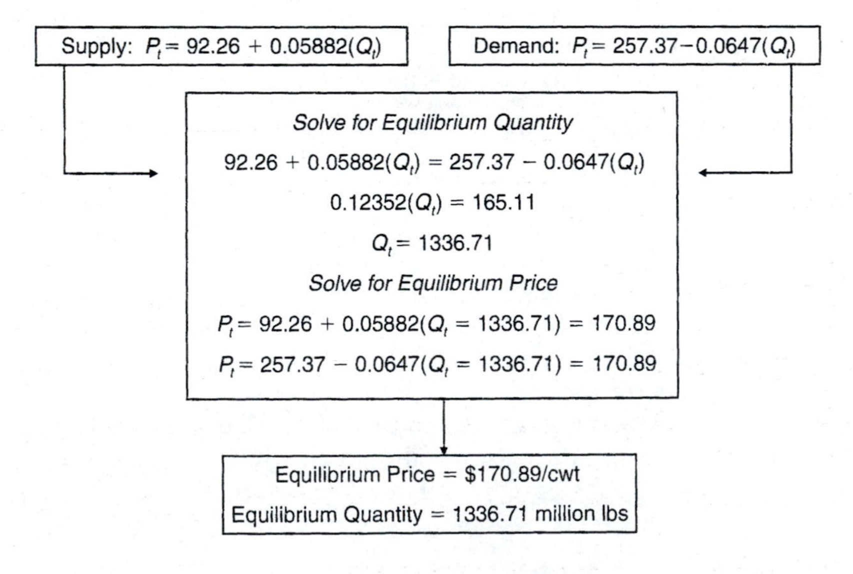

FIGURE 7.10 Equilibrium Price and Quantity and Pork.

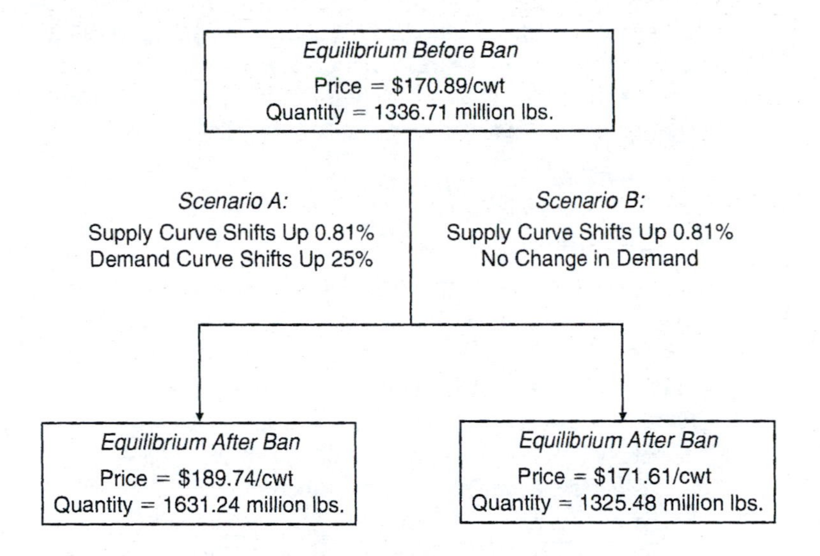

FIGURE 7.11 Effects of Antibiotic Ban.

Even though their pork production costs will rise, consumers will pay a higher pork price to more than offset this cost increase.[7]

Conversely, if Scenario B is correct and demand does not increase in response to the ban, price increases and quantity decreases only slightly. Thus, if Scenario A is correct, then pork producers gain substantially from the ban, whereas if Scenario B is correct, industry profits may or may not fall, depending on whether the increase in price outweighs the loss in quantity. Of course, the fact that the industry has not supported such a ban indicates they feel Scenario B to be the most realistic and that the ban would indeed harm the industry. More precise statements about welfare impacts could be made. Given that we have real supply and demand curves and we know how they shift under different scenarios, we could use the concepts of consumer and producer surplus introduced in Chapter 1 to calculate total surplus before and after the ban. This is beyond the scope of this chapter, however, this chapter does provide the foundation needed for such analysis.

THE LOG-LOG REGRESSION MODEL

In the previous section we estimated supply and demand functions for pork. In Chapter 3 we described how to calculate supply and demand elasticities from these functions. For example, see the demand elasticity formulas in Figure 3.13. Frequently the major reason

[7]Note that this voluntary ban must be enforceable, or else producers could use antibiotics, raise pork cheaper, and sell at higher prices to consumers thinking no antibiotics were used.

economists estimate supply and demand functions is to calculate elasticities, in which case a different type of regression model allows one to obtain elasticity estimates directly.

Take the pork supply function from the previous section and replace each variable with its natural logarithm.

Log-Log Short-Run Supply Function:

$$\ln(Q_t) = a_0 + (a_1)\ln(P_{t-12}) + (a_3)\ln(C_{t-12}) + (a_4)\ln(Q_{t-12})$$

This is called a "log-log" model because the variables on both sides of the equation have been transformed to their natural logarithms. With any function of the form $\ln(Y) = a_0 + a_1\ln(X)$, a_1 turns out to equal the percent change in Y resulting from a 1% change in X.[8] Because the own-price elasticity of supply is the percent change in quantity supplied resulting from a 1% change in price, the coefficient a_1 in the supply function above must be the own-price elasticity of supply! Using the same data as before, if we estimate this function, we get the following, signifying that the short-run own-price elasticity of pork supply equals 0.324. A 1% increase in price one year will cause a 0.324% increase in quantity supplied of pork the following year.

Log-Log Short-Run Supply Function:

$$\ln(Q_t) = 0.380 + (0.324)\ln(P_{t-12}) - (0.182)\ln(C_{t-12}) + (0.737)\ln(Q_{t-12})$$

Price Flexibilities

Frequently this book has discussed the concept of demand elasticities. For example, the own-price demand elasticity indicates the percent change in quantity demanded from a 1% increase in price, and the income elasticity signifies the percent change in quantity demanded resulting from a 1% increase in price. Yet, the production of agricultural goods typically involves a production lag, such that at any given time the quantity of the good is fixed. The resulting price is then the point on the demand curve corresponding to that quantity—the price that clears the fixed supply. In these cases it is not quantity demanded that changes, but the market price. Refer back to the right-hand diagram in Figure 7.6. The supply is perfectly inelastic. A larger supply leads to a lower price. Moreover, a change in any one of the demand shifters (e.g., price of related goods, income) changes the market price but not the quantity. Because it is price that responds to demand changes and not quantity, price flexibilities are used instead of demand elasticities.

---

[8] The proof requires calculus. Consider the function $\ln(Y) = a_0 + a_1\ln(X)$. The derivative of $\ln(Y)$ with respect to X is $dn(Y) = a_1 \times d\ln(X)$, where the "$d$" is the instantaneous rate of change. Note that the derivative of $\ln(X) = dX/X$ and the derivative of $\ln(Y)$ is dY/X. Thus, $a_0 = \frac{dY/Y}{dX/X}$, which is identically the percent change in Y divided by the percent change in X—the elasticity of Y with respect to X!

Price flexibilities are basically the same as elasticities, except that instead of quantity changing in response to a variable, it is price that is changing. For example, the own-price flexibility is measured as

$$Price\ Flexibility = (\%\Delta\ Price)/(\%\Delta\ Quantity\ Supplied)$$

The same log-log transformation can be made to the pork demand function to obtain pork demand flexibilities. After replacing all non-dummy variables in the previous pork demand function with its natural logarithm, our log-log demand function is

$$\begin{aligned} \textit{Log-Log Short-Run Demand Function:}\quad \ln(P_t) = {} & a_0 + (a_t)\ln(Q_t) + (a_2)\ln(B_t) \\ & + (a_3)\ln(O_t) + (a_4)\ln(I_t) \\ & + a_5(DQ1_t) + a_6(DQ2_t) \\ & + a_7(DQ3_t) + (a_8)\ln(Q_{t-12}) \end{aligned}$$

Dummy variables either equal one or zero, and the natural logarithm of zero is undefined, so they cannot be transformed to log form. The coefficient a_1 now indicates the percent change in price resulting from a 1% rise in quantity supplied—the very definition of price flexibility. Estimating this function using the same data as before yields the following, resulting in a short-run demand flexibility of –0.398. For every 1% increase in pork produced in the short run, the pork price falls by 0.398%.

$$\begin{aligned} \textit{Log-Log Short-Run Demand Function:}\quad \ln(P_t) = {} & 1.267 - (0.398)\ln(Q_t) \\ & + (0.290)\ln(B_t) + (0.568)\ln(O_t) \\ & + (0.321)\ln(I_t) - 0.035(DQ1_t) \\ & - 0.058(DQ2_t) - 0.049(DQ3_t) \\ & + (0.022)\ln(Q_{t-12}) \end{aligned}$$

Other empirical facts can be gleaned from this log-log demand function. For every 1% increase in beef and poultry prices, pork prices rise 0.29% and 0.568%, respectively. Pork prices rise 0.321% for every 1% rise in income. Chapter 3 repeatedly illustrated the importance and usefulness of elasticities. They are used to assess the impact of government regulations as well as trade agreements between countries. But where do elasticities come from? They come from real-world data, using the methods described above.

DO BEEF ADVERTISEMENTS WORK?

Each time someone sells a steer or heifer, the government requires them to pay one dollar to an organization called the beef checkoff. Funds collected by the checkoff are then used to stimulate and sustain beef demand. They fund advertisements such as the familiar "Beef: It's What's For Dinner" campaign, that aim is to increase the demand for all beef products. The money is also used to fund beef safety research, consumer education, and beef export efforts.

Why does the government tax cattle producers to fund beef advertisements? The reason is that cattle producers asked the government for this particular tax. Yes, you heard us right; people actually asked for a tax! Most cattle producers want to advertise to consumers so that beef demand is sustained and market share is not lost to pork and poultry. The problem is that no single cattle producer has the incentive to conduct nationwide advertising. If farmer Kevin Yon advertises beef, hoping to realize higher prices for his 100 head of cattle, the benefits of those advertisements go to Kevin and all other cattle producers. Kevin ultimately receives very little of the benefits, because the benefits are spread over all cattle producers. Those other producers are called *free-riders*; they benefit from the advertisement without paying. Due to the free-rider problem, no single producer advertises beef on television. All cattlemen reap the rewards from advertising generic beef, but advertising is not profitable for a single producer.

A free-rider is someone who benefits from a program without paying their share of the cost.

This is a coordination problem. No single producer wants to conduct generic beef advertisements alone, but cattle producers would like to collectively fund advertisements. A method is needed to eliminate free-riders. The simple solution is to have the government make everyone pay. Now free-riding becomes illegal. Periodically, the government conducts a vote among cattlemen to determine if the beef checkoff should continue. Cattlemen should support the checkoff if the advertisements increase their profits. In fact, they do believe in the profitability of the checkoff, because they repeatedly vote to keep the checkoff. However, it would be nice if they could find some evidence that the checkoff raises beef demand. Fortunately, regression analysis gives us a simple but powerful method to make this assessment.

In this section we want to determine whether money spent on beef advertisements increases beef demand. A beef demand function will be estimated much like the pork demand function, except that a variable for money spent on advertising will be included. If the coefficient on this variable has a p-value less than 0.05, then one may conclude that the advertisements do indeed increase beef demand. The beef market is similar to the pork market discussed previously. There is a long production lag of about two years. Thus, the amount of beef for sale to consumers today was determined two years ago and is (relatively) fixed, regardless of the price of beef. The price of beef is then the market price that induces consumers to purchase all the beef available.

Given the implications of the production lag, we specify a demand function similar to the pork demand function, with several differences. First, we are more concerned with the price cattle producers receive than the price of retail beef, so the dependent variable is the price of live-cattle.[9] Live-cattle prices are still influenced, albeit indirectly, by consumer demand (this is a derived demand discussed in Chapter 6). If consumer demand rises, beef prices rise, and beef processors will pay more for live-cattle to meet this rising beef demand. Second, an additional variable is included, a variable equaling the amount of money spent on generic beef advertisements. It seems plausible that it takes time for advertisements to take effect. Money spent on advertisements in the first quarter of the year will have its greatest effect in the second quarter. Third, we will ignore seasonality and consumption habits. They are optional and are ignored here to keep the demand function simple. The following live-cattle demand function is estimated.

---

[9]Cattle that are ready to be slaughtered and processed into beef are also referred to as fed-cattle.

Live Cattle Demand Function: *Live-Cattle Price*

$$= a_0 + a_1(\textit{Quantity of Live-Cattle Supplied}) + a_2(\textit{Pork Price}) + a_3(\textit{Poultry Price}) + a_4(\textit{Disposable Income}) + a_5(\textit{Generic Beef Advertising Expenditures Last Quarter})$$

As always, our prices, income, and advertising expenditure variables are converted to real 1982 dollars. The data used are available at the textbook website and are quarterly data from 1980 to 1997.[10] The following Excel output is provided in Figure 7.12. As indicated by the p-values, all variables significantly affect live-cattle prices. Most of the coefficient signs are as expected. The greater the cattle supply, the lower the price, due to a downward sloping demand curve (recall that the supply at any given time is fixed and was determined by previous prices). As hog and poultry prices rise, the demand for live-cattle rises and increases live-cattle prices. Beef is usually thought to be a normal good, however, the negative sign on income suggests that beef demand falls when income rises, making beef an inferior good. The counterintuitive result may be due to the fact that beef demand had been trending downward due to health concerns and incomes trending upward throughout the sample. Because there is no variable for

SUMMARY OUTPUT				
Regression Statistics				
Multiple R	0.96			
R Square	0.93			
Adjusted R Square	0.92			
Standard Error	2.88			
Observations	71			
ANOVA				
	df	*SS*	*MS*	*F*
Regression	5	6830	1366	164
Residual	65	540	8	
Total	70	7371		
	Coefficients	*Standard Error*	*t Stat*	*P-value*
Intercept	158.1017	9.6271	16.4226	0.0000
Quantity of Cattle Slaughtered	-0.0091	0.0009	-9.6736	0.0000
Real US Disposable Income	-0.0159	0.0021	-7.5532	0.0000
Real Hog Prices	0.1926	0.0712	2.7042	0.0087
Real Poultry Prices	0.4378	0.0761	5.7531	0.0000
Real Beef Advertising Expenditures Last	0.0004	0.0002	2.4918	0.0153

FIGURE 7.12 Live-Cattle Demand Function.

[10]The authors would like to thank George Davis of Virginia Tech University for making these data available.

health concerns, the regression falsely attributed the impacts of health concerns to income, leading to a negative relationship between income and beef prices.[11]

Now to the question of interest: Does generic beef advertising increase the demand for cattle? The coefficient on generic beef advertising is positive, and its p-value is less than 5%. This indicates that advertising does increase beef demand. In fact, after further analysis, agricultural economists have calculated that every dollar spent on these generic beef advertisements generates $9.84 in cattle producer profits (Davis 2005). This should be of great comfort to cattle producers who fund these advertisements, a comfort only regression analysis can provide.

TIME-SERIES ANALYSIS

The chapter has thus far focused on the use of regression for estimating supply and demand. This is referred to as *fundamental price analysis*, where prices are analyzed by supply and demand factors. That is, price is modeled as a function of the variables that shift supply and demand curves. Often, a different type of price analysis is more useful, one we refer to as *time-series analysis*. This type of analysis also uses regression but models price directly as a function of time, like the year or the season, or even past prices.

Fundamental Price Analysis: Where price changes are modeled as a function of supply and demand determinants; the factors shifting supply and demand curves.

Time-Series Price Analysis: Where price changes are modeled as a function of time and past prices; price changes are predicted by trends, and not the factors shifting supply and demand curves.

Seasonal and Time Trend Models

Recall in the last chapter that corn prices follow a seasonal pattern. Corn prices are lowest around harvest in October and November. Prices rise between December and April and fall from April to October. Also, corn prices are trending downward due to better technologies that reduce the cost of corn production. In many cases, we may want to predict corn prices in different seasons and different years. For example, if one is considering storing their corn for three months after harvesting it in November, one might want to predict February prices to see if the increase in prices from November to February justify the storage cost.

Consider the following regression model, where the predicted corn price is stated as a function of the month and a yearly time trend variable.

$$\begin{aligned} \textit{Predicted Corn Price} = a_0 &+ a_1(\textit{January}) + a_2(\textit{February}) + a_3(\textit{March}) \\ &+ a_4(\textit{April}) + a_5(\textit{May}) + a_6(\textit{June}) + a_7(\textit{July}) \\ &+ a_8(\textit{August}) + a_9(\textit{September}) + a_{10}(\textit{October}) \\ &+ a_{11}(\textit{November}) + a_{12}(\textit{Yearly Time Trend}) \end{aligned}$$

The first 11 variables are monthly dummy variables. For example, if the current month is March, the variable *March* equals one and all other monthly dummy variables equal zero. Recall that when your dummy variables describe every possible state

[11]Variables acting as a proxy can and are frequently used to account for demand shifts due to health concerns.

of the world, you must exclude one of them. Here, there are 12 months, so we must exclude the dummy variable for one month. The dummy variable for *December* was chosen to be excluded. The yearly time trend variable increases by one unit each year. Our corn data covers the years 1960 through 2005, and the time trend is constructed to equal 1 in 1960, 2 in 1961, . . . , and 46 in 2005. Thus, the variable *Yearly Time Trend* equals the year minus 1959.

The regression is estimated as follows and is illustrated in Figure 7.13. First, nominal corn price data and consumer price index (CPI) data are retrieved. The nominal prices are converted to real prices using the formula $real\ price = (nominal\ price/CPI) \times 100$. The CPI has a base year of 1982, which means that the real prices are stated in terms of 1982 dollars. Next, the monthly dummy variables and the time trend variable is constructed in Excel. See Figure 7.13, and note how the month variables are created so that *July* equals one in the month *July* and zero for all other months. Then, use the Excel regression tool to estimate a regression with price as the dependent variable and the monthly variables and year variable as the explanatory variables. Running the regression then yields the following prediction equation.

$$\begin{aligned} \textit{Predicted Corn Price} = {} & 4.20 + 0.09(\textit{January}) + 0.10(\textit{February}) + 0.11(\textit{March}) \\ & + 0.11(\textit{April}) + 0.15(\textit{May}) + 0.17(\textit{June}) + 0.16(\textit{July}) \\ & + 0.14(\textit{August}) + 0.06(\textit{September}) - 0.04(\textit{October}) \\ & - 0.10(\textit{November}) - 0.07(\textit{Yearly Time Trend}) \end{aligned}$$

To understand how to interpret this regression, suppose we wish to forecast prices for February of 2007. The variable *February* equals one, the variable *Yearly Time Trend* equals $2007 - 1959 = 48$, and all other variables equal zero. The predicted price is then $4.20 + 0.10(1) - 0.07 \times (2007 - 1959 = 48) = \$0.94$. If we perform this same calculation for every month, the predicted prices will be as shown in Figure 7.14. If you are familiar with current corn prices, they will seem low, but remember these are prices stated in 1982 dollars. Notice that the dummy variable for each month illustrates the premium or discount for selling corn that month, relative to December, because we excluded the December dummy variable. The coefficient of 0.09 for January says that prices are \$0.09 per bushel higher in January than December, whereas the coefficient of -0.07 in November says prices are \$0.07 lower in November than December.

Given this estimate, suppose that you harvest your corn at the end of November, and it costs you \$0.05 per bushel (in real 1982 dollars) to store corn each month. By storing corn throughout December and selling in January, you incur costs of \$0.05 per bushel but should realize a higher price of \$0.93 compared to \$0.84. It is indeed profitable to store corn through December and sell in January—on average. Should you perhaps store through January as well? The answer is no. You only receive an increase in price of \$0.01 from selling in February instead of January, yet you pay \$0.05 per bushel to store corn throughout January. The answer is clear: harvest your grain in November, store it throughout December, and sell it in January.

Date	Real Corn Price in 1982 Dollars	January	February	March	April	May	June	July	August	September	October	November	Year
Jan-60	$ 3.41	1	0	0	0	0	0	0	0	0	0	0	1
Feb-60	$ 3.44	0	1	0	0	0	0	0	0	0	0	0	1
Mar-60	$ 3.47	0	0	1	0	0	0	0	0	0	0	0	1
Apr-60	$ 3.59	0	0	0	1	0	0	0	0	0	0	0	1
May-60	$ 3.66	0	0	0	0	1	0	0	0	0	0	0	1
Jun-60	$ 3.68	0	0	0	0	0	1	0	0	0	0	0	1
Jul-60	$ 3.68	0	0	0	0	0	0	1	0	0	0	0	1
Aug-60	$ 3.61	0	0	0	0	0	0	0	1	0	0	0	1
Sep-60	$ 3.58	0	0	0	0	0	0	0	0	1	0	0	1
Oct-60	$ 3.33	0	0	0	0	0	0	0	0	0	1	0	1
Nov-60	$ 3.03	0	0	0	0	0	0	0	0	0	0	1	1
Dec-60	$ 3.08	0	0	0	0	0	0	0	0	0	0	0	1
Jan-61	$ 3.26	1	0	0	0	0	0	0	0	0	0	0	2
Feb-61	$ 3.39	0	1	0	0	0	0	0	0	0	0	0	2
Mar-61	$ 3.39	0	0	1	0	0	0	0	0	0	0	0	2
Apr-61	$ 3.25	0	0	0	1	0	0	0	0	0	0	0	2
May-61	$ 3.42	0	0	0	0	1	0	0	0	0	0	0	2
Jun-61	$ 3.46	0	0	0	0	0	1	0	0	0	0	0	2
Jul-61	$ 3.50	0	0	0	0	0	0	1	0	0	0	0	2
Aug-61	$ 3.48	0	0	0	0	0	0	0	1	0	0	0	2
Sep-61	$ 3.47	0	0	0	0	0	0	0	0	1	0	0	2
Oct-61	$ 3.40	0	0	0	0	0	0	0	0	0	1	0	2

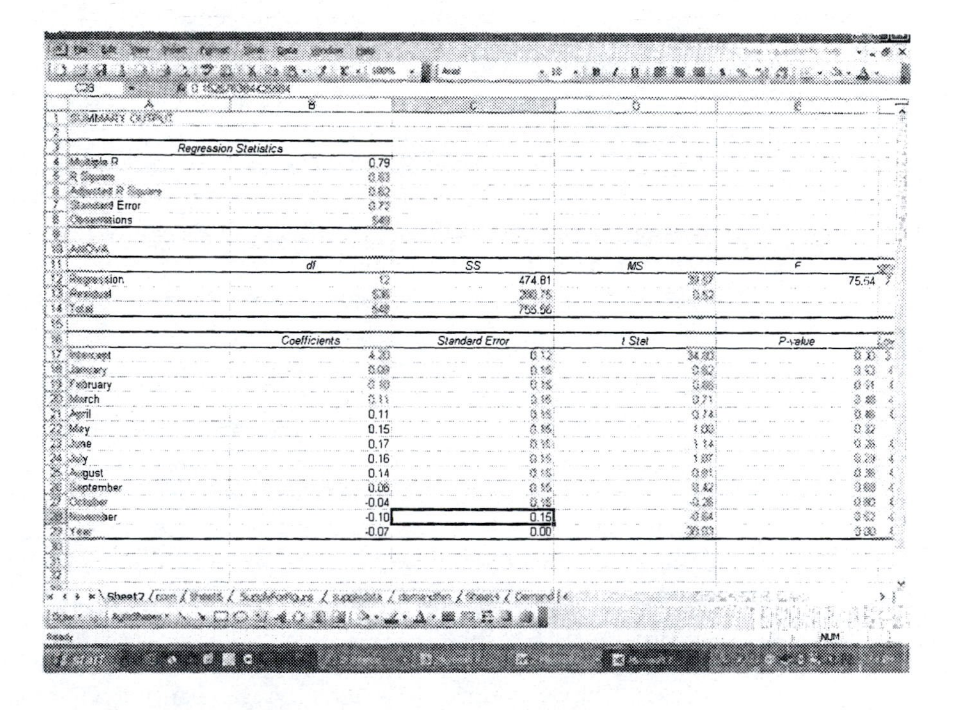
SUMMARY OUTPUT

Regression Statistics	
Multiple R	0.79
R Square	[illegible]
Adjusted R Square	[illegible]
Standard Error	[illegible]
Observations	[illegible]

ANOVA

	df	SS	MS	F
Regression	12	474.81	[illegible]	75.54
Residual	[illegible]	[illegible]	[illegible]	
Total	[illegible]	755.56		

	Coefficients	Standard Error	t Stat	P-value
Intercept	[illegible]	[illegible]	[illegible]	[illegible]
January	[illegible]	[illegible]	[illegible]	[illegible]
February	[illegible]	[illegible]	[illegible]	[illegible]
March	[illegible]	[illegible]	[illegible]	[illegible]
April	0.11	[illegible]	[illegible]	[illegible]
May	0.15	[illegible]	[illegible]	[illegible]
June	0.17	[illegible]	[illegible]	[illegible]
July	0.16	[illegible]	[illegible]	[illegible]
August	0.14	[illegible]	[illegible]	[illegible]
September	0.06	[illegible]	[illegible]	[illegible]
October	-0.04	[illegible]	[illegible]	[illegible]
November	-0.10	0.15	[illegible]	[illegible]
Year	-0.07	0.00	[illegible]	[illegible]

FIGURE 7.13 Time-Series Analysis of Corn Prices.

Month	Predicted Price
January	4.2 – 0.07(48) + 0.09 = 0.93
February	4.2 – 0.07(48) + 0.10 = 0.94
March	4.2 – 0.07(48) + 0.11 = 0.95
April	4.2 – 0.07(48) + 0.11 = 0.95
May	4.2 – 0.07(48) + 0.15 = 0.99
June	4.2 – 0.07(48) + 0.17 = 1.01
July	4.2 – 0.07(48) + 0.16 = 1.00
August	4.2 – 0.07(48) + 0.14 = 0.98
September	4.2 – 0.07(48) + 0.06 = 0.90
October	4.2 – 0.07(48) – 0.04 = 0.08
November	4.2 – 0.07(48) – 0.10 = 0.74
December	141.79 – 0.07(48) = 0.84

FIGURE 7.14 Predicted Corn Price by Month for 2007.

Modeling Price Trends with Autoregressive Models

Trends occur in every market. In the beginning of the twenty-first century, pork and beef prices trended upward partly due to the popularity of high protein diets. Many people believe this to be a fad, and over time drawbacks to these diets may become known and its popularity will fade. In the last chapter we also discussed price cycles and how beef and pork prices follow upward and downward trends. The basic idea is that past prices can help predict future prices. A simple regression model can be used to predict future prices based on current prices. The model is called an autoregressive model and simply uses past prices as the explanatory variables. Consider an example of forecasting live-cattle prices. Let P_t be the current price where t denotes the month. This makes P_{t-1} one month prior to time t, and P_{t-12} 12 months prior to time t. These are *lagged variables*, meaning they "lag" behind the explanatory variable. An example of an autoregressive model is

$$\textit{Predicted } P_t = a_0 + a_1P_{t-1} + a_2P_{t-2} + a_3P_{t-3}$$

Technically, the above model is referred to as an AR(3) model because it is an autoregressive model using three lagged variables. If the prices did not exhibit seasonality, the AR(3) would be an appropriate model, but we know that live-cattle prices are higher in some months and lower in others. To reflect this, we will also include a seasonal variable P_{t-12}. This model is denoted $AR_{12}(3)$, because it contains three lagged variables and a seasonal lagged variable 12 months prior. The formula for this seasonal autoregressive model is

$$\textit{Predicted } P_t = a_0 + a_1P_{t-1} + a_2P_{t-2} + a_3P_{t-3} + a_4P_{t-12}$$

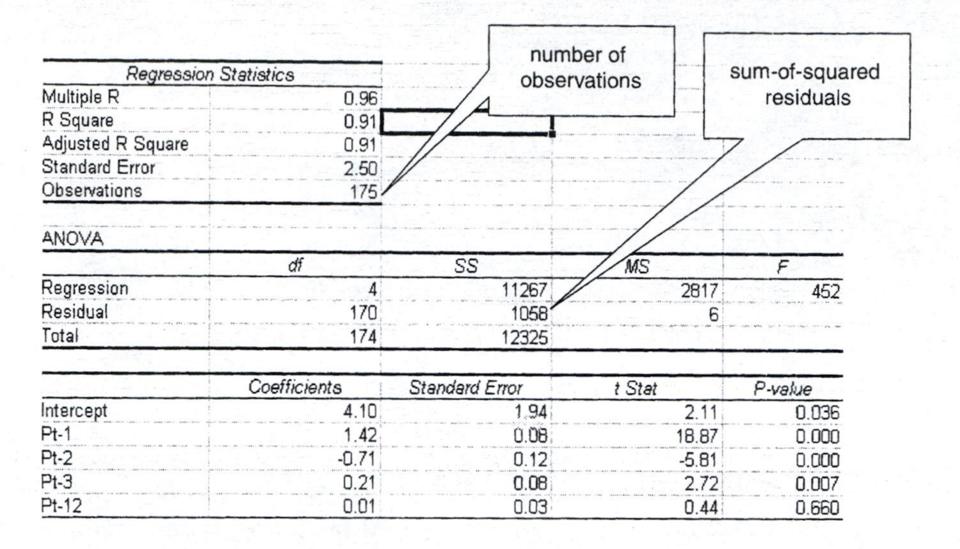

Regression Statistics	
Multiple R	0.96
R Square	0.91
Adjusted R Square	0.91
Standard Error	2.50
Observations	175

ANOVA

	df	SS	MS	F
Regression	4	11267	2817	452
Residual	170	1058	6	
Total	174	12325		

	Coefficients	Standard Error	t Stat	P-value
Intercept	4.10	1.94	2.11	0.036
Pt-1	1.42	0.08	18.87	0.000
Pt-2	-0.71	0.12	-5.81	0.000
Pt-3	0.21	0.08	2.72	0.007
Pt-12	0.01	0.03	0.44	0.660

FIGURE 7.15 Estimating Seasonal Autoregressive Model ($AR_{12}(3)$ Model).

The regression estimates from Excel are shown in Figure 7.15.

$$\textit{Predicted } P_t = 4.10 + 1.42P_{t-1} - 0.71P_{t-2} + 0.21P_{t-3} + 0.01P_{t-12}.$$

Unlike the regression models we estimated before, the actual coefficients have little meaning, and we usually ignore p-values. The seasonal lags are determined by knowledge of the process underlying prices. In this example, we know that it is less expensive to raise calves born in winter, so this leads to prices that are higher in certain months. In the above example three lags were used. The actual number of lags to use is the difficult part of autoregressive modeling, and there are two strategies to pursue. One is to simply use three or more lags and don't look back. The other is to choose a number of lags that maximizes the information contained in the regression. For now, let us just keep our $AR_{12}(3)$ model.

Autoregressive models are almost always used for forecasting into the future. Consider a setting where the current month is July 2004, and we wish to forecast prices for August and September of 2004. Actual prices are shown in Figure 7.16. To forecast prices for August 2004, we simply plug in July 2004 price for P_{t-1}, June 2004 prices for P_{t-2}, May 2004 prices for P_{t-3}, and August 2003 prices for P_{t-12}, as shown in the figure.

$$\begin{aligned}\textit{Predicted August 2004 Price} = {} & a_0 + a_1(\textit{July 2004 Price}) + a_2(\textit{June 2004 Price}) \\ & + a_3(\textit{May 2004 Price}) \\ & + a_4(\textit{August 2003 Price})\end{aligned}$$

However, if we use the same formula to predict September 2004 prices, we run into a problem.

Predicted $P_t = 4.10 + 1.42P_{t-1} - 0.71P_{t-2} + 0.21P_{t-3} + 0.01P_{t-12}$

Current Month = July, 2004	*Forecasted Price for August 2004:*
Price in July 2004 = $84.53	*Predicted Price* = 4.10 + 1.42(84.53) – 0.71(89.14)
Price in June 2004 = $89.14	+ 0.21(87.69) + 0.01(80.30) = **80.06**
Price in May 2004 = $87.69	*Forecasted Price for September 2004:*
Price in September 2003 = $88.08	*Predicted Price* = 4.10 + 1.42(**80.06**) – 0.71(84.53)
Price in August 2003 = $80.30	+ 0.21(89.14) + 0.01(88.08) = 77.37

FIGURE 7.16 Forecasting with an $AR_{12}(3)$ Model.

$$\textit{Predicted September 2004 Price} = a_0 + a_1(\textit{August 2004 Price}) + a_2(\textit{July 2004 Price}) + a_3(\textit{June 2004 Price}) + a_4(\textit{September 2003 Price})$$

The problem is that the August 2004 price is an explanatory variable, but it is currently July 2004, so we do not know that price. In this case, we simply substitute the predicted August 2004 price for the true price, as shown in Figure 7.16. This model uses three lagged explanatory variables: P_{t-1}, P_{t-2}, and P_{t-3}. As mentioned before, the number of lagged variables one should use is not obvious. One can either just pick a particular number of lags (though three or more should be used) and run with it, or one can employ more sophisticated regression techniques.

For reasons too technical for this book, we rarely use p-values to determine how many lagged explanatory variables to include. Instead, one must use information criteria, which are measures of the amount of information contained in a regression. In particular, this section will employ the *Akaike Information Criterion* (AIC) to determine how many lagged explanatory variables to include. For any one regression, the formula for the AIC is

$$\text{AIC} = \ln\left(\frac{SSR}{N}\right) + 2\left(\frac{\text{number of explanatory variables} + 1}{N}\right)$$

The variable N stands for the number of observations used in the regression. In Figure 7.15, one can see that N for the seasonal autoregressive model equals 175. The acronym *SSR* stands for sum-of-squared residuals. Recall that the regression model provides a "best-fit" line for the relationship between the explanatory variables and the dependent variable. The name "best-fit" line indicates there is some prediction error. A residual is just a single prediction error, so the *SSR* is simply the sum of all squared prediction errors. In the seasonal autoregressive model, it is the sum of all 175 squared prediction errors. As shown in Figure 7.15 there are four explanatory variables (P_{t-1}, P_{t-2}, P_{t-3}, and P_{t-12}), and the *SSR* is 1058. Thus, for this model the AIC value is

$$\text{AIC} = \ln\left(\frac{1058}{175}\right) + 2\left(\frac{4 + 1}{175}\right) = 1.8565$$

The lower the AIC value, the greater the amount of information contained in the regression. Next, let us see if including one more lagged explanatory variable (P_{t-4}) will increase the model's information. After reestimating the extra explanatory variable, the model's AIC value is 1.8412. This is lower than the former model, which tells us an $AR_{12}(4)$ is better than an $AR_{12}(3)$ model. Then, we could keep adding an extra lag until the AIC value stops decreasing and then use the model with the lowest AIC value to form predictions of live-cattle prices.

Compared to the regression models using supply and demand information, this autoregressive model seems unsophisticated. It would seem that information on the supply and demand of cattle would predict cattle prices better than autoregressive models. It turns out this is not the case. Autoregressive models are surprisingly accurate predictors, usually outperforming supply and demand models. In fact, one of the authors compared the performance of time-series models to supply and demand models for forecasting cattle prices up to six months into the future. Regardless of the forecast horizon, the time-series models performed best, a finding discovered by many other authors. Anytime prices exhibit trends, or patterns that are difficult to pinpoint using seasonal or supply and demand information, autoregressive models provide accurate projections of the future. Businesses use them frequently to forecast sales, inventories, and prices of other products.

HEDONIC PRICE ANALYSIS

Throughout this chapter we have discussed the general goods of pork and beef as if there was only one type of "pork" and one type of "beef." But clearly, a T-bone steak is better than ground beef, and pork chops are better than pig's feet. There is a supply and demand for pork and within it a supply and demand for various types of pork cuts. For a general good, like meat, there are often many specific goods, or varieties of goods, with their own equilibrium price.

Varieties of a general good differ according to their attributes. Beef can be described by its marbling (fat deposits), tenderness, flavor, and so forth. T-bone steaks tend to be higher in marbling and flavor than filet mignons, but lower in tenderness. T-bone steaks tend to sell for higher prices (on a per pound basis) than filet mignons, indicating that consumers place a higher value on marbling and flavor than tenderness. The value of a good then depends on the attributes comprising that good and the value of each attribute. Hedonic price analysis uses regression to measure the value of these individual attributes. Hedonic price analysis is conducted in regression by making the price of a good the dependent variable and the attributes comprising the good the explanatory variables. Thus, a hedonic price analysis does not describe just supply or just demand, but the price resulting from supply and demand for different varieties of a general good. For example, instead of using regression to describe the supply and demand for "pork" in general, regression is used to articulate how pork prices differ according to the pork cut, tenderness, fat content, and package appearance.

Hedonic Regression: A regression where the price of a product variety is the dependent variable and the attributes describing that variety comprise the explanatory variables.

Consider a good you are more familiar with: graduates from agricultural colleges. There is a supply and demand for college agricultural graduates and an equilibrium price. Yet, no two college graduates are the same. Some have agribusiness degrees and others have animal science degrees. Some are male, some are female, some have master's degrees, and some have work experience but others have none. These factors are indeed attributes that describe any particular college graduate. The salary (i.e., the price) each graduate receives depends on the attributes describing them and the supply and demand for those attributes. Below, we use regression to predict an employee's salary based on their attributes, thereby decomposing the determinants of salary, revealing what individual attributes employers value most.

Andrew Barkley is a professor of agricultural economics at Kansas State University who conducted a survey of graduates from the Kansas State University College of Agriculture. The survey elicited people's salary, degrees, gender, marital status, and years of work experience. Figure 7.17 shows a variety of variables collected from this survey, from dummy variables describing the individual's major and gender, to salary, to years of experience. By using regression to predict salary

salary = salary in 1997 dollars per year

agecon = dummy variable equaling one if the person has an undergraduate degree in agricultural economics, or agribusiness management, or related field

agronomy = dummy variable equaling one if the person has an undergraduate degree in an agronomy or related field

ansi = dummy variable equaling one if the person has an undergraduate degree in animal science or related field

bakesci = dummy variable equaling one if the person has an undergraduate degree in milling, baking, or feed science

female = dummy variable equaling one if person is female

married = dummy variable equaling one if person is married

experience = years of work related experience

noms = dummy variable equaling one if person does not have a master's degree

	Coefficients	*Standard Error*	*t Stat*	*p-value*
Intercept	42900.19	3244.34	13.22	0.00
agronomy	−9277.86	2516.59	−3.69	0.00
ansi	−6154.30	1789.80	−3.44	0.00
bakesci	9179.24	2400.46	3.82	0.00
female	−8313.06	2157.95	−3.85	0.00
married	6282.70	1889.58	3.32	0.00
experience	1357.42	123.84	10.96	0.00
noms	−9221.05	2503.99	−3.68	0.00

FIGURE 7.17 Hedonic Price Analysis of Agricultural College Graduates.

as a function of other attributes, we can develop an understanding for the value employers place on each major, experience, and even gender and marital status. Thus, *salary* will be the dependent variable. To capture the effects of major on degree, the dummy variables *agecon*, *agronomy*, *ansi*, and *bakesci* are available. However, all respondents have at least one of these degrees. Remember that when a set of dummy variables encompasses all possibilities, one must be excluded from the regression. Let us drop *agecon*. Including variables for gender, marital status, experience, and absence of a master's degree results in the following regression model.

$$\begin{aligned} \textit{Predicted Salary} &= a_0 + a_1(\textit{agronomy}) + a_2(\textit{ansi}) + a_3(\textit{bakesci}) + a_4(\textit{female}) \\ &\quad + a_5(\textit{married}) + a_6(\textit{experience}) + a_7(\textit{noms}) \end{aligned}$$

Consider what the value of a_0 signifies. If a person is an agricultural economics or agribusiness major (hereafter, simply agribusiness), their values of *agronomy*, *ansi*, and *bakesci* equal zero. If the person is also male and unmarried, does not have a master's degree, and has no experience, the remaining explanatory variables also have a zero value. All that is left in the equation is *predicted salary* $= a_0$. Thus, a_0 is the salary for the person described above. The coefficient a_1 then tells us the premium/discount associated with having an agronomy degree, *relative to* an agricultural economics or agribusiness degree. The premium or discount paid to females relative to males is given by the value of a_4, and a_7 is the difference in salary without a master's degree than with that degree. Finally, a_6 shows the change in salary from an additional year of work experience. Using the survey data (which is provided at the class website) to estimate the regression yields the estimates in Figure 7.17. As the figure illustrates, all explanatory variables have low p-values and do in fact significantly impact salaries paid.

$$\begin{aligned} \textit{Predicted Salary} &= 42900 - 9278(\textit{agronomy}) - 6154(\textit{ansi}) \\ &\quad + 9179(\textit{bakesci}) - 8313(\textit{female}) + 6283(\textit{married}) \\ &\quad + 1357(\textit{experience}) - 9221(\textit{noms}) \end{aligned}$$

The results indicate that, relative to agricultural economics majors, agronomy majors make $9,278 and animal science majors make $6,154 less, but those in the baking science major make almost $10,000 more. Females make $8,313 less than males. Why females receive a lower salary is not known for certain. It may be that females are discriminated against, or their value is less to the company because they are more likely to take maternity leave or quit to raise children. Employees who are married make $6,283 more than their unmarried counterparts, though we cannot tell you exactly why. Perhaps the responsibilities of marriage create more motivated workers. Or, perhaps the individuals willing to make the sacrifices necessary for a successful marriage will also make similar sacrifices for a successful business. Employers highly value workers who get along well with others, and these people are also more likely to marry and stay married. Finally, each year of experience increases one's salary by about $1,357 per year and the absence of a master's degree reduces salary by $9,221.

The reader should bear in mind that these are only averages, actual salaries will differ than what is predicted here, sometimes by a lot. They are not perfect predictions, though they are the best available. This is just one example of a hedonic price analysis. Other examples abound. Hemlock trees in New England have recently come under attack from a pest named the hemlock woody adelgid. Large sections of hemlock forests have been decimated by the pest. Should the government step in with an eradication program to limit this damage? The answer partly depends on the value people place on hemlock trees. If people truly value the trees, they will pay more for houses located close to these trees. An economist by the name of Elizabeth Murphy conducted a hedonic price analysis of property values in New England, where price was modeled as a function of the house size and type, location, and proximity to hemlock forests. The regression revealed that property values did increase when located closer to healthy hemlocks, so people do value the hemlock trees (Murphy 2004). Whether they value it enough for government to enact costly pest eradication measures is another question. Other researchers have used similar methods to demonstrate how living close to a swine farm depresses one's property values. And yes, the regression shows that living close to a swine farm depresses property values (who would have thought?) (Palmquist, Roka, and Vukina 1997). As long as the equilibrium price of a good can be decomposed into individual attributes defining that good, hedonic price analysis can be a useful mechanism for estimating the relative value of those individual attributes.

SUMMARY

Many important questions boil down to asking what is the effect of one variable on another. How does a 5% rise in pork production costs affect pork prices? How do beef advertisements influence beef demand? What is the difference in corn prices from February to March? Do males get paid different salaries than females? These questions can be answered using regression analysis. This chapter discussed how to conduct regression analysis assuming no prior knowledge of regression. By the end of the chapter, we were doing some pretty fancy stuff! Do not be deceived; good regression analysis requires more than what was covered in this chapter. However, mastering the material covered here is a necessary first step in learning more advanced regression analysis techniques.

CROSSWORD PUZZLE

For answers with more than one word, leave a blank space between each word.

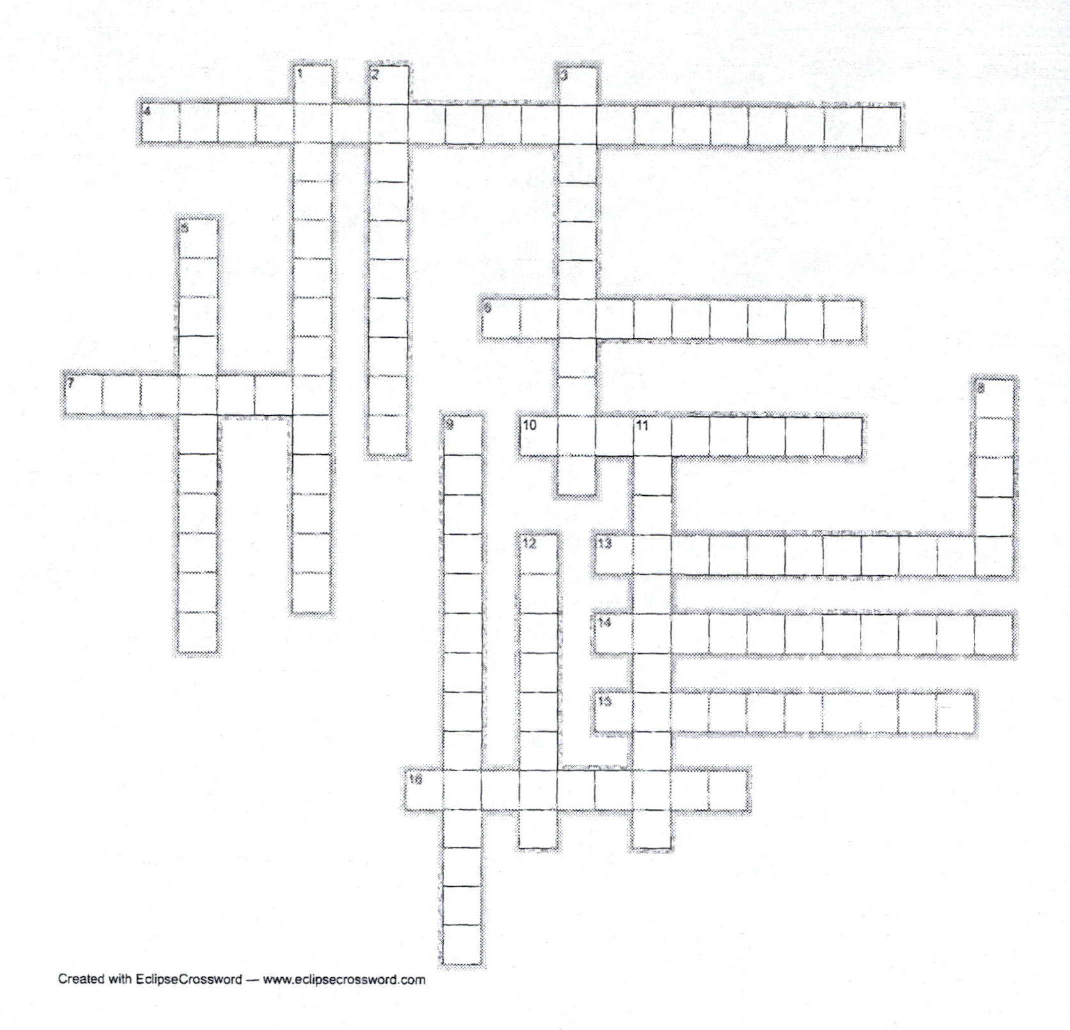

Across

4. An index used to convert nominal prices to real prices.
6. Someone who benefits from an activity without paying their fair share.
7. The _______ indicates the probability that an explanatory variable does not truly impact the dependent variable value.
10. In regression, independent variables are used in an equation to predict the value of a _______ variable.
13. The price _______ illustrates the percent change in price resulting from a 1% increase in the supply of a good.
14. The Akaike _______ Criterion is a measure of the information contained in a model that can be used to select the number of lagged prices in an autoregressive model.
15. Hedonic price analysis is conducted using regression where the price of a good is the dependent variable and the _______ describing the good comprise the explanatory variables.
16. If a supply function contains the lagged quantity as an explanatory variable, it must be a _______-_______ supply function.

Down

1. _______ models predict future prices based on past prices.
2. In _______ analysis, computer programs are used to calculate the unknown coefficients of a prediction equation in order to minimize the sum-of-squared prediction errors.
3. Information on past prices, seasonality, or time trends are used to predict prices in _______-_______ price analysis.
5. Information on the supply and demand of a commodity is used to predict prices in _______ price analysis.
8. A variable that equals either zero or one.
9. In the equation $\ln(Y) = a + b\ln(X)$, b equals the _______ _______ in the value of Y resulting from a 1% change in the value of X.
11. Another word for an independent variable is a(n) _______ variable.
12. The _______ indicates the percent of variation in the dependent variable explained by the explanatory variables.

STUDY QUESTIONS

1. On the class website there is an Excel spreadsheet titled "Steak WTP." The data were collected in a grocery store where subjects were given a regular steak and were then allowed to bid in an auction to upgrade to a "guaranteed tender steak." Column A has the bid for each person, which reveals how much more each person is willing to pay for a guaranteed tender steak over a regular steak. You are a company wishing to market these guaranteed tender steaks and wish to know your target market. Using the female dummy variable and regression, demonstrate whether males or females are willing to pay more for the guaranteed steaks, or if there is no real difference between their willingness-to-pay. Calculate how much more females or males are willing to pay, if any. Show your exact regression equation, and use p-values to justify your answer.

Suppose you estimate the following supply and demand functions. Assume an agricultural good with a production lag of five time periods. Thus, quantity supplied today depends on prices and costs five time periods ago, and price today depends on the fixed supply determined five time periods prior. Q_t and P_t refer to the quantity and price of the good in time t, C_t is the price of a production input, and S_t is the price of a substitute good.

$$\textit{Short-Run Supply Function:}\quad Q_t = -200 + 4(P_{t-5}) + 0.5(C_{t-5}) + 0.2(Q_{t-5})$$
$$\textit{Short-Run Demand Function:}\quad P_t = 1000 - 5(Q_t) + 0.5(S_t) + 0.05(Q_{t-5})$$

2.a. Solve for the long-run supply and demand curves.

2.b. Suppose that the average value of $C_t = 20$ and $S_t = 100$. Calculate the long-run supply and demand curves.

2.c. Calculate the long-run equilibrium price and quantity.

Suppose you estimate the following log-log supply and demand curves.

$$\textit{Log-Log Supply Function:}\quad \ln(Q_t) = a_0 + a_1\ln(P_{t-5})\, a_2\ln(C_{t-5})$$
$$\textit{Log-Log Demand Function:}\quad \ln(P_t) = b_0 + b_1\ln(Q_t) + b_2\ln(S_t)$$

3.a. What is the own-price elasticity of supply?
3.b. What is the demand flexibility?

At the textbook website, find the spreadsheet titled "Cotton Yields." This sheet contains data on average South Carolina cotton yields from 1953 to 2006. Yield is measured in pounds of cotton per acre. Each year, technological advancements allow producers to obtain more cotton from each acre. Thus, each year, we would expect the average cotton yield to rise. To reflect technological advancements made each year, we often use a "time trend" variable like the one in the spreadsheet. Notice that the time trend variable equals 1 in 1953, 2 in 1954, and 54 in 2006.

4.a. Estimate the relationship between time and yields using the following regression model: $Yield = a_0 + a_1(Time\ Trend)$
4.b. Based on the p-value of a_1, do cotton yields truly increase over time?
4.c. By how much do cotton yields increase each year?
4.d. Using this regression, what is the predicted cotton yield in 2010?

CHAPTER EIGHT

International Trade in Agriculture

Free trade, one of the greatest blessings which a government can confer on a people, is in almost every country unpopular.

—*Thomas Babington Macaulay*, British politician, historian, and writer

No country was ever ruined by trade.

—*Benjamin Franklin*

INTRODUCTION

A wheat buyer in North Carolina is free to purchase wheat from a Kansas farmer. They are free to strike a deal, without any government interference. A wheat buyer in Costa Rica is not so free. She must pay a tax (referred to as a tariff) to purchase from the Kansas farmer. This tax limits trade of wheat between the two countries. In 2005 the U.S. government passed a Central American–Dominican Republic Free Trade Agreement (CAFTA), which made trade between the United States and Central American countries almost as easy as trade between Kansas and North Carolina (except for the longer distance, of course). The agreement basically lowers the cost of importing and exporting to Costa Rica, the Dominican Republic, El Salvador, Guatemala, Honduras, and Nicaragua. The agreement ultimately passed, but in July 2005 it was unclear whether it would. Some groups supported it, and among those were most U.S. grain and meat producers. Some opposed it, and among those was the powerful sugar lobby. The debate was called the "trade fight of 2005," according to U.S. Trade Representative Chris Padilla (Schuff 2005a, 2005b).

This is a typical story. One group is for free trade, because they profit from it, and another opposes free trade because they would lose money. What you probably do not know is that 99% of all economists would say free trade is good for society in almost every circumstance. And economists rarely agree to this extent. The purpose of this chapter is to illustrate why economists advocate free trade, and why passing free trade agreements like CAFTA can be so difficult. A discussion of the importance of trade to agriculture is also provided. Specifically, the chapter objectives are

1. illustrate the concept of comparative advantage and gains from trade
2. define and discuss a country's trade balance and its meaning
3. cover the controversial topic of creative destruction
4. review the role of exchange rates in trade
5. study various barriers to trade
6. develop a mathematical model of international markets

Trade is important to agriculture. Around 21% of U.S. agricultural production is exported to other countries. These countries, in return for our agricultural exports, send goods to our country. We export corn to Japan and they export cars to us; we trade our poultry for Russia's oil. Around 24% of U.S. crop production is exported. Roughly half of coarse grains like corn and barley is marketed abroad. In fact, the United States has been a net exporter of agricultural commodities since 1959. About 7% of beef and 18% of poultry is exported, but exports of other animal products like hides, tallow, and fish are 30% of production (Economic Research Service 2002). The United States is the leading exporter of agricultural products, but it is also the leading importer (Jerardo 2004).

Because so many agricultural firms market their commodities overseas, agricultural industry profits hinge critically on changes in other countries. A recession or an exchange rate change can lead to a boom or bust in the U.S. agricultural industry. For these reasons, it is imperative that agriculturalists understand the economics of international trade. It is perhaps more important to understand trade of agricultural products in the context of trade in all products. International trade, in general, is a good thing, increasing wealth for all trading parties. And when trade flows change in response to market changes, such as an increase or decrease in U.S. exports, those changes usually benefit one party and harm another. The overall impact, however, tends to be positive, as illustrated below.

HOW TO GET SOMETHING FOR NOTHING

Free trade between countries benefits both countries. Both can become richer without working more. To see how, we need to use our imagination. Suppose we have two countries: the United States and Cuba. Due to U.S. trade policies, there is virtually no trade of goods and services between the two. What if, suddenly, the two countries were allowed to freely import and export to one another? There is good reason to believe both countries would become richer. To illustrate this, we will tell a story of two clans. The story is simple, but there are numbers involved so it will take some focus on your part. The story of two clans trading is an excellent metaphor for two countries trading and illustrates perhaps the most profound economic concept ever discovered.

A Story of Two Clans

Forget countries, let us think smaller. There are two primitive clans: Wu Tang Clan and No Tang Clan. Each clan produces and consumes only grain and salmon. Both work 10 hours per day. If Wu Tang Clan fishes, it can catch 1 salmon per hour, and if it gathers grain, it can produce 1 bushel of grain (hereafter, bushel) per hour. This means the opportunity cost of 1 salmon is 1 bushel. If it chooses to fish for 1 hour, it gives up the 1 bushel it could have produced during that hour. For the same reason, the opportunity cost of 1 bushel is 1 salmon. No Tang Clan can also catch 1 salmon for each hour it spends fishing, but it only gathers a half bushel of grain for each hour spent gathering. They simply are not as talented at gathering grain. Thus, the opportunity cost of salmon for No Tang is a half bushel, and the opportunity cost of 1 bushel is 2 salmon. To gather 1 bushel takes 2 hours, which could have been used to catch 2 salmon. Being primitive clans, they do not use money. Instead of dollar prices, the opportunity costs describe the number of salmon and bushels given up. But if you think about it, that is all money is anyway. If a CD is $15 and a candy bar is $1, the cost of one CD is 15 candy bars. For now, suppose these two clans do not trade with one another.

For the 10 hours they work, each clan must decide how many hours to fish for salmon and how many hours to gather grain. There are many different possibilities, and if we graph these possibilities, we get something called a *production possibility frontier* (PPF), as shown in Figure 8.1. If No Tang Clan spends all 10 hours gathering grain, since it gathers a half bushel per hour, it will gather 5 bushels. If it spends all 10 hours fishing, since it catches 1 salmon per hour, it will catch 10 salmon. Or, if it spends 5 hours gathering grain and 5 hours fishing, it will gather 2.5 bushels of grain and catch 5 salmon. Each of these three possibilities are shown as points in Figure 8.1. A similar PPF can be drawn for Wu Tang Clan. If Wu Tang works 10 hours fishing, it will have 10 salmon, and if it spends all 10 hours gathering grain, it will have 10 bushels. Or, it might spend 5 hours fishing and 5 hours gathering, at which point

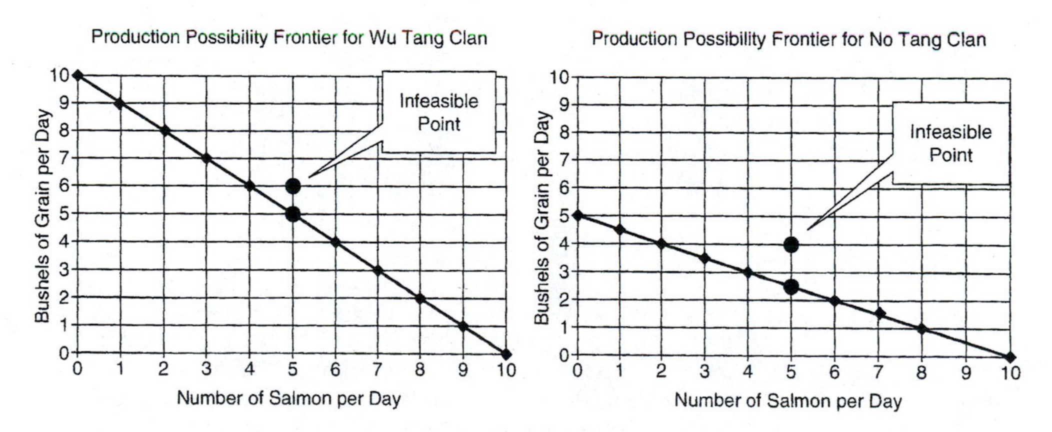

FIGURE 8.1 Production Possibility Frontiers.

	Opportunity Cost of **One Salmon**	Opportunity Cost of **One Bushel of Grain**
Wu Tang Clan	1 bushel of grain	1 salmon
No Tang Clan	1/2 bushel of grain	2 salmon

FIGURE 8.2 Opportunity Cost of Production.

it will have 5 salmon and 5 bushels. Observe the PPF in Figure 8.1 for Wu Tang Clan. It can produce and consume any combination of salmon and grain on this line. The same can be said for the PPF for No Tang Clan.

The region above the PPF is called the *infeasible region*, as neither clan can consume or produce in that region. Figure 8.1 shows an infeasible point for each clan. Wu Tang Clan cannot produce 5 salmon and 6 bushels of grain—it is simply not possible assuming they work 10 hours per day. Similarly, No Tang Clan cannot produce 5 salmon and 4 bushels; if it produces 5 salmon, the most bushels it can gather with the remaining 5 hours is 2.5. Clans would like to consume at infeasible points—we all want more goods for free—but they cannot. However, we will now show that if the two clans engage in trade, both can consume at these infeasible points. Through trade, both clans can get something from nothing. To illustrate how, we must first learn the concept of *comparative advantage*. Figure 8.2 shows the opportunity cost of salmon and grain for the clans.

Comparative advantage: One group is said to have the comparative advantage in the production of a good if it can produce that good at a lower opportunity cost than another group.

Notice two things. No Tang Clan can produce salmon at a lower opportunity cost. It gives up only a half bushel to obtain a salmon, whereas Wu Tang Clan gives up a whole bushel. This means No Tang Clan has the comparative advantage in salmon. This is like saying No Tang can produce salmon cheaper. Similarly, Wu Tang Clan has the comparative advantage in grain production, because it produces grain at a lower opportunity cost (gives up less salmon than No Tang Clan). The concept of comparative advantage leads to one of the most important economic results: If groups have a comparative advantage in certain goods, they can engage in a mutually beneficial trade. To realize this gain from trade, they should follow the rule of trade shown in Figure 8.3.

To illustrate, let Wu Tang Clan produce only grain, because they can produce grain at a lower opportunity cost. If Wu Tang produces only grain, they will produce 10

Gains from Trade: If two groups have a comparative advantage in particular goods, they can strike a mutually beneficial trade. Each group should produce the good for which they possess a comparative advantage and trade them for the goods for which they do not have a comparative advantage, so both groups can become wealthier without working more.

FIGURE 8.3 Important Economic Result.

		Production	Exports	Imports	Consumption
Wu Tang Clan	Salmon	0	0	5	5
	Grain	10	4	0	10−4 = 6
No Tang Clan	Salmon	10	5	0	10−5 = 5
	Grain	0	0	4	4

FIGURE 8.4 Mutually Beneficial Trade Between Clans.

bushels.[1] Similarly, if No Tang Clan produces only salmon, they will produce 10 salmon. The clans then strike a deal. Wu Tang trades 4 bushels of grain for 5 of No Tang's salmon. Wu Tang exports 4 bushels and imports 5 salmon, and No Tang exports 5 salmon and imports 4 bushels. As Figure 8.4 shows, Wu Tang now consumes 5 salmon and 6 bushels of grain, and No Tang has 5 salmon and 4 bushels of grain. Here is the magic of trade: Both of these are points we said were infeasible in the absence of trade. Look back at Figure 8.1. Suppose that before trade, both clans produced and consumed at the point on the PPF line. They would like to reach the infeasible point, but that point is simply not possible without trade. After trade, Wu Tang consumes 5 salmon and 6 grain. Wu Tang receives 1 extra bushel without giving up any salmon. It is almost as if they got 1 bushel of grain for free! The same can be said for No Tang Clan. If No Tang was consuming 5 salmon and 2.5 bushels (shown in red), with trade they can consume 5 salmon and 4 grain: they receive 1.5 bushels of grain for free!

Absolute Advantage Does Not Matter

There are some things Belize can produce that the United States cannot, like bananas. There are also things the United States can produce that Belize cannot, like hops. This provides an obvious motivation for the two countries to trade—both can produce an item the other cannot. But let us ignore this for the time being. Assume both countries can produce all goods. Picture the United States with its ability to produce in large quantity virtually any item. Then, picture the country of Belize, where mules still draw plows and the tallest building in the country is a temple built by the Mayans somewhere between 300 and 900 AD. The United States has an *absolute advantage* over Belize in the production of almost everything. We can produce more cars, more corn, more everything than Belize, even on a per-person basis. Some would interpret this to imply that because the United States has an absolute advantage in everything, the United States could never gain from trading with Belize. How could an advanced country benefit from trade with a developing country?

A country has an absolute advantage over another country in the production of a good if it can produce more of that good.

Refer back to Wu Tang and No Tang Clan. Wu Tang can produce just as much salmon as No Tang, but more grain. Overall, Wu Tang Clan has an absolute advantage over No Tang Clan. No Tang does not have an absolute advantage in any good. This is akin to a developed country (Wu Tang) with large production capabilities and a

[1]Wu Tang Clan works 10 hours per day and can produce 1 bushel per hour, yielding 10 bushels if they only gather grain.

developing country (No Tang) with limited production capabilities. Still, both countries are made wealthier through trade. The reason is that absolute advantage does not matter. So long as the opportunity costs of production are different, each group will have a comparative advantage in something and can be made better off through trade. So even though the United States may have an absolute advantage in every good over Belize, both countries will have a comparative advantage in something, and both will be made better off through trade. Through this simple example, you can see the falsehood in the argument that free trade will mean developed countries take advantage of underdeveloped countries. Both countries benefit!

The Point of the Story

You may have heard the saying, "there is no such thing as a free lunch." However, we have now seen that with trade there is a free lunch. In our previous story we showed two clans able to strike a mutually beneficial trade—they both became richer without working more. But how realistic is this story? Very realistic. To show this, let us begin questioning assumptions made in the story. The story showed that groups with different opportunity costs of production can both consume more, without working more, by trading. Will groups always differ in their opportunity cost of production? Will they always have a comparative advantage in certain goods? Yes, almost certainly. If they do not, their opportunity costs are identical, and though they cannot gain from trade, they cannot be harmed by trade either—trade will not make either group worse off and will probably make both groups better off.

Does trade between clans really describe trade between countries? Yes. The groups were called clans, but we could just as easily called them towns, states, or countries. The clans bartered for goods. They did not use money, even though countries trade using money. Well, money just represents goods, and opportunity costs can be thought of as a price. The story only showed it was *possible* for trade to make everyone richer, not that it has to. Yet, remember that trade is voluntary. Why would any clan engage in a trade that makes itself worse off? Trade is like magic; it makes goods and services appear out of thin air that could never be obtained without trade. Trade makes people richer without working more. The story above demonstrates a concept that virtually all economists agree on: Free trade between any groups, including countries, will usually make all groups better off.

If you think about it, this is a major reason why you participate in society. Have you ever thought of immersing yourself in a secluded forest, eating and consuming only what you produce? Probably not, because you realize you would be quite poor. Instead, you trade things you produce with society for other things, just like the United States trades goods and services it produces with the world for other things. The authors have a comparative advantage in writing economic textbooks (we hope you agree), so we produce textbooks and trade them for other things like food, clothes, and music. You have a comparative advantage in something else and will spend most of your life producing those goods and trading them with society through that thing we call money.

Next time you hear someone say we should close our borders and produce everything in the United States, ask them the following. If that is a good idea, then should Oklahoma close its borders and not trade with other states? Maybe New York City should produce everything it needs within its city limits. If they say that sounds crazy, ask them how it is any different from closing our borders to other countries. You will soon find their ideas are not based on sound logic. Hopefully, this section provided you with sound economic reasoning that trade is usually mutually beneficial. More logic cannot hurt, though. The story above showed that trade *can* benefit both parties, not that it always does. Although it makes sense that small groups will not engage in trade unless both are made better off, it is harder to see this when the groups are countries and the countries trade through markets. Therefore, in the next section we will tell another story, one where countries trade through competitive markets. Like the last story, we will again find that both countries gain from trade.

TRADE BETWEEN COUNTRIES

When countries trade, they do not send delegates from each country to "strike a deal." They trade through markets. We purchase Japanese cars through a car market, and Japanese consumers purchase our wheat through wheat markets. Like the previous section, we will tell a simple story of trade: trade of rice between the United States and Japan. These countries currently trade little rice due to government barriers. Japan places an import tariff on rice (a tax on all imports of rice into Japan). The United States can produce rice cheaper than Japan (the price in United States is about 20% less than the price in Japan), but the tariff makes it more expensive to Japanese consumers.

Suppose the import tariff was removed. You know from an earlier chapter on the Indifference Principle that rice prices in the two countries will converge. U.S. producers will start selling rice to Japan because their prices are higher. This increases the supply of rice in Japan and lowers Japanese rice prices. There will now be less rice sold in the United States (because more rice grown in the United States is sent to Japan), and so price in the United States will rise. The Japanese price falls, and the U.S. prices rises. This will continue until the price difference equals the transportation costs between the two countries, and U.S. producers are now indifferent between exporting to Japan or selling rice at home.

That was an old story; here is a new one. We want to extend the story to ask: How are the citizens in Japan and United States affected by the elimination of the import tariff? Are they better or worse off? All we need to answer this question is to know how consumer and producer surplus changes. Figure 8.5 shows supply and demand curves for rice in the United States. The left-hand graph depicts a situation where the import tariff prohibits trade. All rice produced in the United States must be sold in the United States.[2] With the tariff, the U.S. price and quantity is P_1 and Q_1, consumer surplus is the area a, and producer surplus is the area b.

---

[2]We are implicitly assuming the United States does not trade rice with any other country either.

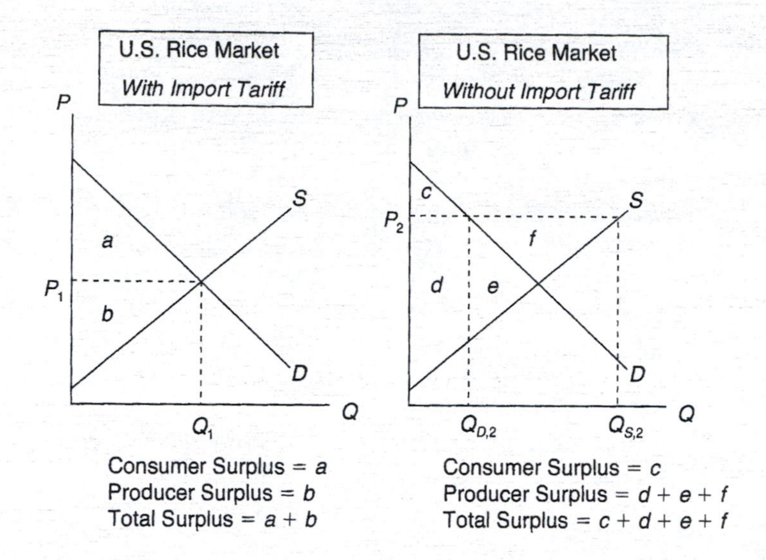

FIGURE 8.5 Impact of Japan Import Tariff Removal on the United States.

Now suppose the tariff is removed and Japan is free to import rice from the United States. Rice starts leaving the United States bound for Japan. Producers in the United States continue to export rice to Japan, and the price of rice in the two countries begins to converge. Eventually, the price in United States equals P_2, where producers are indifferent between selling in Japan or the United States. Consequently, U.S. consumers decrease their rice consumption from Q_1 to $Q_{D,2}$ (the subscript D stands for demand). Consumers are worse off, but by how much? We can measure this by consumer surplus: the area above price and below demand for all quantities consumed. With the import tariff, consumer surplus is area a, and without the tariff is area c. Area c is clearly smaller than a, so consumers are worse off by the area $c - a$ (this is the difference in the two areas, made to be negative to reflect a loss to consumers). How are U.S. producers affected? Recall that producer profits can be measured by producer surplus: the area below price and above supply for all quantities produced.[3] Producers now receive a higher price of P_2 and increase production from Q_1 to $Q_{S,2}$ (the s subscript stands for supply). A total of $Q_{D,2}$ is sold in the United States, and $Q_{S,2} - Q_{D,2}$ is exported to Japan. With the import tariff, producer surplus is area b and without the tariff is area $d + e + f$. Producer surplus rises from the elimination of the import tariff.

We are mostly concerned with how the United States as a whole is affected. Luckily, both consumer and producer surplus are measured in dollars, so we can define total surplus as the sum of producer and consumer surplus and use that as a measure of societal welfare. As you see from Figure 8.5, the elimination of the import tariff improves societal welfare by the area f. Consumers lose and producers benefit,

[3]Recall that producer surplus measures profits above fixed costs.

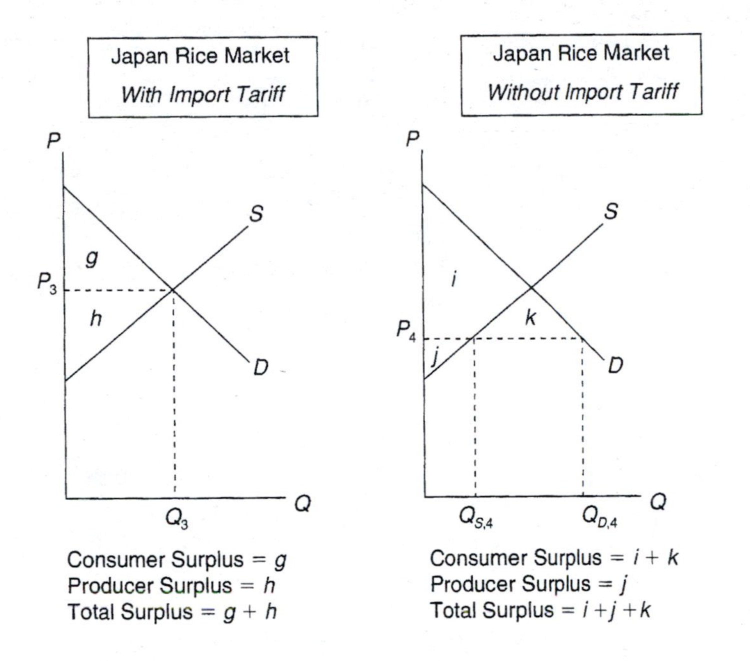

FIGURE 8.6 Impact of Japan Import Tariff Removal on Japan.

but the benefit to producers clearly outweighs the loss to consumers, and society as a whole is made better off. The United States as a whole benefits from eliminating the tariff and engaging in free trade. What about Japan?

Figure 8.6 shows the impacts to Japan. When the tariff is removed, the price of rice in Japan falls from P_3 to P_4. Japanese consumers benefit from the lower price, but producers are hurt. Although Japanese production falls from Q_3 to $Q_{S,4}$, exports more than make up for this production decline. Japan imports $Q_{D,4} - Q_{S,4}$ from the United States, consuming a total of $Q_{D,4}$ rice. Consumer surplus in Japan rises from g to $i + k$ when the tariff is removed, so consumers are better off. Although producer surplus falls from h to j, Japan as a whole is better off. The gains to consumers outweigh the costs to producers, and total surplus increases by the area k. Essentially, Japan is richer by k amount. Societal welfare in Japan improves by eliminating the import tariff on rice.

The Point of the Story

This story makes a profound statement. When governments eliminate barriers to trade between countries, and those countries increase trade through markets, both countries are made better off. Although some people in each country are harmed by trade, others gain and the gains outweigh the losses. This again shows the power of markets to improve the human lot. Also notice that with proper government transfers, there need be no losers from trade. To illustrate, recall that total surplus in the United States rose by the area f in Figure 8.5. Suppose this area equals $1 million.

There is nothing to prevent governments from forcing those receiving the one million in benefits to share some with the losers. In fact, because the size of the pie increases (the pie being total surplus, a measure of happiness), if winners are forced to share with losers, there is nothing to prevent everyone from winning. In reality, this can be difficult to do, because it is not always clear who loses and wins by how much. But what this does show is that a government can pursue free trade and transfer policies together in a way to minimize the burden any one individual experiences from free trade. A perfect government would always be able to pursue trade in a way that makes everyone better off.

BALANCED TRADE

It is not uncommon to see news reports of a "trade deficit" in the United States. Sometimes the reports concern our trade deficit with particular countries, like Japan or China, and sometimes they concern our trade deficit with the world. The report will undoubtedly use words like *worsen*, in order to scare you into thinking something is wrong. For example, in the beginning of 2005 most U.S. newspapers ran a story depicting the horrors of a trade deficit. The stories will contain lines like, "For December, the deficit actually shrank. But at $56.4 billion, it was still the worst monthly showing ever." The word *worst* implies that the trade balance is bad. In reality there is little reason to fear a trade deficit, and no reason to fear a trade deficit with one particular country. The purpose of this section is to illustrate why, so that you can interpret news reports more accurately.

First, what exactly is a trade deficit? It depends on whether you are talking about a trade deficit with a single country or a trade deficit with the world. The *trade balance* for a country is its exports minus its imports. All exports and imports are measured as value, like the value of exports and imports in U.S. dollars. If the trade balance is negative, we import more than we export and a trade deficit exists. If the trade balance is positive, we export more than we import and have a trade surplus. The trade balance with the world equals exports to all other countries minus imports from all other countries. The news report above would make it seem we run a world trade deficit. In reality, we do not, and even if we did run a world trade deficit and could sustain that deficit, it would be a good thing. It would be like winning the million dollar sweepstakes over and over again.

- Trade Balance = Total Exports − Total Imports, measured in U.S. dollars.
- If trade balance > 0, we have a trade surplus.
- If trade balance < 0, we have a trade deficit.
- There is a trade balance with the world, and a trade balance with each country.

The Zero Trade Balance

Suppose we ran a trade deficit with the world. This means we export less than we import. Put another way, other countries are sending us all kinds of goods and services, and we are sending them little in return. Why would other countries send us toys, cars, food, and other stuff without demanding we export goods and services to them? This is like a stranger giving you a car in return for nothing. You "import" one car from the stranger and "export" nothing. The value of your imports are clearly larger than the value of your exports, and you run a trade deficit with this stranger. Clearly, no sane stranger would make this trade, and if they would, you would gain from it.

Using another example, we will demonstrate why the trade balance must be zero (why countries will generally run neither a trade balance or a surplus). Suppose we import all kinds of goods and services from the world. Japan sends us cars, China sends us fireworks, and Germany sends us beer. These are not "countries" sending us things for free; they are businesspeople from these countries selling things to the United States. Americans pay for goods and services using one currency: the U.S. dollar. We pay for cars with dollars, fireworks with dollars, and beer with dollars. What do the exporters from Japan, China, and Germany do with these dollars? These dollars cannot buy anything in their country, so they must spend it in the one place that takes U.S. dollars—the United States.[4] Thus, every dollar spent on foreign imports must return to the United States in the form of exports. So do not worry about hurting U.S. workers the next time you drink a Heineken. Sure, you could have chosen an American beer, but American workers as a whole are not hurt. The money you spent on the Heineken—an import—will eventually come back to the United States in the form of a foreigner purchasing our exports. Imports must equal exports, as long as countries do not give us things for free, and we do not give other countries things for free.

Perhaps this may seem like an overly simplistic story. When we purchase goods from China, we do not actually give them U.S. dollars; we go to a foreign exchange market, exchange dollars for yuan (the Chinese currency), and purchase from Chinese businesses using the yuan. More accurately, Honda may establish a car dealership where Americans purchase cars with dollars. The owners of Honda live in Japan. The profits they make in the United States are profits in dollars, and the Japanese owners exchange the dollars for yen. They then spend the yen in their own country. Although a foreign exchange market is used, the story really does not change. When U.S. dollars are brought to the exchange market to exchange them for yen, there is someone on the other side of the transaction exchanging yen for dollars. And what can you do with dollars? You spend them in the United States on goods and services. Again, every dollar spent on foreign imports comes back to the United States in the form of exports. Again, we find that exports must equal imports.

The One Exception: Foreign Aid

Previously we argued that the trade balance must always be zero. Neither trade deficits nor surplus can exist. That is not exactly true when there is foreign aid. When we offer foreign aid to developing countries, we are essentially delivering them goods and services and asking nothing in return. We are exporting more than we are importing. The United States and most developed countries regularly give foreign

[4]This is not literally true. Other countries have adopted the U.S. dollar as their currency as well. However, you get the same result accounting for these anomalies, so we will ignore it here.

aid, which means we regularly export more than we import, which means we regularly run a trade surplus. In reality, the United States runs a trade surplus through foreign aid, but the amounts are usually so low that we ignore them and say we run a zero trade balance.

Are the Newspapers Wrong?

Previously we argued that countries must run a zero trade balance, except when foreign aid exists. Because foreign aid is relatively small, we simply say all countries run a zero trade balance. However, the news story cited above clearly stated the United States runs a trade deficit. Somebody must be lying, right? Not exactly. When the government measures exports and imports, they include only goods and services. They do not include investments. Foreigners often sell goods and services to the United States and then use the dollars from those sales to buy American investments like stocks and bonds. The government records us importing goods and services but does not record us exporting investments. We may import $1 billion worth of goods and services, and the foreigners take that $1 billion and purchase one-half billion dollars of U.S. goods and services (our exports) and one-half billion dollars worth of U.S. investments. The government records imports of $1 billion but exports of only one-half billion and concludes our trade deficit equals one-half billion.

Let us return to the question: Is a trade deficit with the world bad? On the one hand, no. There really is no trade deficit. If anything, there is a trade surplus because we give foreign aid. However, foreign purchases of U.S. investments is not counted as an export, so the government measures a trade deficit. On the other hand, when we run a trade deficit, we are essentially borrowing money from foreigners. Suppose Japanese businessmen sell cars in the United States and use the dollars from their sales to purchase Treasury Bonds. When someone buys a treasury bond, they are making a loan to the U.S. government, where the government pays back the principal plus interest at a later date. We essentially borrowed money from the Japanese to purchase cars and must pay back the Japanese at a later date, plus interest. The point is that the trade balance—as the government measures it—indicates net foreign investment. The United States imports more goods and services than it exports, making it a net borrower of the world. Japan, on the other hand, runs a trade surplus, which means they are net lenders to the world.

- Net Foreign Investment (NFI) = Exports of Goods and Services Minus Imports of Goods and Services.
- If NFI > 0, the country is a net lender to the world.
- If NFI < 0, the country is a net borrower from the world.

Is being a net borrower bad? There is no clear answer, but it partially depends on what we do with the money we borrow. Just like a household, if you borrow money to fund current consumption, you will be poorer in the future, but if you borrow money and invest it in a business, and that business is profitable, you will be richer in the future. Whether the United States is borrowing money wisely remains to be seen. Moreover, one reason we run a world trade deficit is that America is a great place to invest your money. The economy is robust, safe, and typically growing. International investors know this and regularly purchase U.S. investments. If the United States runs a world trade deficit it is partially because it is the best place in the world to invest money, and it stands to reason the deficit should not be viewed as a bad thing.

If anything, it is a sign of a stable and strong economy. So far, we have mainly discussed the trade balance between a single country and the world. Let us now discuss the trade balance between two particular countries.

Is Your Trade Deficit with Wal-Mart a Bad Thing?

We have never met you. All we know is that you are college students. Yet, we are willing to bet that you run a trade deficit with Wal-Mart. Ask yourself how much you spend at Wal-Mart each year; these are your imports from Wal-Mart. Ask yourself how much Wal-Mart pays you for goods and services you produce (including your labor); these are your exports to Wal-Mart. If you do not work at Wal-Mart or sell goods to Wal-Mart, your exports to Wal-Mart are zero. Subtract your imports from your exports and that is your trade balance with Wal-Mart. We are willing to bet that your trade balance is negative, meaning you run a trade deficit with Wal-Mart.

Who cares if you run a trade deficit with Wal-Mart? Not the authors. Similarly, there is no cause for concern if we run a trade deficit with one particular country. If we run a trade deficit with Japan (which we do), we must run a trade surplus with some other country, because the trade balance must be zero. The authors run a trade deficit with Wal-Mart, but we run a trade surplus with Oklahoma State University (we sell lots of labor to the university, and the only thing we buy are football tickets; our exports exceed our imports). You run a trade deficit with Wal-Mart, but a trade surplus at the place you work. Always remember two things when reading the newspaper: (1) a country runs neither a trade surplus nor a trade deficit with the rest of the world, unless foreign aid is involved; and (2) a trade deficit or surplus with any one country is not important and should not be given a connotation like "good" or "bad."

CREATIVE DESTRUCTION: WHY FREE TRADE IS OFTEN UNPOPULAR

The discussion above creates a rosy picture of trade. Trade makes everyone better off, we argue, and even if there are some losers, they can be compensated by the winners so that they become winners as well. With a perfect government, trade would always make everyone happier. But, of course, a perfect government is a dream. In reality some people are harmed by trade with other countries, and they tend to yell louder than the people who gain, making trade seem like a bad thing.

Consider the following story. It will seem silly at first but the basic story is true. Before the 1970s most cars purchased in the United States were made in the United States. There was an opportunity cost of car production; there always is. Let us make up some numbers and keep the story simple. America's workforce is divided between car and corn production such that it produces 100,000 cars and 1,000,000 bushels of corn. America's PPF (production possibilities frontier) was such that to produce one car it had to give up 10 bushels of corn.

Then, a spectacular discovery was made. Out in the ocean is an island of magic, a very specific type of magic. If you shipped five bushels of corn to the island, they would magically turn into one car, just like the kind of cars sold in the United States, and sometimes better. The U.S. leaders ran some numbers and discovered what this means. If all workers currently making cars began raising corn instead, and all that extra corn was shipped to the island to be transformed into cars, it would have 200,000 cars—double the number of cars currently produced. Before using the magic island, the country had 100,000 cars and 1,000,000 bushels of corn. After using the magic island, it would have 200,000 cars and 1,000,000 bushels of corn. It could obtain 100,000 extra cars from using the magic island, without giving up anything or working harder, and is clearly richer.

However, many Americans thought this was a bad idea. The reason is that the people in the automobile industry would experience a temporary cost of moving from car builders to corn farmers. No one wants to lose their job. Even though Americans as a whole would be better off, a subset of Americans would be experience a temporary, but perhaps painful adjustment. But because the magic island will be there forever, and the transition cost is temporary, over many years the United States must become richer if they would use the island.

This refers to the concept of *creative destruction*. To become richer, we needed to destroy automobile jobs, and use that free labor to create more corn. One thing had to be *destroyed* for another to be *created*, and that which is created has a greater value than that which was destroyed. After sending that extra corn to the magic island, we would all be richer. We will now let you in on a secret. That magic island is called Japan, and it wasn't really magic that turned corn into cars; it was trade. Ships were loaded with corn and sold to Japan. Japanese pay for corn with their currency, called the yen. We could take these yen and purchase Japanese automobiles. The Japanese learned how to make better cars for less, so in the end, it was in America's best interest to produce fewer cars, more corn, and trade corn for cars. It was cheaper for us to produce corn and trade the corn for cars than to make the cars ourselves. We had a comparative advantage in corn and Japan in cars. Both countries could become richer without working more. Instead of freely trading with Japan, however, there was political pressure to limit imports of Japanese cars. Ronald Reagan even persuaded Japan to voluntarily limit the number of cars sold to the United States. Why would politicians prohibit us from becoming richer through trade with Japan? The reason is that those who would lose from free trade have more incentives to lobby than those who would gain.

Consider sugar in the United States Sugar is obtained by raising either sugarcane in Florida or sugar beets in other states. We could import sugar for less money than it costs to produce it, that is, if sugar imports were not taxed. Lucky for the U.S. sugar producers, they are protected from foreign competition. The government charges high taxes on sugar imports, making imported sugar more expensive than domestic sugar. Because of the import taxes (called import *tariffs*), sugar producers make more money. However, we know that the United States as a whole would be made better off if it eliminated these import tariffs. Although sugar producers would lose about $1,046 million dollars, consumers would gain $1,900 million.[5] For every dollar sugar

[5] 1999 dollars.

producers would lose, sugar consumers (you and I) would gain close to two dollars (Beghin 2001). Why, then, has this program been around for so long? The reason is simple. The number of sugar producers is few, and each producer stands to lose a lot if the import tariffs are removed. If there is even a rumor of eliminating the tariffs, they start beating on their congressperson's door. Compare this to yourself. You just read that the sugar program costs you money, yet you will do nothing about it. You will not write a letter to your congressperson or even shoot off an e-mail. Neither will we. You will do nothing, because, as a single consumer, you stand to gain little from eliminating the tariffs. The benefits are concentrated and the costs are dispersed.

Politicians have good reason to love protectionist policies (policies "protecting" a country from free trade). The benefits are clearly visible to the parties it benefits, but the costs are mostly hidden from the parties it harms. Pursuing protectionist policies allows a politician to make friends without gaining enemies, which is often difficult in the political arena. This can certainly be said for the sugar tariffs. The benefits of the sugar import tariffs are spread among a few producers, and they know they need the tariffs to survive. The costs are spread among all consumers, and if tariffs were removed, we would barely notice it. Although the costs of the tariff outweigh the benefits, the benefit per producer is large, but the cost per consumer is small. Although the total costs outweigh the benefits, producers have more political power due to their small numbers. Producers lobby hard, and with their small numbers attract little attention. They make larger profits and you pay higher prices, and the country as a whole is worse off. We have fewer toys than we could have without the tariff.

In the 1920s, car ownership became a common occurrence. Businesses began regularly shipping goods via automobile. This was a great improvement over the horse and carriage, and even the train, which was limited to railroad tracks. Let there be no doubt: The use of the automobile has made us wealthier and happier. Yet, let there also be no doubt that the horse and carriage industry was hurt. Some horse trainers and some carriage makers had to find other jobs. Their jobs were *destroyed* so automobile jobs could be *created*. Advances in agriculture like mechanization and fertilizer allows us to produce the same amount of food with fewer farmers. There are one-third less farmers now than in 1900, yet farm output is seven times greater. Where did the farmers go? They now produce other things like computers, Playstations, and iPods. Agriculture jobs had to be *destroyed* so that computer jobs could be *created*. From these technological advancements in agriculture arose a wealthier nation, one with more food and more toys.

There are basically two ways to produce a good. One is to directly produce it. The other is to produce an alternative good and trade it with other countries for the good. Sometimes it is cheaper to make things inside our own country, and sometimes it is easier to obtain it through trade. As we have seen, free trade with other countries provides wealth gains that are no different from technological advancements; with both, you get more goods and services for the same amount of work. Both also involve creative destruction. Yet, we are far more open to creative destruction stemming from technological innovation than free trade. Hopefully, this chapter has demonstrated that this bias is unjustified.

EXCHANGE RATES IN TRADE

Trade between countries is more complicated and costly than trade between groups within a country for many reasons. Distance, language, and different laws make striking a deal no easy matter. Another obstacle to trade between countries is currency. To buy goods in another country, we must buy their currency first. If you sell goods in another country, your revenues are in a foreign currency and you must convert them to your currency. These currency conversions take place in exchange rate markets. Let us return to the example of exchanging rice between the United States and Japan. With free trade, the United States exports rice to Japan. In the market for U.S. rice exports, the supply curve refers to the willingness of U.S. rice producers to export rice, and the demand curve refers to the willingness of Japanese consumers to import rice. These supply and demand curves are shown in Figure 8.7.

When we draw supply and demand curves, we usually just label the *y*-axis *P* for price, without stating what currency the price refers to. It is usually not important, but because Japanese consumers and U.S. rice producers use different currencies, it is important. In this case, we decided to state currency in U.S. dollars. This is fine for U.S. rice producers; that is the currency they respond to when determining how much to export. Japanese consumers make purchases using the yen, so Figure 8.7 should be modified to reflect this. On August 2, 2005, the exchange rate for yen was 112 yen/dollar. This means that one dollar will purchase 112 yen, and that one yen will purchase 1/112 dollars in foreign exchange markets. Thus, on this day, one dollar equaled 112 yen. Let us assume that the supply and demand curves in Figure 8.7 refer to this exchange rate.

- If the dollar appreciates in value relative to the yen, one dollar purchases more yen, and one yen purchases less dollars.
- If the dollar appreciates relative to the yen, the yen depreciates relative to the dollar.

What if the dollar *appreciates* in value, meaning it can purchase more yen? If one dollar purchases more yen, then one yen purchases *less* dollars. This means the cost of importing rice into Japan rises for Japanese consumers. Think of it this way. To purchase rice from the United States, the Japanese must first use their yen to

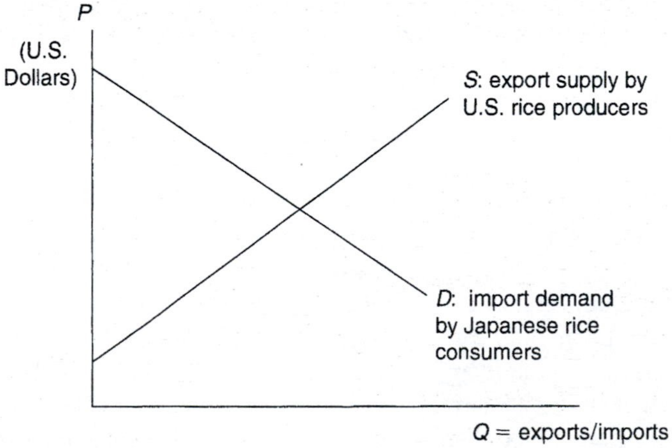

FIGURE 8.7 Market for Exports of U.S. Rice to Japan.

purchase U.S. dollars and they pay U.S. farmers for their rice in dollars. If it takes more yen to purchase those dollars, then the price of rice to the Japanese rises, even if the price of rice in U.S. dollars stays the same. Because it is more expensive for Japan to import U.S. rice, their quantity demanded of U.S. rice falls. Even when the U.S. price in dollars is the same (remember the y-axis is in U.S. dollars), Japan purchases less rice because it takes more yen to obtain one dollar. Suppose the opposite occurred and the U.S. dollar *depreciates* relative to the Japanese yen. This implies the yen appreciates relative to the dollar; one yen can purchase more dollars. This makes the cost of importing U.S. rice cheaper, and Japan responds by increasing their demand for U.S. rice.

This is why exchange rates are important to agriculture. The United States is a net exporter of agricultural products, meaning we export more agricultural products than we import. Roughly 20% of all U.S. agricultural production is exported (Economic Research Service 2006). If the dollar appreciates relative to other currencies, our exports become more expensive to other countries, demand for our exports falls, and so will farm income. For the same reason, if the dollar depreciates, then demand for our exports will rise and farm income will rise with it. This does not mean that it is "good" when the dollar depreciates. When the value of the dollar falls, our exports become cheaper to other countries so we export more, but our imports from other countries become more expensive. Put another way, if the dollar depreciates, we sell more wheat abroad, but German beer costs us more. Because exchange rates determine exports and imports, they settle to a value that sets exports equal to imports. Exchange rates are a price for currency, and they serve an important role of directing trade flows between countries.

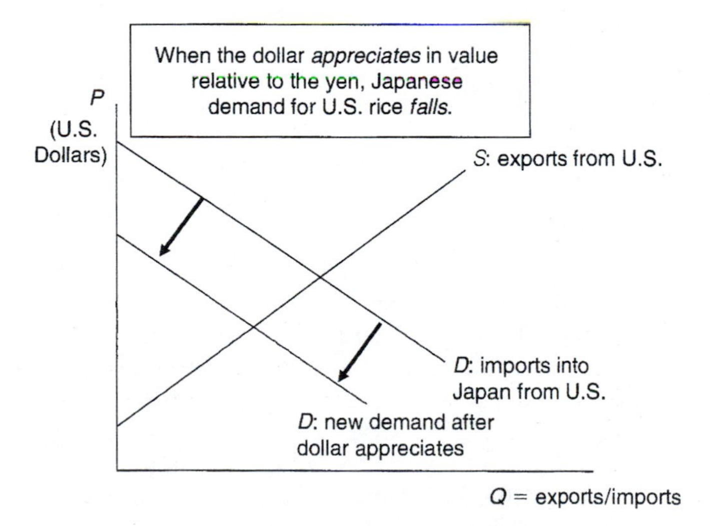

FIGURE 8.8 Impact of Exchange Rate Changes.

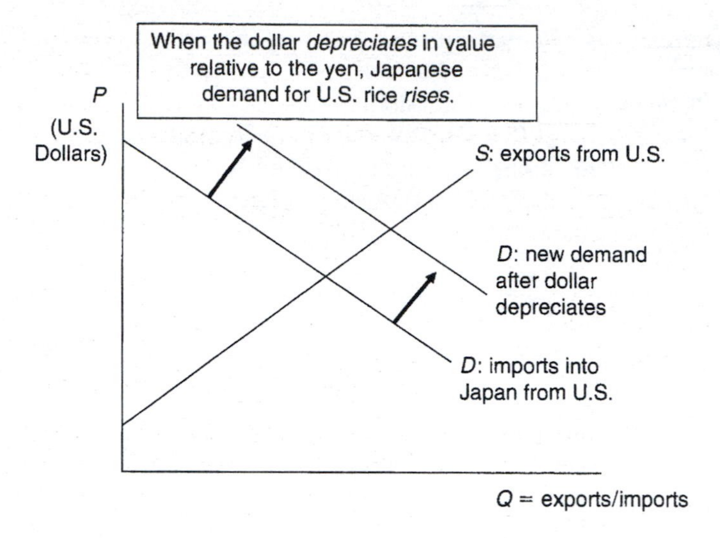

FIGURE 8.9 Impact of Exchange Rate Changes.

Case Study: Exchange Rates, China, and the Declining Dollar

There are numerous currencies in the world, which means there are numerous exchange rates. Many exchange rates are determined in a currency market, meaning the exchange rates are determined by negotiations between buyers and sellers of currency. In other countries, governments fix the exchange rate, and they do it in a rather strange way. The number of Chinese yuan one U.S. dollar can buy, for example, is fixed by the Chinese government. There are many in China who would like to exchange their yuan for dollars, but when they try to, the Chinese government comes in and steals the transaction from them. Imagine every time you try to purchase a car, the government comes in and offers a dollar more and steals your purchase. That is basically what China does when some of its citizens try to purchase dollars.

Here comes the weird part. After the Chinese government has given up all this yuan to purchase all these dollars, it sits on the dollars. They are stored in a vault collecting dust. Why, you say? The government wants foreigners to hold lots of yuan to import Chinese goods, but it does not want its citizens to hold many dollars to purchase U.S. goods. They wish to promote their exports but limit their imports. By keeping the U.S. dollars stored in vaults and prohibiting their sale to Chinese citizens, the dollar becomes more scarce in China. The result is that the dollar becomes "overvalued" relative to the yuan, because the dollar appears scarcer than it really is. This similarly makes the yuan "undervalued" relative to the dollar.

This benefits the United States in one way but harms it in another. When China exchanges yuan for dollars and sits on the dollars, it allows its exports to exceed its imports. China views this as a good thing. To them, this is like "winning" against the

United States because they are selling us more than we are selling them. But think about it, who is really winning? China is exporting many goods and services but is not really asking for much in return. If they continue to sit on these dollars, they are basically giving us goods and services for free. When your exports are greater than your imports, you are giving away more than you take, and that makes you poorer.

Some would argue that China will not continue to sit on U.S. dollars forever. Eventually, their reserves of U.S. dollars will become so large that they will certainly want to spend it. If they do, then all of these dollars will start pouring into the United States to purchase American goods. American exports to China will increase dramatically. Dollars will no longer seem "scarce," and the exchange rate (number of yuan that one dollar can purchase) will fall as the dollar depreciates. If the dollar depreciates, the yuan will appreciate; Chinese exports will fall and imports will rise. This may create an instability in international markets. At first, dollars were tucked away in Chinese vaults. Then, the dollars come pouring into the United States from China. Put differently, at first, U.S. exports to China are small, then they explode. Many will begin to wonder just how valuable the dollar is, and as the exchange rate falls dramatically, people will become uncertain over its future value.

Markets perform better—they serve society better—when it operates in a stable environment. The Chinese approach to managing its currency is viewed by many Americans as introducing unnecessary volatility in international markets. If instead, China just allowed its currency to be determined by markets, perhaps we could avoid this instability. Others contend that markets themselves lead to volatile prices, and that is true. Artificially setting exchange rates, and then letting markets adjust them periodically, can generate more volatility than completely free markets though. Recently, under pressure from the United States, China did release some of its dollar reserves, and the value of the dollar relative to the yuan fell. This will undoubtedly benefit U.S. farmers, because the depreciation of the dollar increases demand for agricultural products.

BARRIERS TO TRADE

It should be obvious by now that the authors are pro-trade. There are many economic theories and much empirical evidence indicating that trade is beneficial for all countries. This is not to say that *all* trade is good. Indeed, one can identify cases where trade brings greater wealth in the short run but lower wealth in the long run. A popular appeal for limiting trade is national security. We import huge amounts of oil, and that oil is happily used by the American consumer. Yet even President George W. Bush warned we were "addicted to oil." Many are calling for research into alternative fuels, like ethanol made from corn, which allow us to produce our own fuel rather than import fuel from abroad. Yet, despite the few exceptions, free trade in general benefits the poor and rich alike in every region of the world. Fortunately, the world is moving towards freer trade, but barriers to trade still exist. Given that politicians have incentives to pursue protectionist policies (policies that erect barriers to trade between countries), they will probably always be with us in some form. It is at this

Barriers to Trade

(1) Import Tariffs
(2) Import Quotas
(3) Voluntary Import Restraints
(4) Quality Restrictions
(5) Export Subsidies
(6) Export Taxes
(7) Trade Embargoes
(8) State Traders
(9) Exchange-Rate Distortions

point we should discuss the arguments for limiting trade between countries and the type of trade barriers used.[6]

Import Tariffs and Import Quotas

To fully understand international trade we must understand the types of barriers to trade and their possible justifications. *Import tariffs* and *quotas* are designed to limit imports into a country. Import tariffs charge a tax for each unit imported, and quotas designate a maximum amount that can be imported. Often, the two are combined so that a certain amount may be imported, but anything over that amount is charged a tax.

Voluntary Import Restraints

Voluntary import restraints usually stem from political pressure. In the 1980s, Ronald Reagan convinced Japan to voluntarily limit the number of cars they exported to the United States. The word *voluntary* is deceiving though. President Reagan threatened to impose import tariffs if they did not voluntarily limit their exports.

Quality Restrictions

Quality restrictions seem justifiable; they prohibit imports that do not meet certain quality standards. The longest running quality restriction was the German Reinheitsgebot (purity) law that began in 1516 and was only recently repealed. This law stated that all beer sold in Germany must be made using only water, barley, hops, and yeast. Many countries, including the United States, include rice as a major beer ingredient. Thus, the German Purity Law prohibited companies like Budweiser and Coors from selling in Germany. Were the purity laws protectionist, or were they meant to ensure consumers high-quality beer? Most self-claimed beer connoisseurs would claim that beer made with rice is indeed of lower quality. However, the vast majority of U.S. beer drinkers consume beer made with rice, and rice makes beer less expensive. Some consumers prefer beer made with rice, and it is relatively easy for a consumer to determine whether the beer includes rice, so it is unlikely that this law protected consumers.

Recent years have witnessed a pronounced increase in nontariff trade barriers related to food quality and safety issues. For example, since 1989, the European Union has banned imports of U.S. beef due to the use of added growth hormones in U.S. beef production. Similarly, the European Union has imposed a moratorium on approving new varieties of genetically modified crops and has implemented a mandatory labeling law on foods containing genetically modified ingredients. Although one

[6]The types of barriers to trade listed here are a minor adaptation from Knutson, Penn, and Boehm (1990).

might argue that the European Union has erected protectionist policies to protect their agricultural producers, their consumers might actually be more concerned about such issues than Americans. It is not as though U.S. citizens are without their fears. For example, concern is often expressed about importing beef from Mexico and Central America, where food safety standards are not as stringent as the United States.

This chapter has drilled home the concept that free trade is better for society. Is this always the case? Not necessarily if there are concerns over food safety and quality. If consumers are concerned about the quality or safety of a product from a foreign country, it is possible that certain trade barriers can enhance domestic welfare. However, one must always be wary of quality restrictions and check to determine whether there are real food safety concerns, or if food safety concerns are artificially generated to protect domestic producers.

Politicians can impose quality restrictions, giving the appearance they are protecting consumers, when their real motivations are otherwise. In March of 2003 Russia placed a ban on U.S. poultry citing unsanitary practices during chicken slaughter. It is no coincidence that this occurred right after President George W. Bush placed higher import tariffs on steel, harming the Russian steel industry. Russia was not really concerned about food safety; they were concerned with protecting their steel industry. Most countries now have trade agreements, limiting the extent to which they erect barriers to trade. These agreements usually allow trade barriers when there are safety concerns, making quality restrictions sometimes the only trade barrier available. In some cases quality restrictions are appropriate. After the BSE outbreak in Britain, countries quickly prohibited imports of English beef. There was a serious safety concern, and quality restrictions alleviated this concern.

Export Subsidies

Export subsidies are one of the most widely used trade barriers, especially by the European Union. Often, politicians want to make sure their farmers receive a higher price than farmers in the rest of the world. For example, wheat may be selling for $3.30 across the world. This means a Kansas wheat farmer will receive about $3.30 if they sell their wheat on the market. The government may want wheat farmers to receive $4.30, instead. Sometimes this is because it costs the farmer more to grow the wheat than they can receive in the market. The government will then either buy the wheat from the farmer at $4.30 and sell it on the market at $3.30 (losing $1.00 in the process), or it will pay the farmer a subsidy of $1.00 per bushel in addition to the $3.30 per bushel the farmer receives from her sales.

The United States and the European Union (EU) were the largest users of export subsidies before 1995. In the 1980s, the two countries fought each other's wheat export subsidies fiercely. If the EU increased its wheat subsidies, that would put the U.S. wheat farmers at a disadvantage, and so the United States increased their subsidies in kind. This went back and forth, using much of taxpayer's money in the process. Eventually realizing this subsidy war served neither country's interest, at the

Uruguay Round Agreement on Agriculture in 1995, they agreed to decrease their subsidies. There are still disagreements over the use of export subsidies around the world, but the general trend is to decrease their use.

Other Trade Barriers

Similar to export subsidies is an export tax, by which a country taxes its exports. In December of 2004, China announced it would tax its exports of textiles. This was done to avoid a confrontation with the United States. China can produce textiles cheaper than the United States, and imports of Chinese textiles had led to a decline in the U.S. textile sector. It was anticipated that the United States was going to limit textile imports from China. To avoid this, China implemented an export tax. This assured U.S. producers that Chinese imports would not run them out of business, and so calls for import barriers ceased.

Have you ever purchased a Cuban cigar? Not legally in the United States. The reason is that the United States has a *trade embargo* placed on Cuba, where neither exports to nor imports from Cuba are allowed. A wheat farmer in Canada can sell her wheat to only one place: the Canadian Wheat Board. One can purchase Canadian wheat from only one place: the Canadian Wheat Board. This makes the Canadian Wheat Board (CWB) something like a monopoly on Canadian wheat. They are not a real monopoly because the CWB faces competitors abroad, but many claim that the market power attained by the board allows them to extract a higher price for Canadian wheat farmers. For this reason, we refer to the CWB as a *state trader*. Imagine a Nebraska wheat farmer who must negotiate wheat prices all by herself. She is a small player in a big market. Then imagine a Canadian wheat farmer who has the CWB negotiating prices on her behalf. The CWB is a fairly big player in a big market; thus, it seems likely that the CWB may give Canadian farmers a competitive advantage over U.S. farmers. But then, if those running the CWB are not competent, Canadian wheat growers may be at a disadvantage.

An earlier section of this chapter discussed how exchange rates can be set by prices or by governments. When governments set exchange rates, it allows the government to give their firms a competitive advantage. Recall how China frequently purchases U.S. dollars and stores them in vaults, preventing them from being spent by anyone. This allows China to temporarily run a trade surplus and gives their businesses that export items like textiles to the United States a competitive, and perhaps unfair, advantage over U.S. producers. This is an example of an *exchange-rate distortion*.

Five Goals of the World Trade Organization

Trade between countries should be . . .

(1) Without discrimination
(2) Freer
(3) Predictable
(4) More competitive
(5) More beneficial for less developed countries

The Future of Trade

Society seems to becoming increasingly aware of the benefits of trade. Economists have been virtually unanimous in their support of free trade, and others are increasingly agreeing with them. Still, the incentive for politicians to erect trade barriers is present. Most politicians know trade is good for society, but they also know getting

Unfair Trade Barriers

- Children's polyester sweaters are hit with a 35% import tax, but mink coats can be imported duty-free.
- Tariffs increase the price of orange juice by 40%, but French Perrier water faces a miniscule 0.8% tax.
- Mothers who buy imported infant formula must pay 17% more, but those who can afford lobster can buy it duty-free.
- Each American is allowed to consume only one teaspoon of foreign ice cream, one teaspoon of foreign butter, one ounce of foreign dried milk, one pound of foreign cheese, and four pounds of foreign beef.
- Even though families with incomes below $10,000 spend three times as much of their disposable income on milk as families with incomes of $35,000, U.S. quotas raise the price of dried milk by 161%.
- Even though families in the bottom fifth of the income distribution spend almost four times as much of their income on clothes as do those in the top fifth, clothing quotas and tariffs increase retail prices by as much as $40 billion per year.

Source: Bovard, J. "Americas Unfairest Taxes: Tariffs and Quotas." National Center for Policy Analysis. Policy Report No. 171. May 1992.

reelected requires support from key groups. Fortunately, the general trend is towards greater trade liberalization. Perhaps the most important step was creating the World Trade Organization (WTO) in 1995. For countries to move together towards free trade, they need a place to hold forums, set of rules for fair trade, and a place to settle disputes. This place is the WTO. It is not "headed" by any single country; it is the creation of many countries and was created with five general goals in mind. These are the goals that all member countries agreed upon and agreed to make steady advances towards, with certain exceptions (World Trade Organization 2005).

The first goal is that countries should not discriminate among countries in their trade practices. If you erect trade barriers to one country, you must do so for all countries. If you freely trade with China, you should freely trade with Mongolia. There are exceptions. Countries are allowed to discriminate through the creation of free trade agreements, like NAFTA, CAFTA, and the European Union. A country is allowed to discriminate towards a country if that country is competing unfairly. When the United States placed an import tariff on steel, Russia retaliated by placing an import embargo on poultry from the United States. Russia did not violate the WTO agreement because the United States was not playing fair with their import tariff, which put Russian steel producers at a disadvantage. Steel producers in the European Union were harmed as well. But then, why was the United States allowed to place the import tariffs in the first place? The WTO allows trade restrictions to *temporarily* protect domestic industries. Finally, as one would hope, countries are allowed to restrict trade with another country for safety concerns, like when countries banned imports of British beef due to BSE concerns.

WTO member countries have agreed to pursue *freer trade*. Each country still has barriers to trade. The European Union still uses export subsidies and the United States still has a sugar quota. However, all countries have agreed to dismantle these barriers over time and have already taken steps to do so. The goal is to pursue free trade gradually and through negotiations. Another goal is *predictability* of trade. If countries arbitrarily erect trade barriers with no warning, this disrupts all economies, causing wealth losses to all. A strong economy requires substantial capital investment, and firms will not invest in a business venture that relies on exports or imports unless the market is stable. Trade should be *more competitive*, meaning producers in each country should compete on a level playing field. When the European Union subsidizes its farmers, they are not on a level playing field with the United States. The first part of this chapter illustrated the gains from trade. It showed how all countries are better off from trade. But these gains only occur if firms in all countries compete on a level field.

Finally, WTO members have agreed that trade should be *more beneficial for less-developed countries*. Developing countries are allowed more time to meet trade agreements than developed countries. The pursuit of freer trade and more open societies around the world has become known as globalization. People often oppose globalization believing that it harms developing countries, but the real effect is otherwise. Globalization has served to lift millions out of poverty. Some are concerned that globalization "makes the rich richer and the poor poorer," but the evidence points otherwise. True, not every group has benefited from globalization, but those who do not benefit (like those in sub-Saharan Africa) are just bypassed from it, not harmed by it (Sachs 2005).

THE MATHEMATICS OF INTERNATIONAL MARKET EQUILIBRIUM

International markets were described above using supply and demand diagrams. Yet it is often desirable to capture markets in mathematical form. When organizations calculate the impacts of trade barriers or trade agreements, they need numbers, and the numbers come from and are used in mathematical models. Using an earlier example, suppose that the United States cannot trade rice with Japan because of Japan's excessively high import tariffs. Consequently, the price of rice is higher in Japan. Suppose also that these tariffs are expected to be eliminated soon. We know the Japan rice price will fall and the U.S. rice price will rise, but by how much? Answering this question requires mathematical models of supply and demand, very similar to those in covered in Chapter 3. Suppose you know the supply and demand curves in both countries, and they are reported below. The curves are reported in U.S. dollars for Japan, which means they assume a particular exchange rate. In the next section you will learn how to calculate the supply and demand change if exchange rates move.

Using these supply and demand curves, we want to solve for the equilibrium price of rice if countries are allowed to freely trade rice (no import tariffs). We also want to know how much rice will be exported from the United States into Japan. For simplicity, we will assume the cost of transporting rice from the United States to Japan is so small we can ignore it, meaning after free trade ensues the price will be bid up in the United States and down in Japan until the prices are equal.

There are three general steps to calculating the new international price of rice and trade volume.

Step 1: Calculate the equilibrium price of rice in both countries if trade does not occur.
Step 2: Using these prices, find the export supply curve and the import demand curve.
Step 3: Set the export supply curve equal to the import demand curve and solve for the equilibrium price and quantity. The price will be the international price.

U.S. Rice Market	Japan Rice Market (at exchange rate 112 yen/dollar)
Supply: $P_{Dollar} = 200 + 1(Q)$	Supply: $P_{Dollar} = 500 + 1.5(Q)$
Demand: $P_{Dollar} = 400 - 1.2(Q)$	Demand: $P_{Dollar} = 800 - 1(Q)$
P_{Dollar} = price in U.S. dollars	P_{Dollar} = price in U.S. dollars
Q = tons	Q = tons

FIGURE 8.10

Step 1: Solve for domestic prices and quantities without trade.

If the United States and Japan did not trade, market prices would be set by the supply and demand in each country. Using the supply and demand curves for each country, let us calculate this equilibrium, as shown in Chapter 4.

Step 2: Find export supply and import demand curves.

This is the easy part. The export supply and import supply curves are shown in Figures 8.11 and 8.12, but how do we know those are the correct curves? Think about

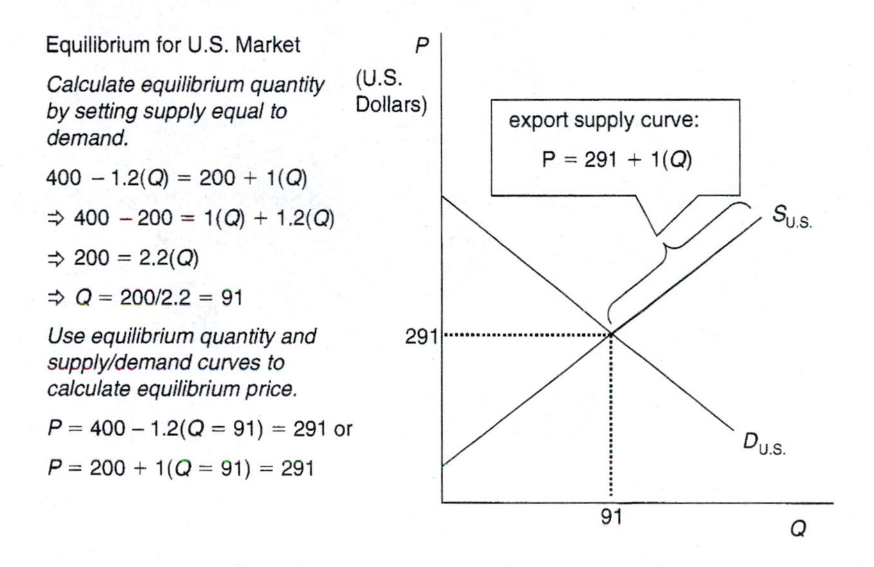

FIGURE 8.11 No Trade Equilibrium and Export Supply Curve for United States.

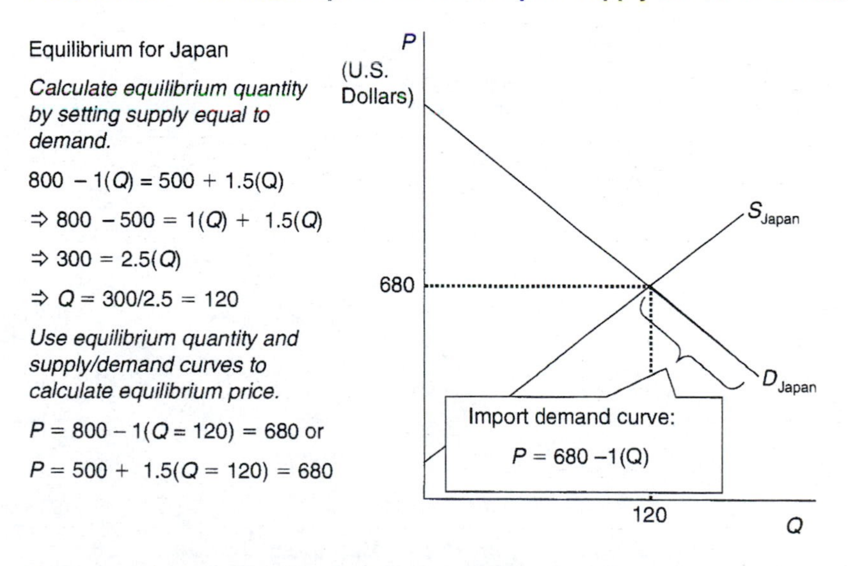

Figure 8.12 No Trade Equilibrium and Export Supply Curve for Japan.

it this way. For the United States to produce more rice for export to Japan, it must receive a price higher than \$291. Thus, the supply curve intercept is \$291. If the price is lower than \$291, it will not export any rice. If the price rises to \$300, how much more will it produce? The U.S. supply curve has a slope of 1. So when the price rises by one, the quantity supplied will increase by 1. Therefore, the export supply curve is $291 + 1(Q = \text{Exports})$. Similarly, Japan has no reason to buy rice from the United States unless it can get a price lower than 680. Thus, the import supply curve has an intercept of 680—once price falls below 680 they will purchase imports. The slope of the Japan demand curve is negative one, so for every one-dollar decrease in price they will import one additional ton of rice. Through this logic, we know the Japan import demand curve is $680 - 1(Q = \text{Imports})$.

Step 3: Solve for price and quantity that sets export supply equal to import demand.

After trade, the price in the United States rises from 291 to 485.5, and the price in Japan falls from 680 to 485.5. This new common price of rice is often referred to as the international or world price of rice. However, we should caution you that the equilibrium trade quantity is not 194.5. That is, the level of exports and imports after trade is not 194.5 units. Exports and imports (which must equal one another) will be greater than this. As the world rice price rises, price moves up the U.S. export supply curve and both U.S. exports and production rise. However, as the price rises, U.S. consumers are also consuming less. For example, if domestic production rises by

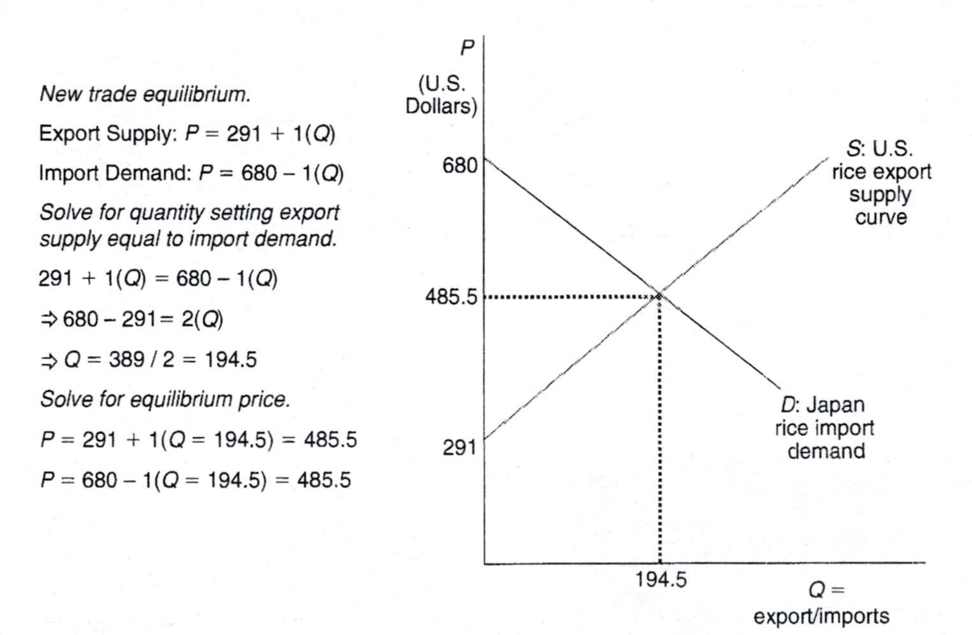

FIGURE 8.13 International Market Equilibrium.

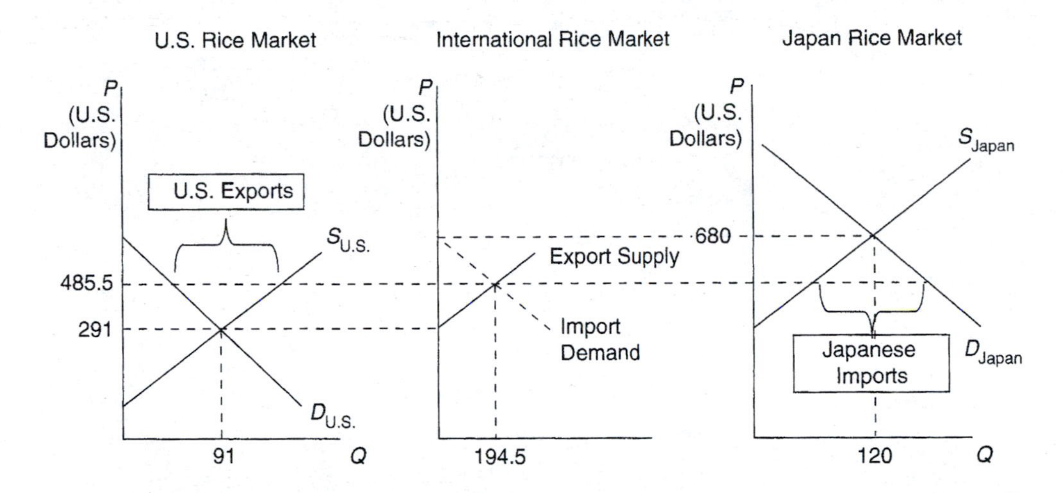

FIGURE 8.14 Three-Panel Diagram of Trade Equilibrium.

50 tons and domestic consumption falls by 20 tons, then total exports are 50 + 20 = 70 tons. Similarly, as the rice price falls in Japan, Japanese consumers are purchasing more and Japanese rice producers are producing less. The actual imports are greater than the increase in consumption. Imports equal the increase in consumption plus the decrease in production.

Figure 8.14 reiterates the international market equilibrium, but shows the big picture in a three-panel diagram. This is one of the most familiar trade diagrams in economics. Before trade, the price in Japan was higher than in the United States. If allowed, U.S. rice producers would gladly export to Japan at a higher price, and Japan consumers would gladly import at a lower price. When trade is allowed, assuming no transportation costs, the price in both countries will equal. This new "international" or "world" price of 485.5 is where the export supply curve and import demand curves cross. At this price, U.S. exports of rice equal Japan imports of rice. The price of rice in each country now depends on market conditions in both countries.

The Mathematics of Exchange Rates

The United States currency is the U.S. dollar, and the Japanese currency is the yen. In reality, supply and demand in each country is stated in each country's currency, as shown in Figure 8.15. However, to calculate the international market equilibrium, we need to convert all supply and demand curves to a common currency. This is easier than you might think. We could convert the U.S. curves to yen, or the Japanese curves to dollars. Let us do the latter. When writing this section, the newspapers reported the exchange rate between the dollar and the yen to be 112 yen per dollar. This means that one dollar was equivalent to 112 yen, and one yen was equivalent to 1/112

Japan Rice Market (Yen Currency)

Supply: $P_{Yen} = 56{,}000 + 200(Q)$

Demand: $P_{Yen} = 89{,}600 - 112(Q)$

P_{Yen} = price in Japanese Yen

Q = tons

Japan Rice Market (U.S. Currency)
at exchange rate 112 yen/dollar

New Supply: $(P_{Yen} = 56{,}000 + 200(Q)) / 112$

New Demand: $(P_{Yen} = 89{,}600 - 112(Q)) / 112$

New Supply: $P_{Dollar} = 500 + 1.79(Q)$

New Demand: $P_{Dollar} = 800 - 1.00(Q)$

P_{Dollar} = price in U.S. Dollars

FIGURE 8.15

dollars. To convert yen to dollars, all we need to do is divide the number of yen by 112. More importantly, if we divide both sides of the Japan supply and demand equations by 112 (preserving the equations' relationship), we convert their equations to the U.S. currency, which can be compared with supply and demand equations in the United States.

What if the dollar appreciates in value and the new exchange rate is 140 yen per dollar? How will the demand for rice change? Using the formulas above, we know that the new demand curve will be $P_{Dollar} = (P_{Yen} = 89{,}600 - 112(Q))/ 140 = 640 - 0.8(Q)$. The Japan demand curve intercept decreases from 800 to 640, indicating a decrease in Japanese rice demand. Japanese demand for rice falls; the world price of rice falls; the U.S. exports and Japan imports less rice.

SUMMARY

A popular *New York Times* columnist Thomas Friedman wrote a book called *The World Is Flat*. The title means that the barriers that separated countries, and more importantly economies, in the past are withering away. We trade freely with India, and U.S. workers compete with Indian workers. Trade between countries present of host of issues, few of which can be debated among people with civility. Yet economists, despite their tendency to disagree with one another on a host of topics, are surprisingly united on trade issues. For the most part, free trade between countries is good, and it is difficult to find an example otherwise.

Think of all the ways you cooperate and trade with other people. You study with people. You work at one place and use the money you make to buy products from another place. All our great wealth is derived from the fact that we work where we have a comparative advantage and trade with others. This chapter sought to demonstrate the benefits of trade through several examples. It then went on to describe how market prices change once international trade ensues and describe the role of foreign exchange markets.

CROSSWORD PUZZLE

For answers with more than one word, leave a blank space between each word.

Across

2. The rate at which one can trade one currency for another.
3. An import _______ is a tax placed on imports.
6. A production possibility _______ shows the combination of goods a group can produce.
9. A country possesses this when it can produce more of a good than another country.
10. Exports to all other countries minus imports from all other countries.

Down

1. A country possesses this when it can produce a good at a lower opportunity cost than another.
4. If the U.S. dollar can now purchase more francs, we say the dollar _______.
5. This term refers to the fact that as trade or technological developments occur, jobs in one industry often must be destroyed to create jobs in another.
7. A point lying above the production possibility frontier curve is said to be what?
8. A limit set on the amount of imports of a good from a region.

STUDY QUESTIONS

Use the following information for Questions 1–4.

Two countries, the United States and Cuba, do not trade due to trade embargos. Both countries produce beer and cigars. Suppose that the production possibility frontier for each country is as presented in Figure 8.16.

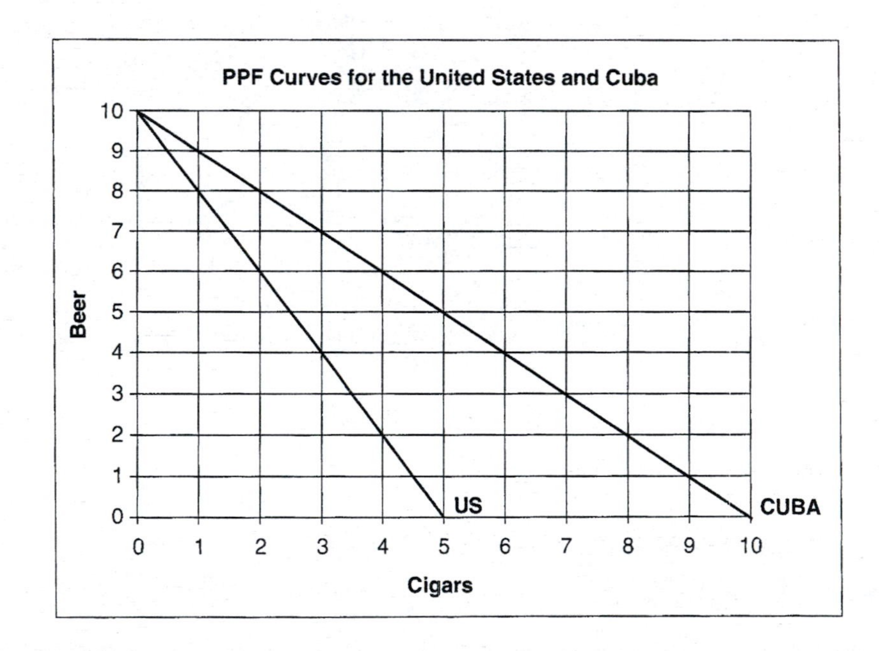

FIGURE 8.16

1. Fill in the table on page 249 showing the opportunity cost of production for the United States and Cuba for beer and cigars.

	Opportunity Cost of Beer	Opportunity Cost of Cigars
U.S.	_______ cigars	_______ beers
Cuba	_______ cigars	_______ beers

2. Which country has the comparative advantage in beer? Which country has the comparative advantage in cigars?
3. For the two countries to gain from trade, how much of each good should each country produce? Answer by filling in the table below.

	Beer Production (# beers produced)	Cigar Production (# cigars produced)
U.S.		
Cuba		

4. If a newspaper reports that the United States is running a trade deficit with the rest of the world, this indicates that
   a. foreigners are purchasing U.S. investments like stocks and bonds
   b. the United States is borrowing money from the rest of the world
   c. the U.S. net foreign investment is negative
   d. all of the above
5. Explain why, if we count the sale of investments as an export, if a country gives foreign aid to other nations and receives no foreign aid, its trade balance must be positive.
6. Explain why politicians love protectionist policies despite overwhelming evidence that trade promotes growth for all countries.
7. The United States and Cuba does not currently trade. Suppose that the price of similar quality cigars is higher in the United States than Cuba. If free trade ensues, what will happen to the price of cigars in both countries?
8. Consider the supply and demand for a general good in both the United States and China. Suppose the two countries are allowed to engage in free trade and the transaction costs of trade are zero. Graph China's excess supply and the U.S.'s excess demand curve in the middle graph. Indicate the world price resulting from trade, exports from China, and imports to the United States.

   World Price = _______

   U.S. Imports = _______

   Chinese Exports = _______

Refer to Figure 8.17 for Questions 9–12.

9. If the dollar appreciates relative to the Chinese currency (the yuan), the U.S. excess demand (*circle one*) FALLS / RISES and U.S. imports (*circle one*) FALLS / RISES.

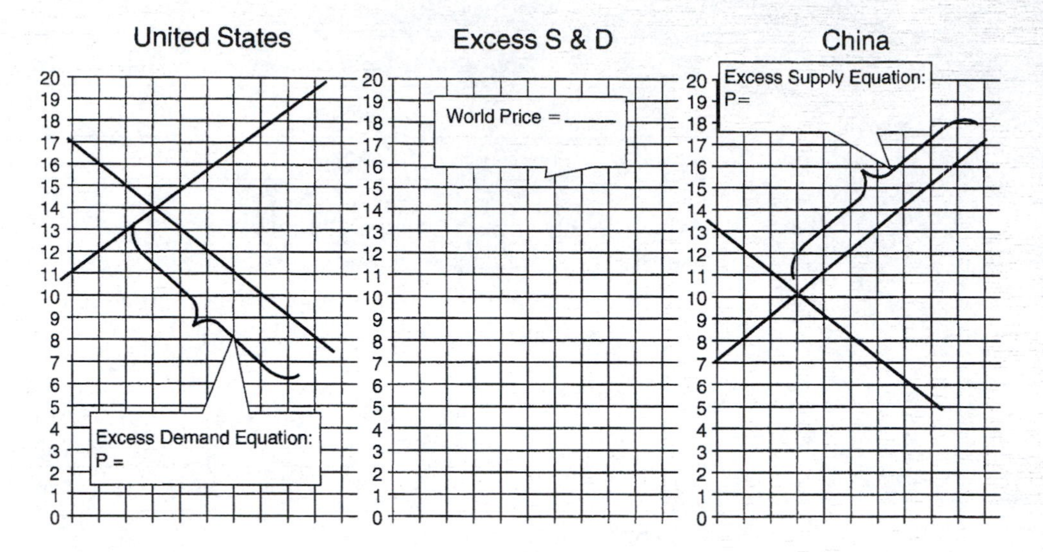

FIGURE 8.17

Note: The *y*-axis is in U.S. dollars and the *x*-axis refers to quantities.

10. If the dollar depreciates relative to the Chinese currency (the yuan), the U.S. excess demand (*circle one*) FALLS / RISES and Chinese exports (*circle one*) FALLS / RISES.
11. Use the three-panel diagram in Figure 8.18 to fill in the blanks below.

United States

Consumer Surplus Before Trade ________________ (use letters in Figure 8.18.)

Producer Surplus Before Trade ________________ (use letters in Figure 8.18.)

Total Surplus Before Trade ________________ (use letters in Figure 8.18.)

Consumer Surplus After Trade ________________ (use letters in Figure 8.18.)

Producer Surplus After Trade ________________ (use letters in Figure 8.18.)

Total Surplus After Trade ________________ (use letters in Figure 8.18.)

Japan

Consumer Surplus Before Trade ________________ (use letters in Figure 8.18.)

Producer Surplus Before Trade ________________ (use letters in Figure 8.18.)

Total Surplus Before Trade ________________ (use letters in Figure 8.18.)

Consumer Surplus After Trade ________________ (use letters in Figure 8.18.)

Producer Surplus After Trade ________________ (use letters in Figure 8.18.)

Total Surplus After Trade ________________ (use letters in Figure 8.18.)

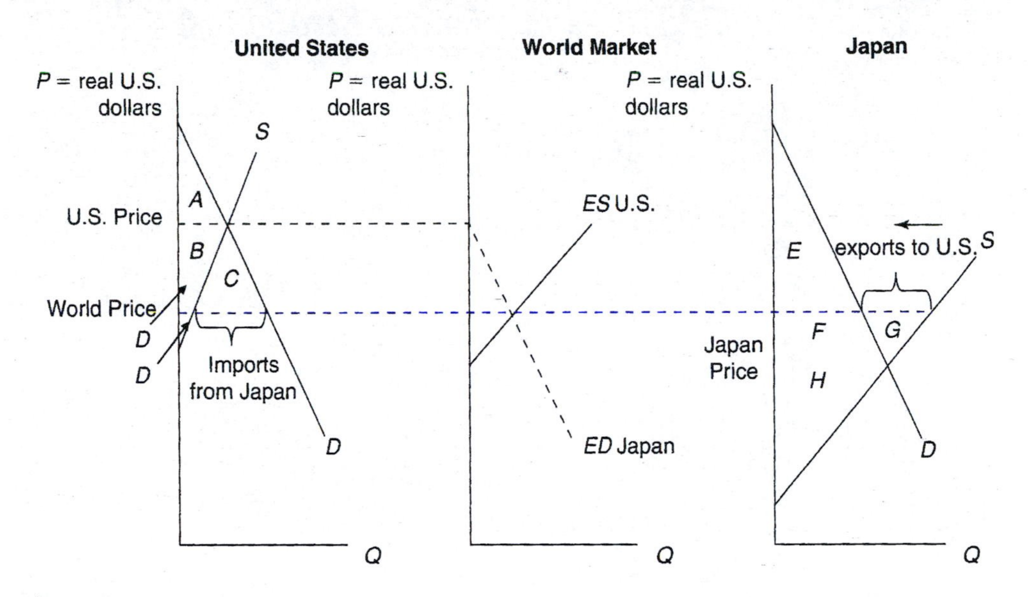

FIGURE 8.18

Final Welfare Analysis

Total Welfare (Total Surplus) Change for United States ____________________ (use letters in Figure 8.18.)

Total Welfare (Total Surplus) Change for Japan ____________________________ (use letters in Figure 8.18.)

CHAPTER NINE

Managing Price Through Futures Markets

How We Lost Money Gambling

People regularly gamble on football, basketball, and horse races, whether it is legal or not. What if you like to gamble but are not crazy about sports? Luckily, a company in Ireland allows you to gamble on political events like whether Hillary Rodham Clinton will be the Democrat presidential nominee in 2008 or whether the bird influenza strain H5N1 will be confirmed in the United States before a certain date. The company is called Intrade. When George W. Bush nominated Harriet Myers for the Supreme Court justice, the authors believed she would make it through the nomination process. We were so sure, we wanted to bet money on it, so we went to Intrade and gambled. The gamble was conducted through a futures market. For $3.60, we purchased a futures contract that paid out $10 if Harriet Myers became a Supreme Court Justice and zero if she did not. Needless to say, we lost $3.60.

It turns out that the price of the Harriet Myers contract can be used to predict the probability of her becoming a Supreme Court justice. If the market price of the contract (that pays $10 if she is confirmed) is $3.60, the market believes that there is a 36% chance she will become a Supreme Court justice. The authors thought the probability was greater than 36%, so we purchased the contract for $3.60. The market was right, and we were wrong. It turns out that futures contracts are very good predictors of events. The University of Iowa conducts a futures market for U.S. political elections where you can buy contracts that pay out money if a certain person wins an election. The price of the contract can be used to predict the probability of a candidate winning, and these predictions consistently outperform polls. The Pentagon even considered using futures markets to predict whether terrorists would strike on U.S. soil. Futures markets are more than just gambling centers. They provide useful information about certain events, like whether the bird influenza will be discovered in the United States. Futures markets are useful for agricultural purposes as well. As this chapter describes, they provide accurate forecasts of future agricultural prices and allow producers to lock in a price for their crop long before the crop is harvested.

INTRODUCTION

In agriculture there is a delay between the time production decisions are made and the final product is ready. Winter wheat is planted in October but not harvested until July. It takes two years from the time a cow is bred until its offspring is ready for slaughter. This delay is called a production lag. Some producers wait until the product is ready before it is sold. The wheat producer may harvest the wheat and then sell it at the local wheat purchasing center, and the cattle producer may raise the steer and take it to the local auction to be sold to the highest bidder. In these cases, the producer receives the *spot price* for their commodity, which is the price of the good at the present time and location. Markets, such as auctions, which sell commodities where the price is determined at the point of exchange, are known as *spot* or *cash markets*.

Often, however, producers wish to sell the product before the production process is finalized. Other times they wish to lock in a price before the commodity is sold or establish a minimum price they will receive, like insurance assuring them a certain amount of revenue. The same goes for buyers of the commodity. The major cost for flour producers is wheat cost. If the price of wheat skyrockets, a firm could suffer large losses. To protect against such losses, many flour producers wish to set what they feel is a "reasonable price" well in advance of the actual purchase.

The purpose of this chapter is to introduce you to marketing tools provided by agricultural futures markets. Specifically, the objectives are to

1. instill an understanding of futures markets
2. describe how to use futures markets to hedge commodities
3. describe how to use futures markets to predict future prices
4. describe how to use options to establish a minimum selling or maximum buying price

Both buyers and sellers often wish to establish a price in advance of the actual exchange. This can be accomplished several ways. One is to enter a *forward contract*, which may be formal or informal. A forward contract is an agreement stipulating the amount to be exchanged and the exchange price, where the actual exchange is made at a later date. When Bruce Willis announced a $1 million reward for information leading to the capture of Osama Bin Laden, that was an informal forward contract because nothing legal was signed. If your roommate purchases two tickets to the Superbowl and you agree to pay him $500 for one of the tickets in two months, you entered into a forward contract established through oral agreement. Some forward contracts are more formal requiring legally binding contracts, signed by both parties. As discussed in Chapter 6, many hog producers enter forward contracts (i.e., marketing contracts) by which they agree to sell a certain number of hogs at a later date and at a particular price. In other situations, the forward contract does not specify a particular price, but a formula dictating how the price will be established. The cattle market often employs "top-of-the-market pricing," where a cattle producer agrees to sell a particular number of cattle to a specific buyer, and the buyer agrees to pay the highest spot price observed between the time the contract is signed and the exchange takes place.

Futures contracts are a highly standardized form of forward contracts (a) that trade on an organized exchange center like the Chicago Board of Trade or Chicago Mercantile; (b) that specify a particular good, delivery date, delivery mechanism, and exchange price; (c) where the payment to and from the buyer and seller is backed by the exchange; and (d) where the agreement is backed by a good-faith deposit called a margin. There are futures contracts for numerous types of agricultural products from wheat to pork bellies to milk. Futures contracts even exist for nonagricultural commodities like natural gas and foreign currencies.

A few examples of futures contracts are shown in Figure 9.1. One may enter futures contracts to buy or sell corn at the Chicago Board of Trade, where each contract refers to 5,000 bushels. One may enter futures contracts to buy or sell lean hogs at the Chicago Mercantile, where one contract refers to 20,000 lbs. of lean hogs. For every contract exchanged there is a buyer and seller. Here is how it works. Suppose that I raise live-cattle and you purchase live-cattle for a beef processor.[1] It is currently August 10, 2005, and I plan to have a large group of cattle ready for slaughter in December 2005. Coincidentally, you plan to purchase cattle in December. Neither of us wish to wait until December to set the exchange price. We wish to set a price now. I call my futures broker and inform her that I wish to sell one December cattle futures contract in a certain price range. You call your broker and inform her you wish to buy one December cattle futures contract in a certain price range. Both brokers send a representative to the trading floor seeking to make the exchange.

If the price ranges overlap, I sell the futures contract and you purchase the futures contract at a price negotiated by the brokers' representatives. Suppose the negotiated price is $84.00/cwt. By selling one December cattle futures contract, I agree to deliver 40,000 lbs. of live-cattle in December and receive $84.00/cwt for the

Commodity	Exchange	Contract Size
Corn	Chicago Board of Trade	5,000 bushels
Wheat	Kansas City Board of Trade	5,000 bushels
Feeder-Cattle	Chicago Mercantile	50,000 lbs.
Live-Cattle	Chicago Mercantile	40,000 lbs.
Soybeans	Chicago Board of Trade	5,000 bushels
Lean Hogs	Chicago Mercantile	20,000 lbs.

FIGURE 9.1 Futures Contract Examples.

[1]Reminder: Live-cattle refer to cattle that are ready to be slaughtered and processed into beef.

sale. By purchasing one contract, you agree to accept delivery of 40,000 lbs. of live-cattle in December, and you pay $84.00/cwt. Thus, the futures contract is simply a standardized forward contract. It allows you to set a buying price and me to set a selling price in advance of the actual exchange. The commodity exchange, whether it is the Chicago Mercantile or the Chicago Board of Trade, simply designs the contract and provides a place for you and I to enter the contract. It also backs the agreement. If you do not have the money to pay for the cattle, the exchange will suffer the loss and I will still receive my selling price. Of course, the exchange has many precautions in place to protect them from such losses. The point is that I do not have to worry about your integrity, and you do not have to worry about mine. Whenever a futures contract is "traded," one party sells the contract and one party purchases the contract. The seller agrees to sell the good at the contract price and the buyer agrees to purchase it at that price.

Every business day, thousands of futures contracts for numerous commodities are bought and sold. The prices at which they are traded are reported in daily newspapers. See Figure 9.2 illustrating how the futures contract prices are reported. This figure is taken directly from the major newspaper in Tulsa, Oklahoma. This contract is for live-cattle at the Chicago Mercantile (hence, the CME acronym in the figure) and refers to 40,000 lbs. of live-cattle. The left-hand side shows the various live-cattle contracts that are traded. One may exchange contracts for cattle in August 2005, October 2005, and so on, to August 2006. At the top is the column heading Open, High, Low, Settle, and Chg., and the body of the table refers to futures contract prices for contracts traded on August 10, 2005. The Open price for an October 2005 contract was $81.05 per cwt. This means that the first October 2005 live-cattle futures traded that day traded for $81.05. The seller of that contract agreed to sell 40,000 lbs. of

	Open	High	Low	Settle	Chg.
CATTLE (CME) **40,000 lbs.– cents per lb.**					
Aug 05	80.00	80.22	79.75	79.92	+.12
Oct 05	81.05	81.30	80.57	80.65	–.07
Dec 05	83.80	84.15	83.50	83.62	–.12
Feb 06	86.00	86.35	85.85	85.97	–.20
Apr 06	84.50	84.75	84.15	84.42	–.12
Jun 06	80.20	80.35	79.60	79.60	–.30
Aug 06	79.52	79.52	79.52	79.52	–.27

Est. sales 11,282. Mon's sales 15,866
Mon's open int. 131,490. +309

FIGURE 9.2 Live-Cattle Futures Contract Prices as Reported in Newspapers.
Note: This figure was created to mimic the price reporting format in traditional newspapers. These are real prices, as reported by *Tulsa World*, Section E-4, August 10, 2005.

live-cattle in October for $81.05, and the buyer agreed to purchase it at that price. As the day progressed, many other futures contracts were traded. The highest and lowest prices observed were $81.30 and $80.57, respectively. The last contract traded that day went for a price of $80.65 and is referred to as the settlement price. The "Chg." column of −0.07 indicates that the settlement price was $0.07 lower than the previous day's settlement price.

In any one day contracts may trade for many different prices, but we usually say "the price" of the contract for that day equals the settlement price. Notice that the price of a June contract is $79.60, but the price of an April contract is $84.42. This indicates that cattle to be exchanged in April are trading at a higher price than cattle to be traded in June. As of August 10, April cattle are worth more than June cattle. This may reflect a higher anticipated demand in June or a lower anticipated supply. Each contract refers to a forward contract for cattle in a different month. Thus, each contract refers to a different market, the market for cattle, but the market for cattle at different time periods.

To clarify the difference between a spot price and a futures contract price, refer to Figure 9.3. This graph shows the spot price of cattle at Dodge City, Kansas, from May 1999 to June 2000. The spot price in July of 1999 was around $66, which means the price of cattle in July was around $66. To be more specific, the price of cattle in July of 1999 *to be exchanged* in July of 1999 was $66. The price of a June 2000 futures contract is also depicted. The futures contract is a June 2000 contract, which means

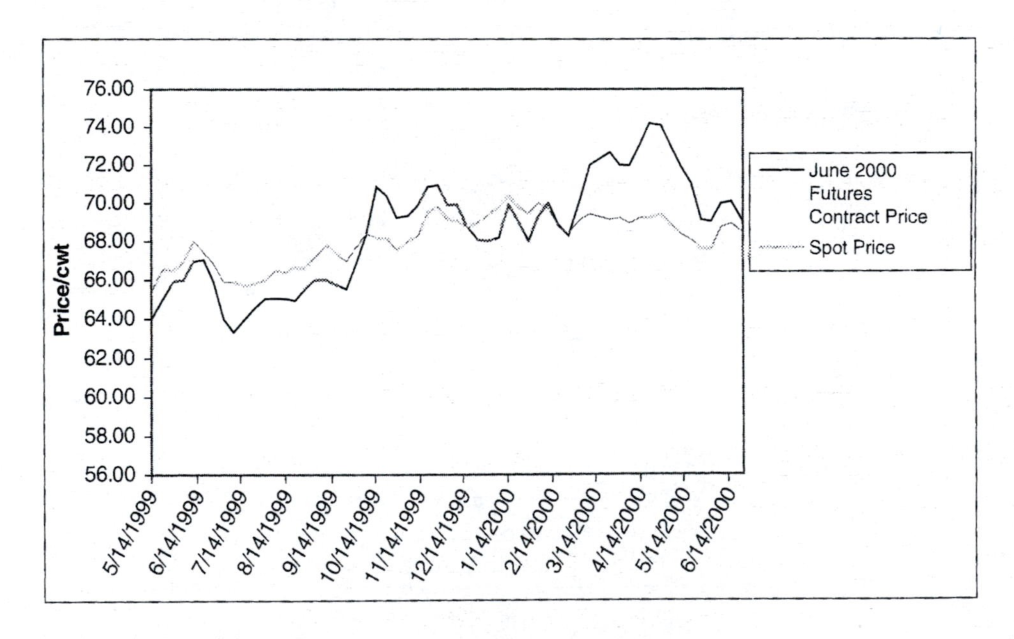

FIGURE 9.3 Spot and Futures Prices for Live-Cattle.

Note: Spot prices are the weekly prices at Dodge City for finished steers between 1,100 and 1,300 lbs. Futures prices refer to settlement prices at the Chicago Mercantile.

it refers to cattle to be exchanged in June of 2000. In July of 1999, the price of this contract was $63.00, which is $3 less than the spot price. The prices are different because they refer to cattle to be exchanged in different time periods. The futures price of $63.00 indicates that in July of 1999, buyers and sellers of cattle agreed they would exchange cattle in June of 2000 for a price of $63. Remember, the spot price at any time refers to the price of cattle at that time, whereas the futures price at any time refers to the price of cattle at contract expiration (a June 2000 futures contract expires in June of 2000; a May 2003 futures contract expires in May of 2003).

At contract expiration, the futures price and spot price should be approximately equal.

Notice what happens to the price series in June of 2000. The spot price and futures price are approximately equal. The reason is that the two prices now refer to the price of cattle traded in the same time period. In June of 2000, the spot price refers to the price of cattle to be exchanged in June of 2000. Coincidentally, in June of 2000, the price of a June 2000 futures contract also refers to the price of cattle to be exchanged in June of 2000. Remember the Law of One Price: The difference between prices of two identical goods in two different regions should not exceed transportation costs. Otherwise, arbitrage would occur, forcing the price difference to be within transportation costs.

This can be proven by contradiction. Suppose that in June of 2000, the futures price was much higher than the spot price. Smart investors could then purchase cattle in the spot market and sell them in the futures market for a higher price. Many investors would begin doing this, placing upward pressure on spot prices and downward pressure on futures prices. Investors would keep arbitraging until the prices were close to one another and profit opportunities no longer exist. At this point (where the Indifference Principle holds), the difference between the spot price and futures price—at contract expiration—is less than transportation costs. Similarly, cattle producers will see the higher futures market prices and will sell to the futures market instead of the spot market. The increase in supply in the futures market lowers price and the supply decrease in the spot market raises the spot price, forcing the futures and spot price closer to one another. This continues until the price difference does not justify transporting cattle to the futures market exchange. Eventually they will be within transportation costs. Commodity exchanges have delivery points scattered across the United States, ensuring the price differences for commodities is small.

Offsetting Futures Contracts

Up to this point, we have made it seem like people who sell futures contracts actually deliver the product, and people who buy futures contracts actually accept the product at a delivery point. In reality, less than 1% of buyers and sellers deliver or accept delivery. The reason is that there is a much easier way to fulfill one's contract obligation. Suppose that I purchase a used car from you for a price of $5,000, but then turn around and sell it for $6,000. I made a profit of $1,000 and no longer own the car. Similarly, if I purchase a futures contract, I can fulfill my obligation by selling a futures contract. This is referred to as "offsetting" one's futures contract. By buying

	May 12, 2004	November 5, 2004
November 2004 Soybean Futures Contract Price	$7.54/bushel	$5.20/bushel
Soybean Spot Price	$9.56/bushel	$5.36/bushel

FIGURE 9.4 Soybean Futures and Spot Prices.
Source: Articles from *Tulsa World* and the Livestock Marketing Information Center.

and then selling a wheat futures contract, I do not have to accept delivery of wheat, but I do earn (or lose) the difference between the buying and selling price. Consider the following example, using the price data reported in Figure 9.4. These are real spot and futures market prices. Notice that in May, the difference between the spot and November futures soybean price is rather large. However, in November at contract expiration the prices are close to one another, reflecting the Law of One Price. Suppose that you purchase a November soybeans futures contract in May at a price of $7.54 per bushel. If you do not offset your contract, in November when the contract expires, you would have to accept delivery of 5,000 bushels of soybeans and pay $7.54 for each bushel. An alternative to accepting delivery is to offset your futures contract by selling a November 2004 soybean futures contract, and you can offset anytime before the contract expiration. Contracts usually expire in the middle of the expiration month. If you offset in November by selling a November soybean futures contract at a price of $5.20, your profits are (selling price − buying price)(bushels per contract)(number of contracts) = (5.20 − 7.54)(5,000)(1) = −$11,700. You lose close to $12,000. You basically bought soybeans for $7.54 and sold them for $5.20. This is like buying a car for $20,000 and then selling it for $10,000—you lose money.

If you sell a futures contract, you may offset by buying the same type of futures contract.

If you buy a futures contract, you may offset by selling the same type of futures contract.

When you offset, your contract obligations are fulfilled, and you neither need to accept or make delivery of the product.

To summarize, if you initially purchase a futures contract, you may fulfill your contract obligation by selling another contract, and your profits are the selling price minus buying price times the number of units under contract. Similarly, if you initially sell a futures contract, you may fulfill your contract obligations by purchasing another contract, and your profits go by the same formula. In keeping with the previous example, suppose you sell a November 2004 soybean futures contract in May of 2004 at a price of $7.54. If you do not offset, you must deliver 5,000 bushels in November and will receive a price of $7.54. Alternatively, you may offset by purchasing a November 2004 soybean futures contract. If you offset in November, you purchase the contract at a price of $5.20. Your profits from offsetting are (selling price − buying price)(bushels per contract)(number of contracts) = (7.54 − 5.20)(5,000)(1) = $11,700. The futures transactions make you close to $12,000 because you bought soybeans at a low price and sold them at a high price.

To drive the point home, consider one additional example depicted in Figure 9.5 where Nature Boy Ric Flair and Macho Man Randy Savage trade September corn futures contracts. On January 7, Ric buys two of these contracts and Randy sells two. These are the same contracts, so they trade for the same price of $2.42. In July, Ric

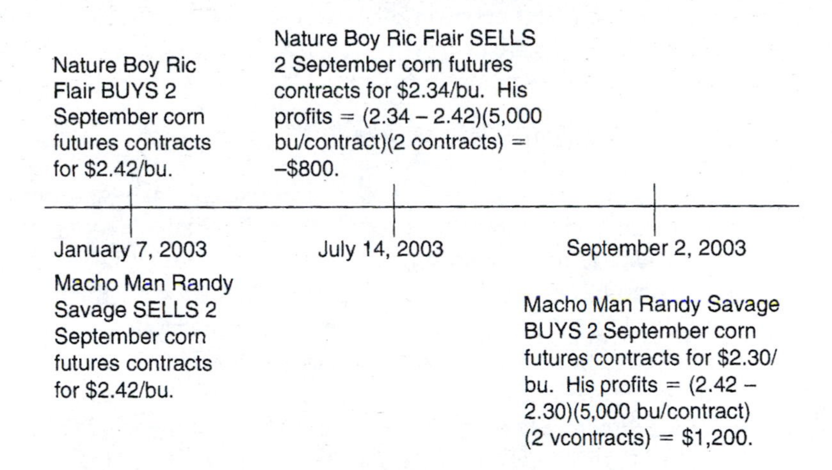

FIGURE 9.5 Hypothetical Corn Futures Transactions.

decides to offset and fulfill his futures obligations by selling two of the same futures contracts. The price of the contract has now fallen to $2.34, so Ric loses $800 from his transactions. Randy holds onto his contracts longer and does not offset until August, at which time the contract is trading at a price of $2.30. After offsetting by purchasing two September corn contracts, Macho Man Randy Savage makes $1,200. Neither Ric nor Randy ever grew corn and would not know what to do with corn if they had it. Yet, they were both able to trade corn futures contracts.

Profits from Offsetting Futures Contracts = (Selling Price – Buying Price) (Units per Contract)(Number of Contracts)

It is interesting that you may trade soybean futures contracts without ever owning soybeans, having the capacity to take delivery of soybeans, or even knowing what a soybean looks like. Plenty of futures traders deal only on paper and never actually see the commodity. This may be difficult to visualize, but the key to understanding it is to realize that buyers and sellers do not negotiate with each other, but through the exchange. Consider the following hypothetical story. I sell a soybean futures contract and you purchase the same contract. So far, we are the only buyers and sellers of this contract. In reality, I am selling the soybeans to you, and I agree to deliver 5,000 bushels of soybeans to you. Yet I do not deliver the soybeans to you, but to the commodity exchange at a particular place. You then go and pick up the commodity at that place. You and I never meet. Now, suppose that after selling the soybean futures contract, I offset by purchasing another contract. For me to purchase a soybean contract, there must be a seller, and this seller named Sam agrees to deliver 5,000 bushels of soybeans to the exchange. I am now "out of the market" because I both sold and purchased 5,000 soybeans to myself. If you sell soybeans to yourself, it does not matter if you own any soybeans or not. The only remaining contract holders are you and Sam. Although I am now out of the market and will not deliver soybeans, Sam will deliver soybeans and you may still accept delivery of 5,000 bushels. But neither of you may want to deliver or assume delivery of soybeans. Both of you offset. You initially purchased a futures contract, so you offset by selling a futures contract. Sam sold a contract so he offsets by purchasing a contract. Now everyone is out of the market and no soybeans are delivered. This is how futures markets really work.

Speculating in Futures Markets

Figure 9.5 depicts a hypothetical story of Macho Man Randy Savage and Nature Boy Ric Flair trading corn futures contracts. Ric Flair lost money; Randy Savage made money. Had Ric known the futures price would fall between January and July, he would not have bought the contracts. If Ric Flair knew the price would fall, he would have sold contracts like Randy Savage and made money from that information. When one sells contracts anticipating the contract price will fall, they take a *short position* in the futures market. Conversely, if one thinks prices will rise, they take a *long* position and purchase contracts in hopes of making money. This brings up an important point: If one knows the direction of futures prices, they can make profits by trading futures contracts. This is *speculation*, much like trading stocks based on information about whether stock prices will rise or fall.

If a speculator thinks the price of a futures contract will fall, she takes a *short* position (sells the contract).

If a speculator thinks the price of a futures contract will rise, she takes a *long* position (buys the contract).

Indeed, speculators comprise a large component of all futures traders. Some people make their living speculating on the futures market. In general, these speculators are serious students of agricultural markets and are quite good at predicting price movements. They have spent years studying the economics of agricultural markets and thousands of dollars in sophisticated computer equipment. Yet, in their efforts to make profits, speculators change price in a way to eliminate those profit opportunities. Recall the three I's of economic theory: incentives, interactions, and indifference. Figure 9.6 presents a hypothetical situation in the futures market. It is January 7, and the current price of a December corn futures contract is $2.42. Corn will be planted in April and harvested in December. A new weather news report forecasts dry

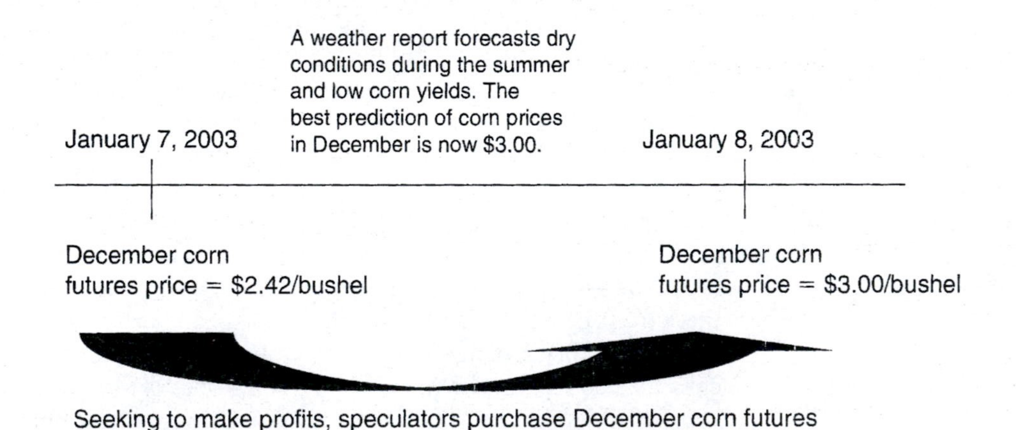

FIGURE 9.6 Futures Market Prices Are the Best Predictors of Future Spot Prices.

conditions this summer, suggesting the corn harvest will be small. A small harvest will lead to high corn prices in December, and the best forecast of corn prices in December is now $3.00 per bushel. When we say the "best forecast," we mean that the people who know corn markets the best and can predict corn prices the best forecast a December spot price of $3.00.

The people who know corn markets best are also people who speculate on the corn futures market. If they were not the best forecasters, they would have lost tons of money long ago and would not be able to buy and sell futures contracts. These speculators believe the spot price in December will be $3.00, yet they see the price of a December corn contract is only $2.42. Profits can be made by buying futures contracts. The speculator can buy December corn contracts, accept delivery in December paying $2.42 per bushel, and then sell the corn on the spot market for $3.00. Alternatively, remember that the futures price at expiration must equal (or be close to) the spot price. Thus, if the spot price of corn in December is $3.00, the futures price of a December corn contract must also be close to $3.00. The speculator can therefore purchase December corn contracts now at $2.42, selling the contracts in December at a price close to $3.00, making profits.

Speculators have every reason to believe they can make large profits by buying December corn contracts, so they purchase all they can. Yet, the more they try to purchase, the higher they bid the price. Speculators keep purchasing additional contracts as long as the contract price is below $3.00 and keep bidding the price up. Eventually, the price will be bid up to $3.00, and no more profit opportunities will exist. At this point, they are indifferent between purchasing a contract or not, because it neither makes more nor loses money. By pursuing profits and making profits, they eliminate profit opportunities for others. Speculators follow *incentives* by purchasing contracts when purchasing is profitable. Every buyer needs a seller, and the *interactions* that result when speculators try to buy more contracts bids the price of the contracts up. More and more contracts are purchased until speculators are *indifferent* between purchasing or not.

A similar story would occur in Figure 9.6 if the contract price in January was higher than $3.00, the best forecast of December spot prices. Now, speculators can make profits by selling contracts at a high price and buying contracts later at $3.00. As speculators sell more and more contracts, they bid the price down until it hits $3.00 and there are no more profits to be had. The point is this: Regardless of the starting futures price, if the best forecast of the spot price at contract expiration is $3.00, the futures price will quickly change to equal $3.00. Because speculators like money and because they are good at forecasting future prices, futures prices are always the best prediction of spot prices at contract expiration.

The futures price is the best forecast of spot prices at contract expiration.

To understand exactly what this implies, refer back to Figure 9.2. On August 10, 2005, many cattle producers were wondering what cattle prices would be in the future. What will the price be in February 2006? The best prediction is the settlement price of a February 2006 contract, which is $85.97. What will the price be in August 2006? The best prediction is the settlement price of an August 2006 contract, which is $79.52. Many studies have sought a forecasting method that outperforms the futures market, yet few are successful. One of the authors devoted their entire master's thesis trying to outperform the futures market and failed. The evidence is clear: The best forecast of future spot prices can be found in futures contract prices. This does not imply that the futures price is always correct. Very rarely does the futures contract price predict perfectly. There is a 50% chance it will overshoot and a 50% chance it will undershoot spot prices in the future. Futures markets are not perfect predictors of future spot prices, but they are the best predictors.

Futures markets provide the best predictions of future spot prices. The reason is that futures market speculators are adept at assimilating all market information accurately and quickly. Figure 9.6 depicts a situation where one news report triggers a rise in corn prices in just one day to reflect new information on growing conditions. In reality, prices move in seconds in response to new information. Consequently, we refer to futures markets as *efficient markets*, meaning the market incorporates all new information quickly and accurately into futures market prices.

HEDGING IN FUTURES MARKETS

It is April, and a corn farmer will soon begin planting. The corn will not be harvested until December. Wondering what price to expect for corn in December, she consults the futures market and finds the price of a December corn futures contract is $2.60 per bushel. This is the best prediction of December spot prices. That price is based on all the information available at the present time. Before December, many things could happen to change this price. Lots of rain will lead to a good harvest and lower prices. A shortage of corn in China will increase demand for U.S. exports and increase corn prices. Something will happen, and the spot price will not be exactly $2.60, but $2.60 is the still the best prediction available at the time. The farmer sees the $2.60 price and decides she would be happy with this price. If possible, she would like to "lock in" this price. This is possible by hedging in the futures market.

The Hedging Concept

After years of not making bowl games your college is playing in the Rose Bowl for the national championship. If you win, you will go crazy—Clemson University crazy—rolling over cars and setting fire to local establishments. If you lose, you will become despondent and will not talk for days. Your life now becomes risky; it will either improve dramatically or start sucking big time. If only there was some way to keep your happiness just as it was before the game—there is a way! Simply bet a large amount of money (say $500) against your team. If you win, you will be happy over becoming national champions but sad over losing $500. The two cancel out and you are neither happy nor sad. If you lose, you anguish over the football loss but revel in the monetary gain. The two cancel out and you are neither happy nor sad. That is the concept of a hedge. A hedge is an instrument designed to leave you no better or worse off than your current state. The hedge instrument is designed to give you money in bad times and take your money in good times. A hedge is like car insurance. It costs you money when things go your way and you do not wreck, but it gives you money when the unfortunate event occurs and you are rear-ended by a blonde talking to her sorority sisters on her cell phone about the recent *OC* episode (that's just a joke, the authors have cell phones and watch the *OC* too, but we do not have sorority sisters).

Hedging in Agriculture for Sellers

Refer back to our corn farmer. It is April, and she will soon plant her corn to be harvested in December. The price of a December corn futures contract is $2.60, and the farmer wishes to "lock in" this price by hedging. To hedge, the corn farmer should sell December corn futures contracts now, at planting in April, at the $2.60 price. One contract refers to 5,000 bushels, so if the farmer expects to harvest 20,000 bushels, she should sell four contracts.[2] Come December when the corn is harvested, she could simply deliver the corn to a designated delivery location and receive the $2.60 price, but this incurs transportation costs. Instead, she could offset by buying four of those same contracts. This relieves her of any contract obligations, and she can just sell the corn at a local spot market. The net price she will receive is something close to $2.60, and something predictable. To see this, we need to invoke some math.

Let P^F_{April} be the December corn futures contract price at planting and P^F_{December} be the price in December at corn harvest. Similarly, let P^S_{December} be the spot price in December. The net cash price the farmer receives from her corn equals the spot price plus or minus any profits or losses in the futures market. The per bushel futures profit always equals the selling price minus the buying price. She sold futures at price P^F_{April} and bought futures at the price P^F_{December}, yielding futures profits of $P^F_{\text{April}} - P^F_{\text{December}}$ per bushel. The price received from her corn in the spot market is P^S_{December} per bushel. Putting the two together, the hedge price received for corn is

$$\textit{Hedge Price} = \textit{Spot Price} + \textit{Futures Profits} = P^S_{\text{December}} + (P^F_{\text{April}} - P^F_{\text{December}}).$$

If she made money in the futures market, she receives something higher than the spot price, whereas if she lost money, she receives something less. Rearranging this equation yields

$$\textit{Hedge Price} = P^F_{\text{April}} + (P^S_{\text{December}} - P^F_{\text{December}}) = P^F_{\text{April}} + \textit{Basis}$$

Basis: Spot Price – Futures Contract Price at Expiration

The last term, referred to as the *basis*, equals the spot price minus the futures contract price at expiration. (Warning: Some textbooks will define the basis as the futures price minus the spot price, and some will use the term *cash price* in place of *spot price*.) Both prices refer to the same commodity in the same time period, and according to the Law of One Price, these two prices should be similar. The basis is constrained by transportation costs; it cannot exceed the transportation costs between the farm and the nearest futures delivery point. This makes the basis fairly predictable, and in many areas, small in absolute value. Although spot prices and futures prices vary wildly across time, the basis is much more stable. Locking in this hedge price reduces price risk. If

[2]Or, if she only wants to lock in a price for half of her crop, she should sell two contracts.

you relied on spot prices only, prices could plummet between planting and harvest, but the basis will remain stable. The hedge price will remain stable.

Notice how the hedge works. High corn prices are good for the corn farmer, and low corn prices are bad. The farmer hedges by selling corn contracts at planting. If corn prices fall, she receives a lower spot price but makes more money from her futures transaction. If corn prices rise, she receives a higher spot price but loses money in the futures market. Hedging by selling corn futures is like betting against your football team; it makes you money in bad times but loses you money in good times.

Hedging in Agriculture for Buyers

Buyers also have incentives to lock in a price by hedging. You are a hog producer who purchases large amounts of corn for hog feed. High corn prices are bad for your business. It is currently April, and come December you plan to make a large corn purchase. Observing the December corn futures price of $2.60, you decide this is a good price. You could wait until December and purchase from the spot market, but prices may have risen since then. Rather than risk higher prices in December, you wish to lock in this $2.60 price now. One way is to buy December corn contracts at $2.60 and accept delivery of the corn in December. But this incurs transportation costs, so the buyer is better off just offsetting.

The buyer can execute a hedge in April by buying December corn futures contracts at the $2.60 price. When December arrives, the buyer then offsets her futures positions by selling futures contracts and then purchasing the needed corn from the spot market. This will produce a hedge price that is fairly predictable, and in many areas results in a purchasing price close to $2.60. Again, to see this result, some math must be used. Let P^F_{April} be the price of a December corn futures contract in April and P^F_{December} be the price in December at contract expiration. The corn buyer purchases contracts in April and sells contracts in December, so her per bushel futures profits are $P^F_{\text{December}} - P^F_{\text{April}}$. Similarly, let P^S_{December} be the spot price in December. The net price the buyer pays for corn equals the spot price minus any profits from the futures transaction. If the futures transactions made money, this helps to reduce the cost of corn. If it lost money, this just adds to the amount of cash paid for the corn.

$$\textit{Hedge Price} = \textit{Spot Price} - \textit{Futures Profits} = P^S_{\text{December}} - (P^F_{\text{December}} - P^F_{\text{April}})$$

Rearranging the equation yields

$$\textit{Hedge Price} = P^F_{\text{April}} + (P^S_{\text{December}} - P^F_{\text{December}}) = P^F_{\text{April}} + \textit{Basis}$$

The formula for the buyer's hedge price is exactly the same as the seller's hedge price.

The buyer first purchases a futures contract and then sells it close to expiration. If the contract price declines over that time period, the buyer loses money in the futures market, but gains by purchasing corn at a lower price. Conversely, if corn prices rise during this time, the buyer makes money in the futures market but must pay more for corn. Regardless, the bad offsets the good such that the corn buyer's position is not improved from the day the hedge is executed.

- A seller executes a hedge by selling contracts now and buying the contracts later when she sells the good in the spot market.
- A buyer executes a hedge by buying contracts now and selling the contracts later when she purchases the good from the spot market.
- For buyers and sellers, the hedge price they receive is:

  Hedge Price = Futures Price When Hedge Is Executed + Basis

FIGURE 9.7 Hedging for Buyers and Sellers.

Hedging in Practice

What follows is how you would execute a real hedge, assuming you are a soybean farmer located in eastern Missouri. It is May 12, 2004, and you will soon begin planting soybeans. The soybeans will be harvested in November. Picking up the paper that morning, you see that the settlement price for a November 2004 soybean futures contract is $7.54 per bushel. If you hedge, your hedge price will be this futures contract price plus the basis. To determine what basis to expect, you go to the Agmanager website at Kansas State University (www.agmanager.info). Agricultural economists have posted the map shown in Figure 9.8, illustrating the average soybean basis across the Midwest regions. For eastern Missouri, the map shows the average soybean basis is $0.02. If you hedge, you will receive a hedge price close to $7.54 + $0.02 = $7.56. The hedge price will not be exactly $7.56, but the basis has a low variability. It may be $7.50 or $7.60, but it will be close to $7.56.

This is a good price. Come harvest, the spot price may be higher, but then it may be lower. Rather than risk a lower price, you decide to lock in the hedge price of $7.56. You plan to harvest 100,000 bushels of soybeans. However, you wish to hedge only half of this. That is, you want to lock in a price close to $7.56 for half of your crop and take a chance on the other half that prices will rise. This is similar to purchasing collision insurance for one car but not your other car. Deciding to hedge 50,000 bushels, you sell 10 November 2004 soybean futures contract for $7.54 (one contract corresponds to 5,000 bushels). Throughout the growing season you hold onto this contract. Once the soybeans are harvested in November, you offset your futures position by purchasing 10 November 2004 soybean futures contracts and sell your soybeans in the spot market. The price you will receive for the 50,000 bushels you hedged will be close to $7.56. For the remaining 50,000 bushels you did not hedge, you will receive the spot price, which may be higher or lower than the hedge price depending on market conditions.

A buyer wishing to hedge their purchase of 50,000 bushels will execute her hedge similarly. In May 2004 she buys 10 November 2004 soybean futures contracts. Come November when she makes her purchase, she offsets by selling 10 of those same contracts and purchasing her soybeans in the spot market. The buyer's hedge price will also be close to $7.56. In observing the basis map, it is clear that soybean producers in the western portions of Texas, Oklahoma, Kansas, and Nebraska will receive a lower hedge price than those in the eastern portions. The basis is −$1.08 in western Kansas compared to −$0.29 in

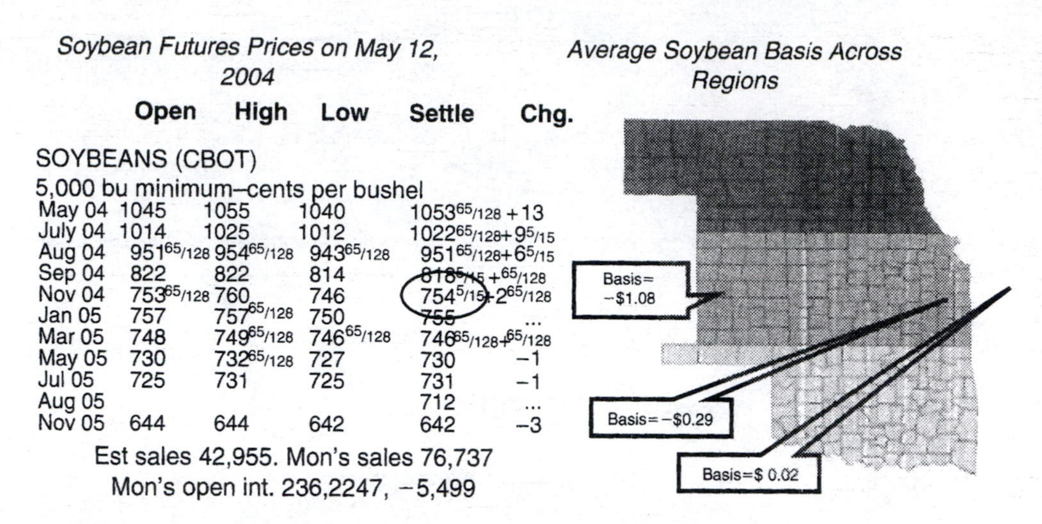

Soybean Futures Prices on May 12, 2004

SOYBEANS (CBOT)

5,000 bu minimum–cents per bushel

	Open	High	Low	Settle	Chg.
May 04	1045	1055	1040	1053 65/128	+13
July 04	1014	1025	1012	1022 65/128	+9 5/15
Aug 04	951 65/128	954 65/128	943 65/128	951 65/128	+6 5/15
Sep 04	822	822	814	818 5/15	+65/128
Nov 04	753 65/128	760	746	754 5/15	+2 65/128
Jan 05	757	757 65/128	750	755	...
Mar 05	748	749 65/128	746 65/128	746 65/128	+65/128
May 05	730	732 65/128	727	730	−1
Jul 05	725	731	725	731	−1
Aug 05				712	...
Nov 05	644	644	642	642	−3

Est sales 42,955. Mon's sales 76,737

Mon's open int. 236,2247, −5,499

FIGURE 9.8 Soybean Futures Prices and Basis Information.

Source: *Tulsa World*, May 12, 2004. Department of Agricultural Economics, Kansas State University, Agmanager website.

eastern Kansas. This, however, does not imply that hedging is less desirable in western Kansas. The reason is that spot prices are also lower in western Kansas by the same amount as the basis is lower, so no matter whether a western Kansas soybean producer hedges her crop or simply sells it on the spot market, she will receive approximately ($1.08 − $0.29) $0.79 per bushel less than her eastern counterparts.

A final note should be given about using futures markets to reduce price risk. Both producers and buyers face price risk, meaning market prices fluctuate in ways that are out of buyers' and sellers' control. Prices may be favorable or unfavorable. Hedging is one method to reduce your price risk. *Once you have hedged*, you have a much better idea of the price you will receive or pay than someone who has not hedged because the basis is predictable. Recall the hedge price is the futures price when the hedge is executed plus the basis. That futures price is set in stone, in an official contract. The basis is determined by market prices but due to the Law of One Price is predictable. This makes the hedge price predictable. However, *before one has hedged*, there is just as much price risk in hedging than the spot market. Both futures prices and spot prices fluctuate wildly, so before hedging you do not know the exact hedge price you will receive. But once you execute a hedge and buy or sell at a particular futures price, you know you will receive or pay that price plus the basis, and the basis will be relatively small and stable.

Cross-Hedging

Sorghum is a feed grain, meaning sorghum is processed into a feed given to livestock. Sorghum is just one feed grain produced in the United States. Other feed grains are corn, soybeans, wheat, barley, oats, and canola. The major feed grain is corn. Suppose that you are a sorghum former planting your sorghum in April and planning to harvest it in October. You would like to hedge your sorghum crop, but there is no sorghum futures contract. What can you do?

You could cross-hedge using corn futures contracts. Sorghum and corn are substitutes for livestock feed, and the Indifference Principle tells us that, on average, livestock producers should be indifferent between feeding sorghum and corn. If corn prices rise, sorghum suddenly becomes the preferred feed grain. Livestock producers increase their demand for sorghum, thereby bidding up the sorghum price. The sorghum price keeps rising until producers are again indifferent between feeding sorghum or corn. Similarly, if corn prices fall, corn becomes the preferred feed grain. The demand for sorghum falls and so does its price. Eventually, the sorghum price becomes so low that producers start using it again and are indifferent between feeding sorghum and corn. The point is that, because corn and sorghum are feed grain substitutes, their prices move in tandem. If corn prices rise, so do sorghum prices. If sorghum prices fall, so do corn prices.

Remember the concept of a hedge. A hedge is an instrument that makes you money when prices are favorable and loses you money when prices are unfavorable. For a sorghum farmer to hedge, she must find a futures transaction that makes her money when sorghum prices are low and loses money when sorghum prices are high. The answer is simple: Sell corn futures when sorghum is planted and buy them back when the sorghum is harvested. The actual sorghum harvested is then sold in the spot market. Between planting and harvest, if sorghum prices fall, we know that corn prices will fall as well. If you sold then bought corn futures, this makes you money. If sorghum prices rise between planting and harvest, corn prices rise as well, losing you money. Thus, a sorghum farmer can hedge her crop by "pretending" to be a corn farmer. She hedges by purchasing a corn futures contract that expires sometime in the fall. She offsets by purchasing the same number and type of corn futures in the fall and sells her sorghum crop in the spot market.

Buyers can cross-hedge as well. A sorghum buyer will hedge by initially buying corn futures, offsetting by selling those corn futures later, and purchasing sorghum directly from the spot market. If sorghum prices rose during this period, that is bad for sorghum buyers, but the hedge made money. The good and the bad cancel each other out. If sorghum prices fell during this period, that is good for sorghum buyers, but the hedge lost money and once again the bad cancels out the good.

There are other commodities besides sorghum for which cross-hedging is an option. Cull cows are mother cows that are at the end of their productive life. Some can no longer conceive and some produce too little milk for the calf. These cows are "culled" and replaced with younger mothers. Large livestock operations may sell numerous cull cows each year and may want to hedge those sales. The Chicago Mercantile has a boneless beef futures contract, where each contract refers to 20,000 lbs. of boneless beef of a particular quality. As the price of boneless beef rises, so should the price of cull cows, because boneless beef is made partly from cull cows. Boneless beef and cull cow prices do not move together perfectly, but they do move in the same general direction. Thus, just like a sorghum producer can cross-hedge using corn futures contracts, a seller or buyer of cull cows can cross-hedge using boneless beef futures contracts.

OPTIONS IN FUTURES MARKETS

Think back to the analogy of hedging in football games. Betting against your team gives you money when your team loses but costs you money when your team wins. The purpose of the hedge is to give you money in bad times and cost you money in good times. Hedging locks in a given level of happiness. Hedging in agricultural futures markets locks in a price. If you are a seller of hogs, a hedge makes you money when times are bad and hog prices are low, but costs you money when times are good and hog prices are high. A hedge precludes you from benefiting from a price rise. Similarly, if you are a buyer of hogs, a hedge precludes you from benefiting if prices fall. Options are an alternative price setting tool that protects you from bad times but allows you to benefit in good times. Basically, options limit the extent to which a bad price movement can hurt you. For sellers of hogs, options provide a price floor, a guaranteed minimum they will receive for their hogs. For buyers of hogs, options provide a price ceiling, a guaranteed maximum they will pay for hogs.

Options are really more like insurance. Imagine a hog seller purchases an insurance plan for a fixed amount. If spot prices fall below a certain level, say $50/cwt, the insurance company pays the difference between the spot price and the minimum price of $50/cwt, guaranteeing you a price of at least $50. For example, if the spot price is $45, the insurance company pays $5 for each cwt of hog insured. When prices are above $50, the insurance company pays nothing. Similarly, a buyer purchases an insurance plan for a fixed amount that pays out if prices rise above a certain level, say $60/cwt. If prices rise above $60, the insurance company pays the difference between the spot price and $60. If prices do not rise above $60, the insurance company pays nothing. That is basically how an option works. There are two types of options, *call options* for buyers and *put options* for sellers.

One may purchase call options and one may purchase put options. There are sellers of put and call options, yet we will ignore the sellers, concentrating only on how to use options to lock in a minimum selling or maximum buying price. A call option gives the owner of the option the right *but not the obligation* to buy a futures contract at a fixed price referred to as a *strike price*. One can purchase a call option and sit on it, meaning they never buy futures at the strike price. Or, one could "exercise" the option and buy futures at the strike price. A put option gives the owner the right but not the obligation to sell a futures contract at a fixed price, also known as a strike price. As with call options, one may sit on the option and never sell futures at the strike price, or one could exercise the option and sell futures at the strike price. To see how these options guarantee a minimum price to sellers and a maximum price for buyers, let us run through some examples.

Put Options for Sellers

It is January 11, 2006, and a hog producer named Douglass will have hogs ready to sell in the spot market in April. A good manager, Douglass wants to protect himself in case hog prices fall to low levels in April. Douglass does not want to hedge though, because he wants to be able to sell his hogs at high prices if the spot price is high in April. To provide a price floor for his hogs that will be sold in April, Douglass purchases a put option for the April 2006 lean hog futures contract at a strike price of $58. What, exactly, does this mean? Once the option is purchased, anytime before the contract expires in mid-April, Douglass may sell an April 2006 lean hog futures contract at a selling price of $58 (i.e., can sell at the strike price). This option price is $0.78 per cwt of hogs in the April 2006 contract, and one contract covers 200 cwt, so the option costs him $156. The option price must be paid regardless of whether Douglass exercises his option to sell futures contracts at the strike price. Recall the formula for the hedge price: Hedge Price = Futures price when hedge is executed + Basis.

The put option allows Douglass to execute a hedge anytime between January and April. If the option is executed, then Douglass sells hog futures at the strike price of $58/cwt. He can sell those futures at this price anytime between when the put option was purchased and the option expiration date. Like a hedge, if Douglass exercises his option by selling futures, come April he will offset by buying the same number of futures and then sell his hogs in the spot market. Thus, if Douglass exercises his put option, he essentially hedged, receiving a hedge price equal to the strike price plus the basis. Of course, the put option was purchased at a price as well, which reduces the net amount received for each hog.

By purchasing a put option, the price a seller receives for the commodity is:

Price = Strike Price + Basis − Option Price or Price = Spot Price − Option Price, whichever is larger.

Conversely, if Douglass never exercises his option, he never sells or buys futures contracts and simply sells his hogs in the spot market. But he still purchased the put option, so the price of the put option reduces the net amount received from the hog sales. Thus, the put option gives Douglass the right, but not the obligation, to hedge at the strike price. Consider again the two prices received depending on Douglass's actions.

Net Price if Put Option Is Exercised = Strike Price + Basis − Option Price

Net Price if Put Option Is Not Exercised = Spot Price − Option Price

Suppose Douglass purchases a put option in January. Specifically, he purchases a put option for an April futures contract. Suppose the put option for an April 2005 hog futures price and a $58/cwt strike is purchased on January 11, 2006 (see Figure 9.9).

Call Option		Put Option	
Strike Price	Settlement Price	Strike Price	Settlement Price
	$/cwt		
$58	$7.58	$58	$0.78
$60	$6.03	$60	$1.20

FIGURE 9.9 April 2006 Lean Hog Option Price Data on January 11, 2006.
Source: Chicago Mercantile website.

Also, suppose the expected basis is $2/cwt. If the option is exercised, the net price equals $58 + $2 − $0.78 = $59.22. Douglass has guaranteed himself a minimum price of $59.22/cwt. Suppose that come April the spot price is $55. By not exercising the option Douglass would receive $55 − $0.78 = $54.22 (remember, you pay for the option regardless of whether you use it, just like you pay car insurance regardless of whether you get into a wreck). Clearly, a price of $59.22 is preferred to $54.22, so Douglass exercises his option. If instead the price was $65, Douglass would not exercise his hogs and would simply receive the spot price minus the option price.

Call Options for Buyers

Just as producers can use put options to establish a minimum selling price, buyers can use call options to establish a maximum buying price. A call option gives the owner of the option the right but not the obligation to purchase futures contracts at the strike price. In keeping with our hog example, it is January 11, 2006, and a hog processor will purchase hogs in April. The processor is scared prices will be high in April and therefore wants to protect itself from high prices, but the processor also wants to take advantage of low prices in April should they exist. This can be accomplished through the purchase of call options.

If a call option is purchased and the hog buyer exercises this option, he buys futures at the strike price, then offsets by selling the futures, and then makes his hog purchases from the spot price. Essentially, the call option allows the buyer to hedge at the strike price. If the option is not exercised, then no futures are bought or sold, and the buyer simply purchases from the spot market, paying the spot price, plus the cost of the option. Depending on whether the call option is exercised, the net price paid is

$$\text{Net Price if Call Option Is Exercised} = \text{Strike Price} + \text{Basis} + \text{Option Price}$$
$$\text{Net Price if Call Option Is Not Exercised} = \text{Spot Price} + \text{Option Price}$$

Figure 9.9 shows that on January 11, 2006, one could purchase a call option at a strike price of $60 for $6.03/cwt. Suppose the hog buyer purchases this call option. Anytime between then and April, the buyer can exercise the option and buy hog futures contracts at $60/cwt. This allows the buyer to hedge anytime until mid-April and receive a hedge price of $60 plus the basis. The option price must be added to the hog procurement cost as well. If the buyer does not exercise the option, he makes no futures transactions, purchases the hogs on the spot market, and still pays the option price.

By purchasing a call option, the price a buyer pays for the commodity is: Price = Strike Price + Basis + Option Price, or Price = Spot Price + Option Price, whichever is smaller.

Suppose that, come April, hog spot prices are $75/cwt, and the expected basis is −$1.50. By exercising the call option, the buyer will only have to pay a price equal to *strike price* + *basis* + *option price* = $60 − $1.50 + $6.03 = $64.53. If he did not exercise the option, he would have to pay the spot price of $75 plus the option price of $6.03, so he is clearly better off exercising the option. Conversely, if April spot prices were a low of $50, the buyer would simply not exercise the option and buy from the spot market at a price of *spot price* + *option price* = $50 + $6.07 = $56.07/cwt.

PUTTING IT ALL TOGETHER

There is a good chance that some readers are still confused about how to use hedging and options to establish a buying or selling price. Futures markets can be confusing, and we found the best way to understand them is through examples, so let us do one more from the seller's perspective. You are a seller of live-cattle located in western Kansas. It is January 14, 2006, and you have cattle that will be ready for slaughter in June. You could wait until June and receive the spot price, execute a hedge now and lock in a hedge price, or purchase a put option that guarantees you a minimum price. After consulting your newspaper or Internet sites, you see that the price of a June 2006 futures price is $87.07 (that was the real settlement price the previous day). You could hedge, and your hedge price would be the price of $87.07 plus the basis. Kansas State University maintains an excellent website for market information at www.agmanager.info. After perusing this website, you find a paper written by Kevin Dhuyvetter (a professor at Kansas State) about the basis for live-cattle in Kansas. One particular chart in this paper shows that the basis for June contracts in western Kansas varies from −$1 to $6/cwt. The average basis is about $2/cwt. Thus, if you hedge, you will receive a hedge price close to $87.07 + $2 = $89.07.

Alternatively, purchasing a put option guarantees you a minimum price and allows you to sell your cattle in the spot market at a high price should spot prices be high in June. Again, after consulting the Internet or newspapers, you find that the price of a put option with a $90/cwt strike price is $4.00. If you purchase the put option, you are guaranteed a minimum selling price equal to the expected hedge price minus the option price: $90 + $2 − $4 = $88. If spot prices are lower than $92, you exercise your option and receive the price of $88. If spot prices are higher than $92, simply let your option expire (do not execute it) and sell cattle on the spot market. For example, if spot prices are $95, your selling price is $95 − $4 = $91. Always subtract out the option price.

June 2006 Live-Cattle	$/cwt
Futures Price	$87.07
Expected Basis in Western Kansas	$2.00
Call Option Price (Strike Price = $90.00)	$1.13
Put Option Price (Strike Price = $90.00)	$4.00

FIGURE 9.10 June 2006 Live-Cattle Futures and Option Prices on January 13, 2006.

Source: Chicago Mercantile website and the www.agmanager.info website.

Let us recap. It is January and you will sell your cattle in June. If you wait and receive the spot price, there is no way to know what that spot price will be, but according to the futures markets the best guess is around $87.07. However, it is almost certain that the spot price will be higher or lower. You could try to negotiate a forward contract with a buyer, assuring you of what you feel is a reasonable price, if such a contract can be negotiated. If you hedge, you will receive a price close to $89.07. If you use put options, the lowest price you will receive is $88, but you may receive a higher price. That higher price may be $88.50, in which you would have been better off hedging, or $92, in which you would be glad you used your option. It is not clear which marketing strategy is best, and it shouldn't be, as we will see shortly.

The relationship between three of these marketing strategies is depicted in Figure 9.11. Alternatives to receiving the spot price are negotiating a forward contract, hedging, or using options. The end result, the net price you end up receiving, depends on several things. Using a forward contract, the net price is simply the price negotiated in the forward contract. With hedging, the net price is the futures price plus the basis. The use of options lets you buy or sell futures at a fixed strike price; thus, options give you control over that futures price.

So, if you were a live-cattle producer, which marketing strategy would you employ? This is a tough decision; there is no clear-cut winner. Indeed, rarely will there be one marketing strategy that is a clear winner. One can only look at the alternatives and decide what is best for herself. The Indifference Principle is always at play,

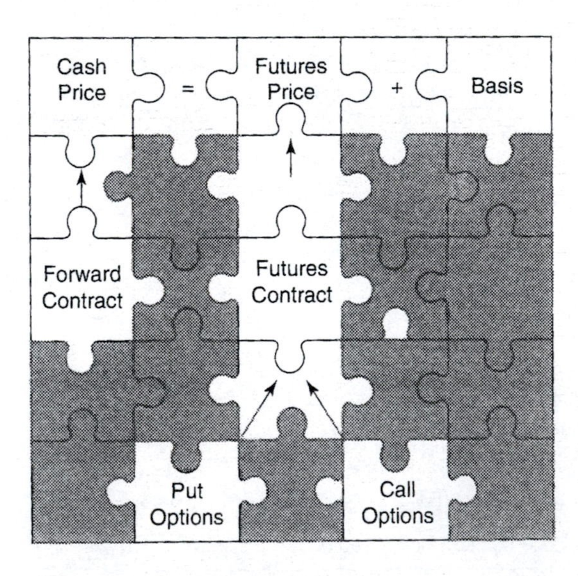

FIGURE 9.11 The Marketing Puzzle and Alternatives to the Spot Prices.
Note: The authors wish to thank Kim Anderson for permitting us to re-create this figure from his extension publications.

and market prices are continually changing to make one indifferent between selling at the future unknown spot price, negotiating a forward contract, hedging, or using options. The reason is simple. It is unlikely you will continually receive a higher price from forward contracts. Why would buyers continually pay significantly higher prices in forward contracts when they could purchase for less in the spot market? If the hedge price was significantly above the expected spot price, speculators could exploit this price difference to make profits, and in the process will realign the two price series. If the minimum price using an option was higher than the average hedge price, all producers would prefer options over hedging. Large groups of producers would cease hedging and would begin purchasing put options. But this drives the price of a put option upward, and the price keeps rising until the minimum price under an option is less than the hedge price.

All things considered, no one marketing strategy will clearly dominate another. Yet people will still find futures markets to be a useful marketing tool. The Indifference Principle assumes that all people have similar tastes. People differ in their attitudes towards risk though. Some people prefer the use of hedging and options because it decreases the risk of an undesirable price movement. Others prefer to sell in the spot market only, taking the risk that prices will be favorable. If people's risk attitudes differ, the Indifference Principle holds only for the average person. On average, people will be indifferent between the various market strategies, but highly risk-averse people will prefer options and hedging, and risk-loving people will not.

Futures markets can be a useful marketing tool, but they should be approached with caution. Fortunately, one does not need to become a futures market expert to market their product using futures markets. Just like there are financial advisors ready and willing to help you prepare for retirement, there are marketing consultants with the competence to help you conduct successful futures markets transactions. We say "success" not to imply they will ensure all your futures market transactions are profitable. If they were that good at speculating in the futures markets, why would they take time to help you? Wouldn't they spend all their time speculating? Instead, these consultants are able to help you manage price risk. Financial advisors are not great stock pickers, but they can help you plan for retirement. Similarly, marketing consultants are not great futures speculators, but they can help you reduce the risk of undesirable fluctuations in market prices.

SUMMARY

Most firms who buy or sell large volumes of agricultural commodities are involved in the futures market. The futures market provides useful predictions of future prices. It allows buyers and sellers to lock in prices long before a commodity actually trades hands via hedging. Futures options allow sellers to lock in a minimum price and buyers a maximum price.

Perhaps the most useful function of a futures market is that it allows producers and buyers of agricultural commodities to lock in a price long before the good is actually bought or sold. By locking in a price, you lock in a profit. Locking in a profit does more than simply make people feel secure and reduce risk; it allows them to borrow money at a lower interest rate. By providing documentation that a firm will turn a profit, regardless of how future prices actually behave, banks are much more willing to loan money. There are many other reasons for trading futures contracts than that described in this chapter. This chapter just skims the surface of futures markets. For those interested, your university most likely has an entire class devoted to futures and options. We encourage those individuals to take such courses.

CROSSWORD PUZZLE

For answers with more than one word, leave a blank space between each word.

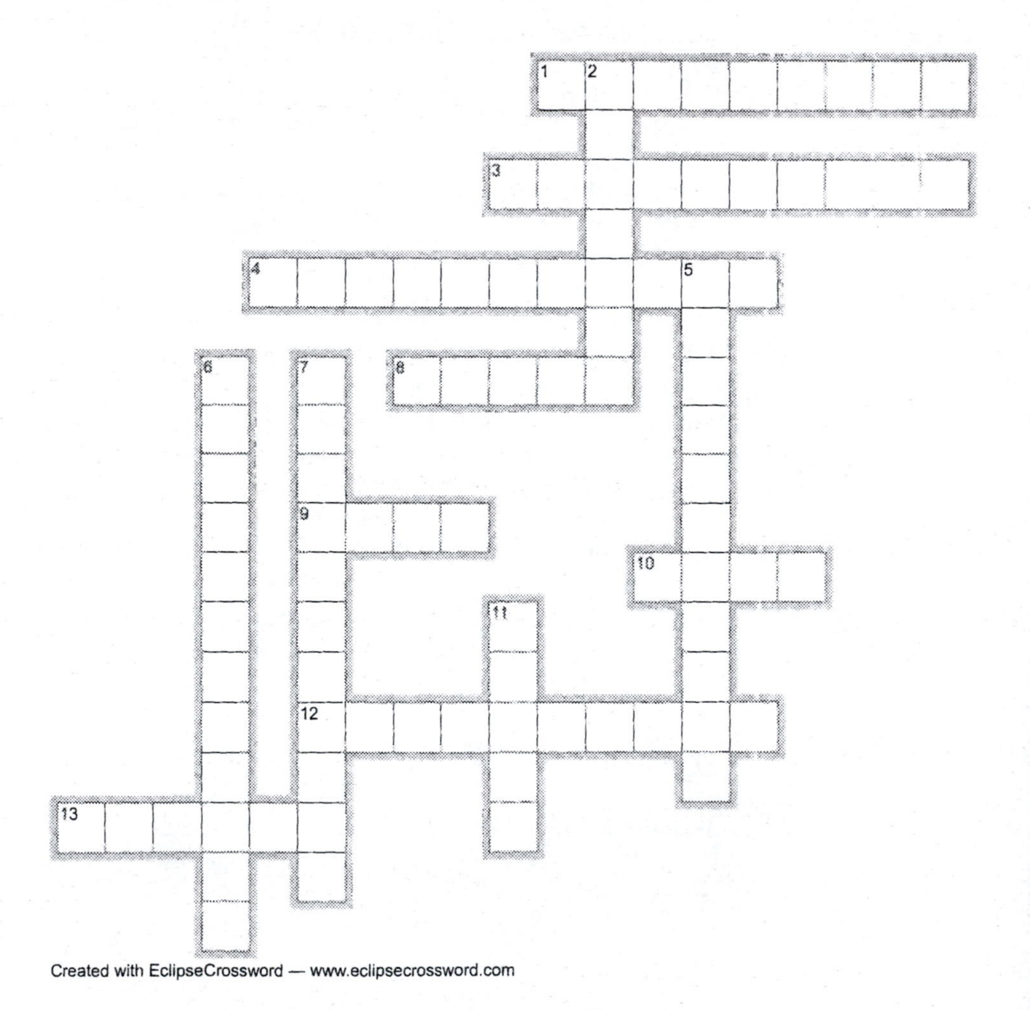

Created with EclipseCrossword — www.eclipsecrossword.com

Across

1. Futures markets are _______ markets, meaning they incorporate all available information quickly and accurately.
3. This gives the owner the right, but not the obligation, to sell futures at a specified strike price.
4. The _______ _______ is the futures price at hedge execution plus the basis.
8. The cash price minus the futures price is referred to as what?
9. A corn farmer who wishes to hedge her crop will initially _______ corn futures contracts, then buy those contracts back later and sell corn in the futures market.
10. The _______ price refers to the price at a market where exchange takes place immediately, at the present time.
12. At contract _______, the cash and futures price should be approximately equal.
13. If one initially sells a futures contract, she offsets by _______ the same futures contract.

Down

2. A highly standardized forward contract backed by an exchange is a(n) _______ contract.
5. This gives the owner the right, but not the obligation, to buy futures at a specified strike price.
6. A put option allows its owner to sell futures at a specified _______ _______.
7. A sorghum purchaser executes a hedge by purchasing corn futures contracts. What is this type of hedge referred to as?
11. An April futures contract traded in February refers to cattle that will be exchanged in the month of _______.

STUDY QUESTIONS

1. Beth purchased two May live-cattle futures contracts in January. One contract is 40,000 lbs. of live-cattle. It is currently March. In March, Beth can fulfill her contract obligations by (*circle all that are correct*)

   (a) purchasing two May live-cattle futures contracts
   (b) selling two May live-cattle futures contracts
   (c) delivering 80,000 lbs. of live-cattle to a futures transaction point
   (d) accepting delivery of 80,000 lbs. of live-cattle at a futures transaction point

Using the following information to answer Questions 2–5.

	May 12, 2004	November 5, 2004
November 2004 Soybean Futures Contract Price (one contract is 5,000 bushels)	\$6.23/bushel	\$7.50/bushel
Soybean Spot Price	\$7.15	\$7.35

2. On November 5, 2004, what was the soybean basis?

3. On May 12, a speculator sells three November soybean futures contracts and offsets those contracts on November 5. What are the speculator's profits? Show your work.
4. Soybean farmers plant in May and harvest in November. A soybean farmer hedges by (*circle one*) BUYING / SELLING November soybean futures in May, (*circle one*) BUYING / SELLING the same number of November soybean futures in November, and selling their grain in the (*circle one*) SPOT MARKET / FUTURES MARKET.
5. Using the data in the table above, if the soybean farmer hedged in 2004, what hedge price did she receive?

Using the futures price information in Figure 9.12 to answer Questions 6 and 7.

	Open	High	Low	Settle	Chg.
CATTLE (CME)					
40,000 lbs.– cents per lb.					
Aug 05	80.00	80.22	79.75	79.92	+.12
Oct 05	81.05	81.30	80.57	80.65	−.07
Dec 05	83.80	84.15	83.50	83.62	−.12
Feb 06	86.00	86.35	85.85	85.97	−.20
Apr 06	84.50	84.75	84.15	84.42	−.12
Jun 06	80.20	80.35	79.60	79.60	−.30
Aug 06	79.52	79.52	79.52	79.52	−.27
Est. sales 11,282. Mon's sales 15,866					
Mon's open int. 131,490. +309					

FIGURE 9.12 Live-Cattle Futures Contract Prices as Reported in Newspapers.
Source: *Tulsa World*, Section E-4, August 10, 2005.

6. On August 10, 2005, what was the best prediction of live-cattle spot prices in February 2006?
7. On August 10, 2005, a buyer of live-cattle decides to execute a hedge for cattle he will purchase in April 2006. In April, he expects the basis to be $2.25/cwt. If he hedges today, about what hedge price does he expect to receive?
8. Markets that quickly and accurately assimilate all available information into market prices are referred to as _______ markets.
9. It is May, and a farmer will soon plant her sorghum. The sorghum will be harvested in November. She wants to hedge her sorghum, but there are no sorghum futures contracts. Explain *how* she can cross-hedge her sorghum and *why* the cross-hedge works. You will be graded on the clarity, accuracy, and completeness of your answer.

Use the information below to answer Question 10.

Contract	Futures Prices on February 6, 2003	Futures Prices on March 1, 2003
	Settlement Price ($ per bushel)	*Settlement Price ($ per bushel)*
March 03 Corn Futures Contract (5,000 bushels)	2.38	2.50
February 04 Corn Futures Contract (5,000 bushels)	2.44	2.62

10. Suppose Dave Chappell bought two March corn contracts on February 6, 2003. It is currently March 1, 2003, the month in which the March contract expires. He can fulfill his contract obligation by (*circle all that are correct*)[3]
    (a) selling two March corn contracts at $2.50/bushel
    (b) selling two March corn contracts at $2.38/bushel
    (c) buying two March corn contracts at $2.50/bushel
    (d) buying two March corn contracts at $2.38/bushel
    (e) accepting delivery of corn and paying $2.38/bushel
    (f) accepting delivery of corn and paying $2.50/bushel
11. Answer true or false: On February 6, 2003, the price of a March corn futures contract must equal (or at least be very close to) the spot price of corn. In one complete sentence, describe why you answered true or false.
12. More than 99% of all futures contract obligations are met by
    (a) offsetting
    (b) accepting or making deliver
    (c) canceling the contract

Use the following table for Questions 13–15.

Futures Contract	Futures Prices on February 6, 2003	Futures Prices on March 1, 2003
	Settlement Price ($ per bushel)	*Settlement Price ($ per bushel)*
March Corn (5,000 bushels)	2.38	2.50
December Corn (5,000 bushels)	2.44	2.62
July Wheat (5,000 bushels)	3.41	3.10

[3]Class notes and worksheet on 2/8/05.

13. Suppose it is March 1, 2003. Between March and December, the spot price of corn is expected to
    (a) rise
    (b) fall
    (c) remain the same
    (d) cannot tell from the information given
14. Suppose it is February 6, 2003. The best prediction of corn spot prices in March is _______.
15. If Pistol Pete purchases one July wheat contract on February 6 and offsets the contract on March 1, how much total money will he receive or pay? Please show your work.

Use the following table for Questions 16–18.

Date	July 2004 Winter Wheat Contract Futures Price (5,000 bushels/contract)	Spot Price in Your Town
	Settlement Price ($ per bushel)	*($ per bushel)*
January 10, 2004	3.30	3.60
July 10, 2004	3.50	3.45

16. You are a wheat farmer who plans to harvest and sell wheat in July. You decide to hedge all of your wheat by selling wheat futures contracts on January 10, 2004. You expect the basis in July to be –$0.10. What is your *expected* hedge price on January 10? Show your math.
17. (*continued from previous question*) Come July 10, you offset your futures position and sell in the your local spot market. What is your *realized* hedge price? Show your math.
18. You purchase a put option for a July 2004 winter wheat futures contract at a price (the option premium) of $0.40 per bushel. The strike price for this option is $4.00 per bushel. If the expected basis in July is $0.25/bushel, what is the minimum expected price you will receive for your wheat by using this option? Show your math.

CHAPTER TEN

Strategic Price Setting

Business is a game, the greatest game in the world if you know how to play it.

—*Thomas J. Watson, Sr.*

We don't want to start a bloodbath, but whatever the competition wants to do, we'll do.

—*Anheuser-Busch*

INTRODUCTION

Many of the goods you purchase are from industries best described by oligopolistic or monopolistic competition. An oligopoly refers to a market where there are only a few sellers of identical goods. Five firms control 80% of beef processed in the United States, and each firm produces similar beef. Thus, the beef processing sector can be described as an oligopoly. Anheuser-Busch, Miller, and Coors control over 80% of the beer market, and in blind taste tests consumers cannot distinguish between the beers produced by these brewers. Coke and Pepsi dominate the soda market, and although each person has their preferred brand, the two are certainly close substitutes for one another.

Why would only three firms dominate the beer market if consumers cannot distinguish between different beers? The answer partly has to do with advertising. Although beer varieties are indistinguishable in blind taste tests, most consumers are not blind. They have been exposed to million dollar advertising campaigns by the three brewers. NASCAR fans who support Dale Earnhardt Jr. will only drink Budweiser, because Dale Jr. drives the Budweiser car. Even though Bud Light tastes very similar to Miller Light or Coors Light, advertising has distinguished Bud Light as a beer with its own unique identity. Viewed in this light, these three brewers do not produce an identical product, but a *differentiated product*. Product differentiation allows a firm to possess market power and charge higher prices. Markets where product differentiation exists, but where the individual brands are still in competition with one another, are referred to as monopolistic competition.

A good example of monopolistic competition is the movie industry. No two movies are alike (though many seem to follow the same Hollywood formula), but most movies are substitutes for one another. Consider the two movies *Star Wars: Return of the Sith* and *Brokeback Mountain*. The former is about a galactic struggle between good and evil, full of fights, death, and betrayal. The latter is about two homosexual cowboys. These movies are anything but identical. Yet, they are in competition with one another because they are both varieties of a general good called movies. Both movies are available to rent at the video store. Chances are, if *Star Wars* had never been made, video rentals of *Brokeback Mountain* would be higher. Granted, there are many people who would rent one but not the other. But so long as there are some people who would watch both, but have limited money and time, the two movies are competitors.

So, to review, an oligopoly is an industry with a few sellers of an identical or very similar product, and monopolistic competition describes an industry with producers of differentiated products who compete with one another for customers. Typically, we say that oligopolies possess market power because barriers to entry limit the number of competitors. There are four major beef processors. You are free to become a beef processor yourself, but you must first raise millions of dollars to build a processing plant—if you wish to be competitive in terms of costs. Because the four existing firms easily meet the market demand, there is little room for another firm. Knowing this, you will have a hard time finding an investor to loan you the money. Firms in monopolistic competition possess market power because they produce a unique or differentiated variety. Guinness beer sells at a premium because no one else produces a beer like Guinness. Budweiser sells at a premium because only Budweiser sponsors Dale Earnhardt Jr. and evokes the image of Clydesdale horses. There are few barriers to entry in monopolistic competition, but firms still possess pricing power due to product differentiation.

In many cases, an industry possesses characteristics of both oligopolistic and monopolistic competition. All gas stations sell virtually the same type of gasoline. Yet some are able to charge a higher price than others. The reason is that the demand at each gas station depends on more than just the gasoline quality. Convenient locations are able to charge a premium because consumers will pay more for the same quality gasoline if it saves them time. From this view, the station sells a differentiated product—convenient gasoline. In one sense, there are few barriers to entry in the gasoline industry. Anyone can open their own gas station. Yet, the number of convenient locations is limited and are typically already filled by another business. The price of land at convenient locations is expensive. So even though there are few barriers to entry for gas stations in general, there are barriers to entry for convenient gas stations.

Throughout this section, we explore industries that display characteristics of both oligopolistic and monopolistic competition, although we will refer to the industry as an oligopoly. We assume that each firm produces a differentiated product of a general good. Each good is an imperfect substitute for another. Imperfect substitutes are like

Coke and Pepsi. People have their preference but will quickly switch to their second choice if the competitor's price is considerably cheaper. Also, we assume barriers to entry. Other firms can enter the industry, but there are obstacles such as high start-up costs or proprietary technology. Each firm has some control over its price. It can raise price without losing all its consumers. However, the presence of close substitutes limits the firm's ability to raise prices. In these cases, firms must make decisions taking into account how their competitors may react. Firms engage in a strategic game, a game for which many outcomes are possible. The firms could collude and charge high prices. Or, if one firm sees the other charging a higher price, it may charge slightly lower prices, steal that firm's customers, and enhance their profits at their competitor's expense. Another possibility is that a fierce price war will ensue, with each firm charging prices so low that no firm makes any money.

Even though there are numerous outcomes in an oligopoly, many times we can make useful predictions regarding expected pricing strategies. An entire field of economics called *game theory* is devoted to studying how people and firms behave in strategic competition. Game theory can be used in numerous settings but is especially useful for describing price-setting behavior in oligopolies. The purpose of this chapter is to

1. introduce the basic concepts of game theory
2. use game theory to discuss price-setting strategies when demand for a firm's product depends on prices set by other firms
3. discuss settings and strategies when a firm can engage in tacit price collusion
4. describe price-setting strategies to limit competition by other firms

GAME THEORY

In the movie *A Beautiful Mind,* John Nash (a brilliant mathematician) is played by Russell Crowe. Nash is at a bar with two friends. In walks four girls; one drop-dead gorgeous and the others rather ordinary. One of Nash's friends suggest they all go after the gorgeous girl, suggesting that intense competition will lead to the best result for everyone. Nash notes otherwise. If they all go after the gorgeous girl and ignore the three ordinary girls, each blocks the other from the gorgeous girl and the ordinary girls become angry at being ignored. The result? No one gets a girl. Conversely, if each goes after a single ordinary girl and no one pursues the gorgeous girl, the three ordinary girls will be flattered—each man gets a girl. Although the gorgeous girl may be offended and no one gets the gorgeous girl, obtaining one of the ordinary girls is better than no girl at all.

This is a game where the outcome for each guy depends on the actions of other people. The study of how interdependent people make decisions, when aware that their actions affect one another, is referred to as *game theory*. In the movie, it is in this scene where Nash supposedly develops a mathematical technique in game theory that would later win him the Nobel Prize in Economics. Although the movie scene is mostly fictional, John Nash is not, and he did win the Nobel Prize for his work in game theory.

Game Theory: The study of how interdependent people behave when aware that their actions affect one another.

The One-Shot Price-Setting Game

The behavior of firms in an oligopoly are best studied using game theory, because the profits realized by each firm depend critically on the prices set by other firms and vice versa. Firms in an oligopoly are playing a game, a game with large monetary stakes. To illustrate game theory, suppose we have a duopoly market, which is a market with only two firms. The two firms are named ADM (Archer Daniels Midland) and Ajinomoto, and both make lysine, an additive used in corn feed to improve animal growth. If the firms both charge high prices, they split the market and both make large profits. If they engage in a price war and both charge low prices (again splitting the market), both make low profits. However, if one firm charges a high price and the other charges a low price, the low-price firm gets all the market and makes the largest profits possible, while the high-price firm receives the lowest profits possible.

This game is depicted in Figure 10.1. If both charge high prices, each receive profits of $50 million. If they engage in a price war and both set low prices, they receive only $30 million each. Yet, if Ajinomoto sets high prices and ADM sets low prices, ADM gets most the customers. ADM receives high profits of $60 and Ajinomoto receives low profits of $10 million. The reverse happens if ADM sets high prices and Ajinomoto low prices. This is a "one-shot game," meaning it is only played once. In a one-shot game, firms can set their strategies without worrying how their actions affect future interactions.

At this point we should discuss the technical details of a "game." Games consist of players, actions, information, and strategies. The *players* in this game are the two firms ADM and Ajinomoto, and their *actions* are either to charge high or low prices. The *information* describes how much each firm knows, and in this game we have common knowledge, meaning each player possesses all possible information and knows that the other players do as well. The *strategies* describe the actions taken by

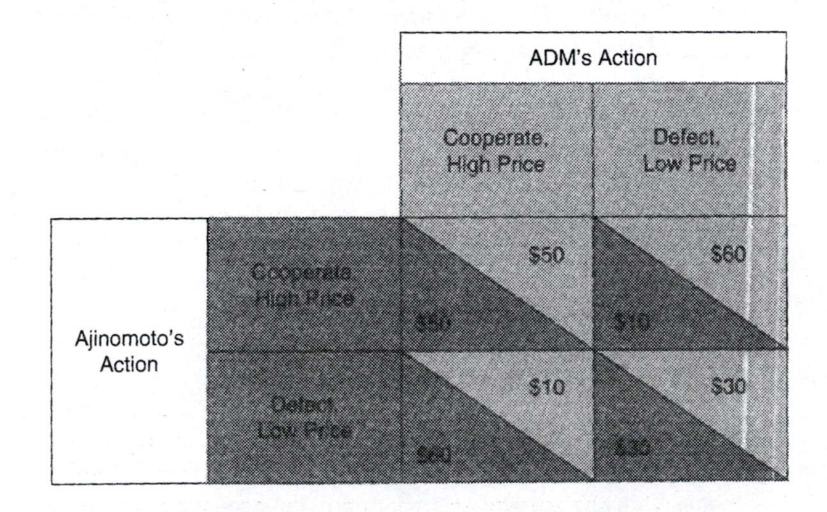

FIGURE 10.1 Price-Setting Game (Profits in Millions of Dollars).

each player, and the *payoffs* show the outcome realized by each player contingent upon the players' strategies.

Dominant Strategy: A strategy that yields the highest payoff regardless of the other players' actions.

Game theory consists of identifying the strategies players will pursue. Suppose that this game were played only once. If you were ADM, which action would you take? Would you charge a high or low price? If Ajinomoto charges a high price, you are better off charging a low price—your profits are $60 instead of $50. If Ajinomoto charges a low price, you are also better off charging a low price—your profits are $30 instead of $10. No matter what action Ajinomoto chooses, you are better off charging a low price. Charging a low price is a *dominant strategy*, because it yields the highest payoff no matter what action Ajinomoto takes. Now pretend that you are Ajinomoto. Charging a low price is also your dominant strategy. Thus, both firms will pursue their dominant strategy and charge a low price. It is the best they can do no matter what their competitor does. Both charge low prices and fail to reach the maximum possible profit.

This is a plausible outcome. Just because a market is dominated by two firms does not mean consumers will face high prices. However, the opposite outcome of high prices is also possible. It is clear from Figure 10.1 that both firms lose from pursuing a dominant strategy. The game we looked at was a noncooperative game, but what if ADM and Ajinomoto could work together? If both firms could collude and agree to set high prices, they both benefit in the form of higher prices. For this collusion to work, each firm needs to be sure the other party will stick to the collusion agreement. If ADM and Ajinomoto agree to collude, but ADM defects from the collusion, ADM makes greater profits at Ajinomoto's expense. When this price-setting game is played only once, both firms have incentives to defect and the collusion deal would likely fail. But when the game is allowed to repeat, collusion may emerge.

The Repeating Price-Setting Game

In real markets the price-setting game is played repeatedly. Firms can set their price, observe prices set by other firms, observe the outcome, and change their price-setting strategy appropriately. Let us modify the price-setting game to better reflect reality by allowing it to repeat infinitely. In Period 1 each firm chooses their actions and observes their payoff. Then they move to Period 2 where they observe the actions of other players in the last period, the payoffs that resulted, and change their strategy appropriately. We then move to Periods 3, 4, 5, and so on. At no point does the game end. This type of game has been studied extensively, by biologists as well as economists. The reason is that the game closely describes the game of altruism in evolution. Many animals live in groups. Vampire bats feed on the blood of other animals at night. Not all hunts are successful though. Many times bats will not find a victim, but when they do find a victim, they typically suck more blood than they need. This presents an opportunity for the bats to cooperate by sharing their blood on successful nights with other bats who were unsuccessful. The group as a whole benefits from such altruism because they all experience a regular supply of blood.

However, so long as there are many altruistic bats within the group, each individual bat has the incentive to defect from the cooperative agreement. Consider a single vampire bat named Dracula who lives with a large group of bats with altruistic genes. Dracula does not possess the altruistic genes like the others. Dracula possesses the jerk gene. When Dracula's hunts are unsuccessful, he still receives blood from those in the group that share the fruits of their successful hunt. Yet, when Dracula's hunts are successful, he hides the fruits of his hunt and does not share. By being a jerk, Dracula receives more blood and therefore more nourishment than the others. Dracula grows to be a bigger and healthier bat and will therefore have more offspring than the others.

His offspring also possess the jerk gene (and not the altruistic gene), allowing them to produce more offspring than bats with the altruistic genes. Eventually, Dracula's descendants dominate the population of bats, and the altruistic gene disappears. In the bat population all bats eventually contain the jerk gene, and the group as a whole is made worse off. They no longer share the fruits of their night hunts and must rely solely on themselves for food. As a result, each bat receives less nourishment than it did when the altruistic gene was present.

This is just like the infinitely repeating price-setting game. Both firms are better off if they cooperate and collude to set high prices than if they both set low prices. Yet each individual firm has the incentive to defect and charge a low price. Just like we have reason to believe altruistic genes could not persevere in bat populations, we have reason to believe that firms will defect from their collusion agreements. However, vampire bats do indeed share blood with fellow bats. The altruistic gene is present in vampire bats and many other species including humans. How many times have you given blood expecting nothing in return? Moreover, we observe firms that do form successful collusion agreements.

So what is the dominant strategy in an infinitely repeating price-setting game? To answer this, the political scientist Robert Axelrod in 1980 issued a challenge for experts in game theory to play the repeating price-setting game. These experts were asked to submit a programmed strategy, meaning a set of rules dictating which action—"cooperate" or "defect"—would be chosen in any given round. In the price-setting game, to cooperate means to choose a high price, and to defect means to choose a low price. Some strategies submitted were *unforgiving*, meaning "cooperate" was chosen in the first period but if the opponent ever defected, the player would always choose "defect." Otherwise, *unforgiving* would always choose "cooperate." Other strategies were forgiving or "nice." For example, one expert named Anatol Rapoport entered a *tit-for-tat* strategy. In the first period, he chose "cooperate." In the next period, he chose whatever strategy the opponent chose in the previous period. If the opponent chose "defect" in the first period, he would choose "defect" in the second period. If the opponent chose "cooperate" in the first period, he would choose "cooperate" in the second period. The *tit-for-tat* revenge only lasts one period. After that, *tit-for-tat* simply mimics the opponent's action the previous period. Thus, *tit-for-tat* wants to engage in cooperation but is quick to punish defectors. Cooperative the first period, then *tit-for-tat* thereafter; whatever you chose the last period, he chose the current period. A third strategy might be *defect* where the player always chooses "defect" regardless of what happened in previous periods.

Axelrod then paired each strategy with one another in a computer simulation and calculated the total payoffs over many, many periods. The total payoffs represented the "score" of the game, where a higher score is better. For example, if an *unforgiving* strategy was paired with a *defect* strategy, from the second period forward both players would choose "defect" and both would receive a low score. A *tit-for-tat* matched with *defect* would also score low, whereas *tit-for-tat* matched with *unforgiving* would perform well because both players would cooperate in all periods.

The game proceeded as follows. Each strategy was matched with every other strategy to form many separate pairings. The price-setting game was played over and over, and the score for each strategy-pair in a period equaled their combined profits for that period. Each strategy-pair was then allowed to reproduce a number of times. The number of times a strategy-pair could multiply depended on its profits. The more profits, the more it could multiply. In biology, this is akin to producing more offspring. The game is just like allowing bats with jerk genes to reproduce with other bats possessing the jerk gene or the altruistic gene. In economics, this is like a firm expanding its production or creating similar firms (like increasing the number of franchises) if it is profitable. Wal-Mart is a highly successful retailer and, because of its success, has expanded into virtually every town in America, just like animals with successful genes spread their genes throughout the species' population.

After 1,000 simulations, it was clear which strategy dominated the population of strategy-pairs: *tit-for-tat*. Remember that *tit-for-tat* is a strategy where you cooperate the first period, but then copy whatever strategy your opponent chose in the last period. The reason *tit-for-tat* performed so well is that it cooperates well with "nice" opponents but is not subject to exploitation by "mean" opponents. If *tit-for-tat* encounters an opponent willing to cooperate, both firms charge high prices and receive high profits. If *tit-for-tat* encounters an opponent that likes to defect, it does not allow that strategy to profit at its expense. That is, in Figure 10.1, if the opponent likes to defect, *tit-for-tat* also defects, making sure it receives profits of $30 million instead of $10 million.

It turns out that vampire bats also display this *tit-for-tat* strategy. Each bat remembers who shared blood with them in the past. When they see a bat who helped them in the past without blood, they share whatever blood they have. And if a bat never shared its blood with anyone, seldom would others share blood with it. Vampire bats are nice to other bats that were nice and punish those who are not (Dawkins 1999).

Trigger Pricing

Trigger Pricing: Set high price if competitors set high price; low price if competitors set low price.

The *tit-for-tat* strategy seems to be the dominant strategy in infinitely repeating price-setting games. This strategy is regularly observed in animals, including humans. In the vampire bat story told above, it turns out that vampire bats form "friendships" with other bats. They do not share their blood with just any hungry bat; they tend to share their blood with bats who fed them in the past. Firms too have been observed to display the *tit-for-tat* strategy, which more formally is called *trigger pricing*. A firm makes it known that if other firms set high prices, they will also set a high price. But if their competitors set a low price, the firm makes it known they will also set a low price. This changes the price-setting game considerably. Consider again the duopoly where ADM and Ajinomoto are the only two firms selling lysine. If Ajinomoto announces this trigger-pricing strategy, it eliminates two possibilities from the game. No longer can the firms charge different prices, and the game modifies to that shown in Figure 10.2. Ajinomoto has announced its strategy: Set high prices if ADM sets high prices and low prices if ADM sets low prices. ADM must now decide its dominant strategy. If ADM sets a low price, it knows Ajinomoto will as well and its profits will be $30 million. If ADM sets a high price, Ajinomoto will too, leading to higher profits of $50 million. It is clear, ADM will set a high price and both firms realize profits of $50 million. With the trigger-pricing strategy in place, cooperate now becomes a dominant strategy and the firms' joint profits are maximized.

Consider the difference in the price-setting game when it is played one time versus repeatedly. When played only once, both firms set low prices (their dominant strategy) and receive low profits of $30 million. When played repeatedly, the dominant strategy is *tit-for-tat* and both firms receive higher profits of $50 million. How does a firm exactly pursue a *tit-for-tat* strategy? One is to simply collude with each other. Meet face to face, agree to charge high prices, and make it clear you will drop your price if the other does not live up to the agreement. ADM (Archer Daniels Midland) and Ajinomoto are real firms, and they really sell lysine, and they really colluded. The lysine market is an oligopoly made up of the firms ADM, Ajinomoto, and three smaller firms. These firms met in person and agreed to set identical high prices. This is called price-fixing. A firm can enhance its profits by price-fixing, so long as there are barriers prohibiting new firms from entering the market and charging low prices. Price-fixing is illegal though and only makes you money if you do not get caught. ADM did get caught and was fined $70 million (*Feedstuffs* 2004).

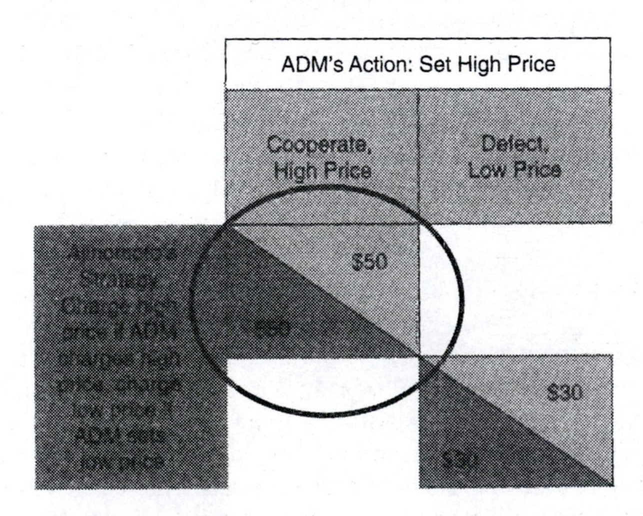

FIGURE 10.2 Trigger Pricing.

However, oligopolies can maintain high prices using the *tit-for-tat* strategy in a legal fashion. Firms cannot meet with competitors and conspire to charge high prices. However, it can make its *tit-for-tat* or trigger-pricing strategy evident simply through the prices it sets. Anheuser-Busch is the leading U.S. brewer with a 44% market share in 1996. There is a tendency in oligopolistic industries for the largest firm to set the price and other firms follow. In 1953 Anheuser-Busch experienced an increase in costs due to a union wage agreement and increased its beer prices in response. Even though many other brewers increased their prices in turn, some midwestern competitors did not. To punish the defectors, Anheuser-Busch responded by slashing its prices and stealing their competitor's market share. Two years later when Anheuser-Busch raised its prices again, their midwestern competitors did so as well, fearing another *tit-for-tat* response if they did not. Consider another example from the brewing industry. Coors and Miller cut the price of some of their beer in 1988, hoping to steal market share from Stroh and Heileman. They ended up stealing from Anheuser-Busch as well though. Within 18 months of the price cut, Anheuser-Busch announced "we don't want to start a bloodbath, but whatever the competition wants to do, we'll do" (Tremblay and Tremblay 2005). Once a firm announces in advance that it will pursue a trigger-pricing policy, they are essentially announcing a low-price guarantee, which can be an effective way for oligopolies to raise prices.

Low-Price Guarantees

Office Depot, Staples, and OfficeMax form an oligopoly in the office supplies market. In May of 2003, OfficeMax ran advertisements saying if customers found a lower price at any competitor's store, OfficeMax would match that lower price plus slash an extra 15% off. This advertisement sends a clear signal to Office Depot and Staples that OfficeMax engages in *tit-for-tat* pricing strategies. Given OfficeMax's commitment to offer low prices if their competitor's prices are low, the dominant strategy for each firm is to set high prices, giving the three firms higher profits. Referred to as *low-price guarantees*, this pricing strategy can be viewed as a legal form of price-fixing. However, a low-price guarantee could also be a strategy of undercutting the competitors and stealing their market share. Low-price guarantees can promote competition and lower prices or promote collusion and raise prices. Theoretically, it is ambiguous whether low-price guarantees raise prices or lower prices.

Low-Price Guarantee: A promise made to consumers that the firm will meet or beat a competitor's price for the same product.

In 1986, Winn-Dixie in Raleigh, North Carolina, ran an advertisement in the local newspaper stating it would match Food Lion's supermarket prices for specific products. Winn-Dixie and Food Lion combined had a 50% market share in the area, giving it market power and the ability to collude. If Winn-Dixie's low-price guarantee was intended to engage Food Lion in a price war, prices should fall after the announcement. If it was meant to facilitate collusion, price should rise after the announcement. Luckily, data were collected on the items for which the low-price guarantee was made before and after the low-price guarantee announcement. Prices rose after the announcement, indicating the low-price guarantee facilitated collusion. It was meant to make it clear that Winn-Dixie would pursue a *tit-for-tat* strategy in the price-setting game, leading both supermarkets to raise prices and accrue higher profits (Hess and Gerstner 1991). Prices even rose in supermarkets not participating in the low-price guarantee. These supermarkets simply saw the higher prices and realized that a price-setting game was in progress. Noting that supermarket profits are highest when all firms collude and set high prices, these other supermarkets also set high prices (Hess and Gerstner 1991). This is commonly referred to as *tacit collusion* (tacit means not spoken but understood).

Tacit Collusion

It is possible for firms in an oligopoly to collude even without any communication. In game theory, there is a mathematical proof called the *Folk Theorem* that shows if the players are rational, in repeated games players will develop cooperative strategies, even if they are ultimately competitors. People naturally display *tit-for-tat* strategies in everyday life. Experiments have repeatedly shown that people readily cooperate with others who have a reputation for cooperation but will quickly punish noncooperators, even at their personal expense. One of the most famous games is the ultimatum game played by two people. The game begins by giving one player a fixed amount of money, say, $10. This person is called the allocator. This player must decide how much to share with the second player called the receiver. The receiver either accepts the offer or rejects it, in which case neither receive any money.

From one point of view, the allocator should offer the receiver a small amount, like $1.00, or $0.25, or even a penny. Consider a low offer of $1.00. Once the $1.00 is offered, the receiver either accepts the dollar or rejects it. If they reject, the receiver gets no money. If the receiver prefers $1.00 to no dollars, the receiver should accept the $1.00 offer. Knowing this, the allocator only offers the receiver $1.00, leaving $9 for herself. There is a problem with this logic. It assumes people get no pleasure from punishing unfair offers. If you were the receiver and were offered only $1.00 of the initial $10 sum, you would probably feel this was unfair. The allocator didn't earn the $10 and doesn't deserve it any more than you. Rather than accept the $1.00, many of you would choose to punish the allocator by rejecting the offer. The allocator now receives nothing, and that is her punishment for being greedy.

Allocators take this into consideration. Knowing they will be punished for unfair offers, many allocators offer an equal split of the money. This game has been repeated across many cultures. On average the allocator offers 40% to 50% of the money to the receiver, and the offer is usually accepted. Offers less than 20% are consistently rejected (Hessell, Sloof, and Kuilen 2004). This illustrates the *tit-for-tat* strategy. If the allocator is nice, the receiver will be nice in response. If the allocator gets greedy, the receiver will respond with punishment. The low-price guarantee discussed in the previous section is like a punishment for failing to collude. It is well known in the brewing industry that Anheuser-Busch will punish fellow brewers for setting prices too low; Anheuser-Busch has said it themselves. But often people do not need to be told that they will be punished for failing to collude. People expect other people to follow a *tit-for-tat* strategy. There is a tacit understanding that all firms should set high prices, so firms do set high prices, and collusion occurs without even any communication.

School systems in Texas can purchase milk from a variety of vendors. Milk sales at different locations are coordinated using a sealed bid auction. Each vendor submits a sealed bid to a particular location, and the lowest bidder gets a one-year contract. Vendors as a whole can benefit from all agreeing to submit identically high prices. There must be a large degree of trust for the agreement to work, because one vendor could submit prices slightly lower than the others and capture the entire market at a reasonably high price. To prevent collusion among the vendors, everyone submitting bids had to sign a noncollusive affidavit, saying they have not and will not coordinate prices with any other person. Although communication between the vendors was prohibited, milk prices in one particular region appeared abnormally high relative to the other regions.

An economic analysis concluded that tacit collusion had occurred. Bidders pursuing *tit-for-tat* strategies were able to raise prices by punishing those who bid low. A vendor who historically sold to a particular school district would submit relatively high bids to that school district but extremely high bids in the other districts. This was done hoping other vendors would do the same. Other vendors returned the favor. The result was that each vendor sold to the same districts at a high price. In a sense, the vendors had a tacit agreement that gave each individual vendor a monopoly in certain school districts. If Vendor A defected and submited low bids in Vendor B's district, Vendor B responded in kind by submitting low bids in Vendor A's territory the following year. The threat of punishment was strong enough that few defections occurred. When they did, the punishment in response brought those defectors back to the tacit arrangement (Lee 1999).

In other settings, tacit collusion can be difficult to maintain. If several or more firms are setting high prices, a rebel firm can always set slightly lower prices, taking over the market and earning high profits in the process. Each firm has a strong incentive to defect from tacit collusion arrangements, and they often do. The more firms in a market, the less likely tacit collusion will occur. Researchers have employed laboratory experiments to determine the effectiveness of tacit collusion without communication under different market settings. In duopolies (two firms only), tacit collusion without communication readily occurs. However, once a third firm enters the market, tacit collusion without communication becomes difficult (Muren and Pyddoke 1999). Thus, although tacit collusion can arise in oligopoly markets, it is probably the exception rather than the rule.

Nash Equilibrium

So far we have only discussed games where dominant strategies exist, but not all games have a dominant strategy. For a long time this limited the usage of game theory in economics. Then along came John Nash, whose life was portrayed by Russell Crowe in the movie *A Beautiful Mind*. Nash discussed a particular type of equilibrium that exists in many games. Referred to as *Nash Equilibrium*, it is undoubtedly the

A Nash Equilibrium exists when all players are employing their best strategy given the strategies of all other players.

most important concept in game theory. A Nash Equilibrium exists when all players are following their best strategy, given the strategies pursued by the other players. Refer back to Figure 10.1 and assume a one-shot price-setting game. The point *cooperate/cooperate* is not a Nash Equilibrium. Given that your opponent is cooperating, your profits are higher if you defect. At *cooperate/cooperate*, neither player is satisfied with their strategy given the opponent's strategy, so *cooperate/cooperate* is not a Nash Equilibrium. Now consider the point *defect/defect*. At this point, neither firm gains from changing their strategy to *cooperate*. At *defect/defect*, both firms have no incentives to change their strategy, so the point *defect/defect* is a Nash Equilibrium. The concept of Nash Equilibrium allows us to develop a universal market model, one that encompasses both perfect competition and monopoly and everything in between (assuming all firms produce the same good). This model is commonly referred to as the Cournot Model and is described below.

THE COURNOT MODEL OF IMPERFECT COMPETITION

First, we will issue a warning. This is probably the most difficult concept in this textbook and requires the most math. What follows is the simplest presentation of the Cournot model we could construct. If you are not able to follow the math, simply assume that the price in the Cournot model is somewhere between the price in a monopoly and the price in perfect competition, and the more firms in the market, the lower the price. For example, if a monopoly would charge \$100, but a competitive market would produce a price of \$10, then in a duopoly, the price would be closer to \$100, say \$80. With three firms, it be lower than \$80, and with four firms even lower. As the number of firms becomes very large, price will eventually equal \$10, the perfectly competitive price. That is a basic description of the Cournot Model outcome. The Cournot model is described more fully below. If you are scared of the math, you can just proceed to the section "Beating the Cournot Model."

The price-setting game describes the basic game played between firms in imperfect competition. In one way, firms fight against each other for consumers and profits. In another way, firms as a whole can make themselves better off by working together. The price-setting game has the advantage of simplicity, because it clearly demonstrates the competition/cooperation trade-off, but the disadvantage that it provides an overly simplistic view of markets. In reality, firms can do more than set a high or low price; they have a wide array of prices to set. Moreover, there are often more than two firms in the market. This section describes the most useful and most general model for oligopolies: the Cournot model. It is also the model that provides the best predictions about oligopoly behavior. Due to its greater power, it is also more complicated.

The Cournot model is based on several assumptions. First, assume there are N firms alike in every manner. They face the same constant marginal cost of c. That is, for every unit produced, cost increases by exactly c, regardless of how many units have already been produced. Each firm produces the same identical product and therefore receive the same price. Also, we assume that each firm produces the exact same output, and that output level is chosen to maximize the firm's profits, given the output level of other firms. Given that the firms are identical, they look at the market conditions and come up with the same conclusion about how much they should produce.

Nash Equilibrium in Cournot Model with Identical Firms Requires

(1) All firms choose quantities to maximize profits, taken other firms' quantity levels as given.
(2) All firms produce the same quantity.

Although not proven here, the assumption of identical output for all firms basically assumes a Nash Equilibrium. The Nash Equilibrium is where all firms are happy with their output, given the output of all other firms. If firms are identical, facing the same marginal cost and receiving the same price, the Nash Equilibrium requires that all firms produce the same output. Proving this requires a bit more math than we wish to expose students to in this textbook. Thus, we motivate the Nash Equilibrium condition of identical output by all firms by appealing to the fact that all firms are identical. If Firm A is identical to Firm B, and both face the same cost and price, how does it make sense that Firm A decides to produce 100 units but Firm B decides to produce 50 units? It doesn't make sense, and mathematically you can prove it.

Let production by each firm be given by q. The market demand curve goes by the formula $P = a - b(Q)$ where Q is industry output, where $Q = Nq$. Next, we solve for the optimal quantity q for each firm. First, substitute Nq for Q in the demand equation to get $P = a - b(Nq)$. Next, notice that the term Nq can be written as $(N - 1)q + q$. The first term $(N - 1)q$ refers to the output by all firms except one, and q refers to output by the excluded firm. Let this excluded firm be called Firm A and denote its output as q_A. Notice we can rewrite the demand curve as $P = [a - b(N - 1)q] - bq_A$. This function shows how price changes if Firm A increases or decreases its output. The term $[a - b(N - 1)q]$ is the intercept and "$-b$" is the slope of Firm A's demand curve. Thus, the presence of N firms is like giving Firm A a monopoly, but at a lower demand. If Firm A was the only firm, it would face the demand curve $P = [a] - bq_A$ and would choose its output level so that marginal revenue equals marginal cost as described in Chapter 4.

Recall that if the demand curve is $P = [a] - bq_A$, the marginal revenue curve is $MR = [a] - 2bq_A$. It then follows that if the demand curve is $P = [a - b(N - 1)q] - bq_A$, the marginal revenue curve for Firm A is $MR = [a - b(N - 1)q] - 2bq_A$. Thus, a single firm in Cournot Competition acts just like a monopoly with a lower demand. The firm will choose its output level so that marginal revenue equals marginal cost, taken as given the output level of other firms. Then, because we assume all firms produce at the same output, we assume that $q_A = q$. Finally, by solving for the q that sets the firm's marginal revenue and cost equal, we have solved for the Nash Equilibrium. This is illustrated in Figure 10.3. Notice the two central features driving the mathematics of the Nash Equilibrium. First, each firm is assumed to choose the profit-maximizing quantity taken as given all other firms' output. Second, all firms choose the same output level.

The Nash Equilibrium tells us that the market price will be: *Cournot Model Market Price with N firms* $= a - [N/(N + 1)](a - c)$. As more firms enter, the term $N/(N + 1)$ becomes larger, reducing the market price. Let us assume that $a = \$100$ and $c = \$10$. Notice what happens when there is one firm. The single firm behaves as a monopoly and charges a high price of \$55. When a second firm enters, forming a duopoly, the market price falls to \$40. Once a third firm enters the market,

Model Setup: N identical firms with identical marginal cost c; market demand curve $P = a - b(Q)$ where Q is total production by all firms.

Solve for q_A^*: Set marginal revenue and marginal cost equal

$MR = [a - (N-1)bq] - 2bq_A^* = c;$

- Use the Nash Equilibrium condition that $q_A^* = q$; all firms produce the same output.

  $[a - c] = 2bq + (N-1)bq = 2bq + Nbq - bq = (N+1)bq$
- Thus, in Cournot-Nash Equilibrium:

  $q = (a - c)/[b(N+1)]$
- The market output (output by all N firms) and market price is then

  $Q = Nq = [N/(N+1)][(a-c)/b]$

  $P = a - [N/(N+1)](a-c)$

FIGURE 10.3 Solving for Profit-Maximizing Quantity for a Single Firm with *N* Competitors.

we have an oligopoly, and prices fall to $33. When a fourth and fifth firm enters, the market is still an oligopoly, but prices keep falling as new firms arrive. When a large number of firms exist, the market can be said to be perfectly competitive. With 100 firms the price is $11, which is only $1.00 more than marginal costs of $10. If we keep increasing the number of firms, the price will converge to marginal cost of $10. The appeal of the Cournot model is its ability to capture almost any market structure. When only one firm exists in the market, the Cournot model tells us the price will be $55, which is indeed the monopoly price. When the number of firms is large, the Cournot model tells us price falls to marginal cost, which is what we expect from the perfectly competitive market. Now we can make precise predictions for any number of firms, as shown in Figure 10.4.

Profits for each firm equals price minus marginal cost, times the firms' output. As can be seen from Figure 10.4, profits indeed fall as more firms enter the market. When the number of firms becomes large, price equals marginal cost and profits are zero. But remember these are economic profits, not accounting profits. A firm can make economic profits of zero but still have enough money to pay the manager and all its employees. Economic profits of zero do not imply the firm is not making money, it only implies the firm is not making more money than it could in its next best alternative.

The Cournot Model in the Laboratory

The appeal of the Cournot model is that it predicts both price and quantity for most market settings: monopoly, oligopoly, and perfect competition. It can easily be modified to accompany oligopsony and monopsony as well. For this reason, the Cournot

Number of Firms (N)	Cournot Price: $P = a - [N/(N+1)](a-c)$ $a = \$100$; $c = \$10$	Cournot Quantity for Each Firm $q = (a-c)/[(N+1)b]$	Market Output $Q = Nq$	Profits per Firm $(P-c)(q)$
1, monopoly	\$55	45.00	45.00 × 1 = 45.00	(\$55−\$10)(45.00) = \$202.50
2, duoploy	\$40	30.00	60.00	\$900.00
3, oligopoly	\$33	22.50	67.50	\$506.00
4, oligopoly	\$28	18.00	72.00	\$324.00
5, oligopoly	\$25	15.00	75.00	\$225.00
100, perfectly competitive market	\$11	0.89	89.11	\$0.89

FIGURE 10.4 Cournot Price as Number of Firms Rises ($a = \$100$; $b = \$1$; $c = \$10$).

model is like the grand or unified theory of market prices (when firms sell identical goods). But just because the Cournot model has nice mathematical properties does not imply that it reflects real-world behavior. People are not equations, and the construction of mathematical economic models requires us to view the world in a more simple manner than it really is. Yet, the basic ideas behind the Cournot model seem sensible. Ultimately, any economic model is judged by its ability to predict real behavior. Ideally, we would collect information on market demand, firms' cost of production, and firms' decisions and calculate how close the firms' prices and quantities conform to the Cournot model predictions. Alas, market demand can be difficult to estimate and rarely does one know firms' cost of production. So, to study these issues economists construct their own markets inside of a laboratory. In a laboratory, the economists can impose consumer demand and cost of production exactly. Human subjects are brought in to act as firms and are paid according to how much profits they earn. Inside the laboratory we can measure exactly how close the experimental markets conform to the predictions of the Cournot model.

Many such experiments have been carried out. The human subjects are typically students and are recruited by paying them money. When the students arrive at the experiment, they are told they will play the role as a firm. The student is told how much it costs to produce a single unit of a good and are also given information about the consumer demand curve and the number of competing firms. Without communicating with other students, or even knowing who their competitors they are, the subjects choose how much they want to produce. Based on the production level chosen by all firms, a market price results and firm profits are earned. This is repeated over and over, without letting the subjects know when the game will end.

These experiments show the Cournot model does a pretty good job of predicting real human behavior. One way of measuring the distance between the experiment results with Cournot model predictions is the ratio of actual market production to the production level predicted by the Cournot model. This is referred to as the *Cournot Realism Ratio,* because it indicates the extent to which the Cournot model reflects reality. For example, if "firms" in the experiment produce 90 units but the Cournot model predicts production of 100 units, this ratio is 90/100. A group of economists studied 19 such experiments where the number of firms varied from 2 to 5. Across all 19 experiments the average Cournot Realism Ratio was 0.998. Thus, the Cournot model is a powerful predictor of actual behavior. The Cournot Realism Ratio also tended to increase with the number of firms. As shown in Figure 10.5, a duopoly tends to lead to greater collusion than predicted by the Cournot model. Actual market production is less than the Cournot model predicts, indicated by a realism ratio less than one, which implies that actual prices are higher than the Cournot price. As more firms are added to the market, the realism ratio is greater than one and tends to rise with the number of firms. The greater the ratio becomes, the more competitive the market becomes. With five firms, market production is greater and market price is lower than the Cournot model would predict. Still, the Cournot model does a surprisingly good job at predicting human behavior (Huck, Normann, and Dechssler 2004).

Beating the Cournot Model

The Cournot model provides a good indicator of the price and profits one should expect in an oligopoly. In a sense, the outcome of a Cournot model is somewhere between monopoly (or a really good collusion) and perfect competition. Unlike perfect competition, firms do make economic profits in the Cournot model. However, firms could enhance their profits by colluding with one another. If firms joined together and acted like one single firm, a monopoly, and then split the profits, each firm would make more money. Indeed, in some settings firms do behave collusively. Employing tacit collusion or some other method, they are able to set high prices and therefore obtain high profits. Other times firms engage in fierce competition driving prices and profits down. Of all the outcomes that can result from oligopolies, we can generalize them into one of three groups. The *Nash Outcome* is the price, quantity, and profits predicted by the Cournot model. Remember the solution to the Cournot model assumes a Nash Equilibrium. Firms view the output of their competitors and,

Number of Firms (N)	Cournot Realism Ratio = (actual output)/(output predicted from Cournot Model)
2, duopoly	0.927
3, oligopoly	1.027
4, oligopoly	1.029
5, oligopoly	1.050

FIGURE 10.5 Cournot Realism Ratio.

Source: Huck, Normann, and Oechssler (2004).

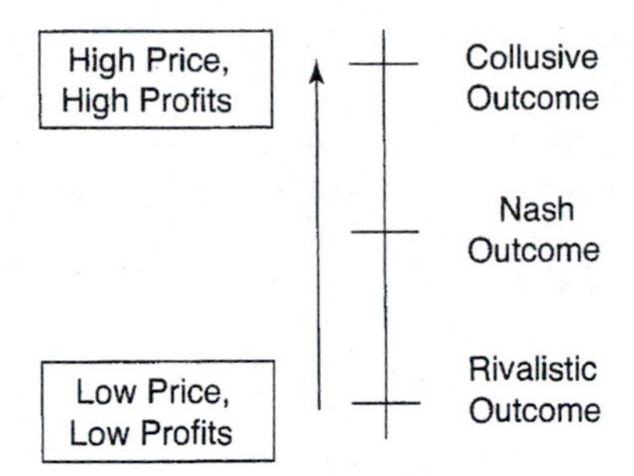

FIGURE 10.6 Three Outcomes of Oligopolistic Competition.

given that output, produce their profits maximizing output. Firms continually change their output decisions as their competitors change strategies and eventually reach a Nash Equilibrium.

Other times firms are able to collude through the use of tacit collusion, trigger pricing, low-price guarantees, or illegal means. They set high prices and receive high profits, leading to the *collusive outcome*. In other cases they seek to drive each other out of business. Each firm produces a large amount and charges low prices, hoping their firm's profits will be low enough to drive them out of business. This latter case where prices and profits are low is the *rivalistic outcome*. If possible, a firm will want to beat the Cournot model, realizing a collusive outcome and high profits. This is not easy, especially when there are more than three firms. Yet there are some market settings and firm strategies that increase the probability of a collusive outcome. Understanding these settings and strategies will help guide a firm in knowing when the Cournot Model can be beat and how it can be beat.

The type of laboratory experiments described previously provide a wealth of information regarding when a collusive outcome can be realized through tacit collusion. The beauty of a laboratory experiment is that we can focus on one feature differentiating markets, hold everything else constant, and determine the impact of that feature on the market price. Below are four features that increase the probability of a collusive outcome through tacit collusion, some of which can be controlled by the firm and some of which cannot. These features are the result of numerous economic experiments and summarized by Huck, Normann, and Oechassler (2004).

1. Stable Competitors Facilitate Tacit Collusion. Tacit collusion only works when all firms have a mutual understanding that all firms will set a high price. If one firm defects from the coalition and charges a slightly lower price, the defecting firms reap enormous profits and the other firms make little. Yet, when a firm defects, other firms quickly punish that firm by charging even lower prices. Remember the case of tacit collusion of milk sales to Texas school systems. If one milk seller charged a low price in a given area, others quickly charged an even lower price, taking away the defector's market. Some firms may have even taken a loss to punish the defector, but the punishment worked. Over many years, without communicating with one another (that we know of), the milk sellers reached a collusive outcome and high profits.

The tacit collusion only worked because the milk sellers stayed the same year after year. Previous experience among the sellers taught them to set prices high or be punished. We also saw cases where Anheuser-Busch would punish competing brewers for failing to charge high prices. The punishment came in the form of Anheuser-Busch charging even lower prices and stealing customers. Remember the quote from Anheuser-Busch at the beginning of this chapter, "We don't want to start a bloodbath, but whatever the competition wants to do, we'll do."

These examples of tacit collusion and trigger pricing only work to promote a collusive outcome if the same firms exist in the market year after year. A new firm entering the market has not participated in the tacit collusion, has not been punished for defecting, and so is more likely to charge low prices in an attempt to steal market share. This presents a very important lesson about oligopolies: Each firm has some incentive to make sure their competitors stay in business, as opposed to being replaced by a new firm. Anheuser-Busch, Miller, and Coors are the three leading brewers. They have developed a mutual understanding over the years by observing their competitors' reactions to their own price changes. We are not saying they behave as a monopoly. Certainly, the fierce advertising wars are a testimony to the fact that these firms are competitors. But it seems plausible that their experience with one another has led to higher prices than would be realized if new firms continually entered and exited the industry. If Coors goes out of business and sells their breweries to a new firm, Anheuser-Busch has a new, unfamiliar competitor to deal with. It is entirely likely that Anheuser-Busch would prefer that Coors stay in business than compete against this new rival.

Imagine if you had to play the price-setting game repeatedly against an opponent. You are asked whether you would like to play against the same person over and over or whether you would like to play against a different person every period. Most of you would choose the former, and here is why. When the price-setting game is played repeatedly against a different person, you are essentially playing the one-shot price-setting game over and over, where the dominant strategy is for both players to defect. This leads to the rivalistic outcome. But if you play against the same person over and over, you are playing the repeating price-setting game, which according to the Folk Theorem tends to lead to the collusive outcome. Thus, the collusive outcome is more likely when there are stable trading partners in an oligopoly.

2. Pre-Play Communication Facilitates Tacit Collusion. Consider again our price-setting game, but suppose you and your competitor are allowed to communicate prior to playing. Nothing in the game has really changed, and if played only once, the dominant strategy is to both defect and set low prices. However, experiments have shown that when the players are allowed to communicate, they are more likely to collude. People simply trust each other more once they get to know each other and feel more guilty defecting against someone they know. The best way for firms to collude is to simply meet with each other, talk, and agree on what prices to set. This is price-fixing though and is illegal. But do not be deceived, businesses do price-fix. William

Shepherd (1997) in his excellent book, *The Economics of Industrial Organization*, recorded the following quotes. One executive stated that "the overwhelming majority of businessmen discuss pricing with their competitors." Another said, "price-fixing has always been done in the business." A third quote is, "It's just the way you do business. There's an unwritten law that you don't compete. It's been that way for 50 years." We say this not to encourage firms to price-fix. We specifically discourage price-fixing or anything illegal for that matter. We mention these quotes because they present a realistic picture of real-world markets.

The low-price guarantee described previously is an obvious form or pre-play communication. A firm announces in advance it will meet any competitors' price. As we previously showed, this can lead firms to collude and charge higher prices. Another regularly used tool is the assignment of a *price leader*. Firms typically do not change their prices at exactly the same time. One firm changes its price, which induces other firms to react accordingly. Often, there is one firm in the industry who makes the first price change—the price leader. If the price leader raises its price, so do all the other firms. Not surprisingly, the price leader is usually the firm with the largest market share. The common knowledge among all firms that the price leader will determine all price changes is an informal method of pre-play communication. *Price leadership* is in some ways similar to all firms acting as a monopoly. Giving one firm the duty of setting price is much like giving that one firm control over all firms. If you think about it, what is the difference between meeting illegally to set prices and the informal understanding that the price leader will set prices? Again, we look to the brewing industry for an interesting illustration of price leadership. When brewing costs rose in 1974, it was in the interest of all brewers to raise their beer prices. When the chairman of Schlitz was interviewed, he said "a price increase is needed, but it will take Anheuser-Busch to do it" (Tremblay and Tremblay 2005). Clearly, Anheuser-Busch was the designated price leader in the brewing industry. Other oligopolies are also noted for having a price leader—including the meatpacking industry (1890–1920) and the cigarette and steel industry (1930–1940) (Shepherd 1997).

3. Experience Facilitates Tacit Collusion. This may not be surprising, but the more experience one has in games like the price-setting game, the greater the likelihood of a collusive outcome. This has the simple but important implication that a firm should draw on experienced employees for help with price determination. Employees who have spent years setting prices and learning how competitors react are more likely to produce a collusive outcome.

4. Firm Homogeneity Facilitates Tacit Collusion. Firm homogeneity describes the extent to which firms are alike. If firms are alike, we say firms are homogenous, whereas if firms display many unique characteristics, we say they are heterogeneous. Firms can differ in many ways. They may have a unique brand, sell differentiated products, or have different production costs. The use of advertising to target specific consumer groups also leads to firm heterogeneity. The more homogenous (the more alike) firms are, the easier it is for them to tacitly collude. If firms produce roughly the same product, then consumers can substitute between each firm's output. By tacitly colluding, the firms limit the degree to which consumers can seek lower prices elsewhere, forcing them to pay a high price. Also, if firms have identical costs, there is little incentive for one firm to try to drive the others out of business, because to do so would require prices below the firms' costs. But if you are a firm with lower production costs than other firms, you will certainly consider the option of setting low prices to drive your competitors out of business.

PRICING TO ELIMINATE RIVALS

The discussion of strategic prices has so far assumed the number of firms is fixed. In most cases though, the number of firms in an industry can change. It is generally thought that if firms in an industry are making profits, so long as there are few barriers to entry, other firms will enter the industry. As an industry consists of more and more firms, the market price will fall and so will each firm's profits. Looking back at Figure 10.4, firm profits go down if the number of firms increases from 1 to 2, or 2 to 3, 3 to 4, and so on. A firm wishing to increase or hold onto its profits will be interested in strategies that limit the number of rivals. Regardless of how many firms are in the industry, all existing firms wish to keep other firms from entering. This section describes pricing strategies to drive competitors out and keep potential competitors from competing.

Limit Pricing

A firm only enters a market if it believes it can make profits. If a potential firm sees existing firms charging a price higher than the cost of production, it will want to enter that market and make profits itself. Suppose that you are a low-cost firm and a potential entrant is a high-cost firm. Your marginal cost of production is only $10 compared to your competitor's cost of $30. If there was no potential entrant, you would maximize your profit as a monopoly by charging a high price of $55. However, you know the potential entrant will see this high price and enter the market. No longer a monopoly, you would have to share the industry profits with this competitor. An alternative is to charge a low price of $30. This equals your competitor's marginal cost, so it has no incentive to enter the market. You maintain your monopoly, albeit at a lower price. This is *limit pricing*, the act of charging a low price with the intent of deterring the entry of a competitor. However appealing limit pricing may sound, most economic analyses find it to be a losing proposition. A more common pricing strategy to thwart rivals is predatory pricing, which is discussed next.

Limit Pricing: Charging a low price to deter potential competitors from entering the market.

Predatory Pricing

With limit pricing the firm charges a low price to discourage firms from entering the market. Predatory pricing occurs when the incumbent firm waits for a firm to enter and then charges low prices to drive the new firm out of business. The concept of

Predatory Pricing: Lowering prices when competitors enter the market in an attempt to drive them out the market.

predatory pricing was popularized by the Standard Oil Company, owned by the famous (or infamous) J. D. Rockefeller. Legend has it that Standard Oil sold oil below cost in certain areas where it had competitors. This forced Standard Oil's competitors to sell below cost in turn. Because the competitors did not have Rockefeller's deep pockets, they would eventually go out of business, leaving J. D. Rockefeller with a monopoly in the area. Once the monopoly was formed, oil prices would rise to high levels and Rockefeller would become even richer. At this point, new firms would not enter these oil markets because they knew Standard Oil would slash prices to drive them out again. As Ida Tarbell (America's first great female journalist) stated, "He [Rockefeller] applied underselling for destroying his rivals' markets with the same deliberation and persistency that characterized all his efforts, and in the long run he always won" (McGee 1958). Standard Oil's alleged predatory pricing received great interest from policymakers. Although this was not the only factor, it was a motivation for passing the Clayton Act of 1914, which made predatory pricing illegal.

The problem is that there is very little evidence that Standard Oil actually participated in predatory pricing. Standard Oil fought its way from having a 10% market share in 1870 to a near monopoly in 1911, but the monopoly was not obtained solely through predatory pricing. It was obtained through mergers and acquisitions. As it turns out, predatory pricing is in many cases unprofitable, even in the long run. Selling products below costs can lead to huge losses. Although those losses might be offset by future profits once the monopoly is formed, in many cases it is far cheaper to build a monopoly by mergers and acquisitions. Of course, antitrust laws prevent a business from obtaining a monopoly by mergers and acquisitions, but not during the time of Rockefeller (McGee 1958).

Similar allegations of predatory pricing were levied at the Gunpowder Trust in 1907. Similar investigations by economists revealed that if predatory pricing was used, it was not as prevalent as believed by courts and the press at the time. The economist investigating the trust reached the conclusion that, "what should be apparent, even to those who still believe predatory pricing might be rational monopolizing behavior under certain conditions, is that mergers and cartels provide a much greater threat to free markets than predatory pricing" (Elzinga 1970).

Obtaining a monopoly through the acquisition of competitors or predatory pricing is illegal today, but that does not mean predatory pricing is not or cannot be used. Price-fixing is illegal, but it still occurs. Similarly, predatory pricing may still be used by firms as long as they can hide it. Consider Wal-Mart, who sells many items at a price far below their competitors. A lawsuit was filed against Wal-Mart in 1993 claiming Wal-Mart pursued predatory pricing on select goods by selling them below cost. Wal-Mart defended this practice by claiming it sold some goods below cost to entice consumers into the store where they would purchase other goods above cost. The court bought Wal-Mart's argument. Whether they really engaged in predatory pricing is uncertain, but the point is that Wal-Mart could have employed predatory pricing under the guise of other marketing schemes (Waldman 2004).

However intuitive predatory pricing is, economic theory and experiments suggest that predatory pricing only works under unique circumstances. What follows is one of those circumstances. A firm could be either a *weak monopolist* or a *strong monopolist*.

A strong monopolist is one who, in the presence of a competitor, can charge a price low enough to drive the competitor out while still making profits itself. A strong monopolist can predatory price and still make money; however, a weak monopolist loses money when it employs predatory pricing. For this reason, a weak monopolist will suffer the losses from predatory pricing only if it believes it can drive competitors out, leaving it with a monopoly in later periods when it will raise prices. The future monopoly profits will then presumably offset the temporary losses incurred from predatory pricing. If a competitor knows you are a strong monopolist, it will exit the industry as soon as you start predatory pricing because it knows it cannot beat you. If the competitor knows you are a weak monopolist, it will not leave the industry. It knows you are taking a loss from predatory pricing and you will not continue doing so forever. It waits on you, you the weak monopolist, to give up, at which point you must share the industry profits. Thus, when there is symmetric information (both firms possess the same information) and the competitor knows you are a weak monopolist, predatory pricing will only lose you money. When there is symmetric information and the competitor knows you are a strong monopolist, then you could predatory price, drive the competitor out, and enjoy the profits from your monopoly. Other potential competitors will see that you are a strong monopolist and will be less likely to enter your market. But remember, if you choose to predatory price, you are engaging in illegal activity. We are not encouraging such illegal activity, but firms do engage in predatory pricing, usually under the guise of more benign intentions. For example, firms who sell multiple products could engage in predatory pricing of a single product, arguing that such pricing is used to get consumers in the door to make money off other products.

See Figure 10.7. describing strong and weak monopolists. Consider the strong monopolist first. If the potential competitor does not enter the market, the monopolist's profits are 10. However, if it enters, the monopolist must choose whether to fight entry by predatory pricing or accommodating entry. When we say "accommodating entry," we simply mean that the firm behaves like the firms in the Cournot Model. In Figure 10.7, if you the strong monopolist fight entry, you make profits of 8, which is higher than profits of 5 if you accommodate entry. The competitor loses money if you fight entry. Given your dominant strategy is to fight entry, you do so, driving your competitor out of business. Once your competitor leaves, you once again enjoy your monopoly and its larger profits.

Now suppose you are a weak monopolist. As before, you are best off if your potential competitor does not enter the market, but if it does, you are best off not fighting entry through predatory pricing. If you do predatory price, your profits are −2, compared to profits of 5 if you accommodate entry. The main difference between a strong and weak monopolist has to do with costs. A strong monopolist can produce at lower cost than its competitor. Thus, it does not have to charge a price less than its cost to drive out the competitor. The strong monopolist can force its competitor to lose money while still making money itself. On the other hand, a weak monopolist has about the same costs of production as its competitor. The only way it can force the competitor to lose money is to charge a price below its own cost and lose money itself. This discussion presents one obvious fact: A strong monopolist is best off employing a predatory pricing strategy to ward off entrants.

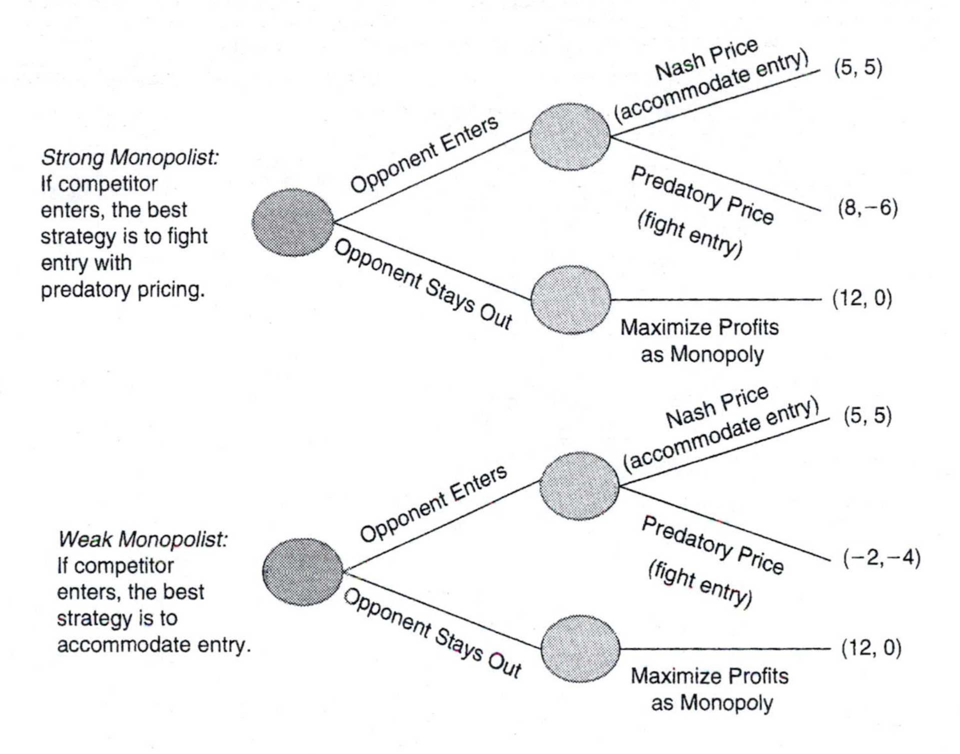

FIGURE 10.7 Predatory Pricing Game (Incumbent Profits, Competitor Profits).

Yet, there are conditions when a weak monopolist should also predatory price. The condition requires *asymmetric information*, where the two firms possess different information. Suppose you are a weak monopolist and you face a potential competitor. Your competitor does not know whether you are a strong or weak monopolist (this is the information asymmetry). If you are a strong monopolist, the competitor leaves the market as soon as you begin predatory pricing. But if it knows you are a weak monopolist, it will not leave the market. With asymmetric information, your competitor does not know whether you are a weak or strong monopolist. Thus, you can predatory price to pretend that you are a strong monopolist. Your competitor sees you predatory price and concludes you must be a strong monopolist. Thinking you are a strong monopolist it knows you will predatory price every period, forcing you to lose money every period. Thus, your competitors' best response is to leave the market. Your predatory pricing scheme worked. Although you lost money during the period you predatory priced, you fooled your competitor into thinking you are a strong monopolist. Now, you do not have to worry about other competitors entering your market anymore.

Asymmetric information exists when some people or firms possess information others do not.

To conclude, there are two instances when predatory pricing is a profitable strategy. One is when there are *asymmetric costs*—one firm can produce at lower costs than others. The low-cost producer is a strong monopolist and uses predatory pricing to drive away competitors. Or, if there is *asymmetric information*, a weak monopolist can pretend to be a strong monopolist, one who produces at a lower cost. Competitors see the weak monopolist predatory price, assume it must be a strong monopolist, and leave the market.

To Predatory Price or to Merge?

Most economists who study predatory pricing come away with the conclusion that mergers or acquisitions will reap higher profits than predatory pricing. Better to buy off your rivals, offering the rival a little bit more than it is worth, than for both of you to engage in expensive price wars. However, two items should be considered before one walks away with this conclusion. First, mergers and acquisitions are public events and receive the attention of antitrust authorities. In fact, mergers require approval of the government, and antitrust authorities will not allow one firm to have a monopoly in a market (except for certain circumstances like the granting of a patent). However, if a firm can weed out all its rivals through predatory pricing, as long as the government believes the firm is a strong monopolist, the firm can attain a legal monopoly. The U.S. government has no problem with a firm attaining monopoly status if it does so by becoming the least-cost producer. Second, predatory pricing and mergers can be complements. If a firm is considering purchasing a rival, it can predatory price first to weaken the opponent. Once the opponent is weakened, it will be willing to sell out at a lower price.

Predatory Pricing in the Sugar Industry

At this point it is useful to discuss a real case of a firm using both predatory pricing and mergers. Before 1887, there were numerous independent sugar refineries. Price-fixing was legal at this time. The firms tried to collude but there were too many firms for the collusion to hold. To solve this problem the firms agreed to merge and act as one firm. Called the Sugar Trust, it was formed in 1887 and controlled 80% of the market capacity (meaning if all firms produced at capacity, the Sugar Trust would have a 80% market share). Not surprisingly, sugar prices rose by 16%. The Sugar Trust later changed its name to the American Sugar Refining Company (ASRC). Claus Spreckels opened a new sugar refinery in 1889 to challenge the ASRC and a fierce price war ensued. The ASRC predatory priced to force spreckels out of the sugar market and it worked. ASRC purchased the weakened rival and several others, now giving it a 95% share of market capacity.

Over the next few years a few small rivals entered the sugar refining market. The rivals were not large enough to cause ASRC much concern. The small firms knew ASRC, being such a larger firm, could easily punish them through predatory pricing, and so ASRC and the small rivals developed an "understanding," meaning they formed a collusive agreement. Then came the big price war. Arbuckle Brothers and Doscher entered the sugar market by constructing their own large refineries. Having none of this, ASRC began serious predatory pricing, leading to a fierce price war.

Following ASRC's lead, all firms began selling below cost, incurring large losses. Someone had to eventually give, and it is rarely the dominant firm. Doscher agreed to merge with two independent refineries, two firms who had an "understanding" with ASRC. This basically means Doscher agreed to follow ASRC's lead and charge high prices. Doscher basically surrendered to ASRC. Arbuckle remained an independent firm, but agreed to cease the price war with ASRC.

The predatory pricing did not force the two firms out, but it did weaken them to the point that they agreed not to try and undercut ASRC's prices. The predatory pricing had another effect too though. It discouraged potential rivals from entering the market. Seeing how ASRC punishes new rivals, potential competitors decided the market was not profitable for new firms. In a study of the sugar market, two economists concluded that although predatory pricing led to negative profits for ASRC in the short run, in the long run it led to higher profits by discouraging future rivals. In fact, for every dollar it lost through predatory pricing, it made $1.058 in future profits from maintaining a larger market share. By all accounts, the predatory pricing was a profitable strategy (Genesove and Mullin 1997, 1998).

SUMMARY

In some industries there exist only a few firms who compete against one another for profits. We call these oligopolies. These settings require something other than the supply and demand model to understand prices. They require game theory. Although an Iowa corn farmer feels no threat from the actions of a Kansas corn farmer, the Miller Brewing Company is directly threatened by the actions of Anheuser-Busch. Oligopoly firms engage in a strategic game where profits depend on each firm's actions. A branch of economics called game theory is used to analyze these firms' strategic behavior.

Oligopoly firms may engage in a price war, leading to low prices and low profits. This is a rivalistic outcome. Other times the firms tolerate each other's presence, allowing each other to make profits, referred to as a Nash outcome. In select cases firms learn how to collude with one another, either by legal or illegal means, and all charge high prices and receive high profits—the collusive outcome. A number of tools can be used to achieve this collusive outcome. Trigger pricing, price leadership, tacit collusion, and low-price guarantees are examples. These tools will not always work, but the greater the homogeneity among firms, experience with price setting, use of the aforementioned tools, and stable competitors, the greater the likelihood firms can collectively charge high prices and enjoy large profits.

CROSSWORD PUZZLE

For answers with more than one word, leave a blank space between each word.

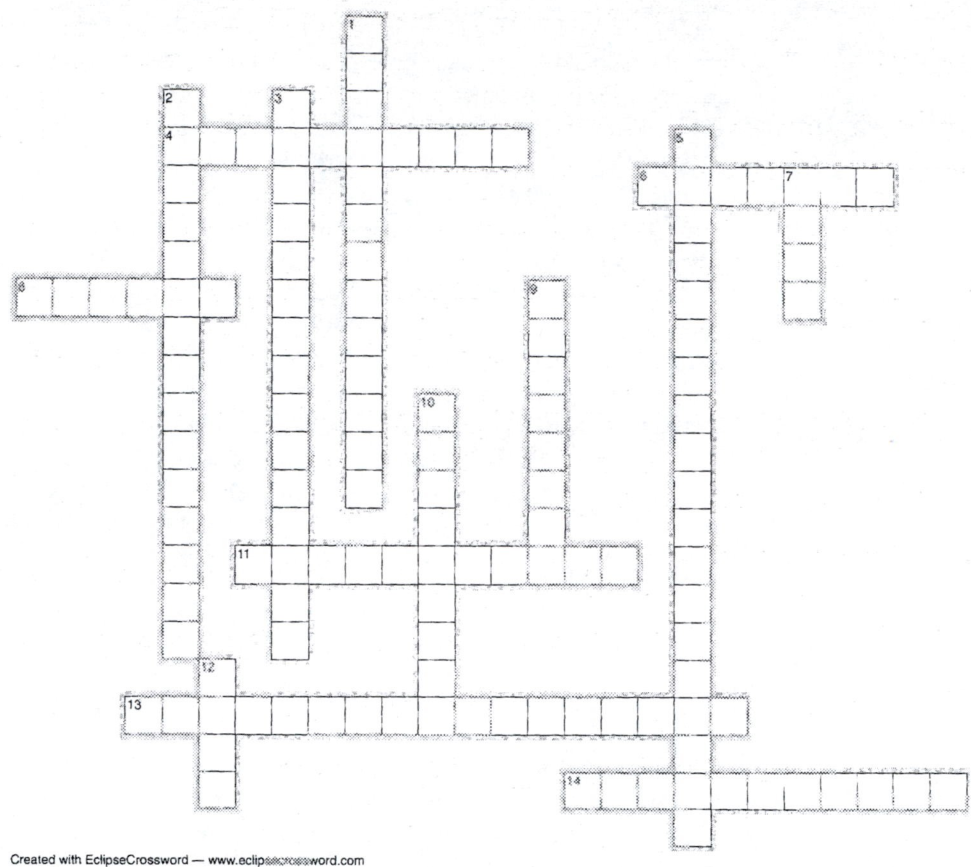

Across

4. Firms who engage in fierce price wars will experience a _______ outcome.
6. The _______ model captures a monopoly market, perfectly competitive market, and all markets in between.
8. A price _______ is one whose competitors allow to set the initial price, after which all competitors charge the same price.
11. In the repeated price-setting game, a strategy where one cooperates in the first period and copies the other player's previous action in each subsequent period.
13. Charge a high price, but if competitors enter the market, slash your prices to drive them out of business.
14. The study of how interdependent people behave when aware that their interactions affect one another.

Down

1. Charging a low price to discourage the entry of potential competitors.
2. A pricing strategy where a firm charges a high price, but quickly slashes its prices if its rivals charge a lower price.
3. An unspoken but understood collusion agreement.
5. A strategy to facilitate collusion where you agree to meet or beat your competitor's price.
7. A _______ equilibrium is where all players are employing their best strategy given the strategies of all other players.
9. A _______ strategy exists when one strategy yields the highest payoff regardless of the competitor's actions.
10. A market with only a few sellers of an identical good.
12. A _______ monopolist is one who incurs losses when they predatory price.

STUDY QUESTIONS

1. In the one-shot price-setting game, both firms are better off setting a low price, regardless of the other firm's price. Thus, "low price" is a _______ strategy for both firms.
2. The equilibrium in games where all players are doing the best they can given the choices of their opponents is referred to as a _______ equilibrium.
3. There are several forms of "pre-play" communication that help facilitate collusion in the price-setting game. One is called trigger-pricing while another is called a low-price guarantee. Choose ONE of these pricing strategies, and discuss exactly how a firm would use it in practice. A concise but clear explanation will do.
4. In the infinitely repeating price-setting game, the _______-_______-_______ strategy performs well.
5. Predatory pricing can be profitable when there are _______ costs or _______ information.
6. Wal-Mart begins predatory pricing and driving its competitors out of business because it can produce at a much lower cost. Even while predatory pricing, Wal-Mart still enjoys a healthy profit. Therefore, Wal-Mart is a _______ monopolist. Sal-Mart enters a different town that does not have a Wal-Mart and begins predatory pricing, setting prices as low as Wal-Mart does in other towns. Sal-Mart is not a strong monopolist, because it loses money when it predatory prices. Firms in this town think Sal-Mart must produce at a lower cost though, since it charges such low prices, and they go out of business. This leaves Sal-Mart alone with a monopoly, and it raises prices to recover its losses. Sal-Mart is a _______ monopolist.
7. Philip Morris and RJR must decide on whether to conduct television advertisements. The payoff from their decision, given the other's choice, is shown in Figure 10.8. This advertising game is played once. In a Nash Equilibrium, Philip Morris will choose to _______ and RJR will choose to _______.

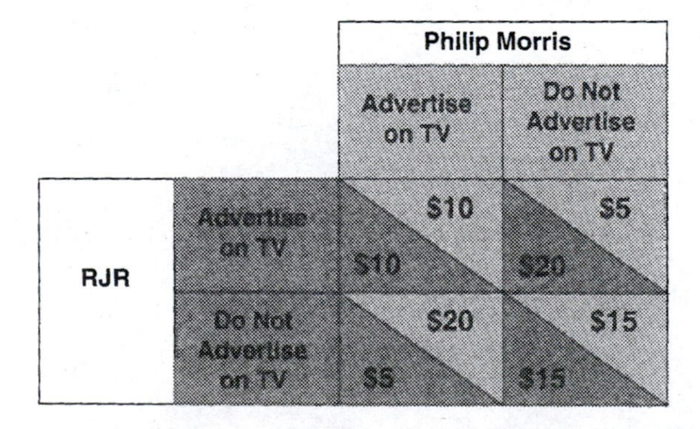

FIGURE 10.8 Advertising Game (Profits in Billions of Dollars).

8. You see your competitor lower prices, so you respond by lowering prices even more, inflicting damage on your competitor. After experiencing this several times, your competitor learns to keep its price high with yours. This type of pricing strategy is referred to as what?
9. In class we said there are four factors that facilitate collusion. Fill in the missing factor.
   1. Stable trading partners
   2. Pre-play communication
   3. Experience
   4. ______
10. Describe a weak monopolist in one to three sentences.

CHAPTER ELEVEN

Pricing Schemes

A customer leaving your store happy is a customer with your money in her pocket.

—*Rober McCormick,* professor of economics, Clemson University in his excellent textbook *Managerial Economics*

[A] fitting testament to the instincts of a Congress that, from the standpoint of the public interest, can't go home soon enough . . .

—A *Washington Post* editorial on the U.S. government's policy allowing milk producers to raise prices through a particular pricing scheme

INTRODUCTION

Consumers encounter many different pricing styles. Go to a matinee and all movie tickets sell for the same price, despite the fact that some movies are clearly more popular than others. But different people pay different prices. Both the elderly and college students typically receive a discount. Within the movie theatre you can purchase popcorn and oversized candy bars at ridiculously high prices. However, you may use the bathroom in the theatre without paying a fee. The theatre gladly pays for the bathroom plumbing and cleaning but makes you pay for your popcorn. Managing a successful business often requires creative pricing schemes, which is what this chapter is all about. This chapter shows why businesses give senior citizen discounts, and it is not always because they care for the elderly. It shows why businesses often sell "bundles" of a good like season rather than individual sporting event tickets. It discusses why a bar may issue a cover charge to enter, selling you cheap beer afterwards, and why candy bars at movie theatres are giant compared to the kind sold in convenience stores.

The act of charging different prices to different people is called *price discrimination*. Given that companies employ so many different pricing schemes, at least one chapter on such pricing is in order. A firm can substantially increase its profits by developing a clever pricing method; some of the pricing schemes discussed in this chapter are

1. all-or-nothing pricing
2. two-part pricing
3. second-degree price discrimination
4. third-degree price discrimination
5. bundling
6. required tie-in sales

Several assumptions will be made throughout this chapter. First, we will assume that the marginal cost of production is constant and equals MC. This assumption is not made because it is realistic (often, however, it is) but because it allows us to construct simple mathematical models. By keeping the marginal cost structure of firms simple, we can focus on consumers and how to devise pricing systems to make the most money. Second, we assume a single firm that possesses significant market power. Whether it be a monopoly or monopolistic competition, this firm can raise prices without losing all its customers to rivals. This simply implies that the firm faces a downward sloping demand curve. Third, instead of constructing market demand curves, we will use demand curves for individuals. Specifically, individuals are assumed to possess market demand curves given by the formula:

$P = a - bq$ The term q stands for the quantity purchased for each identical consumer.

STRIVING FOR AN UNHAPPY CUSTOMER

This chapter concerns how to set prices to maximize business profits, so the title of this section may seem surprising. Surely a business wants customers to leave happy so they will return for future purchases. That is not what we are trying to say. Suppose you sell a bull to a cattle producer for $1,000. The buyer was willing to pay $1,500 for the bull, so she is happy with the purchase. Specifically, she benefits by $500. She was willing to pay up to $1,500 for the bull. If she had paid $1,500 for the bull, she basically traded $1,500 in the form of money for $1,500 in the form of a bull. She would be no better off. But if she trades $1,000 in the form of money for $1,500 in the form of a bull, she is better off by $500. That is a happy customer.

Now suppose that you are able to charge her just a little under $1,500, say, $1,450. You receive just about the highest price you could possibly get. The producer leaves with only $50 of value, so she is not very happy. Yet, the fact that she made the purchase indicates this deal was better than she could find elsewhere. That is what we mean by an unhappy customer. The more money you get from your customers, the more money you make and the less happy is the customer. The idea of a profitable pricing scheme is to extract as much happiness from the consumer while maintaining their patronage.

This begs the question: What is the maximum amount of money you can extract from a consumer? The answer is illustrated in Figure 11.1. The figure shows a per-consumer demand curve (the demand curve for a single consumer, showing the

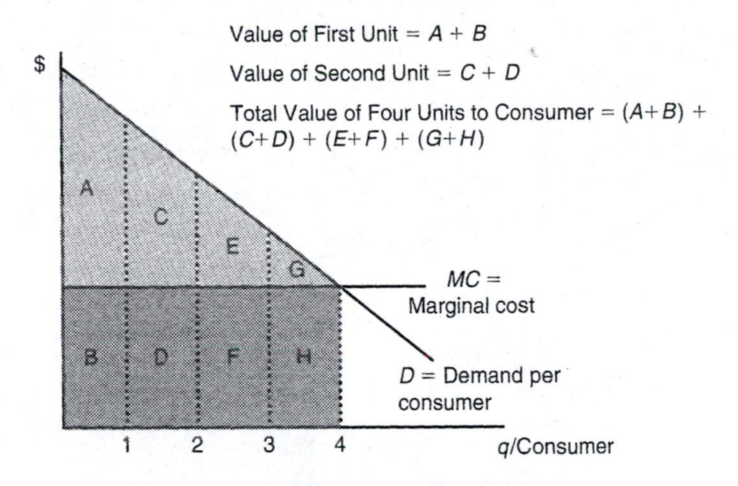

FIGURE 11.1 Per-Consumer Demand Curve.

number of units one consumer will buy at any given price), which can be interpreted two ways, both of which are correct. At any price, the demand curve indicates exactly how many units the consumer will purchase. At any quantity, the height of the demand curve tells us the maximum amount the consumer will pay for that unit. The demand curves we use in this chapter assume that a good can be divided into infinitesimally small units. For example, when moving from 0 to 1 unit on the x-axis in Figure 11.1, this is really like consuming 0.0001 of a unit, plus 0.0001 of a unit, and so on until the sum of those extremely small units equals one. This assumption may seem weird but is made for a good reason. First, it doesn't significantly change the model results. Second, it makes the mathematics much simpler. And if we know students, we know they prefer simpler math!

In Figure 11.1, what is the consumer's maximum willingness to pay for the first unit? It equals the area underneath the demand curve from 0 to 1. This area is the consumer's maximum willingness-to-pay for each infinitesimally small unit on the 0 to 1 interval. This area is $A + B$. The consumer's maximum willingness-to-pay for the second unit is $C + D$. Recalling that value is maximum willingness-to-pay, the value of the third and fourth units equals $E + F$ and $G + H$, respectively. Therefore, the maximum amount the consumer is willing to pay for four units equals $A + B + C + D + E + F + G + H$. The cost of producing each unit equals the marginal cost MC. The cost of producing the first unit equals MC, the cost of the second unit is MC, and so on. Thus, the cost of producing four units equals $4 \times MC$, or, the area underneath the marginal cost curve from 0 to 4. The cost of four units equals $B + D + F + H$. If one takes the consumer's maximum willingness-to-pay for four units and subtracts the cost of four units, the maximum profit one can extract from the consumer equals $A + C + E + G$. You may recognize this as consumer surplus from earlier chapters. Notice that you would never want to sell the fifth unit, because the maximum amount the consumer is willing to pay (given by the height of the

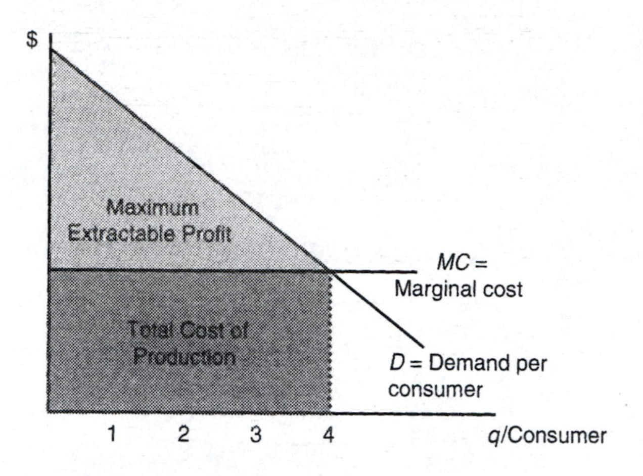

FIGURE 11.2 Per-Consumer Demand Curve.

demand curve) is less than the marginal cost. You would lose money from producing and selling the fifth unit.

This introduces an important concept. Given the per consumer demand curve and marginal cost in Figure 11.1, the most profit a firm can earn is the area underneath the demand curve and above marginal cost. The question then becomes how to set prices so that a firm extracts this maximum profit. This question is addressed in the following sections.

FIRST-DEGREE OR PERFECT PRICE DISCRIMINATION

First-degree price discrimination is simple. Take each consumer and sell her individual units one at a time. Set a price equal to her maximum willingness-to-pay for each unit (or maybe slightly less, just to make sure she purchases the good). Keep selling her more units, charging the maximum willingness-to-pay for each unit until her maximum willingness-to-pay is less than the marginal cost of production. Refer back to Figure 11.1. This means you charge her $A + B$ for the first unit. The cost of producing the first unit was B, so your profits off the first unit equals A. Charge the consumer $C + D$ for the second unit, making a profit of C off the second unit. Charge her $E + F$ for the third unit and $G + H$ for the fourth unit, making profits of E and G, respectively. Do not sell her the fifth unit because her maximum willingness-to-pay is less than the marginal cost of production. The firm would then follow the same method for each consumer. Charge each consumer their maximum willingness-to-pay for each unit until selling more units no longer results in profits. Car dealerships attempt first-degree price discrimination, doing everything they can to charge each consumer as high a price as possible, as long as the price exceeds costs. Two people are likely to pay different prices for an identical car. If you wonder why a salesperson looks at your clothes, asks where you work, and so on, they are simply trying to figure out your income, and consequently, your maximum willingness-to-pay for the car.

After all is said and done, total profits from selling the four units to the consumer equals $A + C + E + G$, which is the maximum total profits one can make. This is sometimes referred to as perfect price discrimination. The consumer gets no benefit from the purchases because she pays a price equal to her value of each unit. Yet the firm receives the maximum profits possible. As you might suspect, first-degree price discrimination is difficult to employ in practice. People do not like negotiating a different price for each unit they purchase, and they do not like haggling over price in general. Consumers prefer simpler pricing systems. For this reason most firms will want a more consumer-friendly method of achieving first-degree price discrimination. Two such methods are discussed below.

All-or-Nothing Pricing

In the example above using Figure 11.1, perfect discrimination worked by charging the consumer $A + B$ for the first unit, $C + D$ for the second unit, $E + F$ for the third unit, and $G + H$ for the fourth unit. The firm would not want to sell the fifth unit because the consumer's willingness-to-pay is less than the cost. We said that this pricing scheme had the undesirable property that the consumer is charged a different price for each unit. Consumers generally do not like that, and besides, charging different prices based on the number of units purchased can be a hassle for the firm. A simple alternative is *all-or-nothing pricing*—instead of selling individual units of the good, a firm sells only bundles consisting of four units. We said that consumers' maximum willingness-to-pay for four units equals $A + B + C + D + E + F + G + H$, so charge them that amount. Because you are charging the consumer a bundle price equal to their maximum willingness-to-pay, they will purchase the bundle of four units. After subtracting the cost of producing four units ($B + D + F + H$) the firm is left with profits of $A + C + E + G$. As we previously stated, this is the maximum profit possible! Moreover, the pricing scheme is simple. Consumers either purchase the bundle of four units at the bundle price or nothing at all. Perfect price discrimination is made easy through all-or-nothing pricing.

All-or-Nothing Pricing: The consumer either purchases a specific number of units at a price or no units.

Consider a numerical example in Figure 11.3. Suppose you offer an individual with the demand curve in Figure 11.3 a bundle consisting of four units. What is the maximum amount she will pay for this bundle? As we have discussed, this maximum is the Area A plus the Area B, or the area underneath the demand curve for all units consumed. Area A is a right triangle, and the formula for the area of a right triangle is the base times the height divided by two. In this example, Area A is $(1/2)(4)(5 - 2) = 6$. Area B is a rectangle whose area equals length times width. In this example, Area B is $4 \times 2 = 8$. Thus, the individual is willing to pay $\$8 + \$6 = \$14$ for a bundle of four units but not a penny more. After charging her a price of \$14, you pay your production costs of $\$4 \times \$2 = \$8$, leaving you with profits of \$6. This is simply the area of the triangle, Area A, which we have said is the maximum profit possible. Bundles larger and smaller than four are undesirable. If the

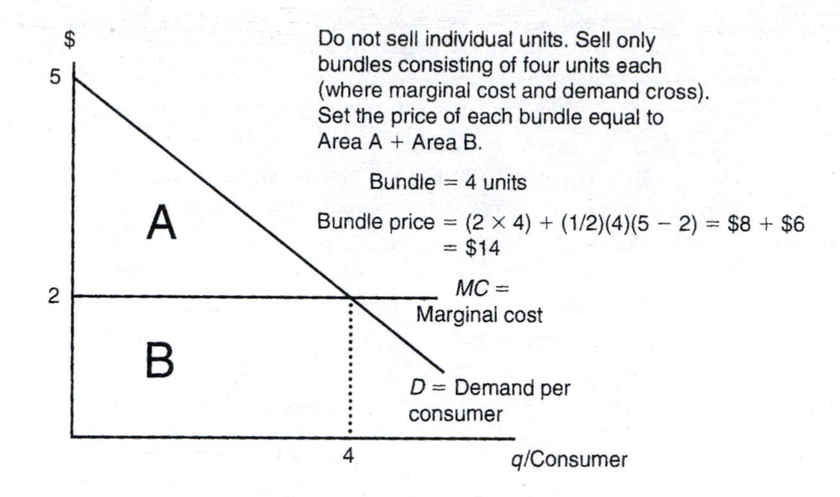

FIGURE 11.3 Example of All-or-Nothing Pricing.

firm sells more than four, it incurs a loss on some units (because consumers' willingness-to-pay is less than the cost for units greater than four). If the firm sells less than four units, it foregoes profits it could have made by selling more. In short, in cases where all consumers have similar demand curves, all-or-nothing pricing is a convenient method of achieving perfect price discrimination and the maximum profits possible.

One example of all-or-nothing pricing are the oversized candy bars at theatres. The theatre could easily sell small candy bars, allowing consumers to purchase more or less. However, profits are higher by bundling two or more regular-size candy bars into one large bar. Another example pertains to hotel rooms during football games. Often hotels require a minimum stay of three nights when a college football bowl game is close by. Again, they could easily charge a per-night rate and allow each person to stay however many nights they wish, but profits are higher through all-or-nothing pricing.

Two-Part Pricing

Two-Part Pricing: The consumer is charged a lump-sum fee plus a price for each unit purchased.

Two-part pricing is yet another way of achieving perfect price discrimination. College students are quite familiar with two-part pricing. Think of all the bars that make you pay a cover charge at the door plus a per-unit price for each beer you drink. There are many reasons for such a pricing system. One is that some consumers go to bars but do not drink beer, so the bar must charge them something to make money off the nondrinkers. Another reason for this two-part pricing system is that it allows firms to conduct perfect price discrimination. Consider Figure 11.4, using a bar as an example. Ignore the cover charge for now. Suppose the consumer enters the bar and is charged a beer price equal to the marginal cost of the beer. In this case, $2 per beer. Given the demand curve the consumer purchases four beers. This is a good deal for the

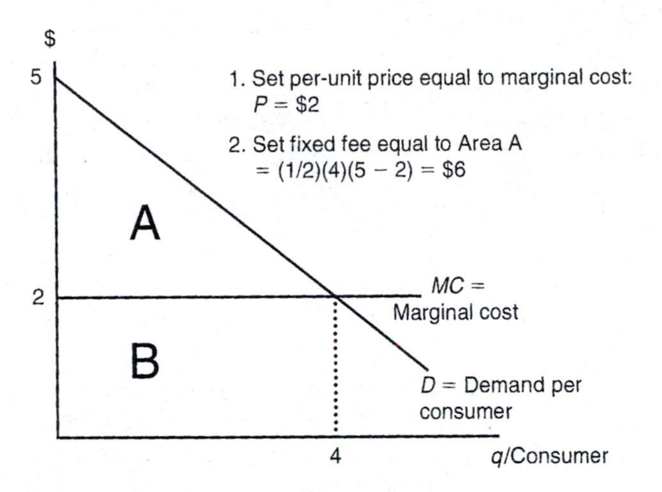

FIGURE 11.4 Example of Two-Part Pricing.

consumer. For each beer he was willing to pay a price indicated by the height of the demand curve, yet for each beer he paid a smaller price. The difference between his maximum willingness-to-pay and the price he really pays for each beer indicates the value or surplus (surplus just being a measure of happiness) he extracts from the exchange. Summing this difference over all four beers you can see the value of being able to purchase four beers at a price of $2 per beer is Area A (the area below the demand curve and above price for all quantities consumed, referred to as consumer surplus in earlier chapters).

Area A is a right triangle, the area of which is the base times the height divided by two. Using the numbers provided Area A equals (1/2)(4)(5 − 2) = $6. The consumer extracts $6 of surplus from being able to purchase beer at $2 per beer. This implies that the consumer is willing to pay up to $6 for the right to purchase $2 beer. The bar owner can then set a cover charge equal to $6 (or a little less). The consumer pays $6 at the door and purchases four beers for $2 a piece. The bar owner's profits equals the revenues minus the costs. Revenues from the consumer equal the $6 cover charge plus the $8 from beer sales (4 beers at $2 per beer). However, the marginal cost of each beer was $2, so the cost of serving the consumer equals $8 (4 beers at $2 per beer). The bar owner ends up making $6 from the patron, which is the area of triangle A in the figure. This two-part pricing system (a fixed fee plus a per-unit charge) yields the maximum profits possible and is therefore a case of perfect price discrimination.

Night clubs frequently employ two-part pricing; so do wholesale clubs like Sam's Club and BJ's Warehouse. When Disneyland first opened it charged a fixed entry fee plus a price for each ride taken. Eventually, Disneyland learned that the marginal cost of selling an additional ride is close to zero. Now Disneyland simply charges a fixed entry fee and each ride thereafter is free.

SECOND-DEGREE PRICE DISCRIMINATION

First-degree price discrimination is referred to as perfect price discrimination in the sense that it achieves the maximum profit possible. Thus far we have assumed that all consumers are identical. For this section let us relax this assumption. Assume now that some consumers have a high demand for a product and others have a lower demand. Ideally, one would identify the high- and low-demand customers and conduct different forms of perfect price discrimination for each type separately. For example, Figure 11.5 illustrates two consumer types. One has a relatively low demand for the product and the other a much greater demand. If the firm can distinguish which group a consumer belongs to, it can charge them an appropriate price. Suppose the low-demand consumers are senior citizens and the high-demand group are teenagers. One can simply sell bundles of the good. On display are bundles of 15 units at a bundle price of $C + D$. Then, advertise a senior citizen discount whereupon proof of age a senior can purchase a bundle of 5 units at a bundle price of $A + B$. By noting differences in demand across groups and establishing different pricing schemes across groups, the firm can ensure maximum profits. If the firm did not target different consumer groups and simply sold bundles of 15 units at $A + B$ each, it would lose all potential profits from senior citizens.

Unfortunately, a business cannot always distinguish between low- and high-demand customers. Different consumer types often look and behave similarly. In these cases perfect price discrimination cannot be used for different groups. The inability to distinguish between consumer types and alter pricing schemes accordingly will result in lower profits, but profits can still be made, even substantial profits. In cases where demand differs across consumers but the firm cannot distinguish between consumer types, the firm can employ second-degree price discrimination to enhance profits.

Second-degree price discrimination involves offering discounts to consumers who purchase more units of the good.

Second-degree price discrimination entails giving all consumers the same price schedule but offering quantity discounts. The assumption is that the firm cannot

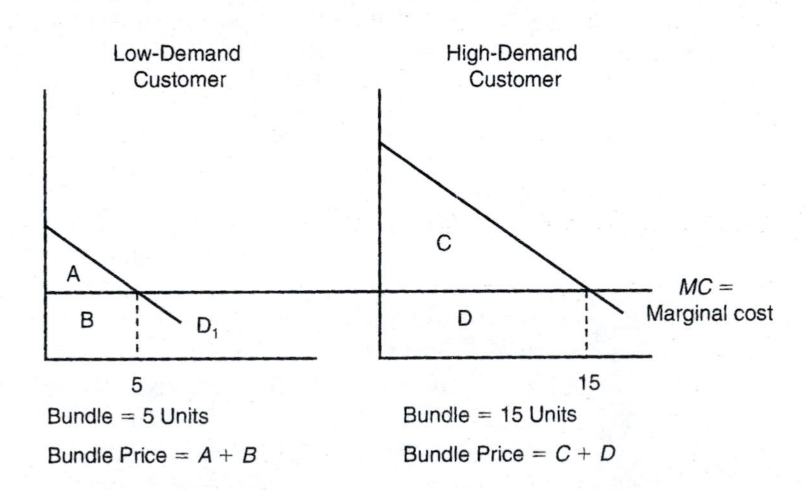

FIGURE 11.5 Perfect Price Discrimination (via All-or-Nothing Pricing) with Different Consumer Types.

distinguish between consumer types, so the firm has no choice but to offer all consumers the same pricing scheme. However, if designed cleverly, the pricing schedule will force consumers to reveal their type and allow the firm to price accordingly. This type of pricing can be quite complicated, so we will only review the basic features here, using the example in Figure 11.5. Suppose that 5 units form a bundle and the bundle price of $A + B$. Then, sell a second bundle at a discounted price. This discounted price must be less than $A + B$, but greater than the cost of producing the bundle (greater than B). The low-demand group will not purchase the second bundle because their willingness-to-pay is less than marginal cost. However, if the price is appropriately set, the high-demand group will purchase the second bundle and you will profit from the second bundle. By selling the first bundle at one price and the second bundle at a discounted price, you induce customers to reveal whether they are high- or low-demand customers. More importantly, your profits are higher than if you simply sold all bundles at the same price.

As we said before, implementing a perfect second-degree price discrimination scheme can be complicated. It depends on the exact demand curves for different consumer groups. However, there are two key rules one should use in designing an optimal second-degree price discrimination scheme. First, the pricing scheme should maximize profits from the low-demand consumers. Notice that this was achieved in the example above. Selling bundles of 5 units at a price of $A + B$ extracted the maximum possible profits from the low-demand group. Next, offer quantity discounts that entice the high-demand group to purchase more than the low-demand group. The exact quantity discount is difficult to identify and depends on the particular setting. At times there may be multiple consumer types—such as low-demand, medium-demand, and high-demand types. In these situations one should follow the same two rules. First, set the price to maximize profits from the low-demand group. Then, offer quantity discounts so that the medium- and high-demand types purchase more than the low-demand group.

You might have noticed a flaw in the preceding discussion. We first assumed that the firm cannot distinguish between different consumer types, and then we instruct you to employ a pricing scheme that maximizes profits from the low-demand type. But if you cannot tell which customers are the low-demand type, how can you set a pricing scheme to maximize profits from these customers? You cannot. However, in some instances the firm may be able to estimate demand by low-demand consumers fairly accurately and can use this estimation to aid in the formation of pricing schemes. If a reasonable estimate is not available, all the firm can really do is to develop a pricing scheme that involves quantity discounts. Try one type of discount and then change it to see if profits rise or fall. Through experimentation with different discounts one can identify a profit-maximizing pricing scheme.

THIRD-DEGREE PRICE DISCRIMINATION

There are many buyers and sellers of milk, but the milk market is not perfectly competitive. The Agricultural Marketing Act of 1937 allows farmers to organize and act something like a monopoly. Within a region in the United States, dairy producers can vote on whether to create a milk marketing order. If more than two-thirds of dairy producers approve, the order is created. At this point, all milk produced within the region must be sold through the marketing order, essentially giving the marketing order something like a monopoly on milk sales. It is not exactly a monopoly because the marketing order cannot control how much milk is produced. But it does get to set milk prices, and that gives it market power to raise profits. See Figure 11.6 illustrating the different milk marketing orders created in the United States.

Most marketing orders use their pricing power to develop a classified pricing scheme. This scheme involves selling fluid milk (the milk you drink) to milk consumers at one price, and selling milk to food processors who make things like cheese and ice cream at another price. It turns out that fluid milk consumers have a more inelastic milk demand than food processors. There are relatively few substitutes for fluid milk but substantial demand. This may be changing with the introduction of milk made from soybeans and rice, but thus far consumers' willingness to substitute soymilk for real milk is unknown. Food processors have substitutes if milk prices increase. They can incorporate artificial flavors that reduce the need for raw milk and can utilize lower-quality milk if high-quality milk becomes too expensive without sacrificing too much taste. Thus, when milk prices rise, food processors decrease their purchases by a larger percentage than milk consumers. Consumer purchases of

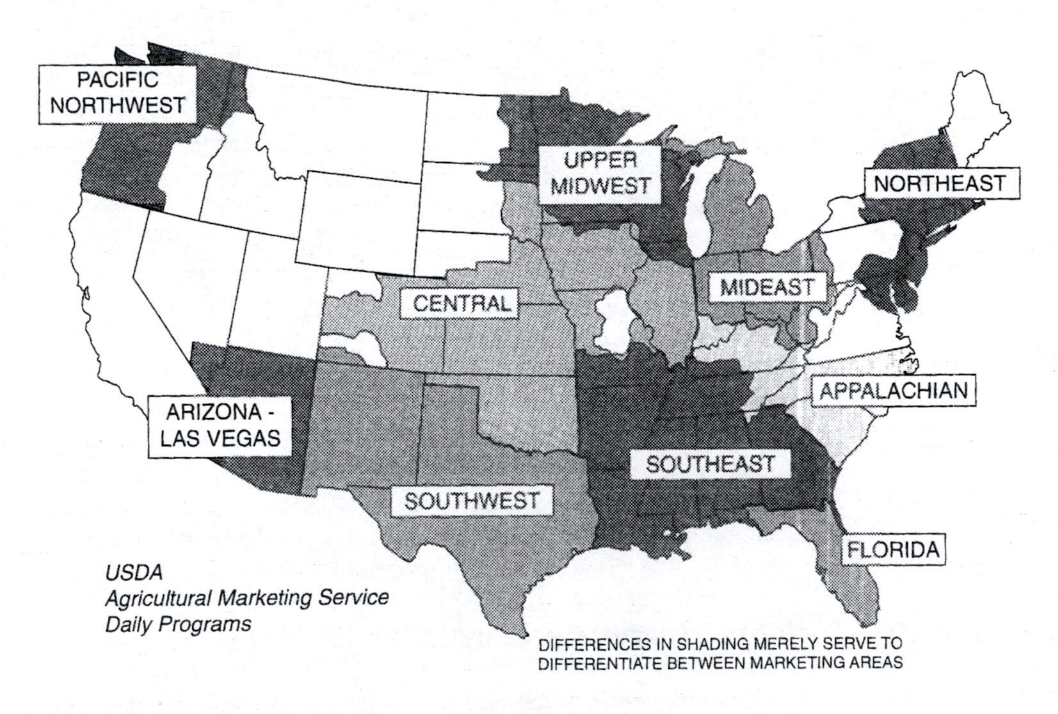

FIGURE 11.6 Federal Milk Marketing Order Areas.
Source: United States Department of Agriculture.

fluid milk are less sensitive to price changes; the demand for fluid milk is more inelastic. Consequently, fluid milk consumers are charged a higher price.

Third-degree price discrimination involves charging consumers with different demand elasticities different prices.

This is an example of *third-degree price discrimination*. A firm is able to distinguish between different consumer types. Each "consumer type" has a different shaped demand curve; one is more elastic than the other. The firm then proceeds to set different prices for each consumer type. The more inelastic the demand of each type, the higher the price it pays. Examples of third-degree price discrimination abound—including senior citizen discounts, student discounts, and airline tickets. Airline tickets tend to be cheaper if they include a Saturday overnight stay. This is because tourists' demand is more elastic than the demand of business travelers. Tourists tend to want to stay overnight on weekends but businesspeople on business trips do not. Tourists are typically looking for a "good deal" whereas businesspeople will typically pay more for a plane ticket. By charging a lower price for tickets with a Saturday overnight stay, the airlines charge tourists one price and businesspeople another, and their profits are higher for doing so.

To understand the nature of third-degree price discrimination we must revisit the optimal price when firms face a downward sloping demand curve. Suppose that marginal cost is constant, as we have throughout this chapter. Additionally, suppose that all consumers can be broken into one of two groups. The first group has a relatively elastic demand and the second group's demand is inelastic. Using our milk example, the first group would be food processors and the second group would be fluid milk consumers. Because marginal cost is constant, the number of units sold to the first group in no way impacts the cost of selling to the second group, so we can treat the two consumer groups as two completely separate markets. Recall the optimal price-setting policy when firms face a downward sloping demand curve. The marginal revenue curve lies underneath the demand curve (and if the demand curve is linear, it falls twice as fast). The firm sells a number of units that sets the marginal revenue for each group equal to marginal cost and charges a price on the demand curve corresponding to this quantity.

Group 1 has an elastic demand and Group 2 has an inelastic demand, as illustrated in Figure 11.7. Notice that when we follow the optimal-pricing rule described above, it must be that the group with a more inelastic demand is charged a higher price. This is why grocery store shoppers pay a higher milk price than food processors, and why businesspeople pay higher plane ticket prices than tourists. Senior citizens are routinely given discounts, but this is not likely due to the benevolence of the firm. Senior citizens have less income because most are retired. Plus, they have more time to shop around for better deals. For this reason, senior citizens are less likely to pay high prices than a working mom who makes lots of money but has hardly any time to shop. As you are well aware, most college students can be properly described as poor. They have little time to work, tuition is always rising, and of course top-notch textbooks such as this one cost money (we apologize). If movie tickets rise, students decrease their moviegoing much more than professors. Thus, theatres routinely charge students a lower price via student discounts. Again, this is not likely a firm who "feels the students' pain." Students have a more elastic demand for movie tickets than college professors, and so they are charged a lower price.

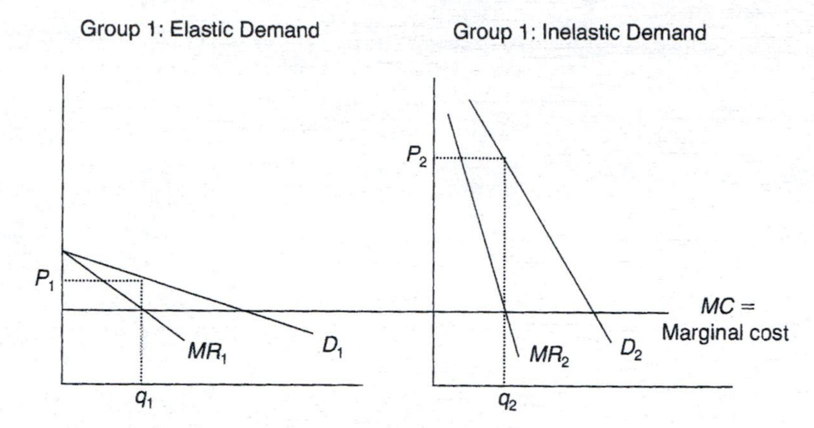

The firm charges a higher price to the group with the more inelastic demand.

FIGURE 11.7 Third-Degree Price Discrimination.

It is important to note that a firm can make more profits using a properly administered third-degree price discrimination scheme than it could make charging a uniform price to all consumers. Looking back at Figure 11.7, suppose we wanted to charge a single price to both groups. This would require either increasing the price charged to Group 1, decreasing the price charged to Group 2, or both. Regardless of which price changes, at least one price will be different from the profit-maximizing prices given by P_1 and P_2. Consequently, if any one price deviates from its profit-maximizing level—by definition—profits fall.

There are other more indirect ways of employing third-degree price discrimination. Retailers routinely offer coupons in newspapers. Some consumers are always looking for a good deal and even enjoy the shopping process. Call these people shoppers. Others hate shopping and do not mind paying a higher price if it requires less shopping. Call these people the nonshoppers. Shoppers have a more elastic demand than nonshoppers. If the price rises, the shoppers will quickly shop around for a better deal whereas the nonshopper will "suck it up" and make the purchase anyway. A firm knows nonshoppers will pay a higher price than shoppers, and it uses coupons to distinguish between the shoppers and nonshoppers. Nonshoppers are less likely to peruse the newspaper looking for coupons. Shoppers peruse all the coupons of every Sunday newspaper. The shoppers bring the coupons to the retail outlet, but the nonshoppers do not, providing a straightforward method for charging a higher price to nonshoppers (who have a more inelastic demand) than shoppers (who have a more elastic demand).

The use of coupons forces consumers to overcome a hurdle to obtaining a lower price. To get a lower price, the consumer must peruse the paper for coupons, cut it out, and bring it to the store. Think of the many other hurdles firms make consumers jump over for a lower price. A prime example can be found the day after Thanksgiving—the biggest shopping day of the year. Despite the fact that demand for retail products is highest on this day, retailers frequently offer huge discounts on select items. However, supplies are limited. To take advantage of this discount, you must arrive by 5 A.M. so that you are one of the first in line when the store opens at 6 A.M. This is a hurdle not many, including the authors, are willing to jump, and it efficiently separates the shoppers from the nonshoppers. This is referred to as *hurdle model of price discrimination*. Consumers are forced to pay a personal cost (i.e., overcome a hurdle), usually in terms of their time, in order to obtain a price discount.

Notice that the main difference between the shoppers and nonshoppers is the opportunity cost of time spent shopping. Shoppers have a relatively low opportunity cost, either because their wages from working are low, they have more time, or they personally enjoy shopping. Nonshoppers have a high opportunity cost of shopping. The hurdle separates those with a high and low opportunity cost of time spent shopping. The key to implementing a successful hurdle is to not make the hurdle too large or too small. Suppose you run an retail outlet and wish to issue coupons for 20% off of all sales in order to price discriminate. If you place the coupons right by the register, the hurdle is too small. Everyone at the checkout line will take the coupon and receive the discount. This is not price discrimination because everyone pays the same low price. If you place the coupons in a very expensive magazine, few people have a chance to use the coupons. In this instance, the hurdle is so high no one has access to the coupons. The best strategy is to place the hurdle just high enough that those with a high opportunity cost of shopping time will not utilize the discount, but low enough that those with a low opportunity cost will.

When Third-Degree Price Discrimination Works

Third-degree price discrimination can only be used in certain settings. To see when it cannot be used, consider the following extreme and silly examples. Suppose the National Football League (NFL) institutes a new pricing policy where individuals making more than $100,000 per year pay a high ticket price but all others pay a low ticket price. This policy is doomed to failure, because those with a low income will buy the tickets initially and then sell it to the wealthier individuals at a higher price, but lower than what the NFL offered the wealthy. *The firm must be able to prevent reselling between customers.* There is another reason why this policy will not work. How can the NFL determine which individual's income exceeds $100,000 and which does not? The only proof would require tax returns, and that is private information. So if the NFL really tried to implement this policy, it must rely on people's word, and many wealthy people would simply say they make less than $100,000 to obtain a lower price. *The firm must be able to identify and separate different consumer groups.* Although wealthy people would pay more for a Superbowl ticket, the NFL cannot exploit this fact because people can resell football tickets and because the NFL cannot distinguish between people with different incomes.

Conditions for Third-Degree Price Discrimination

(1) Firm can identify and separate all consumers into distinct groups.
(2) Reselling between consumer groups is prevented.
(3) The different groups have different elasticities of demand.
(4) Must be able to avoid prosecution by antitrust authorities.

Consider another example where a barber charges different prices to blondes and brunettes. The barber can tell who has what hair color, and one cannot resell a haircut. However, what does the barber possibly have to gain from charging people with different hair color different prices? Does the barber really think blondes will pay more or less than brunettes? To successfully third price discriminate, the different consumer groups *must have different elasticities of demand*. Remember, the firm should change a lower price to the group with a more elastic demand. Given that there is no reason to think blonde people have a more or less elastic demand than brunettes, price discrimination will not enhance profits. In addition, this form of price discrimination is illegal. The Clayton Act of 1914 only allows the charging of different prices to different people if the price reflects differences in cost or quality. Grocery stores can charge more for organic food because it costs more to produce. Grocery stores can also charge more per pound for T-bone steaks than flank steaks because they are of a higher quality. To our knowledge, officially, most senior citizen discounts are illegal. The movie is the same for everyone so one cannot use quality as a reason for charging different prices, and it is hard to argue that it costs less to provide senior citizens with a movie experience.

This brings up another very important point about price discrimination. *Price discrimination is illegal, so the firm must be able to avoid prosecution by antitrust authorities.* Even though senior citizen discounts may be illegal, no government official or politician would dare take a firm to court over it. It seems benign. So do student discounts, and that is why firms get away with it. Car dealerships routinely price discriminate, charging people with less negotiating power a higher price. Yet, the sticker price is the same for all, so this essentially disguises price differences. What about grocery coupons? If you took grocery stores to court, they might argue that coupons entice housewives into the store, and housewives tend to shop during working hours. This helps spread out customer patronage throughout the day. Housewives shop between 8 A.M. and 5 P.M., and working men and women shop after 5 P.M. This smoothes out the number of customers in the store at any one time, providing a more pleasurable shopping experience to all. It also reduces the costs of serving the customers. Because the use of coupons lowers costs, it does not violate the Clayton Act. This line of reasoning certainly seems plausible, plausible enough that it would be difficult to prove price discrimination in court. Firms regularly price discriminate despite its illegality because it is so hard to prove.

Let us return to the example of third-degree price discrimination in the milk market. Why is it so successful for milk producers? Running down the list, first, the milk marketing orders can easily separate its buyers into two groups: fluid milk consumers at the grocery stores and food processors who transform milk into products like cheese and ice cream. These two groups are easily identifiable and distinct. Second, it is illegal for food processors to resell the milk it purchases to grocery stores. Third, fluid milk consumers and food processors have different elasticities of demand—fluid milk demand is less elastic. Finally, the Agricultural Marketing Act of 1937 protects milk marketing orders from antitrust authorities. The act makes price discrimination legal for marketing orders. The practice of third-degree price discrimination is widespread and profitable for many firms. But it is not a profitable practice for all firms. Only under certain settings is it feasible, profitable, and legal. The milk industry is one of the industries in which price discrimination abides.

BUNDLING

Bundling: Selling two or more goods together at a single price.

The discussion of pricing schemes has thus far concerned only a single good. Most firms produce more than one good, so it is natural to ponder whether two distinct goods can be combined and sold as one bundle to increase profits. The answer is yes, and this practice is referred to as *bundling*. To illustrate how bundling can increase profits, consider Figure 11.8. A firm sells its patented herbicide and insecticide. It could sell them individually or could sell them as a bundle, where you either purchase both the herbicide and the insecticide or neither. Farmer Tina Turner and Farmer Ike Turner place different values on the pesticides. Tina is willing to pay up to \$100 for the herbicide whereas Ike will pay up to \$120. Tina will pay up to \$50 for the insecticide and Ike will pay up to \$40. These numbers are often referred to as *reservation prices* or maximum willingness-to-pay. At any price below or equal to the reservation price the consumer will make the purchase. Suppose that the cost of producing each pesticide is well below the lowest reservation price.

First, consider a case where the products are not bundled. Each pesticide is sold separately. The maximum price the firm can charge for the herbicide where both Tina and Ike will purchase it is \$100. The maximum price the firm can charge for the insecticide, which again will entice both to purchase it, is \$40. What are the revenues from these sales? The firm makes $\$100 \times 2 = \$200$ from selling herbicides and $\$40 \times 2 = \$80$ from the insecticide sales, providing total revenues of \$280.

Now, consider a case where the herbicide and insecticide are bundled. The firm sells both the insecticide and the herbicide together for one price. If we sum the reservation price for the herbicide and the insecticide, we have a reservation price for the bundle. Tina will pay a maximum of \$100 for the herbicide and \$50 for the insecticide, so her reservation price for the bundle is \$150. Similarly, Ike's reservation price for the bundle is \$160. Thus, the maximum price the firm can charge for the bundle that still entices both to purchase the bundle is \$150. When both make the purchase, revenues are $\$150 \times 2 = \$300$. This is \$20 more than the maximum revenues when the items are sold separately. The cost of producing the pesticides are the same regardless of how they are sold, so the use of bundling increases profits by \$20!

If the two pesticides can only be purchased as a bundle, this is referred to as *pure bundling*. If the pesticides can be purchased individually or as a bundle, the firm is employing a *mixed bundling* approach. Firms usually employ mixed bundling, and

	Reservation Price for Herbicide	Reservation Price for Insecticide	Reservation Price for a Bundle of Herbicide and Insecticide
Farmer Tina Turner	\$100	\$50	\$150
Farmer Ike Turner	\$120	\$40	\$160

FIGURE 11.8 Bundling Example.

economists have shown that the reason is that it is as least as profitable as pure bundling. In the pure bundling example above the bundle sold for a higher price than the sum of the two prices if they were sold separately. With mixed bundling this cannot be the case. In mixed bundling, the price of the bundle must be less than the price of the individual components, or else no one would purchase the bundle. Think of all the mixed bundling you see in the food industry. Wendy's sells combos consisting of a burger, fries, and drink. You could purchase the burger, fries, and drink individually if you want, but it would cost more than the combo. Low-price Mexican restaurants almost always offer combinations of burritos, tacos, and enchiladas, although each could be purchased individually as well. A final word about bundling: Bundling increases profits because it exploits differences in consumer reservation prices. Different people place different values on burgers, fries, and drinks at Wendy's. This allows Wendy's to increase their profits via mixed bundling. For this reason it is an alternative to price discrimination. In a world where all consumers have the same demand for each product, bundling the products will not enhance profits (Adams and Yellen 1976).

REQUIRED TIE-IN SALES

Under pure bundling a firm only allows consumers to purchase a good if they also purchase another good in the bundle. This is an example of a *tie-in sale,* where consumers are forced to purchase one good before purchasing another. There are different forms of tie-in-sales. One is a *required tie-in sale,* where a purchase commits the consumer to also buy complementary goods from the same firm. An example is fast-food franchises. To open your own Wendy's you must do two things. First, you purchase the franchise rights for a fixed fee. Then, you must purchase all your food supplies from Wendy's. Wendy's makes money from the franchise fee and the food supply sales. This is a required tie-in sale. If you buy the franchise fee, you must also buy food supplies from Wendy's. As you might suspect, Wendy's uses a tie-in sales requirement because it makes them more money.

Like bundling, tie-in sales have the ability to increase firm profits only when the firm's customers have different demands and the firm is unable to distinguish between the different demands. To illustrate, suppose there exists a fast-food franchise called Rick James' Burgers (RJ Burgers). Rick James owns the franchise but leases rights to open franchise restaurants for a fee. For simplicity, assume the RJ Burger restaurant takes a fixed batch of inputs (ground beef, hamburger buns, and ketchup) to produce a single burger. Each batch of inputs used to produce one hamburger costs $2 in a competitive market. RJ Burgers has become a brand name. All individuals who want to open an RJ Burgers restaurant must purchase the franchise license from Rick James. Let's call these franchisees. Rick could just charge a lump sum for the franchise license, letting the franchisees purchase the inputs from the competitive market at $2 per batch, but most franchise restaurants go further. Most franchises charge a lump sum for the right to open a franchise (the franchise license) plus require the franchisees to purchase all their inputs (the batch of inputs used to make burgers) from the franchise. Moreover, they sell the inputs at a higher price than one could obtain elsewhere. This is a required tie-in sale; after buying the franchise license you must also buy all food inputs from the franchise. Part of the reason for the required tie-in sale is that the franchise wants to ensure quality inputs are used and that RJ Burgers at different locations are consistent to protect the brand name. Yet there is another reason. These required tie-in sales can provide higher profits to the franchise than if it sold franchise licenses only.

To illustrate this point, suppose RJ Burgers has two potential franchisees: Charlie Murphy and Eddie Murphy. First consider Charlie. Charlie wants to locate in a place with little traffic, so he expects little store volume. The demand curve for burgers at Charlie's location is $P = 5 - 0.00005(Q)$, as shown in Figure 11.9. If Charlie could purchase each batch of inputs from competitive markets at \$2, his marginal cost of production would be \$2 per burger (there are other inputs of course, like labor, but ignore these to keep the illustration simple). As we discussed in the previous section, the area underneath the demand curve and above price is the maximum amount of profit a firm can extract. Assume Charlie is able to extract the maximum profit available. His profits would be \$90,000. This means that the maximum amount the franchise could charge Charlie for the franchise license is \$90,000. Put differently, Charlie's reservation price for the franchise license is \$90,000.

Now consider Eddie Murphy, who wants to locate an RJ Burgers in a high-traffic area expected to receive a high volume of customers. The demand curve for burgers at Eddie's restaurant is higher: $P = 7 - 0.00005(Q)$. Thus, Eddie can extract a larger profit from the restaurant, \$250,000 at most. This means the franchise could charge Eddie up to \$250,000 for the franchise license. If Rick James knew that Eddie's burger demand would be higher than Charlie's, it would charge them different franchise license prices. Specifically, if the demand curves were known to him, Rick James would charge Eddie \$250,000 and Charlie \$90,000. Unfortunately in this story Rick James cannot tell if one restaurant will have a greater demand than the other, so it cannot charge different prices. Both Eddie and Charlie will say they are a low-volume location. Eddie would be lying but Charlie would not. Given that both say

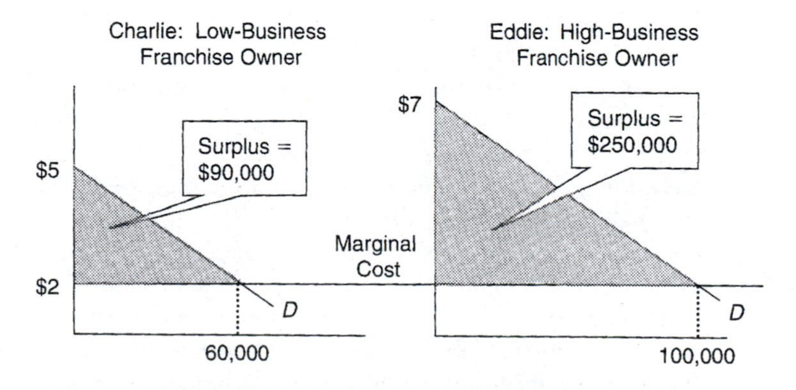

FIGURE 11.9 Required Tie-in Sales Example.

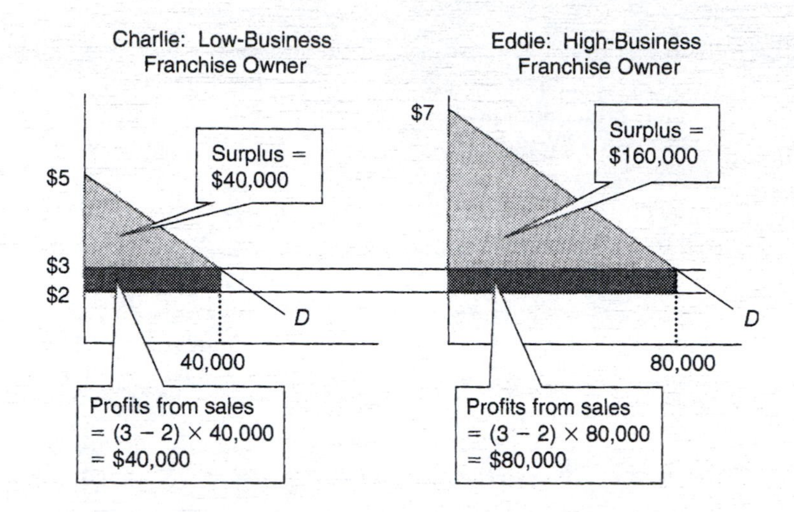

FIGURE 11.10 Required Tie-in Sales Example Continued.

they are low-volume locations, Rick James charges a $90,000 franchise fee to both and will make profits of $180,000.

Now suppose the franchise considers a different pricing scheme—a required tie-in sales scheme. This time it charges both a franchise license fee plus requires both Charlie and Eddie to purchase all their inputs from RJ Burgers. Here is the key: Rick James charges a premium on the inputs. Even though Charlie and Eddie could buy the inputs at $2 per batch from the market, Rick James makes them pay $3 per batch. Figure 11.10 illustrates how profits change with the required tie-in sales scheme. Because Charlie and Eddie must pay more for inputs, they charge higher prices for their burgers and sell less. Consequently, their profits fall. At best, Charlie can make $40,000 and Eddie can make $160,000. Given that RJ Burgers does not know whether Eddie or Charlie will be a high-volume business owner, they both pretend to be a low-volume store and Rick James must charge them the low-volume franchise fee of $40,000. Rick James therefore makes $40,000 × 2 = $80,000 from the franchisees' license fees, much lower than before.

Required Tie-in Sales: Where a consumer agrees to purchase specific components or complements to a good from the same firm it purchases the good.

However, Rick James also makes money off the input sales. Rick buys the inputs from competitive markets at $2 per batch and sells them to Charlie and Eddie at $3 per batch. Making one dollar per batch, since Charlie purchases 40,000 batches and Eddie purchases 80,000 batches, Rick James makes $120,000 from the input sales. Combining this with the franchise license sales, Rick James makes $120,000 + $80,000 = $200,000 in profits. Notice this is an improvement over just charging a franchise license fee, where profits were only $180,000.

There is a simple reason required tie-in sales increase profits; they overcome an asymmetric information problem. The franchise could not tell whether Eddie or Charlie was a high-volume business, so it could not price them according to the

	Franchise License Fee Only	Franchise License Fee Plus Required Tie-in Sales
Profits from license fees	$90,000 × 2 = $180,000	$40,000 × 2 = $80,000
Profits from input sales	$0	$40,000 + $80,000
Total Profits	$180,000	$200,000

FIGURE 11.11 Required Tie-in Sales Example Outcome.

volume of business they will generate. However, the franchise does know that a high-volume restaurant will require more inputs. To sell more hamburgers you need more ground beef. The franchise therefore focuses less on making money through franchise fees and more on making money through input sales. Forcing Eddie and Charlie to purchase inputs through the franchise at a premium, the franchise is able to charge Eddie more money because Eddie purchases more inputs.

Beware, required tie-in sales, or bundling in general, can be illegal. As of October 13, 2006, the Monsanto Corporation was being sued for its required tie-in sales. Monsanto has a patent for a genetically modified soybean seed that is resistant to herbicides with the glyphosate active ingredient. The seed allows farmers to cheaply apply one glyphosate-based herbicide to the entire field of soybeans for weed control. Monsanto also makes an herbicide called Roundup, which contains this glyphosate active ingredient; however, other pesticide manufacturers sell similar herbicides. When Monsanto enters into contracts with farmers to sell their seed, the contracts contain a stipulation that the farmer must purchase only Monsanto's Roundup herbicide. Without this stipulation, farmers would be free to buy close substitutes to Roundup. Because Monsanto has a patent for the genetically modified seed, the required tie-in sale allows them to charge a premium for Roundup. In fact, the price of Roundup is 300% to 400% higher than competing herbicides with the same active ingredients (Fakta 2006).

One of Monsanto's competitors has claimed that this violates the Sherman Act. The Sherman Act specifically states that "Every person who shall monopolize, or attempt to monopolize, or combine or conspire with any other person or persons, to monopolize any part of the trade or commerce among the several States, or with foreign nations, shall be deemed guilty of a felony"[1] Monsanto has a monopoly on the genetically modified seed due to its patent, but it does not have a monopoly on glyphosate-based herbicides. Thus, the required tie-in sale may be seen as an attempt to use their patent to monopolize the glyphosate-based herbicide market as

[1]Sherman Act, Part 2.

well. We shall see what the courts say, but the point is that any form of bundling (including tie-in sales) must not appear to be an attempt to monopolize any particular market.

SUMMARY

This chapter is concerned with a firm that has control over its prices. We are not talking about a cattle producer who simply must take the "going price." We are talking about businesses who either produces a unique product or has a rather large market share. Whether it be product uniqueness or market share that bestows the firm with pricing power, this chapter is concerned with how to use that pricing power. A calculator in the hands of a monkey is of little use, and pricing power is of little use to a manager unless she knows how to exploit it.

Determining an effective pricing scheme involves two important questions: (1) what are consumers' maximum willingness-to-pay and (2) how does the firm charge this maximum willingness-to-pay? Addressing the first question often entails market research, research often as simple as asking people how much they value something. This is covered in the next chapter. If the consumer would purchase more than one unit in a given time period, the question becomes more complex. How much do they value the first unit? How much do they value the second unit? What about the third and the fourth? Although the goal is to charge the consumer a price equal to their value for each unit, we all know this is difficult in practice. Imagine a grocery store trying to charge you a different price for the second ear of corn than the first. Consumers do not like being harassed, especially over the price of an ear of corn. Two pricing schemes were discussed that charge consumers their maximum willingness-to-pay for each unit but in a more consumer-friendly manner. One was all-or-nothing pricing and the other was two-part pricing.

Consumers are not all alike, and when consumer preferences differ, pricing becomes more complicated. The greater the differences among consumers, the more the firm should concentrate on setting different prices for different consumers than trying to maximize profits from any one consumer. When the firm cannot distinguish between consumers with different values for the good, second-price discrimination is appealing. But when the firm can separate consumers into groups with different values, third-degree price discrimination is more profitable.

Finally, businesses are endowed with additional marketing opportunities if they sell more than one product, which most all businesses do. More profits can sometimes be generated by selling two or more items together as a bundle, such as a Burger King combo meal, than if each item was only sold separately. Other times a firm can enhance profits by including a selling clause that if an item is purchased, the consumer must purchase components going with that item from the same firm. This is called required tie-in sales.

Exactly which pricing scheme is best for any business depends on many details, and rarely will it be plainly evident which scheme is best. How well can you estimate the value consumers place on a good? Do consumers' values differ? Do different consumers place different values on alternative products that you sell? As answers to these questions vary, so does the optimal pricing scheme.

CROSSWORD PUZZLE

For answers with more than one word, leave a blank space between each word.

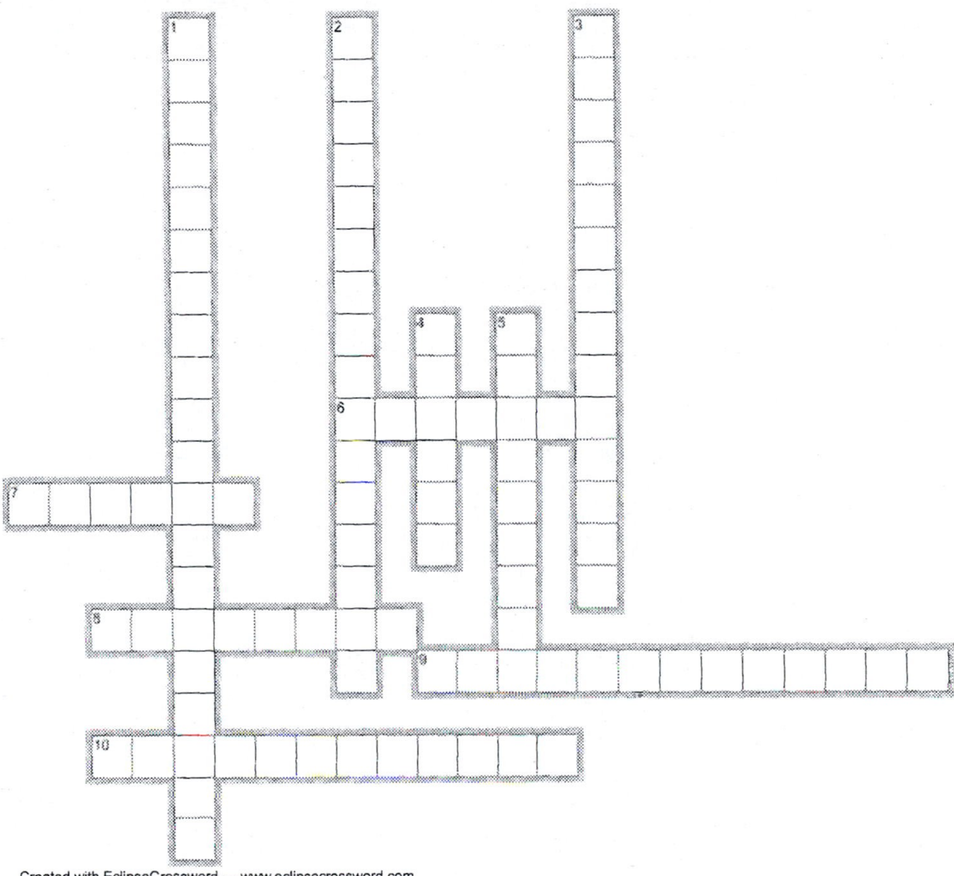

Created with EclipseCrossword — www.eclipsecrossword.com

Across

6. Under _______ price discrimination the firm receives the maximum possible profits.
7. In a required _______ - _______sale, the purchase of good A commits the consumer to also purchase good B from the same firm.
8. When two or more products are sold together.
9. A firm who offers discounts for customers making large purchases is probably using _______ - _______ price discrimination.
10. In __________ - _______ price discrimination, customers are segregated into groups with different demand elasticities and charged different prices.

Down

1. The act of charging different prices to different people.
2. A wholesaler requires one to purchase a membership before one may purchase their items. What type of pricing is this?
3. A price scheme where the firm sells bundles of a good at a single bundle price.
4. The _______ model of price discrimination describes a situation where firms force consumers to pay a personal cost (like gathering coupons) before receiving a price break.
5. In third-degree price discrimination, the consumer group with a more _______ demand is charged a higher price.

STUDY QUESTIONS

Refer to Figure 11.12 to answer Questions 1–3.

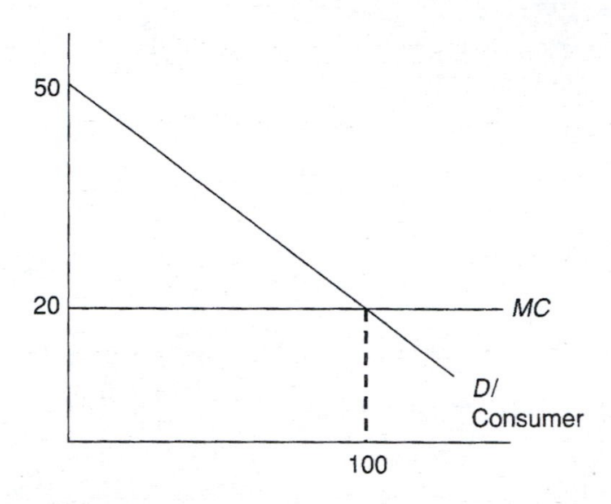

FIGURE 11.12

1. If a firm is able to perfectly price discriminate, what will its per consumer profits be?
2. If a firm employs all-or-nothing pricing, each bundle will consist of _______ units, the bundle price will be _______, and profits per consumer will equal _______.
3. If a firm employs two-part pricing, the fixed fee will be _______ and the per-unit fee will equal _______ for each consumer. Overall profits per consumer will equal _______.
4. A firm is practicing third-degree price discrimination by charging two different groups of consumers two different prices. Call these consumers Group A and Group B. If Group A is charged a higher price relative to Group B, what do we know about its elasticity of demand?

CHAPTER TWELVE

Consumer Behavior and Research

In the News

About 10 years ago, McDonald's introduced a new low-fat burger called the McLean Deluxe. A seaweed derivative was mixed with the meat patty to achieve the lower fat burger. The McLean Deluxe was about 25% lower in fat than the Quarter Pounder and was introduced to appeal to the growing wave of nutritionally conscious consumers. After four months of test marketing, the product was introduced by a massive publicity campaign. What was the result? You know the answer if you've been through a McDonald's lately. You won't see the McLean Deluxe on the menu; the McLean was a McFailure. Despite enormous advertising outlays, McDonald's pulled the McLean Deluxe from their menus in 1996 after only a few years of dismal sales. One franchise owner in New York was quoted as saying, "On a good day I may sell a couple of dozen or more McLeans versus hundreds of Big Macs and Quarter Pounders." In fact, sales of the McLean Deluxe had been so poor that an ABC *Primetime Live* investigation revealed that some restaurants, because of poor sales, occasionally made the McLean Deluxe with regular, higher-fat beef (Prewitt 1992).

How could one of the world's largest and most successful companies make such a blunder? One reason might be attributable to the company's lack of attention to the preferences of their typical customer. A typical McDonald's consumer wants food fast and cheap, not necessarily healthy. Better consumer research and marketing might have saved McDonald's millions of dollars incurred from the failed new product introduction. Another view on the new product introduction is that McDonald's knew the McLean would be unsuccessful with their consumers but wanted to appear healthy to the American public. Indeed, just prior to the introduction of the McLean Deluxe, the president of the National Heart Savers Association placed full-page ads in major newspapers blasting McDonald's for poisoning America with high-fat, high-cholesterol menu items. Perhaps not surprisingly, in 2002 two teenagers sued McDonald's alleging that the chain made them fat and caused health problems. Thus, it might be argued that the introduction of the McLean was a brilliant marketing strategy that would prove useful in future litigation, showing that McDonald's attempted to offer healthy menu items, a move that their customers didn't support. Whether one believes the introduction of the McLean was an ingenious success or monumental blunder, it should be clear that firms need to understand their consumers and the general public to be successful.

INTRODUCTION

This book has thus far focused on standard economic models of consumer and firm behavior. When discussing consumer behavior, the implicit assumption has been that individuals act on a given set of preferences. Preferences are simply a ranking of the relative desirability of competing bundles of goods. Economic models typically focus on observed outcomes, or the actual choices made by consumers, and use these outcomes to define what the underlying preferences of consumers must have been. For example, if a consumer chose a Coke over a Pepsi in a vending machine where the two drinks were equally priced, it must be that the consumer preferred or ranked Coke over Pepsi and preferred the Coke over all other goods that the same amount of money could have bought. Thus, in economic models, preferences are the primary determinant of consumer behavior. Although economic models help us understand consumer response to incentives, changes in constraints, and the like, they are typically silent about where preferences come from, how they change, and factors that influence preferences. Because preferences in economic models are what define choice behavior and vice versa, it would seem that a better of understanding of preferences is in order. This is especially true for those interested in consumer research, a task that aims to better understand consumer psychology, motivation, and decision making in order to improve firm profitability through strategic pricing, promotion, and product design.

It is instructive to consider an example that shows knowing *why* a choice was made, and thus knowing something about preferences, can improve firm decision making. Consider two boys: Jackson and Harrison. Suppose both boys were offered a scoop of ice cream for $1, but both turned down the offer. Apparently neither preferred ice cream over other things that a dollar could buy. The choice for both boys was the same, no ice cream, but the reasons for the choice differed. Since birth, Jackson has been lactose intolerant, and if he eats ice cream, he becomes nauseous and develops a severe stomachache. Harrison, on the other hand, is a bit leery of eating ice cream because he had one too many scoops of pralines and cream prior to a recent roller-coaster ride. Armed with this further information, we can be reasonably sure that a decrease in price from $1.00 to say $0.25 would have no effect on whether Jackson would buy ice cream, but it might temp Harrison to buy. In economic terms, Jackson has perfectly inelastic demand for ice cream; he won't consume it no matter the price. On the other hand, Harrison's demand for ice cream is relatively elastic; he'd consume a whole carton of it if the price was low enough. A defiant defender of typical economic models might argue that all that matters are the outcomes—that neither boy ate ice cream at $1.00, but that Harrison ate when the price was $0.25—regardless of *why* the outcomes were observed. They might further argue that one could simply conduct experiments by varying prices at numerous levels, observe whether a purchase occurred, and never have to know *why* the choices were made. However, it is easy to see from this example that we could do away with the lengthy pricing experiment and ask a simple question about lactose intolerance, which would tell us about Jackson's choice outcomes over *all* price levels. Further, unlike the price-quantity relationship just described, a relationship that is well explained by demand theory, there are many economic outcomes such as decisions made under risk and decisions from bargaining-type situations for which several competing theories or mathematical models can explain the same set of outcomes. In such cases, auxiliary information about preference structures would be of great use.

In this chapter, we first introduce several models to help understand consumer behavior. In particular, this chapter will cover

1. the motivational process
2. the role of values and needs in purchasing behavior
3. attitudes and purchase intention
4. willingness-to-pay
5. models of decision making

After the models of individual decision making are discussed, this chapter then covers research methods that can be employed to better understand the consumer. Topics on consumer research that will be covered include

6. types of consumer research
7. steps in conducting consumer research
8. research methods for measuring values, attitudes, preference, and willingness-to-pay

MODELS OF CONSUMER BEHAVIOR

Understanding consumer behavior is a complex issue that is often left to the realm of psychologists. Nevertheless, as just demonstrated, anyone with an interest in utilizing marketing tools to enhance firm profitability needs have some understanding of consumer behavior. A myriad of factors including an individual's motivation, values, personality, perceptions, learning, attitudes, and the surrounding institutions and culture affect consumer behavior. In what follows, we discuss a few of these issues and present some simple models of how they can affect consumer behavior.

Motivation

Motivation is a multifaceted concept. One simple definition is that motivation is the process that leads people to behave a certain way. Motivation can also be viewed as an internal state or driving force that activates and directs behavior. Based on these definitions, it should be clear that to understand consumer behavior (consumer choice), we need to understand motivation. In many marketing classes, students are told that

Motivation is the internal state or driving force that activates and directs behavior.

the goal of marketing is to satisfy consumers' needs. This is a difficult task if we do not know what those needs are or why they exist. One particular model that relates unfilled needs, wants, and desires to behavior is a model of the motivational process shown in Figure 12.1.

Figure 12.1 shows that behavior is a result of an individual attempting to fulfill a need in order to achieve a particular goal. When a need is unmet, this creates tension the consumer attempts to eliminate. Tension arises from a gap between a consumer's present condition and their ideal condition. The drive strength is simply the degree of arousal or magnitude of tension that exists to satisfy a given need. Once drive strength reaches a certain point, an individual will direct their actions toward a certain behavior, conditional on past learned experiences, cognitive processes, and other factors. Once a behavior is undertaken and the goal met, tension is relieved. For example, around noon every day your stomach probably starts to rumble, identifying an unfulfilled need. The increased rumbling creates tension, which you alleviate by choosing to eat a hamburger, satisfying a number of goals such as maintaining good health. What Figure 12.1 does not necessarily identify is *which* particular behavior will be undertaken; it is simply a model of the process of motivation. That is, the need identified by a growling stomach can be satisfied by eating a hamburger, a salad, or maybe even a piece of cardboard. It is this difference that can help distinguish between a need and a want. A want is a particular manifestation of a need. You may want a hamburger or you may want a salad; both wants can arise from the same need. However, in many cases, different behaviors will be caused by differences in needs and/or goals. In the case of food consumption, eating salad might result from a utilitarian (meaning functional or practical) need such as weight loss or nutritional improvement, whereas eating a hamburger might result from a hedonic (meaning related to pleasure) need such as the need to experience good taste.

This implies that a closer investigation into needs is in order. There are many types of needs and these needs arise from a variety of sources. The sources of motivational needs can be broken down into several general categories. Physiological or

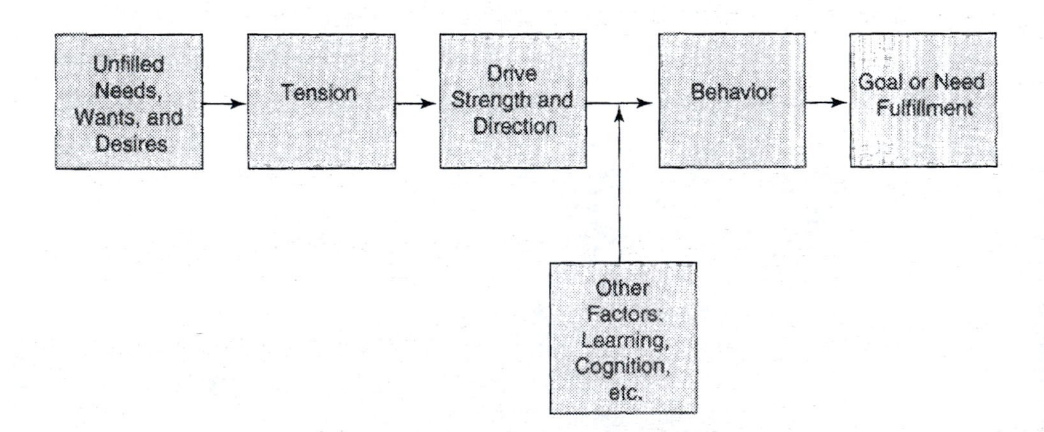

FIGURE 12.1 A Simple Model of the Motivational Process.

biological sources of motivation are those that are based on an individual's physiological condition at a given moment. Most of these needs are activated by involuntary biological responses from a lack of a food, low levels of a particular hormone, and a decrease in body temperature, for example. These are often innate needs with which an individual is born. Another motivational source relates to emotional or affective arousal—the need to feel good or bad, to feel safe, and to maintain levels of optimism and enthusiasm. For example, a person might "go shopping" not because they are motivated to clothe themselves, but because it makes them feel good or it raises self-esteem. A third motivational source stems from cognitive issues that are related to individuals' thought processes. Individuals may be motivated by the need to understand a situation, develop meaning, solve problems, or resolve cognitive dissonance, which arises when two attitudes, goals, or beliefs are in conflict. A final general motivational source of needs relates to environmental or external issues. Such needs arise often due to specific cues in the environment and can result from learned processes. These needs can reflect priorities of a culture, such as individuality, power, risk avoidance, and so on, and will vary from one culture to the next.

Several models have been proposed to classify individual's needs, but none has been more widely received than Maslow's Hierarchy of Needs, which is shown in Figure 12.2 (Maslow 1954). Maslow's Hierarchy of Needs posits that individuals must first meet a more basic need (at the bottom of the pyramid) prior to moving on to a higher need at the top of the pyramid. At the very bottom level are physiological needs such as water, air, and food. Once these needs are met, individuals then focus on safety needs such as security, shelter, order, and stability. Next up the chain of needs are loving/belonging needs, social needs, that relate to individuals need to be

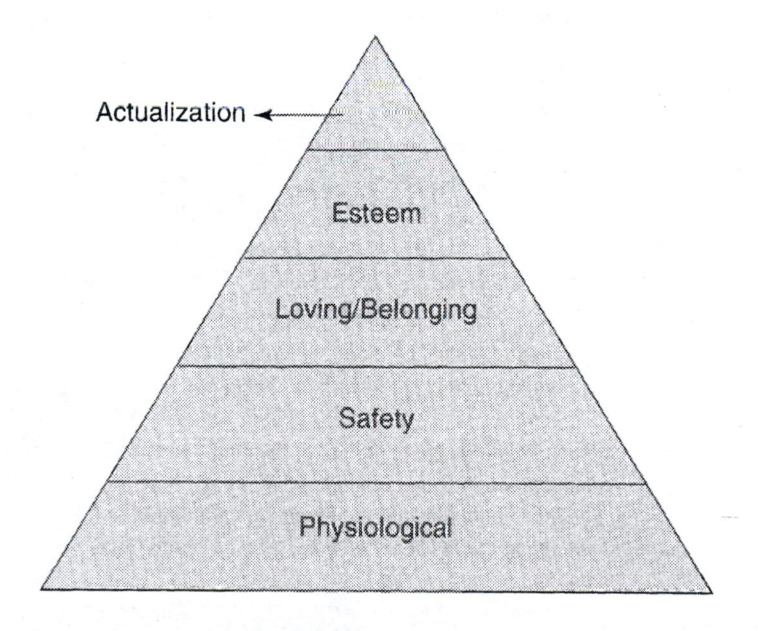

FIGURE 12.2 Maslow's Hierarchy of Needs.

accepted by others. The fourth category of needs is related to esteem or ego needs such as prestige, status, self-esteem, and accomplishment. At the top of the pyramid are needs related to self-actualization where individuals have a need to reach their full potential and relate to needs such as self-fulfillment, creativity, justice, meaning, and so forth. Although Maslow's Hierarchy of Needs provides a useful categorization, it is important to realize that individuals may not necessarily order their needs as in Figure 12.2. For example, an individual often goes through extensive personal hardship without food and shelter to achieve a higher need such as esteem or self-actualization. The "starving artist" is a perfect example.

In addition to Maslow's approach to categorizing needs, other classification systems exist. For example, Achievement-Motivation theory suggests that needs can be placed into three categories: need for affiliation (to be in the company of and accepted by other people), need for power (to control one's environment), and need for achievement. From a marketing standpoint, these and other needs-based classification systems can be helpful in segmenting consumer into different needs-based categories. Consumers with similar needs will likely respond similarly to price changes and promotion, whereas consumers in different segments are likely to behave differently.

Values

In the previous section, we discussed how an individual's behavior is a result of motivation to meet unfilled needs as a way to achieve a goal. However, little was said about the goals that individuals might have and the way that goals influence purchasing behavior. Fortunately, a significant amount of research has focused on investigating individual's values, which are people's broad life goals. In the late 1960s, the psychologist Milton Rokeach (1968, 1973) argued that individuals have two types of values: instrumental and terminal. Terminal values are broad psychological states that represent individuals' preferred state of being. Instrumental values are those actions that lead to preferred terminal values; they are preferred modes of conduct. These values represent the consequences individuals attempt to achieve with their lives. Although values are thought to differ from culture to culture, there is increasing recognition of individual differences in values within a culture. Figure 12.3 lists the original set of 18 instrumental values and 18 terminal values proposed by Rokeach (1968, 1973). Although all the values listed in Figure 12.3 seem like reasonably desirable life objectives, what is important, in terms of identifying differences across individuals, is the relative importance an individual places on the values in Figure 12.3. For example, men tend to ranking "an exciting life" as a more important terminal value that women, who tend to rank the value of "a world of peace" more highly than men.

Values are individuals' broad life goals. Terminal values are individuals' preferred state of being, and instrumental values are those actions leading to terminal values.

One approach to investigating the role of values in purchasing behavior is called *means-end chain analysis*. This approach assumes that individuals make sense of the world by categorizing items in an ordered hierarchy where more abstract items motivate individuals to behave and where more concrete items represent behavioral alternatives from which an individual can choose. Means-end chain analysis represents one such hierarchy where the most abstract components are values and the most concrete components are product attributes. The approach assumes individuals

Means-end chain analysis: the study of how individuals purchase products with the attributes that lead to consequences which fulfill their personal values.

Instrumental Values	Terminal Values
Ambitious	A comfortable life
Broad-minded	An exciting life
Capable	A sense of accomplishment
Cheerful	A world of peace
Clean	A world of beauty
Courageous	Equality
Forgiving	Family security
Helpful	Freedom
Honest	Happiness
Imaginative	Inner harmony
Independent	Mature love
Intellectual	National security
Logical	Pleasure
Loving	Salvation
Obedient	Self-respect
Polite	Social recognition
Responsible	True friendship
Self-controlled	Wisdom

FIGURE 12.3 Values in Rokeach Value Survey.

consume products and the attributes therein, as a means to achieving some end, such as achieving a desired value. Figure 12.4 presents a common means-end chain.

All products represent a bundle of attributes, from the concrete, physical details of the product such as size, shape, and so on to the more abstract attributes such as brand name and quality. For each attribute, the model assumes that consumers associate a particular consequence; that is, consumption of a product and its inherent attributes produces a given set of consequences. Consequences can either be functional (tangible outcomes that are directly experienced) or psychosocial (outcomes

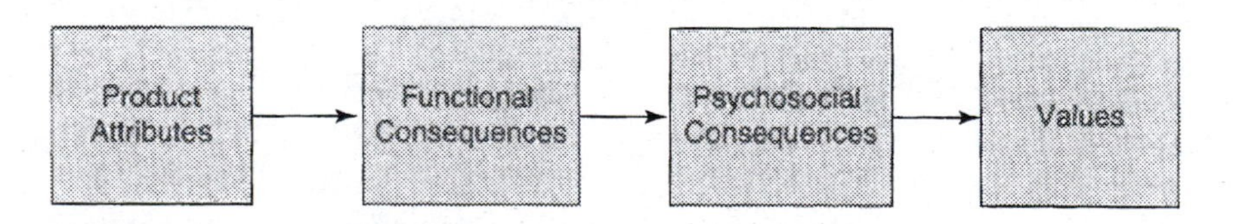

FIGURE 12.4 Means-End Chain.

that are internal and represent how the product makes the consumer and other individuals feel). Consequences are often categorized as benefits, which are desirable consequences, or risks, which are undesirable consequences consumers attempt to avoid. The means-end chain suggests a consumer buys a product with a given bundle of attributes as a means to achieve a particular consequence and value.

To illustrate how a means-end chain might be useful to agribusiness marketers, consider the values that motivate an individual to buy fair trade coffee. Fair trade coffee refers to coffee sold by a marketing channel where the producers, usually in a developing country, are guaranteed a certain portion of the final retail price. Recently de Ferran and Grunert (in press) investigated the values that drove French consumers' purchases of fair trade coffee. They found that individuals had mixed motives and values when buying fair trade coffee. They also found that the typical supermarket consumers bought fair trade coffee for different reasons than consumers who bought fair trade coffee in specialty stores. Figure 12.5 shows that many supermarket consumers associated taste attributes of fair trade coffee with a "good" consequence, which yielded the terminal value of satisfaction. In contrast, the primary chain used by specialty store shoppers was to view the attribute of fair trade as yielding a consequence to either participate in an alternative economy or equality of trade, which yielded a terminal value of a sense of accomplishment or equality between humans. Figure 12.5 also shows that specialty store shoppers were likely to view the attribute of organic as a means of creating a world of beauty; such a chain was not present for any of the supermarket shoppers.

Results from a means-end chain analysis such as that shown in Figure 12.5 can assist marketers in understanding why consumers buy products and can help firms predict how consumers might respond to various promotions. For example, an advertisement

Location	Chain	Attribute	→ Consequence	→ Instrumental Value	→ Terminal Value
Supermarket	1	Taste	Good		Satisfaction
Supermarket	2	Fair trade	Economic Aid	Respectful	Equality between humans
Specialty Store	1	Fair trade	Participate in alternative economy	Responsible	A sense of accomplishment
Specialty Store	2	Fair trade	Equality of trade		Equality between humans
Specialty Store	3	Organic	Respect for environment		A world of beauty

FIGURE 12.5 Prominent Means-End Chains for French Buyers of Fair Trade Coffee.

that focused on individualistic reasons for buying fair trade coffee such as good taste and satisfaction achievement is likely to produce more desirable results with supermarket consumers than those in specialty stores.

It is important to keep in mind that all product attributes do not necessarily produce positive consequences and thus distract from, rather than contribute to, terminal values. For example, consider two individuals who both recognize the same concrete attribute: Potato chips have high fat content. One individual may view this attribute and think of the functional consequence of gaining weight and the psychosocial consequence of feeling bad about their appearance. Such consequences would contribute negatively to the instrumental value of self-control and negatively to the terminal value of self-respect. In other words, eating a potato chip would lead to a lack of self-control and reduced self-respect. Despite this person's disposition, another individual might believe the same attribute of high fat content yields a consequence of good taste, which positively contributes to the terminal values of pleasure and/or happiness.

Because products are essentially bundles of attributes, an individual might have several competing or even conflicting chains from the same product. In such cases, it is difficult to know which means-end chain will prevail in the decision. One way to conceptualize this issue is to focus on consumer involvement. Consumers' product involvement refers to the level of interest a consumer finds in a product or product class. Involvement is a widely accepted variable for explaining variations in consumer behavior. Research has shown that product involvement influences choice behavior, usage frequency, extensiveness of decision-making processes, and response to persuasive messages. Specifically, highly involved consumers tend to think more about specific product categories with which they are highly involved, search more widely for information, process the information obtained in greater depth, and spend more time making purchase decisions relative to less involved individuals.

Although it might not be initially obvious, the concept of involvement can be related to the means-end chain, as shown in Figure 12.6. A consumer's involvement

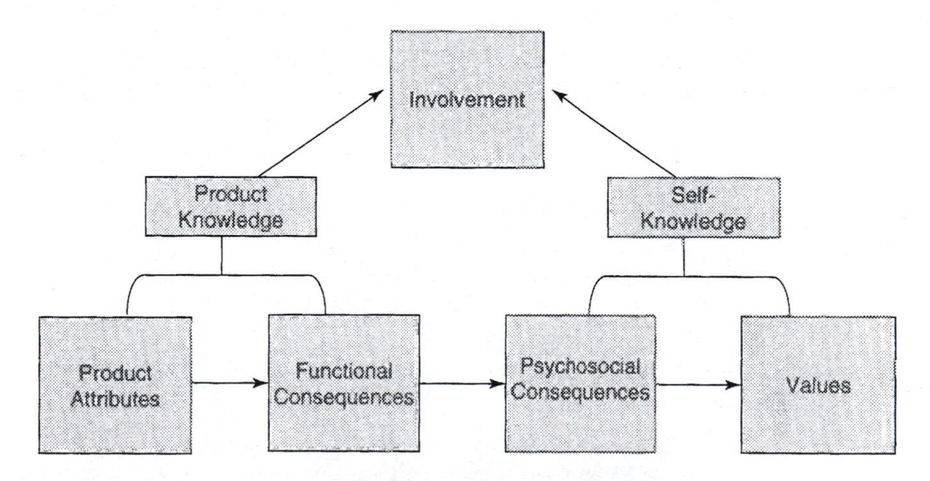

FIGURE 12.6 Effect of Means-End Chain on Involvement.

level is determined by two components of the mean-ends chain: product knowledge and self-knowledge. Product knowledge arises from an understanding of the relationship of product attributes and their functional consequences. Self-knowledge, on the other hand, relates to an individual's understanding of the relationship between psychosocial consequences and values in conjunction with the relative importance of their values. An individual will be more involved with a product if the product's attributes/consequences connect with values that are important to the consumer. By contrast, a consumer will have low levels of involvement with products whose attributes do not yield consequences that contribute to important values. For example, a consumer might construct a means-end chain relating the attribute of baking soda in toothpaste to the consequence of whiter teeth and the value of beauty. However, if (a) beauty is a relatively low ranked value to the consumer, (b) the individual believes there is little relationship between white teeth and beauty, and/or (c) the individual believes baking soda does not improve teeth color, then we can expect the consumer's involvement with toothpaste to be low. Thus, if an individual's level of product involvement can be measured, one can determine the extent to which the individual believes the product will generate consequences that are important to them. One widely used method of measuring involvement proposed by Zaichkowsky (1994) is shown in Figure 12.7, using organic pork chops as an example. Consumers are given a list of adjectives or phrases and their opposite and are asked to circle a number indicating the extent to which the word or phrase best describes their feelings. For example, if a consumer circles number 6 in the first row, they deem organic pork chops to be "important." If they had circled the number 4 instead, the consumer is neutral on the question of whether organic pork chops are important. After a number has been circled for each row, sum up the responses

Important	7	6	5	4	3	2	1	Unimportant
Interesting	7	6	5	4	3	2	1	Boring
Relevant	7	6	5	4	3	2	1	Irrelevant
Exciting	7	6	5	4	3	2	1	Unexciting
Means a lot to me	7	6	5	4	3	2	1	Means nothing
Appealing	7	6	5	4	3	2	1	Unappealing
Fascinating	7	6	5	4	3	2	1	Mundane
Valuable	7	6	5	4	3	2	1	Worthless
Involving	7	6	5	4	3	2	1	Uninvolving
Needed	7	6	5	4	3	2	1	Not needed

FIGURE 12.7 Zaichkowsky Involvement Scale (Using Organic Pork Chops as an Example). To me, organic pork chops are (circle a number corresponding to which word or phrase best describes your feelings).

to obtain an involvement score. For instance, if a consumer is very enthusiastic about organic pork chops (feeling them to be fascinating, important, etc.), they might circle number 7 for each row. This results in the highest possible involvement scale rating of 70.

Consumer Attitudes and Intentions

An attitude is an evaluation or affective (emotional) response to a concept.

Attitudes are one of the most important and studied variables in consumer behavior research. Businesses regularly measure consumers' attitudes as a way to predict and understand behavior. What exactly is an attitude? At some point during your childhood, your mother probably asked that you stop giving her one. What is this concept that so bothered your mother? An attitude is an evaluation or an affective (emotional) response to a concept. Stated differently an attitude represents the extent to which an individual has a favorable or unfavorable opinion of a concept. Attitudes are always toward something. Consumers can have attitudes toward objects and attributes, such as cars, fat content, brands, and so, and even attitudes about behaviors such as shopping for cars, eating a hamburger, choosing a brand, and so on. Although an attitude seems like a rather abstract concept, psychologists and marketers measure attitudes in a relatively straightforward way. For example, an individual's attitude toward the McLean Deluxe burger might be measured by summing an individual's responses to the following three questions.

	Strongly Disagree						Strongly Agree
The McLean Deluxe burger is good.	1	2	3	4	5	6	7
The McLean Deluxe burger is beneficial.	1	2	3	4	5	6	7
The McLean Deluxe burger is pleasant.	1	2	3	4	5	6	7

Psychologists believe that individuals form attitudes about each object or concept they encounter by determining what relationship they have with the concept, whether they like or dislike the concept, whether they think the concept is good or bad, and so on. Typically, consumers do not have to form attitudes each time they encounter a concept; once an attitude is formed, it is presumed to be relatively stable and retrievable from memory. Attitudes can be very product and context specific. For example, you may have a favorable attitude toward ice cream in general, but an unfavorable attitude toward strawberry ice cream. Furthermore, you might have a favorable attitude toward vanilla ice cream at Baskin-Robbins but not vanilla ice cream at Dairy Queen.

So, where do attitudes come from? Attitudes toward an object arise from individuals' beliefs about the attributes an object possesses *and* evaluations of (or attitudes toward) those beliefs. A belief is cognitive knowledge about the object or attribute; a belief represents the attributes an individual thinks a product possesses. For example, consider the attributes of a T-bone steak. Although one could probably list hundreds of attributes if they spend enough time, consumers form attitudes based on a few (typically seven or fewer) salient beliefs—those beliefs that are activated at a specific time and in a specific context. For our purposes, suppose an individual has only three salient beliefs about T-bones steaks:

- T-bones are more expensive than other cuts of meat.
- T-bones taste good.
- T-bones are high in cholesterol.

These statements represent what an individual believes about T-bones. An attitude toward T-bones is formed by combining the beliefs with an individual's evaluation of beliefs. For each of the three beliefs about T-bones, the individual may form the following evaluations:

- High food prices are undesirable.
- Good taste is desirable in food.
- High cholesterol is undesirable.

Quantitatively, if we let b_i represent the strength of an individual's belief that the object has attribute i and let e_i be an individual's evaluation of belief b_i, and there are N salient beliefs about a product, then the individual's attitude about the object (A_0) is:

$$A_0 = \sum_{i=1}^{N} b_i e_i$$

Figure 12.8 illustrates how attitudes toward T-bone steaks could be calculated if an individual were asked to indicate the extent to which they agreed with each of the previous beliefs and belief evaluation statements, where 1 was strongly disagree and 7 was strongly agree.

As can be seen in Figure 12.8, the total attitude toward T-bone steaks is a 69, where a higher attitude value indicates higher acceptance and willingness-to-purchase of the product. Figure 12.8 illustrates how marketers can improve attitudes toward their products. Attitudes toward a product could be made more favorable by correcting or strengthening the individual's existing beliefs, changing evaluations of an important existing belief, or adding new salient beliefs about a product. For example, informing consumers about actual cholesterol content in steaks might change their beliefs about the cholesterol content. Alternatively, advertising about the protein content and healthfulness of beef might cause consumers to add these issues to their list of salient beliefs.

Theory of reasoned action: an individual's attitudes toward engaging in a behavior, together with subjective norms, influence the consumers' intention to undertake the behavior.

Although it is useful to study how attitudes toward products or objects are formed, what is ultimately of interest to the firm is consumer behavior. In this regard, Fishbein and Ajzen proposed the *theory of reasoned action* (1975, 1980). They proposed that an individual's attitudes toward engaging in a behavior, together with subjective norms, influence the consumer's intention to undertake the behavior. Formally, the theory can be expressed as:

$$Behavior \approx Behavior\ Intention = A_{act}w_A + SN_{act}w_{SN}$$

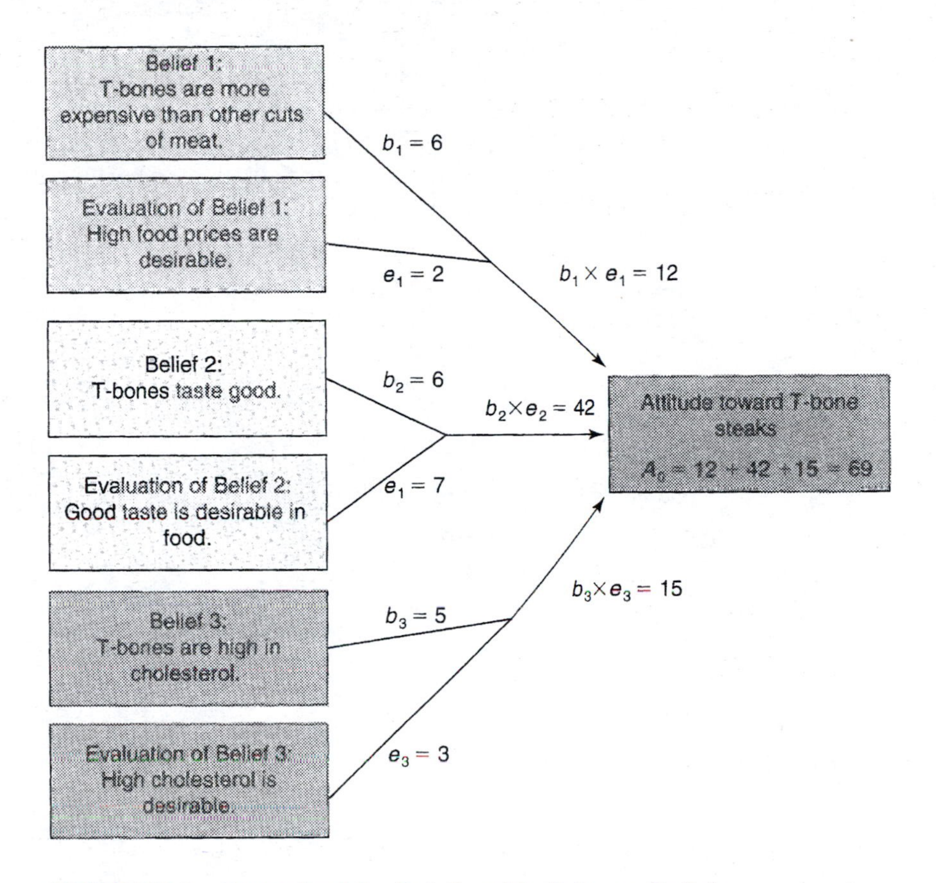

FIGURE 12.8 Example of the Relationship Between Beliefs, Belief Evaluation, and Attitude.

where

$$A_{act} = \sum_{i=1}^{N} b_i e_i$$

= an individual's attitude toward the behavior; b_i is the strength of belief that the action has attribute i and e_i is an individual's evaluation of the outcome of belief b_i

$$SN_{act} = \sum_{i=1}^{M} NB_j MC_j$$

= the subjective norm reflecting whether other people want the consumer to undertake the behavior; NB_j are normative beliefs about what other people want the consumer to do and MC_j is the individual's motivation to comply with the normative belief

The terms w_A and w_{SN} are weights attached to the behavior attitude and subjective norm, respectively. Thus, actual behavior is driven by an intention to undertake that behavior, which in turn is influenced by an individual's attitude toward the behavior and the influence of others. Intentions to undertake a behavior are not always highly correlated with actual behaviors for a myriad of reasons. For example, there is often a period of time that elapses between an intent and an opportunity for behavior during which individuals can receive new information and/or might encounter some unforeseen event. For this reason, attitudes are thought to be more strongly associated with behavioral *intentions* than with actual behaviors. Although the T-bone example discussed previously illustrates how individuals form attitudes toward an object, it is

Column	Variable	Consumer		
		Homer	Marge	Bart
	Behavior			
	Behavior Beliefs (b_i)			
A	By eating French fries I get a food that is convenient[a]	6	6	4
B	By eating French fries I get a food that tastes nice[a]	6	6	4
C	By eating French fries I get a food that is high in fat[a]	6	6	4
	Outcome Evaluation (e_i)			
D	Food that is convenient is desirable[a]	7	4	7
E	Food that tastes nice is desirable[a]	7	6	7
F	Food that is high in fat is desirable[a]	5	1	5
G	A_{act} = *Attitude toward eating fries* $= A \times D + B \times E + C \times F$	114	66	76
	Subjective Norms			
	Normative Beliefs (NB_j)			
H	Dietitians think that I should eat French fries[a]	1	4	1
I	Doctors think that I should eat French fries[a]	1	4	1
J	Fast-food outlets think that I should eat French fries[a]	6	4	3
	Motivation to Comply (MC_j)			
K	In general I want to do what dietitians think I should[a]	1	6	1
L	In general I want to do what doctors think I should[a]	1	6	1
M	In general I want to do what fast-food outlets think I should[a]	6	1	6
N	SN_{act} = *Subjective Norm* $= H \times K + I \times L + J \times M$	38	52	20
	Weights			
O	w_A = Attitude Weight	0.75	0.75	0.75
P	w_{SN} = Subjective Norm Weight	0.25	0.25	0.25
	Intention to Eat French Fries $= G \times O + N \times P$	95	62.5	62

FIGURE 12.9 Intention to Buy French Fries as Determined by the Theory of Reasoned Action.

[a]Responses on a scale of 1 to 7 where 1 = strongly disagree and 7 = strongly agree.

important to recognize that in the theory of reasoned action, it is the attitude toward the behavior that is important. Thus, while A_0 might reflect attitudes toward a T-bone steak, A_{act} might reflect attitudes toward buying or eating a T-bone steak. This distinction is important because an individual might have a favorable attitude toward an object but might have an unfavorable attitude toward a certain behavior involving that object. For example, two people might have a favorable attitude toward the environment. One person might have a favorable attitude toward hiking in the environment, thinking such a behavior would be enjoyable, but the other might view such a behavior unfavorably, believing that hiking will harm the natural wildlife. The final piece of the behavioral intention model is social norm. The model posits that behaviors that are more popular with other people are more likely to be undertaken.

To help illustrate how the theory of reasoned action can be used to understand consumer behavior, consider the data in Figure 12.9, which is loosely based on Towler and Shepherd's study of French fry consumption in England (Towler and Shepherd 1992). Figure 12.9 shows that individuals have three salient beliefs about French fries related to convenience, taste, and fat content. These beliefs couple with outcome evaluations to form an individual's attitude toward eating French fries. We can see from the table that Homer has the most favorable attitude toward eating French fries. Even though Marge has the same beliefs as Homer, her attitude toward eating fries is much less favorable because she has lower evaluations of the outcome associated with each belief. Bart, on the other hand, had the same outcome evaluations as Homer but has different beliefs (perhaps due to his inability to pay attention at school), leading to an attitude toward eating French fries that is more favorable than Marge, but less so than Homer. The latter part of Figure 12.9 calculates the subjective norm associated with eating French fries from each of the consumers' standpoint. Neither Homer nor Bart are motivated to comply with what doctors or dietitians think they should do, but both want to please fast-food restaurants. The final row of Figure 12.9 shows the calculated intentions to eat French fries. Results suggest Homer most intends to eat French fries. The figure also shows that even though Bart has a more favorable attitude toward eating French fries, Marge has a greater intent to actually consume, a result that arises due to differences in social norms. It should be straightforward to see how firms can undertake data from a study like the one just described to improve marketing efforts and increase purchases.

We wrap up this discussion on attitudes by drawing a link between the material in this section and that in the preceding section. In the previous section, we introduced means-end chain analysis to show how consumers evaluate a product by determining how its attributes contribute to consequences and values. In this section, we showed how attitudes are determined by beliefs and evaluations of those beliefs. This suggests a link between values and attitudes. This link is illustrated in Figure 12.10 for one particular attribute and value associated with T-bone steaks. The evaluation of the attribute, high cholesterol in this case, is derived from the evaluations of its end consequence, heart disease and loss of security. Thus, evaluations "flow down" the mean-end chain to form attitudes about an object. Drawing the link between the means-end chain analysis and the theory of reasoned action can help firms better target products to consumers. For example, one strategy might focus on improving the evaluation of an existing strongly held belief about an attribute. Consumers likely believe that T-bone steaks

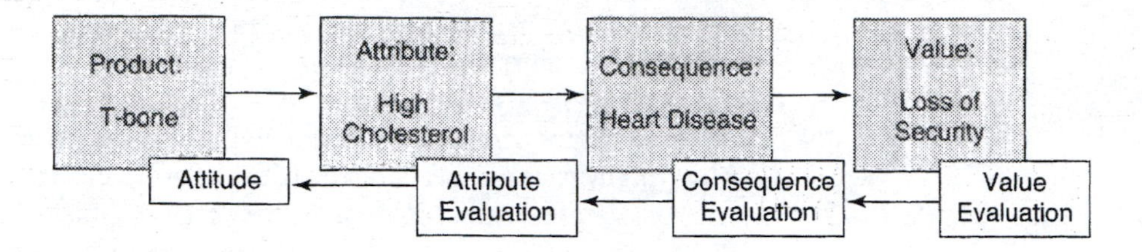

FIGURE 12.10 Link Between Attitudes and Value.

possess the attribute of high cholesterol. As shown in Figure 12.10, the evaluation of this attribute is negative and is associated with the value of lost family security. What consumers may not know is that there are good and bad types of cholesterol and T-bones from certain cattle may have more good cholesterol than bad. If such a product existed, the firm might be able to create a new means-end chain by linking the preexisting attribute of high cholesterol with the value of self-respect, and thus improving the evaluation of high cholesterol and thus attitudes toward T-bone steaks.

Consumer Decision Making

Businesses are ultimately interested in the decisions that consumer make. What brand will they choose? How many units will they buy? Which store will they visit? Each of us makes hundreds of decisions daily; some involve a great deal of thought and consideration, but others are made with virtually no thought at all. How do consumers go about making decisions? This question is the focus of this section.

A variety of theories exists regarding consumers' decision-making processes. On one extreme is the concept of a rational decision maker; this is the decision maker reflected in most economic models. Such a view assumes that consumers accurately gather information about products and accurately rank them in terms of their relatively desirability. The consumer then optimally chooses the most preferred product given budget constraints. Although such a view generates concise mathematical predictions of consumer behavior, it is a bit of a stretch to suggest that most of us approach all decisions in such a rational way. How many of us can really say for sure why we chose one alternative over another or can say that this was the *best* choice given our budget? On the other end of the spectrum is the view that decision making is entirely passive; consumers themselves have very little overt, cognitive control over what they choose. Such a view suggests consumers do not have stable preferences but instead can be easily manipulated by salesmen, external cues, and so on. There is some evidence to suggest that our subconscious mind actually governs many of the decisions over which we believe we choose. A similar view is that consumers choose primarily based on impulses that are generated by emotion. None of these theories are entirely satisfactory by themselves. In fact, there is probably at least a little truth to them all. More likely than not, consumers follow the rational approach when it is required. Consumers can and do give significant thought to many decisions. However, other decisions require little thought and we "choose" to let our choices be governed

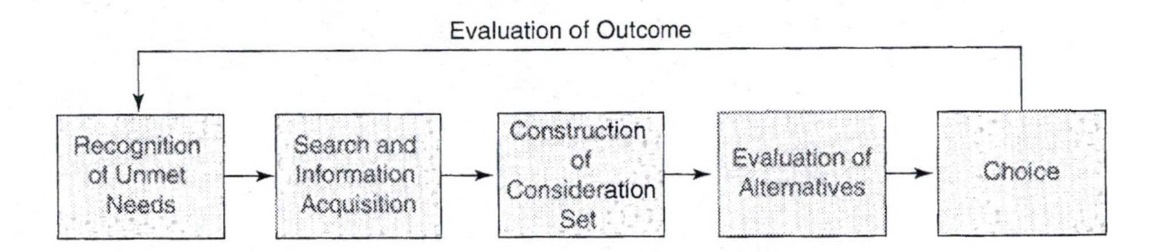

FIGURE 12.11 Decision-Making Process.

by more automated responses. It is this view to which most social scientists subscribe. Most consumers think about decisions and solve problems; however, their decisions may not be perfect or strictly optimal. Individuals likely use heuristics or shortcut decision rules to arrive at a choice rather than strict optimization.

A very simple model of the consumer decision-making process that reflects the view of the consumer as a problem solver (that can be influenced by emotions) is shown in Figure 12.11. As was previously discussed in the section on motivation, the initial step in the choice process is the recognition of an unmet need. A need is recognized when a wedge arises between a consumer's actual state and their desired state and is the result of a desire to achieve a goal or value. Once a need is recognized, the consumer begins a search process to acquire information. The information search might be as simple as recalling from memory past experiences and beliefs or might involve overt processes such as shopping online or talking with friends about products. Some psychologists mischaracterize the economists' view on search and assume that the "economic man" searches until they have complete information about all alternatives; however, work on the economics of information suggests that search is a costly activity and consumers search for additional information up to the point at which they are indifferent to the benefit that extra search would bring and the amount that the extra search would cost.

The outcome of the search process is the construction of the consideration set. The consideration set is the group of alternatives an individual considers or chooses between when making their choice. One way economists have conceptualized this issue is to investigate whether price changes of one product or product category directly affect purchases of another product or product category; if not, the products or product categories are considered separable. Figure 12.12 presents the results of one analysis that was conducted to determine the way that consumers group food purchasing decisions (Eales and Unnevehr 1988). Figure 12.12 shows that consumers apparently separate food and nonfood items when making decisions of what to consume. When considering food items, nonmeat foods are considered separately from meats. This means, for example, that a consumer's preference ranking between an apple and orange does not depend on the existence or price of beef steak. Figure 12.12 further shows that when purchasing meat, individuals make a choice between whole chicken and ground beef, but this choice does not depend on the prices of pork or beef steaks. The point here is that are numerous options available from which a consumer can choose. However, consumers do not usually consider all possible options; they create a consideration set that is comprised of a smaller number of

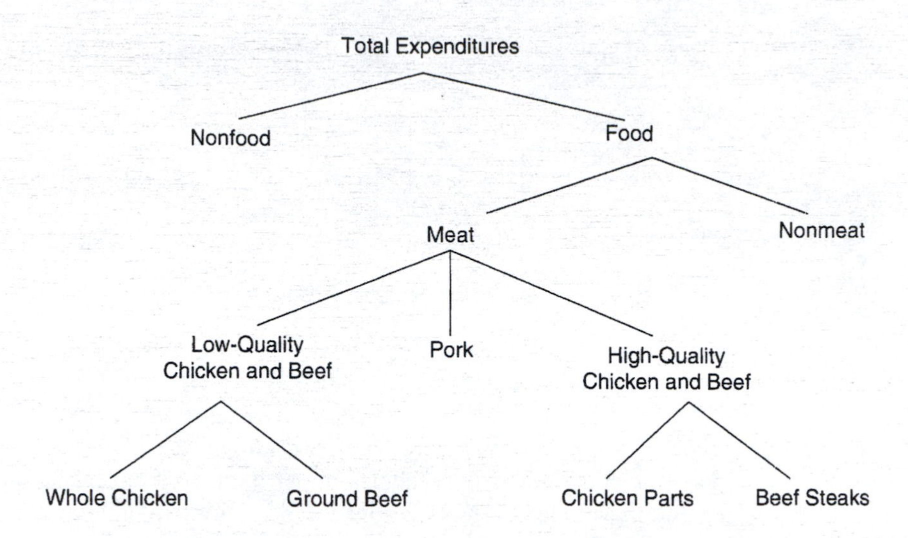

FIGURE 12.12 Structure of Consumers' Consideration Set for Meat.
Source: Eales and Unnevehr (1988).

alternatives. In general, the more knowledgeable the consumer and the more familiar they are with a product category, the larger will be their consideration set.

Once an individual has constructed a consideration set, they must evaluate the options available and determine their relative desirability. The desirability (or utility) of an option is based on the attributes the option possesses. To illustrate, Figure 12.13 shows that a consumer is considering whether to choose to eat a hamburger at

		Alternative 1: McDonald's		Alternative 2: Burger King		Alternative 3: Dairy Queen	
Attribute	Attribute Importance Weight	Belief	Desirability of Attribute Level	Belief	Desirability of Attribute Level	Belief	Desirability of Attribute Level
Price	8	$4.00	1	$4.50	0	$5.00	−1
Speed of Drive-Through	6	5 minutes	−1	3 minutes	1	4 minutes	0
Taste of Burger	10	fair	−1	excellent	1	good	0
Accuracy of Order Delivery	3	90% accurate	0	85% accurate	−1	95% accurate	1
Unweighted Utility			−1		1		0
Weighted Utility			−8		13		−5

FIGURE 12.13 Evaluation of Fast-Food Alternatives.

McDonald's, Burger King, or Dairy Queen. The consumer cares about four attributes possessed by each fast-food restaurant: price, speed of drive-through, taste of burger, and accuracy of order delivery. The consumer attaches a weight associated with the importance of each attribute. The weight for each attribute in Figure 12.13 varies on a scale of 1 to 10. For instance, a researcher asks the person how important is price when determining at which fast-food restaurant to eat, on a scale of 1 to 10, and the person responds "8." However, when posed the same question about the taste of the hamburger, they respond with the rating "10." Speed of drive-through is given a "6" and order accuracy "3." In this case, the consumer cares most about the taste of the burger followed by price. The consumer cares relatively little about ordering accuracy.

The researcher then asks the person about their belief regarding the price, speed of drive-through, taste of burger, and ordering accuracy at each fast-food outlet. For example, the consumer believes that a hamburger at McDonald's will cost $4.00, the burger will be delivered in 5 minutes, the taste will be fair, and there is a 90% chance the consumer will receive the burger they actually ordered. For each belief, the consumer associates a relative desirability of the attribute level. For example, a price of $4.00 is more desirable than a price of $4.50, which is more desirable than a price of $5.00. Thus, the $4.00 is given a desirability level of 1, the $4.50 a level of 0, and $5.00 a level of −1. This is like ranking the three prices from least to most desirable level, where a higher ranking means more desirable, except the rankings are scaled to be between −1 and 1.

Now, the ultimate question is which option does a consumer choose? Where do they buy their hamburger? There are a number of decision rules a consumer might use to make a decision. One decision rule is a *noncompensatory decision rule*, where a consumer rules out all options that have a low desirability on a particular attribute no matter how well it performs on another attribute. For example, in Figure 12.13, suppose the consumer uses a noncompensatory decision rule regarding price. In this case, they would rule out eating at Dairy Queen, no matter how good the burger tasted or how fast or accurately they delivered the food. Another type of decision rule is a *lexicographic decision rule*, where a consumer chooses the product that is best on the most important attribute no matter how poorly it performs on other attributes. Under such a decision rule, our consumer would choose to eat at Burger King because the most important attribute is taste and Burger King scores most highly on this attribute. If the consumer were indeed lexicographic, they would still choose Burger King even if the price was $10.00 or the accuracy rate was 10%. There are also *conjunctive* or *elimination-by-aspects* decision rules where a brand or attribute must meet a certain threshold to be considered. If no brands meet a threshold level on an important attribute, the purchase decision may be delayed. Finally, we consider two *compensatory decision rules*. If a consumer follows a compensatory rule, it means that if a product performs poorly on one attribute, it can compensate by performing well on another attribute, or vice versa. With an *unweighted compensatory decision rule* or *affective decision rule*, individuals simply add up the desirability of each attribute and choose the product with the highest additive desirability. In this case, the unweighted compensatory decision rule would lead to a choice of Burger King because it generated the highest utility level of 1 as compared to a utility level of 0 for Dairy Queen and −1 for McDonald's. The final rule is a *weighted compensatory decision rule*, where a consumer evaluates an alternative based on the summed product of the attribute importance weight and the desirability of the attribute. In this case, the weighted utility of Burger King is $(8 \times 0) + (6 \times 1) + (10 \times 1) + (3 \times -1) = 13$, which is higher than that for Dairy Queen, which is $(8 \times -1) + (6 \times 0) + 10 \times (0) + (3 \times 1) = -5$, or McDonald's, which is -8.

Although some evidence exists for consumers using each type of decision rule in different circumstances, the weighted compensatory decision rule is the one that is most widely used by quantitative marketing researchers to predict choice by employing a research method called conjoint analysis. This particular research method is discussed later in the chapter.

Willingness-to-Pay

Most of the models discussed thus far in the chapter have stemmed from the work of psychologists and marketers. There is an additional theoretical notion that is used by psychologists and marketers but lies primarily within the realm of economics: willingness-to-pay (WTP). If properly measured, WTP represents the *most* an individual would be willing to pay for a good or service; it is the amount of money that, when taken from a person, would make them exactly indifferent to having and not having a good. WTP is an important theoretical notion because, when aggregated across people, it is what makes up the demand curve for a good. For example, consider the individual demand curves for Cheech and Chong in Figure 12.14. At a price of \$10, Cheech will purchase one unit and Chong will purchase two. Another way of interpreting these individual demand curves is to say that Cheech's maximum willingness-to-pay (WTP) for the first unit and Chong's WTP for the second unit is \$10. Similarly, at a price of \$5, Cheech purchases three and Chong purchases eight units.

Economists are often interested in the market demand curve in addition to individual's demand curves. Suppose that Cheech and Chong are the only consumers that make up this market. As Figure 12.14 shows, at a price of \$10 a total of three units

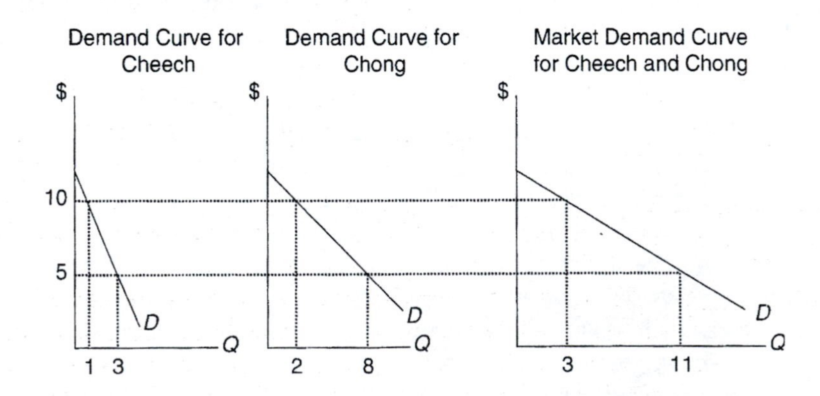

FIGURE 12.14 Market Demand Curves: The Horizontal Summation of Individual Demand Curves.

will be sold to the two consumers Cheech and Chong. At a price of $5 eleven units are sold. These two points are shown on the market demand curve. Notice that to obtain a market demand curve, we simply sum each individual demand curve horizontally. Market demand curves are the horizontal summation of individual demand curves.

Market demand curves are the horizontal summation of individual demand curves.

Firms are often interested in the demand for goods that are not regularly sold in the market. What is the demand for organic beef, environmentally friendly dishwashing liquid, or grass-fed beef? This is difficult to answer because so few stores sell these products. In these cases, one may elicit individuals' WTP for units of a good via consumer research and use these individual WTPs to construct market demand curves. This will be illustrated in a subsequent section.

Phenomena in Consumer Decision Making

A steady stream of research has accumulated in recent years suggesting that consumers exhibit a number of "biases" in decision making. Contrary to the fully rational model of consumer behavior, consumers often behave inconsistently and are influenced by factors that have little objective bearing on the choice task at hand. Economists disagree about the importance of such irrationalities in market outcomes, but at the individual level they appear to be persistent, and as such, firms interested in marketing to consumers might benefit from knowing more about the decision-making processes.

The first behavioral phenomenon we consider is *loss aversion*, which refers to the finding that people tend to value losses more highly than gains. In other words, the loss of $1 tends to hurt worse than a gain of $1 feels good. Loss aversion has a number of important implications for consumer research. For example, it suggests consumers will be more sensitive to a price increase (a loss) than they will be to a decrease in the price of a good (a gain). Loss aversion also implies that people consider the desirability of a good relative to some initial reference point and that people evaluate new goods and opportunities in terms of gains or losses from that reference. As such, research has shown a pronounced "status quo" bias. This results when people highly favor their current condition (or goods they're currently consuming) relative to any change—an effect that relates to loss aversion. This means that choices can often be manipulated by setting one option as the "status quo" and asking people to indicate whether they wish to "opt out." For example, if consumers are told they will be enrolled in health care plan X and are asked whether they would instead like to change to plan Y, research shows many fewer people will change to plan Y as compared to a case when people are simply asked to outright choose between the two health care plans X and Y. In financial markets, loss aversion will cause investors to hold onto stocks longer than what might be considered optimal because they do not want to realize a loss by selling a poor performing stock. Thus, marketers might use the concept of loss aversion to their advantage by carefully considering consumers' reference points and framing changes in products, prices, and advertising in ways that appear as gains rather than losses relative to the reference point.

A second behavioral phenomenon is *probability weighting*. Research has shown that people do not evaluate probabilities in a linear fashion, but instead, low probability events tend to influence choice more highly than what might be expected. For example, suppose someone was given the opportunity to play a lottery with a 1% chance of winning $1 million. If someone was able to play this lottery over and over again, they could expect to earn 0.01 × ($1 million) = $10,000 on average. If someone overweights low-probability events, however, they will not evaluate the probability 1% linearly and give it a weight of exactly of 0.01 but will instead give it a higher weight, say something like 0.02. Thus, rather than viewing the above lottery as yielding $10,000 on average, the person might instead evaluate the lottery as if it would yield 0.02 × ($1 million) = $20,000 on average if played numerous times. This means that consumers will tend to pay more attention to low-probability events like risks from food safety outbreaks, animal growth hormones, genetically modified food, antibiotics, and so on than what might be considered strictly "rational."

Probability weighting can also influence choices when decisions are framed in a slightly different way. For example, combining the concepts of loss aversion and probability weighting, consider the following question, "You are the manager of a food-processing firm. There is currently a 1% chance your food products will make a consumer sick and die. You can invest $1 million to buy a technology that will eliminate the risk. Will you buy the technology?" Research shows that the answers to such a question will be very different than the answer to the following question that has the identical objective outcome, "You are the manager of a food-processing firm. There is currently a 99% chance that everyone who consumes your food products will be safe. You can invest $1 million to buy a technology that will change the probability to 100%. Will you buy the technology?" More likely than not, many more people would answer "yes" they would buy the technology to the first question than they would to the second. The reason is that people will overweight the low, 1%, probability event and because the first question appears as a loss and the second question appears as a gain.

The third behavioral phenomenon is termed an *excessive choice effect*. A common assumption is that more choice is always better. For example, if a restaurant menu has 10 dinner options, it seems natural to think you can only be made better off if a new menu was introduced that had the original 10 options plus one additional option. A person is either as well off as they were before (they can choose one of the original 10 items) or is better off if the 11th option is more preferable. However, research suggests this is not always true. Several studies suggest consumers are, in some situations, less likely to make a purchase when presented with many options to choose from as compared to when they are only presented with a few options. In one example, Iyengar and Lepper (2000) report results of a study where students in an introductory psychology class were given the chance to write a two-page essay for extra credit. Students were divided in two groups. One group was offered 30 topics from which to choose to write the essay, whereas another group was offered only 6 topics from which to choose. Whereas 60% of students presented with 30 topics chose to actually write the essay, 74% of students presented with the smaller set of 6 topics actually performed the extra credit assignment. Further, students who chose from the smaller set of 6 actually scored better on their assignment than students who were able to choose from 30 topics (Iyengar and Lepper 2000). Other studies have shown that retail stores can, in some circumstances, actually increase sales when they reduce the variety of product offerings. Furthermore, Iyengar and Lepper (2000) also show that people are often more satisfied with their purchase when it was made from a smaller choice set than when a purchase was made from a larger choice set. There are a variety of hypotheses regarding why such a phenomenon exists. For example, perhaps people do not like choosing from large choice sets because they are afraid they might regret their choice later given that there were so many other options available that might have been more attractive. Another reason might be related to search costs; when more options are available, it takes more time to determine which option would be most preferred. Regardless of why the effect exists, its implications are clear: More is not always better.

A fourth concept related to consumer behavior is *choice bracketing*. Often, very little consideration is given to how individuals group their choices. In economic theory, it is often assumed that people maximize a utility function defined over all possible goods and outcomes. This view of the consumer may be too all encompassing. Choice bracketing refers to the fact that people often group individual choices together in sets. A set of choices are bracketed together when they are made by taking into account the effect of each choice on all other choices in the set, but not choices outside the set. Narrow bracketing refers to the situation when a person makes decisions from sets that are small, whereas broad brackets refer to the situation where a person makes decisions from sets that are large. The ultimate decision that is made will depend on whether a consumer brackets narrowly or broadly. For example, consider a decision of whether to eat a candy bar. Eating a candy bar may not have a great effect on your body weight. However, eating a candy bar and a rich meal and not exercising may have profound consequences for your weight over the long run. Whether or not you take all the interrelated choices into account will determine which choice will be made. If you only consider the decision of whether to eat the candy bar in isolation, it may seem like a good decision to eat. But if all the choices and their consequences are taken into account, the health consequences can be substantial and may outweigh the satisfaction of eating a single candy bar.

Firms often try to use the concept of bracketing to their advantage by framing decisions in a way that they appear broadly or narrowly bracketed. Advertising slogans such as "Beef: It's What's For Dinner," "Just Do It," and "You deserve a break today" encourage consumers to bracket narrowly, only thinking about beef, shoes, or eating fast food, whereas other slogans such as "Don't Mess with Texas" and "Friends Don't Let Friends Drive Drunk" encourage people to broadly bracket and to think about interrelated decisions of drinking and driving.

A fifth behavioral phenomenon is *time-inconsistent preferences*. All else equal, most of us prefer to have a dollar today as opposed to a dollar tomorrow. This is a concept referred to as discounting. It is often assumed that people discount the future in a constant fashion, but this is not always the case. For example, when offered the choice between $100 now and $101 a day from now, most people will choose the immediate $100; however, given the choice between $100 in one year or $101 in one year and one day, most people will choose to wait the extra day to get $101. This means people tend to discount the future less the more distant the event is to them. Thinking about delaying the satisfaction of eating a delicious hamburger and instead going on a diet today seems pretty unappealing, but going on a diet a year from now does not seem so bad.

When preferences are time-inconsistent, preferences change over time and the choice that is ultimately made depends on timing of the decision. When preferences are time-inconsistent, a person may prefer to constrain their future selves by using precommitment devices, for example, Christmas club savings accounts, payroll withholdings, or joining the army. Businesses and politicians also use precommitment devices by making large capital investments and passing law requiring a supermajority to overturn. Businesses often maliciously use time-inconsistent preferences against consumers by charging exorbitant interest rates on consumer loans.

A final phenomenon we consider here is *hypothetical bias*. Much of the research in psychology and marketing involves asking people hypothetical questions about what they would do in a given situation. Economists, however, have accumulated a wealth of evidence indicating that stated behavior in hypothetical questions deviates, often substantially, from behavior when real money is on the line. For example, it is typically found that when asked to state how much a consumer is willing to pay for a good, they state an amount about twice what they would actually pay when they must depart with their money. In other studies involving choices between lotteries, it is found that people are more averse to risk when real money is on the line as compared to when choices are simply hypothetical. There are a variety of explanations for such behavior, although none is completely satisfying. One answer is that, in a hypothetical setting, people do not suffer any cost from providing answers that deviate from their true preferences. The result is that research participants may strategically try to send a "signal" to researchers to influence future prices or offerings of products. For example, a person may say they like a product less than they really do so that a firm will reduce the price of the product, or a person may say they like a product more than they really do in order to try to convince a firm to offer a new product. In addition to strategic responses, people may simply not put much effort into making hypothetical choices because one choice has the same consequence as the next: nothing. In real-money decisions, however, people are immediately influenced by their choices and as such they need to think more carefully about their decisions. The implications of these findings are that consumer research should take seriously the potential for people to give ill-thought-out or strategic answers to hypothetical questions.

CONSUMER RESEARCH

The preceding sections presented several models of consumer behavior and discussed how the models could be used to inform marketing decisions. Although the models of consumer behavior, in addition to the traditional economic models discussed throughout the book, provide general frameworks for thinking about the consumer, they are unlikely to yield specific insight for particular problems a firm is facing. What is needed, then, is a way to make the abstract models "come to life" with information about the actual problem at hand. Ultimately the goal of marketing is to make and sell what consumers want. A primary tool in achieving that goal is to figure out what consumers want and how to get the product to them at a price they are willing to pay, all while attempting to make a profit. This is where *marketing research* is needed. Marketing research is a multifaceted concept, but at its heart it deals with collecting, analyzing, and presenting *information* to achieve marketing goals. This section will primarily focus on *consumer research,* a type of marketing research that provides information on the needs, wants, attitudes, and preferences of consumers.

Types of Consumer Research and Data

Consumer research comes in all shapes and sizes. The type of consumer research that should be undertaken depends on the particular problem at hand and many times more than one type of research is needed to provide a complete picture.

Exploratory research uses qualitative research to help define the research marketing questions. Confirmatory research uses quantitative methods to answer specific consumer marketing questions.

Exploratory Versus Confirmatory Research. When little is known about a research problem or when interest lies in developing initial insights or identifying alternatives, exploratory research is a useful tool. *Exploratory research* refers to initial research that is conducted to clarify and define a problem and to provide insight to be used in subsequent study. Typically, exploratory research is qualitative—meaning that it cannot be used to statistically test hypotheses about the general population. When exploratory research is conducted directly with consumers, researchers often use methods such as focus groups, in-depth interviews, and projective techniques. In a focus group, a moderator leads a discussion typically among about 10 consumers. The moderator poses questions to the group and provides some structure for the overall conversation; however, the structure is very loose and the purpose is to let ideas and issues be freely discussed. In an in-depth interview, a monitor typically spends 30 to 60 minutes one-on-one with a consumer and asks detailed questions. In-depth interviews are used to construct the means-end chains discussed in the previous section. Projective techniques refer to qualitative research methods that are generally conducted one-on-one with consumers, but where the purpose of the research is not made clear to subjects. Projective techniques use a stimulus to get consumers to project their attitudes or beliefs onto an ambiguous situation. Examples of the technique include word associations, sentence and story completion, cartoon completion tasks, and role-playing. In sum, exploratory research is a very flexible research method that aims to provide structure to future investigations. Often, the results of exploratory research, although potentially providing insights into "why" and in narrowing alternatives, are not directly useful in decision making.

If a decision maker needs to answer questions such as "how many," "which one," "how often," or "how much," they are in need of *confirmatory research.* Confirmatory research, also referred to as conclusive research, aims to provide a specific answer to a specific research problem. Confirmatory research begins with a well-defined objective and attempts to draw generalizable conclusions to be used by decision makers. Confirmatory research is most often quantitative and utilizes mathematical models and statistical techniques to draw inferences. There are countless types of quantitative research techniques; in a preceding chapter we discussed regression analysis and later in this chapter, we will discuss other quantitative research techniques used in consumer research. Some confirmatory research is descriptive in nature, meaning the purpose is simply to describe the composition or characteristics of a group of individuals and perhaps to identify the interrelationship between variables. Some confirmatory research is experimental or causal and seeks to identify cause-and-effect relationships. Causal research attempts to hold all other factors constant (either through statistical regression analysis or through planned experiment) manipulate a potential causal variable, x, and observe the effect on the variable of interest, y.

Primary Versus Secondary Data. Regardless of whether the research is exploratory or confirmatory, data are needed in order to draw inferences. Two general types of data exist. The first is *primary data*. Primary data refer to those data collected firsthand by the researcher by original research designed to answer a specific question. Data collected using some of the qualitative data collection methods previously discussed such as focus groups and in-depth interviews would generate primary data. Other quantitative research methods such as surveys, laboratory experiments, and in-store experiments can also be used to generate primary data. Another type of data is *secondary data*, which refer to information that is typically collected for another purpose, that is already available, and that is often in aggregate form. Private sources of secondary data include scanner data from grocery stores and firms like ACNielsen and sales and trade records of firms. Most secondary data come from public sources including U.S. government agencies such as the Department of Agriculture, Department of Commerce, Bureau of Labor Statistics, and from international agencies such as the World Bank and the Organization for Economic and Co-operation Development (OECD).

The main advantage of primary data is that the researcher can ensure that the collected data are able to adequately answer the question of interest. However, such data are often time-intensive to obtain and expensive to collect. Secondary data, on the other had, are often freely available and can be rapidly used and analyzed. The primary disadvantages of secondary data are that they may not yield ideal tests of the hypotheses of interest and often come in aggregated form, making it difficult to uncover consumer heterogeneity, a key determinant of creating profitable segmentation strategies.

Stated Versus Revealed Preference Research. Another useful distinction is between stated and revealed preference research. Stated preference research refers to those methods that ask individuals, hypothetically, what they would do in a given situation. Stated preference data are most often obtained from mail and phone surveys by asking hypothetical willingness-to-pay or purchase intention questions. Revealed preference research, on the other hand, analyzes what people actually did in a situation that had economic consequences. Revealed preference data can be obtained in mail and phone surveys by asking people to indicate products previously purchased, but such data are also obtained in economic experiments, through observational techniques, and from secondary data sources recording prices and quantities. Both research methods are useful in certain situations and both have their limitations. A primary advantage of stated preference data are that consumers can be asked to evaluate any potential problem or situation—even products that have not actually been developed in situations that have never occurred. This means that stated preference research is very flexible in the types of preferences that can be measured. Another advantage of this approach is that stated preference data can be relatively easy to obtain from a large number of consumers. The primary drawback to stated preference data is that they, as the name implies, are stated. That is, a consumer can give any answer to such a question and suffer no adverse consequence. Of a larger concern, however, is that consumers might answer questions in such a way as to try to benefit themselves later.

For example, suppose a person thinks by saying they are not likely to purchase a new good that they can make a firm believe the good is not very valuable and will therefore sell it at a low price. Clearly, such a person has a strong incentive to understate their preferences for the good. In contrast, suppose a person believes their answer to a survey question might influence whether a company decides to sell a new good. In this situation, a person has a strong incentive to overstate their preferences for the good in order to ensure the chance that they can buy it at a later date. Indeed, a significant amount of research has shown that estimates of willingness-to-pay for a good are drastically overstated in hypothetical questions as compared to purchase questions when real money and real products are on the line (Cummings, Harrison, and Rutström 1995). Thus, the primary advantage of revealed preference data are that they reflect what people actually did; they represent real choices with real money and real goods. A drawback to revealed preference data can be that there is often little variability in variables of interest in the real world. For example, the price of Coke and Pepsi are always priced the same in the vending machine, making it impossible to determine relative price sensitivity of the two drinks. Further, revealed preference data are limited to goods that are actually sold. One exception to this statement is economic experiments such as experimental auctions, in which consumers make nonhypothetical choices or bids in a constructed market where new goods are available.

Steps in Consumer Research

Steps in Consumer Research:
(1) Define the problem
(2) Determine the research design
(3) Analyze and interpret data
(4) Present results

Successful consumer research should be meticulously planned. One does not arrive at useful information by accident. This section suggests several steps to follow when conducting consumer research. Although the steps are numbered sequentially, often researchers will have to back up a step or two before they can achieve their final goal. Research is a dynamic process that often requires trying many ideas until one is successful. Despite this, research should be systematic and should be directed toward achieving a goal.

Step 1: Define the Problem

The first, and perhaps most important, step is to define the problem. In order for research results to have a meaningful impact, they must address a particular problem. An example of problem statements in marketing might be, "Sales of our product have been decreasing for the past year. Should we change the product? How?" Another example is, "If we increase the price of product, what will happen to sales and profit?" Once a problem has been formulated, specific research objectives should be stated. Stated objectives identify what the future research will accomplish and what information it will provide. Research objectives should be phrased in such a way that, at the conclusion of a research project, it is easy to verify whether the objective has, in fact, been accomplished. Examples of a research objective might be "to provide a quantitative estimate of the own- and cross-price elasticities of demand for our product" or "to determine whether adding attribute X to our product will increase sales."

Step 2: Determine the Research Design

There may be several ways to accomplish the same objective. For example, an elasticity of demand could be estimated through analysis of aggregate time-series data or through analysis of stated choice data obtained in a survey. However, in many cases a research objective will point to a specific research design. First, researchers should consider whether the research is exploratory or confirmatory. Once this issue has been settled, the next step is to investigate whether secondary data are readily available that could be used to address the issue at hand. If secondary data are not available or too costly or ill suited to provide an adequate answer to the research question, this would indicate that a primary data collection method would need to be pursued.

If a primary data collection approach is pursued, one must decide on a data collection method: a mail survey, a phone survey, a focus group, a laboratory experiment, or an in-store experiment. Often the research objective will provide some indication of which method is most appropriate. Other times, expert opinion might be sought to help guide the choice. At times, researchers might also decide to pursue more than one approach to determine how sensitive the research findings are to the chosen approach. Once a data collection method has been chosen, the data collection instrument must be developed. For focus groups or experimental approaches, instructions must be written. For mail, phone, and intercept surveys, questions need to be crafted.

Creating an effective survey means paying careful attention to: (1) the phrasing of the questions, (2) the response categories for each survey question, (3) the order of questions in the survey, and (4) the appearance, format, and length of the survey. Surveys should begin with an interesting question that everyone should answer. Sensitive questions such as household income and age should be placed at the end of a survey. When writing survey questions, care should be taken to ask questions for which people do not have a "ready-made answer" but to ask questions to obtain as much information as possible. For example, most people cannot accurately answer questions like "how many times did you eat beef last month" but they can probably answer a question like "do you normally eat beef: (a) never, (b) about once a month, (c) about once a week, or (d) several times a week." In the case of mail and intercept surveys, the survey needs to be visually appealing using uniform font size, spacing, numbering, and response categories to the greatest extent possible.

When conducting a survey, thought must be given about *who* to ask to respond to the survey. Typically, interest lies in estimating some characteristic of a population of individuals. A population could be U.S. citizens, U.S. farmers, U.S. agribusiness people, residents of Oklahoma, and so forth. Regardless of the population, it is likely infeasible to survey everyone in the population. Thus, one obtains a sample from the population. A variety of sampling methods exist. Probably the most common method is random sampling where every individual in the population has the same chance of being included in the sample. One way a random sample is obtained from the U.S. population is by a technique referred to as random digit dialing, where a phone number is randomly created and called. (Note: This approach assumes every individual in the population can be reached by a phone.) Another common sampling approach is called stratified sampling, where some people have a higher probability of being included in the sample than others. For example, perhaps a research question deals primarily with the behavior of high-income farmers; then one might want to "over-sample" people with this characteristic. A stratified random sample might be a useful approach when a researcher wants greater statistical precision with some groups of the population than others or if there is concern that too few people from certain populations will end up in the final sample using pure random sampling techniques.

Because only a portion of the population is surveyed, researchers need to be concerned about *sampling error*. Sampling error is a statistical property directly related to the number of people surveyed. The larger the number of people surveyed, the smaller the sampling error. The next time you see the results of a poll reported on the nightly news, you'll probably see something like "34% of surveyed individuals said they . . ." and in fine print a "margin of error" is often reported. This margin of error is the sampling error. If the sampling error is ±3%, it means the actual percentage of the people in the population who said they believe whatever was asked could be 3% higher or lower than the reported value, typically with 95% confidence.

A major concern with any primary data collection method is nonresponse bias. Nonresponse bias occurs if the characteristics of the individuals that respond to the survey are different than the characteristics of individuals who choose not to respond. For example, suppose a survey was conducted on consumer attitudes toward genetically modified food. More likely than not, people who have strong opinions (both pro and con) are the most likely to respond to the survey; however, these may not be the people of most interest because they have already made up their minds on the issue. As another example, suppose you were interested in conducting a phone survey to determine the attitudes of students on your campus toward binge drinking. Suppose the only list of phone numbers you could obtain were from fraternity membership lists. Would the results of your survey provide an accurate depiction of all students' views on binge drinking? If your answer is no, you understand the consequence of nonresponse bias.

The potential for nonresponse is minimized when (a) the sample is randomly chosen from the population of interest and (b) a large percentage of the sample actually responds to the survey. In the context of mail surveys, a number of steps can be taken to increase response rates including: (1) creating an attractive, well-written, and easy-to-complete survey, (2) sending a pre-letter indicating the importance of the study and the need for a response, (3) sending the survey with a "gift" such as a $1 bill, (4) sending postcard reminders a few days after the initial mailing, (5) resending surveys in a week or two to those who did not respond to the first mailing, and (6) after another few weeks, calling nonresponders or sending them the survey via FedEx or UPS (Dillman 2000). It has been argued that this approach can generate response rates in excess of 75%, meaning three out of every four people sent a survey will fill it out and return it; however, the authors' experiences suggest that it is difficult to achieve response rates higher than 25% for random samples of the U.S. population even if many of the above steps are followed. In cases where low response rates are obtained, it is important to compare the characteristics of the respondents with known characteristics of the population, which may be available from the U.S. Census Bureau. To the extent differences exist, weights can be used to force your sample of respondents to "act" like the population. For example, suppose your sample of respondents consists of more women than men even though we know each group actually represents about 50% of the population. One way to correct for this problem is, when calculating statistics such as the mean response to a survey question with N respondents, to assign a weight to each male respondent higher than $1/N$ and a weight to woman less than $1/N$.

Step 3: Analyze and Interpret Data

Hopefully, when the study began, hypotheses were formulated and models were outlined. Data analysis should focus on testing the original hypotheses of interest and on estimating the proposed models. If a study is well designed, interpretation of the results should be straightforward because the issue was already given a great deal of thought when formulating questions and in hypothesizing. One of the first steps in data analysis, at least for primary data collection, is to turn survey and experimental responses into quantitative information. Data may have been collected to questions with response categories such as "yes," "no" or "male," "female." Although such data are qualitative in nature, they must be turned into quantitative information for statistical analysis, for example, by coding all "yes's" a 1 and all "no's" a 0. A useful second step in any data analysis is to investigate summary statistics of the variables, including such statistics as the mean, median, and standard deviation. Often simple analysis such as creating a histogram to investigate the distribution of responses for a variable or calculating a cross-tab between variables can yield useful insights. More advanced analyses might calculate correlation coefficients between variables or employ statistical procedures such as multiple regression analysis, factor analysis, and cluster analysis.

Step 4: Present Results

No matter how well done the research, it is of little use if results cannot be communicated to decision makers in an effective manner. Often research results are most effectively communicated in charts and graphs. Care should be taken not only to present results but to draw out implications of the results. Finally, limitations of the research should be mentioned. Frequently, the results of a consumer research study will be included in the marketing plan section of a business plan.

METHODS FOR MEASURING VALUE-CHAINS, ATTITUDES, PREFERENCES, AND WILLINGNESS-TO-PAY WITH APPLICATION TO GENETICALLY MODIFIED FOOD

In the first part of this chapter, we outlined several models of consumer behavior; however, the discussion was somewhat abstract. In this section, we discuss a variety of methods to put the models in action. To help illustrate the issues at hand, the methods are discussed in the context of genetically modified food. We focus on genetically modified (GM) food because this is an area where a great deal of consumer research has been conducted; however, it should be clear that the methods can be applied in any context.

Constructing Means-End or Value Chains

Earlier in the chapter, it was discussed that a means-end chain is a theoretical approach that links concrete physical attributes of a product to consequences and then to values. This approach views consumer purchases as a *means to an end*; that is, consumers buy a product because it contains attributes that generate certain desirable consequences that relate to important values to the consumer. Constructing a means-end chain is useful because it helps to understand why someone does or does not buy a product.

The most common method of constructing a means-end chain is called laddering. Laddering is a quasi-structured qualitative interview technique used for linking product attributes to consequences and to values. To illustrate the method, consider the yogurt study conducted by Bredahl (1999) in which interviews were conducted with 50 people in each of four countries: Denmark, Germany, the United Kingdom (UK), and Italy. Initially, participants were "prompted" to think about the attributes that were important in choosing between yogurt varieties. This was done by asking participants to rank four different products according to their preference. The yogurt options differed by fat content, texture, taste, use of additives, and use of GM starter culture. After the products were ranked, participants were asked to give reasons for their ranking. Typically, the answers went something like "this product has attribute *X* whereas the other product had attribute *Y*." Thus, the ranking procedure forced individuals to state salient physical attributes important in choosing between yogurts. Once a set of attributes was elicited for each person, the interviewer took one attribute at a time and asked questions like "why is that important to you?" After an answer was given, participants were, again, asked similar questions to encourage them to give more and more abstract explanations until they could go no further or until they provided an explanation that resembled one of the values in Figure 12.3. Through this series of questioning, the interviewer was able to construct a chain from concrete attributes to functional and psychosocial consequences to instrumental and terminal values. Sometimes in the initial questioning about why one product was ranked above another, a person will provide a more abstract answer that relates to a value or consequence. In that case, reverse laddering was used to figure out the concrete attributes associated with that value or consequence.

One of the challenges in analyzing the data from the laddering technique is that a large number of means-end chains can be constructed. In Bredahl's study, for example, over 640 individual chains were constructed from the interviews with the 50 Danish participants. This means each Danish person constructed about 12.8 chains. The challenge is to summarize the data in a way to glean meaningful insight. The approach typically taken is to try to find the most common chains reported across people and put these in a chart or graph. Bredahl (1999) reported all chains that were mentioned by at least four people.

The most common concrete attributes reported by participants in the yogurt study included additives, fat content, and use of GM. Figures 12.15 and 12.16 show the most prominent chains associated with the attribute GM in Denmark and the United Kingdom. Again, the only chains shown are those reported by at least four people.

There are pronounced differences in the two locations. In particular, the Danish consumers appear to have more elaborate chains, suggesting they are more involved in this food product category and have given more thought to the issue of GM. Although Danish and UK consumers perceived roughly the same amount of consequences with GM, in the United Kingdom, the consequences did not generally follow up the ladder to the value level. In both locations, people associated the attribute of

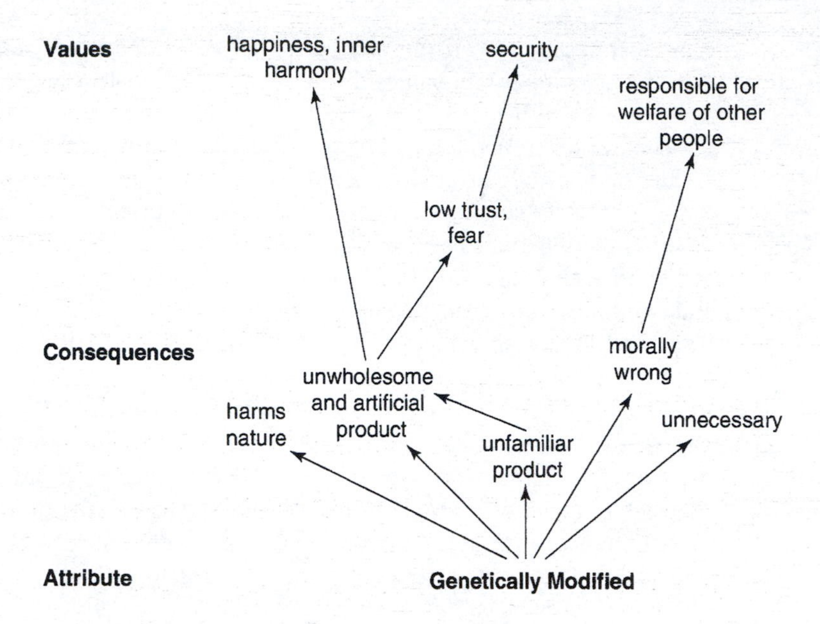

FIGURE 12.15 Means-End Chain for Genetically Modified Yogurt in Denmark.
Source: Bredahl (1999).

GM with the consequences of unfamiliar, unwholesome and artificial, and unnecessary. In Denmark, many of these consequences were *negatively* associated with more abstract values and life goals like happiness, security, and welfare of other people.

The results of such means-end chains analysis can be useful for a variety of purposes. For example, results suggest that even though many Danish consumers considered GM morally wrong, the same was not necessarily true of the British. This finding coupled

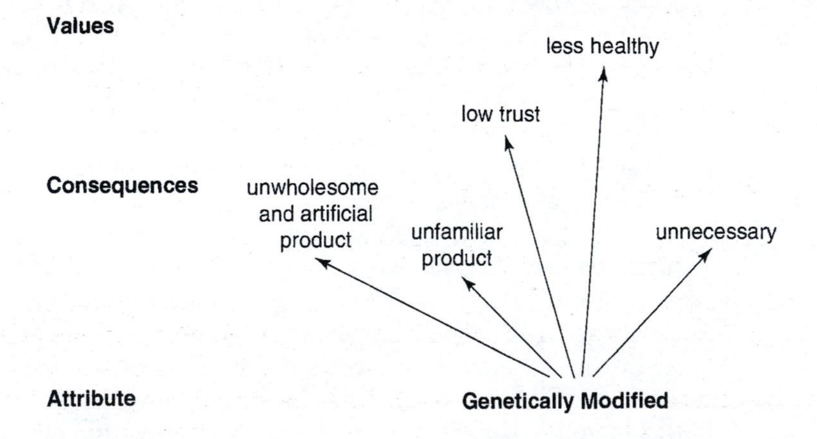

FIGURE 12.16 Means-End Chain for Genetically Modified Yogurt in the United Kingdom.
Source: Bredahl (1999).

with the fact that British consumers did not associate GM with broader life values perhaps suggests they might be more open to the technology and that advertising about the benefits of GM could have a larger impact in the United Kingdom than in Denmark.

Measuring Attitudes. Attitudes are simply someone's opinion (either favorable or unfavorable) about an object or concept. Companies and politicians are constantly measuring attitudes to determine if their strategies are on target. Measuring attitudes can be a relatively straightforward concept. There are generally two methods for measuring attitudes: single-item measures and multiple-item measures. Single-item measures, as the name implies, consist of a single question that is used to represent a person's attitude toward and object or activity. Multiple-item measures, on the other hand, use responses to several questions, which are typically averaged or summed, to represent a person's attitude. There are several motivations for using multiple-item measures such as: (a) the approach allows researchers to test for *reliability* of their measurement device—that is, whether similar results are obtained from repeated measures; (b) attitudes are abstract concepts and it may take several measures to provide a good "picture" of the attitude; and (c) attitudes might be multidimensional and by asking multiple questions, it can be determined whether several different attitudes exist rather than just one.

Regardless of the number of items to include in a scale, there are different types of measurement scales: nominal, ordinal, interval, and ratio. A nominal scale is one in which the measurement is simply an identifier, for example, male or female, yes or no, user or nonuser. For example, the following is a commonly used single-item, nominal scale frequently used to measure attitudes toward the performance of politically elected official, "In general, do you approve or disapprove of the job that George W. Bush is doing as president? Yes, no, or uncertain?" Although such scales are easy for people to answer and for the analyst to summarize, they do not indicate *how much* someone approves or disapproves of the job performance of the president; responses also do not give an indication of job approval compared to other past presidents or elected officials. An ordinal scale is one in which people are asked to rank products/people/services. Keeping with the above example, a single-item, ordinal scale might ask: "Please rank the following people in terms of your approval of the job they did as president: George W. Bush, Bill Clinton, and Ronald Reagan." Ordinal questions provide information on, in this case, job approval, but they do not indicate how much more someone approves of Clinton than Reagan or vice versa. Interval and ratio scales provide such information. An interval scale allows a respondent to indicate how much they prefer/approve/like/agree on a scale typically from 1 to 5 or 1 to 7. For example, another commonly used single-item interval scale used in political polling is the following, "I'd like you to rate your feelings toward George W. Bush as either: very positive, somewhat positive, neutral, somewhat negative, or very negative." Although an interval scale provides information on *how much*, the results cannot be strictly interpreted as saying for example, that "very positive" is twice as positive as "somewhat positive." When such information is needed, a ratio scale can be used. A ratio scale possesses a natural or absolute zero, where there is universal agreement about location (e.g., consider measures of height or weight, where we can legitimately say 200 lbs. is twice as heavy as 100 lbs.).

An example of a single-item ratio scale applied to the political context might be, "Please allocate 100 points to the following people in terms of your approval of the job they did as president: George W. Bush, Bill Clinton, and Ronald Reagan."

Interval scales are probably the most widely used in measuring attitudes. Interval measures are obtained using either Likert or semantic differential scales. In a Likert scale, respondents are typically asked to indicate the degree of agreement/disagreement with a statement on a scale of, say, 1 to 5 where 1 represents strongly disagree and 5 represents strongly agree. In the case of multiple-item measurement, responses to several Likert scale questions are often summed together to represent the measurement of an attitude. A semantic differential scale is often used to measure the meaning of an idea or product, where respondents are asked to indicate their opinion between two polar adjectives. For example, a semantic differential scale might ask people to indicate the extent to which they believe a product is reliable or unreliable (say, on a scale of 1 to 7 where 1 is reliable and 7 is unreliable) or good or bad (say, on a −3 to 3 where −3 is bad, 0 is neither good nor bad, and 3 is good).

Many of these concepts were applied to the concept of GM foods in a study by Saba and Vassallo (2002), who conducted in-person surveys with over 1,000 Italians. The purpose of their study was to measure Italians' attitudes toward eating GM tomatoes and to investigate factors affecting attitudes. They used two questions (e.g., a two-item measure) to measure this attitude using an interval, semantic differential scale. The questions they asked were as follows:

1. My eating of tomatoes produced through gene technology will be . . .

   Extremely Good — Extremely Bad

   1 2 3 4 5 6 7

2. My eating of tomatoes produced through gene technology will be . . .

   Extremely Beneficial — Extremely Harmful

   1 2 3 4 5 6 7

Many Italians had a very unfavorable attitude toward eating GM food. For example, 39.4% answered with a 6 or 7 to question 1 above and 36.4% answered with a 6 or 7 to question 2. Only about 6% of respondents responded with a 1 or a 2 to either one of the above questions. In subsequent analysis, each individual's response to these two questions was summed together and this calculation was interpreted as a measure of their attitude toward eating GM tomatoes.

After establishing that attitudes were generally unfavorable, Saba and Vassallo (2002) were interested in investigating the determinants of such attitudes. They turned to Fishbein and Ajzen's *theory of reasoned action* (1975) discussed earlier in the chapter. In that model, attitudes toward eating GM tomatoes are expected to be influenced by the product of beliefs about eating GM tomatoes and the evaluations of those beliefs. Their model also postulates that attitudes and subjective norms will influence intentions to purchase/eat GM food. Saba and Vassallo (2002) constructed multiple-item measures of beliefs about GM tomatoes and segregated them into positive and negative issues as shown in Figure 12.17. Respondents then

Belief Questions		Extremely Unlikely						Extremely Likely
Positive Issues								
A	How likely do you think that the application of genetic engineering in the production of tomatoes will result in increased food production?	1	2	3	4	5	6	7
B	How likely do you think that the application of genetic engineering in the production of tomatoes will result in possibilities of production in several environments (e.g., arid lands)?	1	2	3	4	5	6	7
C	How likely do you think that the application of genetic engineering in the production of tomatoes will result in reduced use of pesticides?	1	2	3	4	5	6	7
Negative Issues								
D	How likely do you think that the application of genetic engineering in the production of tomatoes will result in ecological damage?	1	2	3	4	5	6	7
E	How likely do you think that the application of genetic engineering in the production of tomatoes will result in reduction of biodiversity?	1	2	3	4	5	6	7
F	How likely do you think that the application of genetic engineering in the production of tomatoes will result in an increase in food allergies?	1	2	3	4	5	6	7

FIGURE 12.17 Questions Used to Measure Italians' Beliefs About Eating Genetically Modified Tomatoes.

answered the questions shown in Figure 12.18, which provided the evaluation of those beliefs. Thus, a measure of attitude toward eating GM tomatoes using the positive issues is constructed by taking the sum-product of the responses to questions in Figures 12.17 and 12.18:

$$\textit{Positive Beliefs} \times \textit{Evaluations} = A \times EA + B \times EB + C \times EC$$

Similarly, a measure of attitude toward eating GM tomatoes using the negative issues is constructed by taking the sum-product of the responses to questions in Figures 12.17 and 12.18:

$$\textit{Negative Beliefs} \times \textit{Evaluations} = D \times ED + E \times EE + F \times EF$$

Behavioral intentions were measured by asking people to respond to the statement, "I intend to eat tomatoes produced by gene technology in the future" on a seven-point scale from extremely unlikely to extremely likely. Finally, subjective norms were measured by asking, "What do you think the attitude of persons important to

Belief Evaluation Questions		Strongly Disagree						Strongly Agree
Positive Issues								
EA	I think that increased food production is desirable.	1	2	3	4	5	6	7
EB	I think that possibilities of production in several environments (e.g., arid lands) is desirable.	1	2	3	4	5	6	7
EC	I think that reduced use of pesticides is desirable.	1	2	3	4	5	6	7
Negative Issues								
ED	I think that ecological damage is desirable.	1	2	3	4	5	6	7
EE	I think that reduction of biodiversity is desirable.	1	2	3	4	5	6	7
EF	I think that an increase in food allergies is desirable.	1	2	3	4	5	6	7

FIGURE 12.18 Questions Used to Measure Italians' Evaluations of Beliefs About Eating Genetically Modified Tomatoes

you would be toward you eating tomatoes produced by gene technology in the future" on a seven-point scale ranging from extremely unfavorable to extremely favorable.

Results of the study indicated that the theory of reasoned action provides a reasonably good depiction of attitudes and behavioral intentions. The correlation coefficient between attitudes, as measured by questions 1 and 2 above, and behavioral intention was 0.54, meaning those with more positive attitudes toward eating GM tomatoes were more likely to intend to eat GM tomatoes. The model also predicts that social norms will influence purchase intentions. The study results confirm his prediction as the correlation coefficient between intentions to eat and social norms was 0.55. Finally, the theory of reasoned action suggests that attitudes are influenced by beliefs and evaluations of beliefs. Saba and Vassallo (2002) found that the correlation between (Positive Beliefs × Evaluations) and attitudes toward eating GM tomatoes, as measured by questions 1 and 2 above, was 0.43 and the correlation between (Negative Beliefs × Evaluations) and attitudes toward eating GM tomatoes, as measured by questions 1 and 2 above, was 0.23. This means that to change attitudes toward eating GM food one might focus on: (a) changing peoples' current beliefs about eating GM food, focusing primarily on positive beliefs as these appear to have the largest impact in this study, (b) adding additional positive or negative beliefs not present in Figure 12.17, (c) change the evaluation of beliefs shown in Figure 12.18, or (d) affect social norms by altering the overall social image of GM.

Measuring Demand and Willingness-to-Pay. In addition to measuring more abstract concepts such as values and attitudes, firms will be interested in more concrete figures like consumer willingness-to-pay (WTP) for a product or a demand curve showing the amount that will be sold at a given price level. Methods for measuring demand and WTP from secondary time-series data typically involve econometric regression models; a topic that is widely covered in many textbooks and discussed in Chapter 7 of this book. Here we focus on several methods for measuring WTP and demand using primary data collection techniques.

This topic is becoming increasingly important in agriculture because many producer groups and agribusinesses are interested in "adding value" to their products by differentiating generic agricultural commodities or developing alternative products or services with new technologies. Considering the fact that tens of thousands of new food products are introduced annually with success rates often as low as 10%, market research into the viability of new products and services is critical.

Depending on how the question is asked, WTP is the maximum amount of money a person is willing to give up to either: (a) obtain a product outright or (b) exchange a product with one type of characteristic (e.g., non-GM) with an otherwise identical product with another characteristic (e.g., GM). Practically, how can agribusiness use these measures? Agribusinesses will typically be interested in the distribution of WTP in a particular market. In particular, these measures can be used to construct inverse demand curves. Consider a study that elicits WTP for one unit of a novel good. It can be assumed that if the price is less than this WTP, the person will choose to buy. Thus, the market demand can be estimated by determining the number (or percentage) of people with WTP higher than a given price. An entire demand curve can be estimated by plotting the number (or percentage) of people with WTP higher than several increasing price levels. To construct such demand curves, it is often assumed a person will only buy one unit of the good during the time period of interest; however, the analysis could be extended by asking people how much they are willing to pay for one unit, two units, and so on.

To measure WTP for a product or service, many methods are available. Probably the simplest method is to ask people to indicate the most they are willing to pay for a good. This is referred to as an *open-ended contingent valuation question*. An example of an open-ended contingent valuation question is, "Assuming the price of a typical bag of corn chips that contains GM ingredients was $2.00, what is the most you would be willing to pay for an otherwise identical bag of corn chips that was non-GM? $______." This approach is valuable in the sense that it provides an exact WTP value for each person. However, research indicates people often have a hard time stating an exact WTP amount and often they do not report what they are willing to pay, but instead report what they think the good costs or what they think the price should be. This question format is also open to strategic responses by individuals. For example, a person might state a WTP amount much lower than they really want to pay so that a company might lower the price of a good. Alternatively, for a new good, a person might say they are willing to pay much more than they really are so that the company will actually introduce the new good and the consumer can decide later if they really want to buy it. Because of these concerns, the open-ended contingent valuation question format is rarely used in economic research.

In contingent valuation, individuals are asked how much they would pay for a good or whether they would purchase a good at a particular price.

Instead, researchers often use discrete-choice questions such as a *dichotomous choice contingent valuation question*. In such a question, people are asked whether they would purchase a product if the price was $X; or they might be asked which product they would buy, if any, if product *X*, *Y*, and *Z* were prided at $X, $Y, and $Z, respectively. For example, consider the following question asked by Lusk (2004) in a survey of Mississippians.

> Imagine you are purchasing rice in your local grocery store. You can choose between two types of rice. One is regular long-grain white rice that has not been genetically engineered. This non-genetically engineered rice does not contain vitamin A. The other rice option is Golden Rice. Golden Rice has been genetically engineered to contain vitamin A. One serving of Golden Rice will satisfy 30% of your daily requirement for vitamin A, as outlined by the FDA. Now, imagine that you are in a grocery store and the price of a 1 lb. bag of regular long-grain white rice is $0.75. Would you purchase a 1 lb. bag of long-grain Golden Rice if it cost $0.75? Yes or No?

Lusk (2004) found that about 60% to 85% of survey respondents said "yes" to this survey question, depending on the survey version employed—meaning that 60% to 85% of people were willing to pay at least $0.75/lb. for Golden Rice. In most dichotomous choice contingent valuation questions, different people are asked whether they would buy the product at slightly different dollar amounts. For example, Lusk (2004) sent several different versions of the survey out so that people randomly received a survey asking if they would buy Golden Rice at a price of $0.55, $0.65, $0.75, or $0.85. As might be expected, fewer people indicated they were willing to buy Golden Rice as the price increased. In one version of his study, Lusk (2004) showed that when the price of Golden Rice was $0.65, 85% said they would buy, but when the price was $0.85, only 62% said they would buy. Thus, even at a $0.15 premium over regular white rice, most people said they would buy the GM product that had been modified to have higher amounts of vitamin A.

Although contingent valuation questions provide useful information, they are limited because they only provide information on someone's WTP for one good or one characteristic of a good. Firms are often interested in someone is willing to pay for one attribute relative to several others. In such cases, researchers often use *conjoint analysis*. Conjoint analysis refers to a technique where consumers either rate, rank, or choose between products that are described by several attributes.

Conjoint analysis asks individuals to compare goods with different attributes by rating the goods, ranking the goods, or choosing between the goods.

A conjoint study typically begins by identifying the attributes that are important to a consumer when choosing which product to buy. Attributes include things like price, package size, ingredients, production methods, and so on. Once one has established the important attributes, then the researcher decides on the number of levels each attribute is to be varied across in the survey. For example, price might be varied between the levels of $1.00 and $2.00 and package size might be varied between the levels of large and small. Then based on these attributes and attribute levels, researchers use an experimental design to create product profiles or product descriptions to show to research participants. For example, one profile might be a large product at $2.00 and another might be a small product at $1.00. Consumers then rate or rank the product profiles in terms of their desirability or they choose which product they would purchase. Finally, regression analysis is used to determine the relative importance of the study attributes and consumers' WTP for each attribute.

To illustrate, consider the study of Baker and Burnham (2001) based on 383 responses to a mail survey administered in the United States. Baker and Burnham (2001) selected cornflakes cereal as their product of analysis and through preliminary research determined to study the attributes of: price, brand, and GM content of the corn. The attribute of price was varied between the levels of $2.75, $3.50, and $4.25, the brand was varied between the levels of Kellogg's and store brand, and the GM content was varied between the levels of GM and non-GM. Given these attributes and attribute levels, there are 12 possible boxes of cornflakes that could be created as shown in Figure 12.19. For example, the first product is a box of Kellogg's cornflakes with GM corn that costs $2.75 and the last product is a box of store brand cornflakes with GM corn that costs $4.25.

Baker and Burnham (2001) requested people to rate each of the 12 products shown in Figure 12.17 on a scale of 1 to 10 where 1 was very undesirable and 10 was very desirable. These ratings or rankings are often interpreted as a measure of a person's utility or satisfaction of an option. Using these rankings, they were able to estimate a regression model (see Chapter 7 for information about regression) for each survey respondent. Their analysis shows that the average model was

$$\begin{aligned} \textit{Utility of option} &\approx \textit{Rating of option} \\ &\approx 10.06 + 1.68 \times \text{Kellogg's} - 1.42 \times \text{Price} - 1.98 \times \text{GM} \end{aligned}$$

The variable "Kellogg's" equals one if the cereal has the Kellogg brand and zero if the store brand. The variable "Price" is simply the price of the cereal, and "GM" equals one if the cereal contains genetically modified ingredients and zero otherwise.

Product Profile	Price ($/18-oz box)	Brand	Type of Corn
1	$2.75	Kellogg's	GM
2	$2.75	Kellogg's	Non-GM
3	$2.75	Store	GM
4	$2.75	Store	Non-GM
5	$3.50	Kellogg's	GM
6	$3.50	Kellogg's	Non-GM
7	$3.50	Store	GM
8	$3.50	Store	Non-GM
9	$4.25	Kellogg's	GM
10	$4.25	Kellogg's	Non-GM
11	$4.25	Store	GM
12	$4.25	Store	Non-GM

FIGURE 12.19 Product Profiles Used in Baker and Burnham (2001) Conjoint Analysis Study of Consumer's Cereal Preferences.

By plugging in different values for Kellogg's, Price, and GM, one can predict how the rating of an option would change as the product attributes are varied.

These results imply that, on average, people prefer eating Kellogg's brand cornflakes to store brand cornflakes; profiles that were Kellogg's brand tended to be rated 1.68 points higher than store brand profiles. The negative sign on the price coefficient (−1.42) indicates for each dollar increase in price, a product's rating falls 1.42. Finally, results indicate that people prefer non-GM to GM cornflakes because GM cornflakes were rated 1.98 less, on average, than non-GM cornflakes.

Overall, these results mean, on average, that GM content is more important than brand name because moving from GM to non-GM changes rating (or utility) by more than changing from Kellogg's to store brand. WTP for an attribute is determined by calculating the price difference that would make a person indifferent (i.e., same rating or utility) between two options that differ in terms of their attributes. Consider one box of cornflakes that is non-GM, store brand, and cost \$3.00; the model predicts such an option would generate a rating (utility) of $10.06 + 1.68 \times (0) - 1.42 \times (\$3.00) - 1.98 \times (0) = 5.8$. Now, consider the anticipated rating (utility) of an otherwise identical product that is GM instead of non-GM: $10.06 + 1.68 \times (0) - 1.42 \times (\$3.00) - 1.98 \times (1) = 3.82$. We can see that the utility of GM product is lower than the non-GM. What if we decreased the price of the GM product by \$1.00 from \$3.00 to \$2.00? The anticipated rating would now be: $10.06 + 1.68 \times (0) - 1.42 \times (\$2.00) - 1.98 \times (1) = 5.24$. The GM product is now more attractive but still not quite as attractive as the non-GM. If we repeat this process a few times, we find that at a price difference of \$1.39 (the non-GM costs \$3.00 and the GM good costs $\$3.00 - \$1.39 = \$1.61$) the average person would be exactly indifferent (i.e., would have the same anticipated rating or utility) between the GM and non-GM options. Thus, the WTP premium for non-GM over GM cornflakes, in this study, is estimated at \$1.39. We can similarly find that the WTP premium for Kellogg's over store brand is about \$1.18.

Although contingent valuation and conjoint analysis are valuable and useful methods to measure WTP, they are subject to one strong weakness—they are hypothetical. At a minimum, people have little incentive to put much effort into thinking about their decisions when the task is hypothetical; at the worst, people might try to provide untruthful answers, at no cost to themselves, in hypothetical questions. Because of this concern, some economists have started to use experimental auctions to obtain better WTP estimates.

In an experimental auction, 10 to 20 people at a time are recruited to participate in an auction session. Often, people are given, for free, a traditional good, say a GM food, and are asked to bid against the other participants in the room to exchange their endowed good for another product that is the same as the endowed good except for one characteristic, say a non-GM food. Bids to exchange the endowed good for the auctioned good can be interpreted as a person's WTP to have the characteristic of interest (e.g., the premium one places on non-GM over GM). In other studies, people are not endowed with a product at all but submit bids for two related goods (with a coin flip determining which good is actually auctioned). In this case, the difference in bids between the two goods is the person's WTP to have the characteristic of interest.

There are a variety of auction mechanisms that can be used, but when interest is in determining WTP, an auction that is *incentive compatible* is needed. An incentive compatible mechanism is one in which a person has an incentive to submit a bid that is exactly equal to their value for a good. One very well-known incentive compatible auction is the second price auction. In a second price auction, each person submits a bid for the good. The highest bidder wins the good but pays the *second highest bid amount* for the good. The second price auction is incentive compatible because bidders cannot influence the price that is paid. If they submit a bid that is lower than their true value for the good, they may not win the auction even though they could have won at a price they were really willing to pay. Conversely, if a person submits a bid that is higher than their true value, they might have to buy the good at a price that is higher than they really want to pay. If a person submits a bid that is exactly equal to their value, they can be assured that they will never have to buy at a price higher than they are willing to pay *and* because if they win, they only pay the (unknown at the time of the bidding) second highest bid amount, they cannot make themselves better off by bidding lower. Thus, the best strategy for a person to follow in a second price auction is to bid an amount equal to their true value for a good. Because real products and real money are exchanged in an experimental setting, participants have a greater incentive to reveal their true value for a good than in a hypothetical survey setting.

In a second price auction, each person submits a bid, the highest bidder wins the good, and the winner pays an amount equal to the second highest bid.

To illustrate the method, consider the experiment conducted by Lusk et al. (2001). In their study, 32 students were recruited to participate in a second price auction. Participants were given an ID number and were endowed with one dollar and a 1 oz. bag of corn chips identified as manufactured with GM corn. Participants were asked to indicate their maximum willingness-to-pay to exchange their bag of GM corn chips for a bag of corn chips not produced with GM corn on a bid sheet with full knowledge that consumption of a bag of chips (GM for nonauction winners and non-GM for the winner) was mandatory upon completion of the auction. Bids were collected and sorted from highest to lowest. The ID number of the highest bidder and the second highest bid amount was written on the board in the front of the room. Then the process was repeated four additional times for a total of five rounds. At the end of the experiment, one of the five rounds was randomly selected as the binding round and the highest bidder in that round paid the appropriate bid amount to receive the bag of corn chips identified as free of GM corn. Because this was a second price auction, the highest bidder won and paid the second highest bid amount to exchange their GM bag of corn chips for the non-GM bag of chips.

In round 1 of the auction, the average bid to exchange the GM bag of chips for the non-GM chips was \$0.035 and this amount increased to \$0.062 by round 5. In round 5, over 70% of students bid exactly \$0, meaning they were not willing to pay anything for non-GM corn chips. These results suggest very little demand for non-GM products, on average, among this sample of consumers. However, this is only part of the story. Almost one fourth of students were willing to pay \$0.25 or more for the exchange. Given that these were only 1-oz. bags of corn chips, this represents a significant WTP

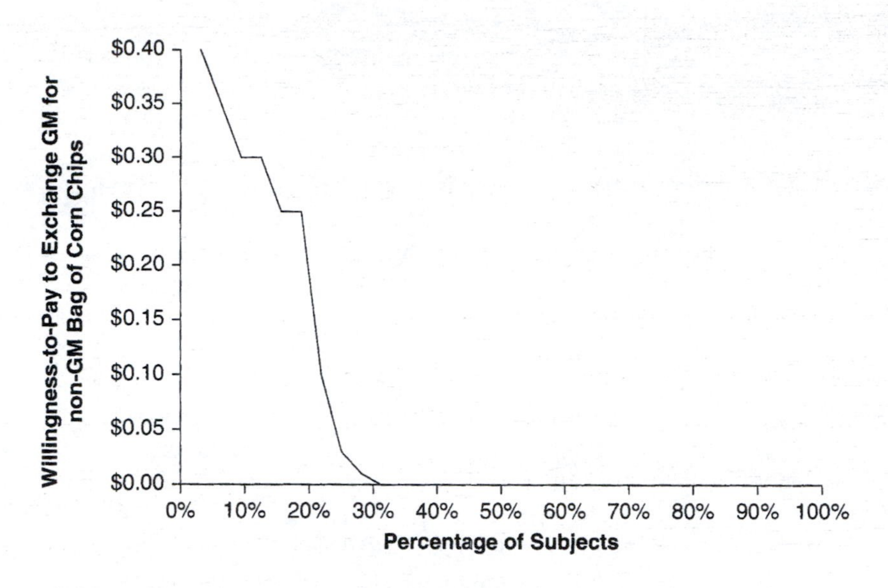

FIGURE 12.20 Distribution of Willingness-to-Pay to Exchange a Bag of GM for Non-GM Corn Chips.

premium among a nontrivial segment of this sample. Figure 12.20 plots the auction bids from round 5 in descending order. Such a graph can be interpreted as a demand curve for non-GM chips assuming each person only buys one and only one bag of chips. For example, at a price premium of $0.30, Figure 12.20 shows that about 10% of subjects would buy the non-GM chips and the other 90% would buy GM.

SUMMARY

This chapter introduced several models, primarily used in psychology and marketing, that are used to characterize consumer behavior. These models provide a basis for thinking about the consumer and provide a framework for analyzing why consumers do what they do. Models related to motivation, needs, values, attitudes, decision making, and willingness-to-pay were covered. In addition to these more abstract concepts, this chapter provided information on how to carry out consumer research. Several types of consumer research were discussed, as were steps in carrying out consumer research. The chapter concluded with information on consumer research methods and showed how the models of consumer behavior could be put into action, paying special attention to consumer research studies that have been conducted on genetically modified food.

CROSSWORD PUZZLE

For answers with more than one word, leave a blank space between each word.

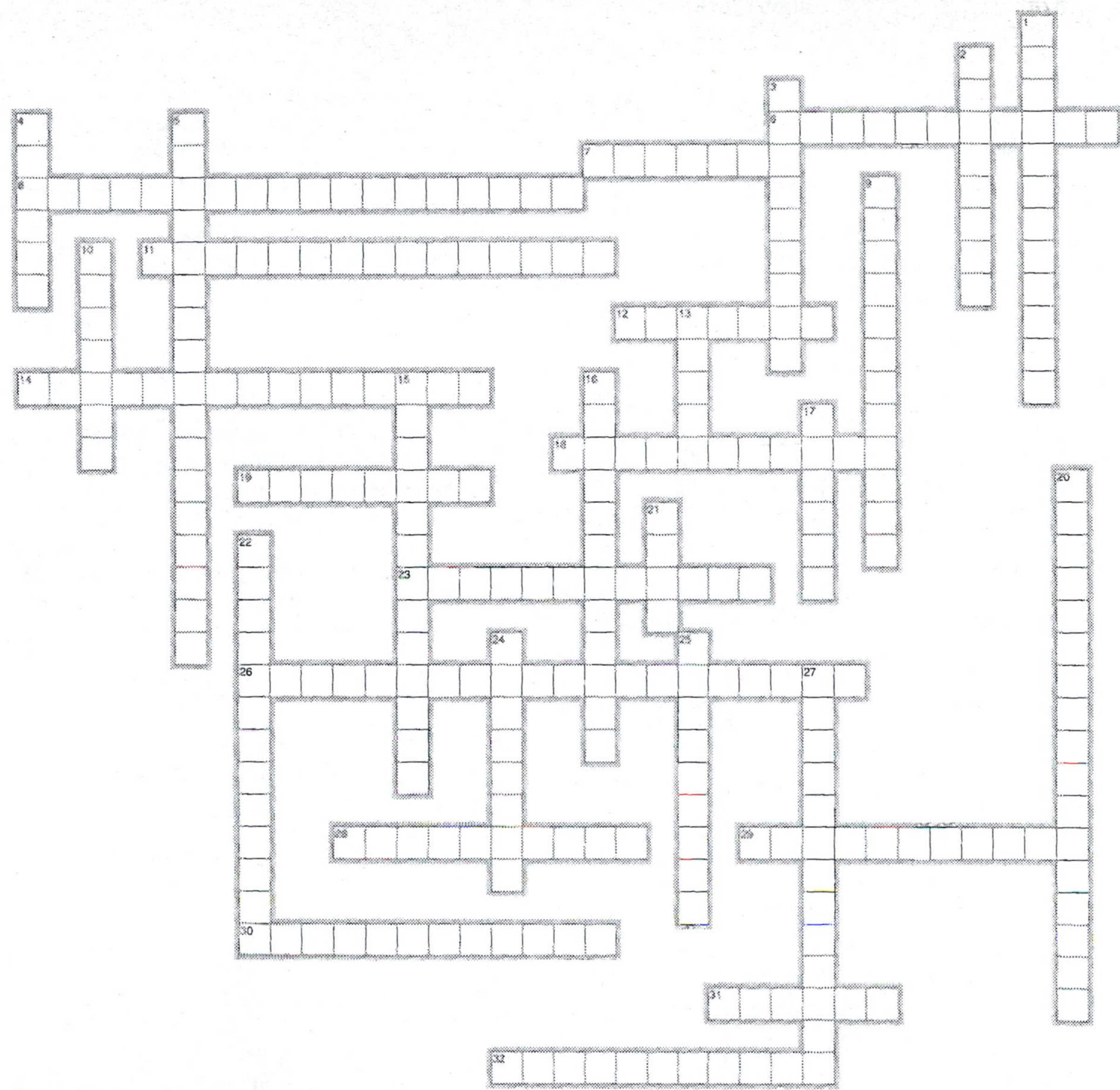

Created with EclipseCrossword — www.eclipsecrossword.com

Across

6. Research that is conducted to clarify and define a problem and to provide insight to be used in subsequent study.
7. Related to pleasure.
8. The amount of money that, when taken from a person, would make them exactly indifferent to having and not having a good.
11. A decision rule where a consumer rules out all options that have a low desirability on a particular attribute no matter how well it performs on another attribute.
12. Data collected firsthand by the researcher by original research designed to answer a specific question.
14. A type of analysis used to investigate the role of values in purchasing behavior.
18. The level of interest a consumer finds in a product or product class.
19. An evaluation or an affective (emotional) response to a concept.
23. Preferred modes (values) of conduct.
26. A mechanism in which a person has an incentive to truthfully reveal their exact value for a good.

28. An internal state or driving force that activates and directs behavior.
29. Attitudes toward an object arise from individuals' beliefs about the attributes an object possesses and _______ of those beliefs.
30. A decision rule where a consumer chooses the product that is best on the most important attribute no matter how poorly it performs on other attributes.
31. Cognitive knowledge about an object or attribute.
32. Functional or practical.

Down

1. A bias relating to the fact that willingness-to-pay is substantially lower when consumers are making decisions with real money.
2. The theory of _______ action suggests an individual's attitudes toward engaging in a behavior, together with subjective norms, influence the consumer's intention to undertake the behavior.
3. Data typically collected for another purpose that are already available and are often in aggregate form.
4. Attitudes are always _______ something.
5. A consumer research technique where respondents rate, rank, or choose between products that are described by several attributes.
9. A decision rule where a consumer evaluates a relative desirability of the attributes, where poor performance of one attribute can be offset by good performance of another attribute.
10. People tend to discount the future less the more _______ the event is to them.
13. Tension arises from a gap between a consumer's present condition and their _______ condition.
15. According to Maslow, the most advanced needs relate to self-_______.
16. Research that aims to provide a specific answer to a specific research problem.
17. In a second price auction, each person submits a bid for the good and the highest bidder wins the good and pays the _______ highest bid amount for the good.
20. The task of better understanding consumer psychology, motivation, and decision making so as to improve firm profitability through strategic pricing, promotion, and product design.
21. A particular manifestation of a need.
22. The most basic needs according to Maslow.
24. Broad psychological states (values) that represent individuals' preferred state of being.
25. A quasi-structured qualitative interview technique used for linking product attributes to consequences and to values.
27. When people value losses more highly than gains.

STUDY QUESTIONS

1. Discuss the advantages and disadvantages of primary and secondary data.
2. What are the four steps in marketing research?
3. Suppose you were interested in determining people's preferences for country-of-origin labels in beef relative to an organic label. Describe how you would conduct a conjoint analysis study to determine the attribute for which people were willing to pay more. How would you use results from a conjoint study to construct a demand curve for organic beef?
4. According to the theory of reasoned action, what strategies could a business pursue to increase a consumer's intent to purchase their product?
5. Using the laddering technique, construct a means-end chain for 2% milk for a classmate.
6. How would you measure someone's attitude toward purchasing genetically modified food? What factors would you expect to affect their attitude?
7. What are the advantages and disadvantages of using experimental auctions to determine willingness-to-pay for a new good?

Part Four: Additional Topics

CHAPTER THIRTEEN

The Firm as a Price Taker

> Mr. Scorpio says productivity is up 2% and it's all because of my motivational techniques, like donuts, and the possibility of more donuts to come.
>
> —*Homer Simpson*

INTRODUCTION

In many industries firms are price takers, meaning they have little control over the price they receive for their output and the price they pay for their inputs. Profits for these firms depend crucially on buying just the right amount of inputs and producing just the right amount of output. This chapter first discusses the production process itself. This production process is then extended to evaluate input decisions by the firm. Basically, we will use information on prices and the production process to answer the question: How much of an input should a firm purchase? The discussion then leads to production costs. It turns out that a firm can be described simply as a set of cost curves. In some ways this is unrealistic, because firms consist of many people and complexities, not just a few curves on a graph. Yet, those curves capture many important features of a firm in an easy-to-understand picture. We give up a little bit of reality for a large amount of understandability. We use these curves and price information to answer the all-important question to a firm: How much should the firm produce? Finally, the costs of production to a firm as they differ across the short run and long run is considered.

To reiterate, the goals of this chapter are to

1. develop a model of production for a firm
2. use this model to study input use by a firm
3. extend these concepts to understand costs of production
4. describe how a price-taking firm maximizes profits
5. distinguish between firm behavior in the short and long run

It is common for students to view the material in this chapter as "a bunch of irrelevant junk." It is easy to think we are oversimplifying agribusiness management. In fact, we agree that we have greatly simplified the task of managing a firm, but for good reason. When engineers design bridges, they first build model bridges to make sure it will work and to understand the things that can go wrong. Before a company mass-produces a new item, they conduct extensive market research on small groups of consumers. Before NASA built the Space Shuttle, you can bet they played around with much smaller versions and simulated the outer space environment first. In a similar vein, this chapter deals with how to maximize profits for a simple price-taking firm where costs of production are known and never change. If you cannot maximize profits in this simple model of the firm, chances are, you will not be able to maximize profits for a really complex firm either. Learn the basics of a golf swing before you compete. Learn the basics of agribusiness management in simple settings before you do the real thing.

THE THREE STAGES OF PRODUCTION

Strip away all the complexities of a firm, and you basically have people turning inputs into outputs. Hop growers take machinery, nitrogen, and hop seed to grow and harvest hops for use in beer. Turkey growers take buildings, corn, water, and medicine to produce turkey meat. Production, the process of transforming inputs into outputs, is an important concept in the study of the firm. A *production function* is a mathematical formula that indicates the output attained from a given level of input.

Imagine you have a plant that has not been watered for weeks. Its leaves are withered and its stem barely able to hold the plant erect. Now you go to water the plant. At first, you give it just a little bit of water. This tiny amount of water is just enough to keep the plant alive but doesn't make it look any healthier. Now you apply a little more, and the plant begins to respond by holding up its leaves again and the stem becoming more erect. This second watering brought the plant from just surviving to recovering. Put differently, you got a bigger "bang for your buck" from the second watering than the first. Then you apply a little more, and the plant responds tremendously. Again, the "bang for your buck" was bigger from the third than the second watering. Its leaves stand out and become strong, and the plant itself stands straight up. From this point on when you apply more water the plant grows, but its health doesn't make the giant leap it did before. And if you keep adding water to the plant, you will eventually drown it. This story describes a firm's production function. The story describes the relationship between any input and its associated output, whether it be fertilizer to corn or rubber to whoopee-cushions.

This story is told to point out that for any production process, whether it be plant production, animal production, or car production, the marginal product goes through three stages. In the first stage the marginal product is positive and rising. The marginal product stays positive but begins declining in the second stage. In the third stage the marginal product turns negative. Consider a slightly more complex example of a tomato canning facility. Many types of inputs are used in this facility.

There is the facility itself, consisting of buildings and land. Forklifts are used to transport crates and tractor-trailers haul tomatoes to the facility. Wage workers operate the machinery. Salaried employees manage the business, perform accounting activities, and coordinate sales. Electricity, fuel, oil, packaging, and machinery parts are just a few of the additional inputs. The firm takes all these inputs, produces canned tomatoes as an output, and sells the canned tomatoes. Assume the canned tomato industry is perfectly competitive, so that the firm is a price taker.

Some inputs are fixed in the short run. These include the facilities, machinery, and salaried employees. Other inputs like electricity and wage labor are variable, meaning they can be easily and quickly increased or decreased. Let us explore the relationship between one of these variable inputs (labor) and output. This will serve as a metaphor between any variable input and output. The units of labor are measured by the number of workers employed each day. A production function tells us the number of canned tomatoes (*total product*) produced for each worker employed each day.

Marginal Product: The additional output realized from increasing the use of an input by one unit.

Suppose we begin with no workers and increase to one worker. The increase in total product from increasing the input level by one is referred to as the *marginal product*. Suppose the addition of one worker over zero workers increases total product from 0 to 5 cans. The marginal product of the first worker is then 5. Suppose a second worker is now added. Before, one worker had to do everything, but with two workers each one can specialize in specific activities. One may concentrate on loading raw tomatoes into the machine, while the other concentrates on unloading canned tomatoes. Specialization allows each worker to be more productive and total product surges from 5 to 20 units. The marginal product of the second worker is 15 and is greater than the marginal product of the first worker. From hiring the second worker, the firm gets "a bigger bang for their buck" than the first worker.

Initially, when more laborers are added, they specialize and productivity for all workers rises. As a result, the marginal product of each additional worker is greater than the previous worker. The marginal product is increasing in the number of inputs used. In Figure 13.1, the marginal product of the first worker is 5 but is 15 for the second worker. If the marginal product increases when additional inputs are employed, we say the firm is in the *first stage of production*. Eventually, as more workers are added, there will come a point where adding another worker increases

Number of Workers	Total Product (TP)	Marginal Product	Average Product	
0	0	—	—	
1	5	5 − 0 = 5	5/1 = 5	Stage One
2	20	20 − 5 = 15	15/2 = 7.5	
3	30	30 − 20 = 10	30/3 = 10	Stage Two
4	35	35 − 30 = 5	35/4 = 8.75	
5	32	32 − 35 = −3	32/5 = 6.4	Stage Three

FIGURE 13.1 The Production Function.

output, but not as much as the previous worker. Production facilities are generally designed to match workers with machines, and a certain number of workers is designed to operate each machine. Once the number of workers exceed this number, those new workers are performing tasks that we could say are "less important." One tomato canning machine may be operated efficiently using two workers, and one forklift is operated by only one worker. Another worker is useful but does not contribute as much as previous workers. The marginal product is positive but is a smaller value. For example, going from two to three workers increases production from 20 to 30 in Figure 13.1. The marginal product of the third worker is 10, which is less than the marginal product of 15 for the second worker. We are now in the *second stage of production*, where the marginal product is positive but falling as the number of inputs increases. Finally, there will come a point where adding a worker actually detracts from production. Imagine we kept adding workers in a factory to the point where it was too crowded to even move. If you cannot move, you cannot work, and those new workers caused production to fall. This is the *third stage of production,* where the marginal product of an input is negative.

The marginal product is positive and increasing in the first stage, positive and decreasing in the second stage, and negative in the third stage of production.

The three stages of production are illustrated in Figure 13.2. It is useful to consider another example to illustrate the production function. Consider the production of wheat using nitrogen as an input. All the other inputs of production (machinery, labor, pesticides) are assumed fixed, and we want to evaluate how corn yield responds to changes in nitrogen use. Without applications of chemical nitrogen to wheat, very little yield will be realized. But wheat responds to the first couple pounds of nitrogen by greatly increasing the plant growth. Another couple pounds and wheat responds more by even greater growth. The marginal product of nitrogen is increasing at low

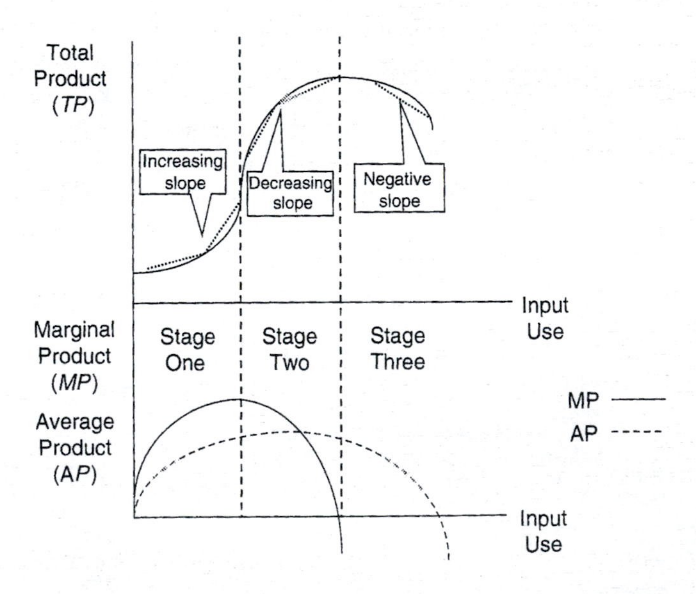

FIGURE 13.2 The Production Function.

levels of use—meaning the marginal product of the fourth pound is greater than the marginal product of the third pound. This is the first stage of production. As you add more nitrogen, wheat will still respond by growing taller, but the response is not as pronounced. The marginal product is still positive but is falling, so you are in the second stage of production. Finally, you will hit a stage where adding nitrogen does not increase wheat yields and may even decrease it. This is the third stage. If you are not familiar with fertilizer and crops, just reconsider our first example of watering a plant. The plant responds to the watering at first, but there comes a point when more water does not increase plant growth.

The presence of the three stages of production give rise to the S-shaped production function in Figure 13.2. In the first stage, marginal product is rising as input use is increased. The marginal product is positive but declining in the second stage. At the third stage, the marginal product is negative and greater input use detracts from total product. The marginal product is really just the slope of the production function at a particular point. To illustrate, the slope between points on the production function is illustrated by the dotted lines in Figure 13.2. The slope tells us how fast production changes with input use. In the first stage, the slope is positive and becomes steeper as more input is used. The slope is still positive but becomes less steep in the second stage. Notice that the slope would be highest at the exact same input level that the marginal product curve is at its peak. This is no coincidence; marginal product is the slope of the production function. Finally, at the third stage, the slope of the production function becomes negative.

In addition to marginal products, the average product is an important concept. As before, the relationship between marginal and average product will be illustrated with an analogy. Suppose we have a baseball team and output is measured by total number of home runs hit by all players during a game. At first we have no players, but then we start adding players. Every new player is better than the previous player, meaning player 2 is better than player 1, player 3 is better than player 2, and so on. Each new player hits more home runs than the previous player. The marginal product of each player is rising, and we are in the first stage of production. Then, after the fifth player is added, we start adding new players such that each new player hits worse than the previous player. Player 6 is worse than player 5, player 7 is worse than player 6, and so on. Each of these new players hit home runs, so they increase output, but they increase output by less than the previous player. This is stage two of production.

Average product is measured as the total output divided by total inputs used. In our baseball analogy, it equals total home runs divided by the total number of players. In stage one, each new player is better than the previous player, and average product is rising. Adding better players increases the average performance of the team. Once we hit stage two, each new player is worse than the previous player. Marginal product starts falling, but as long as the new player is better than the average player on the team, average product continues to rise. At some point, as we continue adding lower-quality players, the new players will be worse than the average player, and average product will begin declining, as shown in Figures 13.1 and the dotted line in the lower graph of Figure 13.2.

OPTIMAL INPUT USE

This is a good point to stop and look at actual production data. Figure 13.3 shows pounds of nitrogen applied to wheat and the corresponding wheat yields for Oklahoma. The yields are taken from actual Oklahoma yield measurements and nitrogen use.[1] We cannot directly measure the marginal product of nitrogen at each level of nitrogen, because nitrogen is not increased in increments of one. It is increased in increments of 20 lbs. However, we can obtain an estimate of the marginal product in a range of input use through the formula $MP = \Delta q/\Delta x$ where Δ means "change," q is total product, and x is input level. Thus, $\Delta q/\Delta x$ is the change in total product divided by the change in pounds of nitrogen. To see how the formula operates, suppose that you work five more hours and your output increases by 100. The change in output is 100 and the change in input use (hours worked) is 5. The marginal product formula yields $MP = \Delta q/\Delta x = 100/5 = 20$. What this means is that for each additional hour worked your output increases by approximately 20 units. That is the very definition of marginal product.

This formula simply takes the slope between two points on the production function (remember, slope equals rise over run, $\Delta q/\Delta x$). For example, when nitrogen is increased from 0 to 20, the marginal product within this range is $MP = \Delta q/\Delta x = (30.5 - 23.0)/(20 - 0) = 0.375$. That is, when using between 0 and 20 lbs. of nitrogen, an extra pound of nitrogen increases wheat yields by about 0.375 bushels per acre.

Using this formula we can see how the marginal product of nitrogen changes with nitrogen use. In Figure 13.3 the marginal product is declining as more nitrogen is used for all nitrogen levels. That is, it appears production is in the second stage of production throughout. What happened to the first stage? Production must have shifted from the first stage to the second stage somewhere between 0 and 20 lbs. of nitrogen, so it cannot be detected in the data. What happened to the third stage? If more than 100 lbs. of nitrogen were applied and one kept adding more and

lbs Nitrogen per Acre	Wheat Yield (Total Product) *bushels/acre*	Marginal Product of Nitrogen	Average Product of Nitrogen
0	23.0	—	—
20	30.5	$(30.5 - 23.0)/(20 - 0) = 0.375$	$30.5/20 = 1.525$
40	35.0	$(35.0 - 30.5)/(40 - 20) = 0.225$	$35.0/40 = 0.875$
60	37.0	$(37.0 - 35.0)/(60 - 40) = 0.100$	$37.0/60 = 0.617$
80	37.8	$(37.8 - 37.0)/(80 - 60) = 0.040$	$37.8/80 = 0.473$

FIGURE 13.3 The Production Function for Oklahoma Wheat.

[1]The data are based off experimental plots but are adjusted to reflect typical nitrogen use rates.

more, there would come a point when wheat yields did not change and would eventually start falling. Even though we cannot observe the first and third production stages in these data, it clearly demonstrates the second stage, which turns out to be the most important stage anyway. The impact of nitrogen on wheat yields declines the more nitrogen that is used. Therefore, there will come a point where increasing nitrogen use is not profitable. At this point, the increase in yield does not justify the additional nitrogen purchase. Let us employ the concept of marginal product to articulate the profit maximizing level of nitrogen use. An additional pound of nitrogen should be used if the value it provides is greater than its cost, just like you do not purchase an item unless you value it more than its price. The value of one more pound of nitrogen is the additional yield it provides (the marginal product) times the value of that yield (the market price of wheat). Not surprisingly, we refer to this as the *marginal value* of input use. The market price for wheat is usually around \$3.25/bu, so the marginal value of nitrogen can be calculated as the marginal product times \$3.25, as shown in Figure 13.4. The marginal cost of nitrogen is roughly \$0.15.

The marginal value of an input equals the marginal product of the input times the price of the good being produced. It is the increase in the value of output from increasing input use by one.

Using Figure 13.4, suppose the farmer initially plans on applying 20 lbs. of nitrogen. Should more than 20 lbs. be used? At 20 lbs., the marginal value of one more pound is \$1.22. That is, by increasing nitrogen use from 20 to 21 pounds per acre, revenue increases by \$1.22 per acre. The marginal cost of increasing nitrogen use from 20 to 21 lbs. is only \$0.15, so profits increase by \$1.22 − \$0.15 = \$1.07/acre. The verdict is clear: More than 20 lbs. should be applied. This reveals a very important production concept: An additional input should be used whenever the marginal value is greater than the input price. Consulting Figure 13.4, the marginal value of nitrogen is greater than the input price up to 60 lbs. per acre. If nitrogen could only be purchased and applied in increments of 20 lbs. per acre, the farmer should apply 60 lbs., and the optimal yield (i.e., profit maximizing yield) is 37.8 bushels per acre.

An additional unit of input should be used whenever the marginal value is greater than the input cost.

This is an important concept. The profit maximizing level of output is *less* than the production maximizing level of output. Put differently, firms do not want to produce at their maximum production level. At 60 lbs. of nitrogen per acre, one could increase yield by using more nitrogen, but this would lower the firm's profits.

lbs Nitrogen per Acre	Wheat Yield (Total Product) *bushels/acre*	Marginal Product of Nitrogen *bushels/acre*	Marginal Value of Nitrogen (Wheat Price = \$3.25/bu) *\$/acre*	Marginal Cost of Nitrogen (Nitrogen Price = \$0.15/lbN)
0	23.6	—	—	—
20	30.5	0.375	0.375 × 3.25 = \$1.22	\$0.15
40	35.0	0.225	\$0.73	\$0.15
60	37.0	0.100	\$0.33	\$0.15
80	37.8	0.040	\$0.13	\$0.15

FIGURE 13.4 Marginal Value and Cost of Input Use.

When Should Cattle Be Sold?

Feedlots are in the business of purchasing adolescent cattle, feeding them a high energy diet, and selling them when they are big enough for slaughter. Cattle within a feedlot are referred to as live-cattle (also fed-cattle). Feedlot profits depend critically on selling cattle at the optimal weight. Each day cattle are on feed, they gain weight, which increases the revenue from each head of cattle sold. If the market price for live-cattle is \$0.75 per pound, then each extra pound increases revenues by \$0.75. However, for each day cattle are kept in the feedlot they must be fed, and that feed costs money. It costs roughly \$1.41 for each additional day a cow is fed. To maximize profits, the manager keeps cattle in the feedlot as long as the value of the extra pounds produced outweigh the costs of feeding the cattle. Our input is "*days on feed*," and the output is pounds per cow.

One of the authors collected data on cattle weights and the number of days cattle were on feed to determine the marginal product of days on feed. Using the regression analysis technique discussed in Chapter 7, a formula for marginal product was calculated as

$$\textit{Marginal Product } (MP) = 4.36 - 0.0157(DOF)$$

where DOF stands for *days on feed*. The first day cattle are brought into the feedlot $DOF = 1$, and the 50th day of being in the feedlot $DOF = 50$. The marginal product equation is interpreted as follows. Suppose a lot of cattle has been held in the feedlot for 25 days, and the manager is considering keeping them one day longer. She can expect each animal to gain $MP = 4.36 - 0.0157(25) = 3.9675$ lbs. from that extra day. If the cattle have been on feed for 200 days, one day of extra feed only results in $MP = 4.36 - 0.0157(200) = 1.22$ lbs. in extra weight. Eventually, there will come a point where the value of an extra pound is less than the cost of feed. At that point, the feedlot manager should sell the cattle. As before, the marginal value of *days on feed* is the marginal product times the market price of live cattle, denoted P_{LC}. In 2004, the average live-cattle price was around \$0.75/lb. The marginal cost or input price of *days on feed* is denoted p_{DOF} and is about \$1.41. That is, keeping cattle on feed one additional day costs \$1.41 per head. This includes all the feed, labor, veterinary, and other costs associated with feeding an animal one additional day. Profits are maximized by keeping cattle on feed as long as the marginal value of *days on feed* is greater than the price. The greater the *days on feed*, the lower will be the marginal value because cattle put on less weight as they age. Eventually, the marginal value will just equal the price, and the cattle will be sold.

The input decision is demonstrated in Figure 13.5. For low input levels, the marginal product is high, higher than the input cost. As long as marginal value is greater than the input price, more of that input is used and profits rise. For example, when cattle have been on feed 50 days, the marginal product is 3.575 lbs. Holding cattle one more day increases cattle weights by 3.575 lbs. per cow. At a price of \$0.75 per lb., this

Set Marginal Value of Input = Input Price

Assume output price $= P_{LC} = 0.75$; Input Price $= p_{DOF} = 1.41$

$$P_{LC}[4.36 - 0.0157(DOF^*)] = p_{DOF}$$

$$0.75\,[4.36 - 0.0157(DOF^*)] = 1.41$$

$$DOF^* = (3.27 - 1.41)/0.011775 = 158 \textit{ days on feed}$$

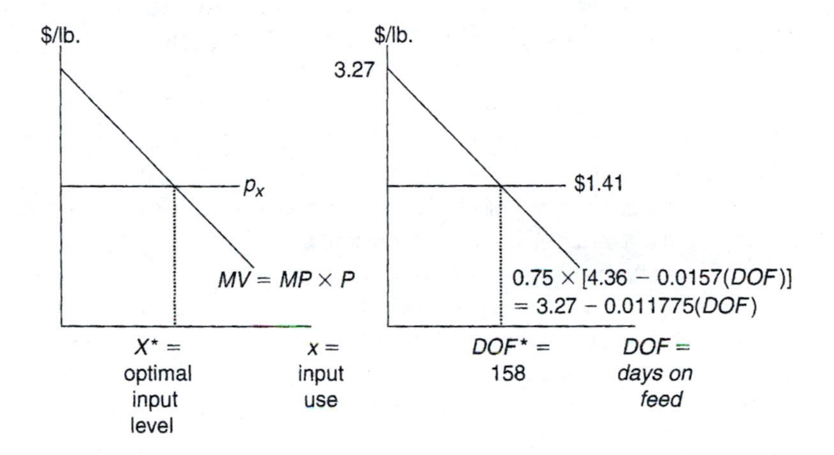

FIGURE 13.5 Solving for the Optimal Number of *Days on Feed*.

translates into $2.68 per head of greater revenues. The cost of holding cattle one more day is only $1.41, so increasing *days on feed* from 50 to 51 days increases profits by $2.68 − $1.41 = $1.27 per head. Using the same logic, the feedlot manager should keep increasing *days on feed* until the marginal value equals the input price. Using the calculations shown in Figure 13.5, profits are maximized by keeping cattle on feed approximately 158 days.

Input Demand by Firms

Notice that the optimal *days on feed* is given by the intersection of the marginal value curve and the input price. Thus, the marginal value curve is the input demand curve for the firm. This relationship can be used to study how input use changes. First, consider the obvious case where the input price rises. To equate marginal value with the input price, the firm must use less of the input. As less of the input is used, marginal product rises, aligning marginal value with the new input price. Using the same logic, if the price of an input falls, the quantity demanded by the firm rises.

Now consider the case where there is technological progress, and more output can be produced using the same amount of input. An example is the use of growth hormones in cattle, which increases the rate at which cattle convert feed to muscle. This is another way of saying that growth hormones increase the marginal product at all input levels. Given that the marginal value equals marginal product times cost, if marginal product rises, the marginal value curve shifts upward. The intersection of input demand and input price is now at a higher input level. The firm buys more of the input.

Consider again the business of producing live-cattle, where *days on feed* is an input. Some cattle are administered regular, low doses of antibiotics in their feed and water to promote growth. Because this poses a health hazard, some lawmakers have considered banning this practice. What would happen to the optimal *days on feed* if these antibiotics were eliminated? Without these antibiotics, cattle would grow at a slower rate, putting on fewer pounds each day. This is just another way of saying the marginal product would fall. If the marginal product falls, then the marginal value curve shifts downward. The firm's demand for *days on feed* falls, and live-cattle are sold and slaughtered at a younger age. This suggests that a ban on antibiotics not only hurts antibiotic makers and cattle producers, but corn producers as well. As the marginal product of *days on feed* falls, feedlots feed their cattle fewer days, reducing their demand for feed and lowering their corn purchases. Finally, suppose that the price of live-cattle rises. Each additional pound gained by cattle now translates into greater revenues. Because the demand curve is marginal product times the output price, the firm's demand for *days on feed* rises and cattle are slaughtered at an older age.

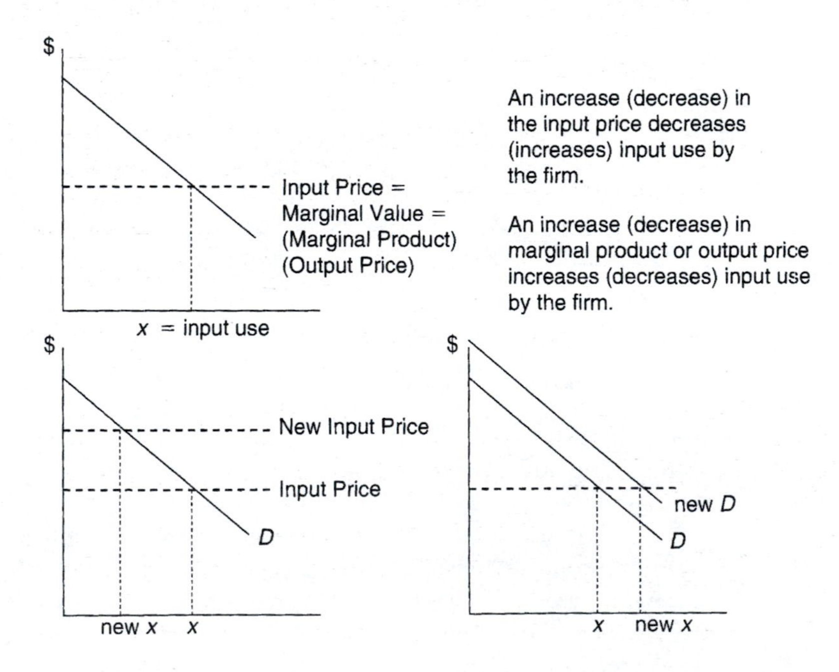

FIGURE 13.6 Input Demand.

COSTS OF PRODUCTION

Short Run: A period of time over which at least one input is being held fixed.

Long Run: A period of time after which no inputs are held fixed.

Production is the process of turning inputs into outputs. Inputs come at a price, so the ability of a firm to produce at a low cost depends on how well they use their inputs. Increasing production requires the use of more inputs, yet a firm does not always have control over the level of all inputs. A milk processing facility uses expensive buildings, machinery, milk, labor, and raw materials as its inputs. If a milk processing facility wishes to process greater volumes of milk, it can increase the use of all those inputs. However, it takes time for a new building to be built. Even though the firm can immediately vary its use of milk, labor, and most raw materials, it cannot immediately increase or decrease the number of its processing facilities (like milk tanks, building, piping). In a period of time, some of its inputs are held fixed and cannot be varied. We refer to this as the *short run*, when at least one input is fixed. Given time, the firm can build more facilities. Given time, the firm can increase or decrease the level of *all* inputs used. The length of time it takes to vary all inputs is referred to as the *long run*. As an example, suppose the milk processing facility can freely vary all inputs at any time except its processing facility (buildings and milk processing machinery). It can easily increase or decrease labor, volume of milk, and raw materials. However, it takes two years for a new processing facility to be built. The short run is then less than two years and the long run is greater than two years.

At any point in time, some inputs can be varied and some inputs may be fixed. Increasing production requires greater use of those variable inputs, and the costs of those variable inputs are referred to as *variable costs*. Variable costs are costs that vary with production. When production rises, so do variable costs. Conversely, *fixed costs* refer to the cost of inputs that do not change with the level of output. Using our milk processing facility example, recall that in the short run the number of processing facilities cannot be changed. The firm likely took a loan to pay for the building, and each month it must make a payment to the bank. That payment is the same regardless of how much milk is processed, making it a fixed cost.

In this section we wish to build a simple economic model illustrating how input use translates into production costs. Let production or total product produced by the firm be denoted q. Total costs equal variable costs plus fixed costs. Denote fixed costs as FC. Suppose that the only variable cost (VC) is labor. The amount of labor used (# of hours) is given by L and the price of labor is the wage rate w ($ per hour). Variable costs in this example equal labor use times the wage rate: variable costs $= wL$. Fixed costs equal FC regardless of the output level or number of hours worked.

$$\text{Total Costs } (TC) = VC + FC = wL + FC$$

In previous chapters we extensively relied on the concept of *marginal cost*, which is the additional cost incurred from increasing production by one unit. The formula for measuring marginal cost is the change in total cost divided by the change in output: $\Delta TC/\Delta q$ where Δ means "change." To illustrate the formula, suppose increasing production by one increases costs by 10. The marginal cost is then $\Delta TC/\Delta q = 10/1 = 10$. If increasing production by two increases costs by 16, a measure of marginal cost is $\Delta TC/\Delta q = 16/2 = 8$. It costs \$8 for each additional unit produced. In our example, the only way to increase production is to increase labor, as all other inputs are held fixed. The change in total costs equals the change in labor times the wage rate. If the wage rate is \$10 per hour and 50 more hours are employed, costs rise by $w\Delta L = (\$10)(50) = 500$.

$$\textit{Change in total costs} = \Delta TC = w\Delta L$$

Then, noting that marginal cost is the change in total cost divided by the change in quantity produced, we obtain marginal cost by dividing both sides of the previous equation by Δq.

$$\textit{Marginal Cost (MC)} = \frac{\Delta TC}{\Delta q} = w\frac{\Delta L}{\Delta q} = \frac{w}{(\Delta q/\Delta L)} = \frac{w}{MP}$$

Notice that the term $(\Delta q/\Delta L)$ is simply the marginal product. Marginal cost can simply be stated as the input price divided by the marginal product of that input. This makes sense. The higher the price of inputs, the greater the cost of purchasing those inputs and producing a product. The higher the price of labor, the greater the cost of using laborers to take raw vegetables and package them into ready-to-eat salads. The formula also tells us that the greater the marginal product, the lower the marginal cost. The more productive your laborers are, the less your cost of using those workers to produce a product. At this point we should revisit our previous discussing of marginal product and the three stages of production. The marginal product curve is re-created in Figure 13.7. Marginal cost is inversely related to marginal product. In the first stage of production marginal product is rising, which means marginal cost must be falling. Marginal product starts falling in the second stage, and so marginal cost starts rising. Therefore, the marginal cost curve falls in the first stage of production then rises in the second stage.

The concept of average cost is also important. First, consider average variable costs (*AVC*), which is simply variable costs divided by output. Using some algebra, we obtain the relation

$$AVC = \frac{VC}{q} = \frac{wL}{q} = \frac{w}{(q/L)} = \frac{w}{AP}$$

Variable Cost: A cost that changes with production.

Fixed Cost: A cost that remains the same regardless of the production level.

Average variable costs are simply the input price divided by the average product. Before, we showed that at low levels of input use average product is rising. Inputs are becoming more productive, so average variable costs naturally fall as a result. However, at some point the average product will begin falling, which increases average variable costs. Average variable costs are then inversely related to average product, as shown in Figure 13.7.

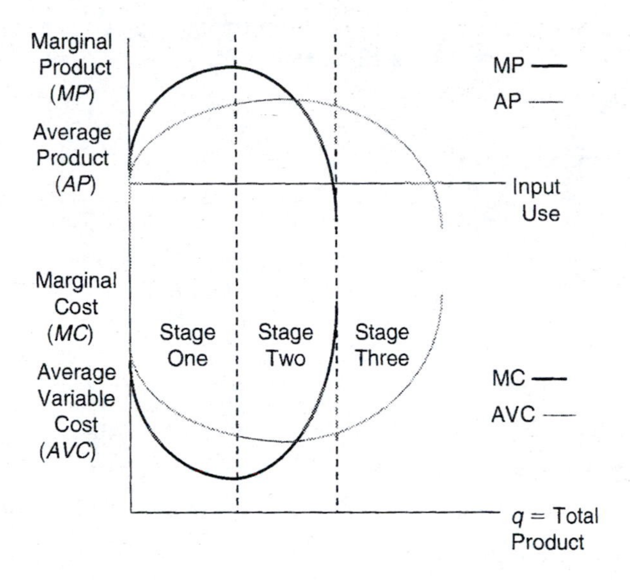

FIGURE 13.7 Marginal and Average Cost Curves.

Finally there are average total costs (*ATC*), which are total costs divided by total product. The only difference between average variable costs and average costs are average fixed costs.

$$ATC = \frac{VC}{q} + \frac{FC}{q} = \frac{wL}{q} + \frac{FC}{q} = \frac{w}{(q/L)} + \frac{FC}{q} = \frac{w}{AP} + \frac{FC}{q}$$

Fixed costs (*FC*) are constant; so as output rises (q becomes large), the average fixed costs (FC/q) will become small. For large levels of output, the difference between average variable costs and average costs is small. As illustrated in Figure 13.8, at zero output

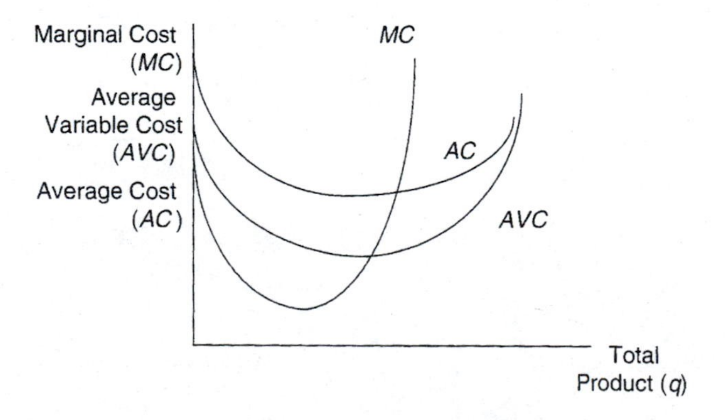

FIGURE 13.8 Marginal and Average Cost Curves.

there is a significant difference between the average variable cost and average cost curve. This difference is fixed costs. However, as total product becomes large, the two curves converge.

A Numerical Example

To help illustrate, consider a numerical example. Refer back to the hypothetical production function in Figure 13.1, which is re-created below. The input is labor, which we will assume costs $5 per worker. Fixed costs are $15. Variable costs are calculated simply as the number of workers times the per worker cost of $5, and total costs are variable costs plus the $15 fixed cost. Marginal cost, average variable cost, and average cost are calculated as described in the formulas above and re-created below.

$$\text{Marginal Cost} = (\text{change in total product})/(\text{change in total costs})$$
$$\text{Average Variable Cost} = (\text{variable cost})/(\text{total product})$$
$$\text{Average Cost} = (\text{total cost} = \text{variable cost} + \text{fixed cost})/(\text{total product})$$

Marginal cost is decreasing at first and then starts to increase. Average variable costs follow a similar pattern, although its minimum is at a higher quantity. The average

Number of Workers	Total Product (*TP*)	Variable Costs	Fixed Cost	Total Costs	Marginal Cost	Average Variable Cost	Average Cost
0	0	$0	$15	$15	—	—	—
1	5	$10	$15	$25	($25 − $15)/ $(5 − 0) = $2	$10/5 = $2	$5
2	20	$20	$15	$35	$0.66	$1	$1.75
3	30	$30	$15	$45	$1	$1	$1.5
4	35	$40	$15	$55	$2	$1.14	$1.57
5	32	$50	$15	$65	—	$1.56	$2.03

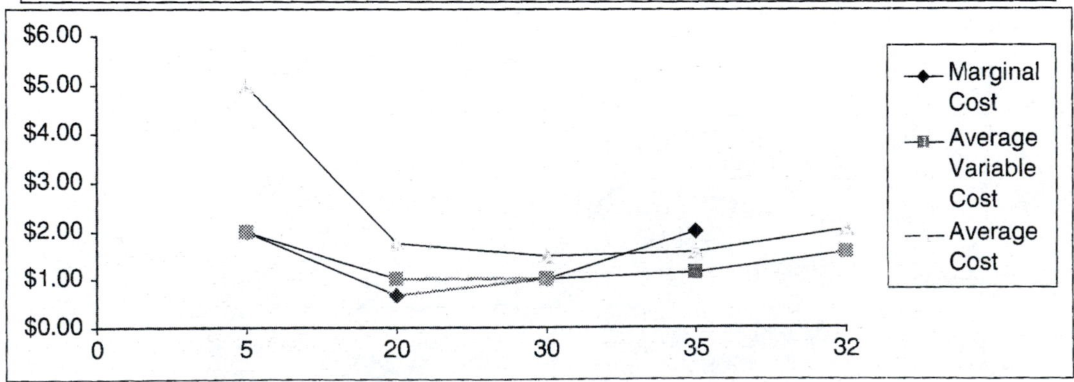

FIGURE 13.9 Hypothetical Firm (Cost per Worker = $10; Fixed Costs = $15).

cost curve lies above the average variable cost curve. For the remainder of this chapter we will focus exclusively on these curves, but note that their shape is derived from the three stages of the production function and input prices.

MAXIMIZING PROFIT

In the previous sections we saw that firms will continue to use more inputs until the marginal value of the input equals its price. Although we concentrated on one input at a time, in reality managers seek to equate marginal value and price for *all* inputs. This maximizes the firm's profits. In this section, we want to study how a firm maximizes profits when it is a price taker. The firm cannot control prices, neither the price it receives for output nor the price it pays for inputs. We assume that however much the firm is producing, it is producing that amount at the least cost possible. The only variable left to discuss is how much the firm should produce. That is, given a marginal cost, average variable cost, and average cost curve, what is the profit-maximizing quantity? The answer turns out to be surprisingly simple, following the two rules below.

Determining the Profit-Maximizing Output Level.

Step 1: If price is greater than the minimum average variable cost, go to step 2. Otherwise, produce nothing.

Step 2: Find the quantity where price equals marginal cost and produce that quantity.

Consider the equation for profit: $Profit = Revenues - VC - FC$, where VC is variable and FC is fixed costs. We can multiply and divide VC by the output level q. This leaves the equation unchanged, as $(VC/q) \times q = VC$. However, noting that VC/q is simply average variable costs, it allows a convenient representation for profit.

$$Profit = Revenues - \left(\frac{VC}{q}\right)(q) - FC = Revenues - (AVC)(q) - FC$$

If we then note that revenue equals price times quantity, $(P)(q)$, profits can be rewritten as

$$Profit = (P)(q) - (AVC)(q) + FC = [(P) - (AVC)](q) - FC$$

If the firm does not produce anything, making $q = 0$, profits simply equal $-FC$. However, as long as it can produce some quantity where average variable cost is less than price, it can make profits greater than $-FC$ and should do so. In the equation above, if the firm can produce at some quantity where $[(P) - (AVC)]$ is positive, it can make more profits than just $-FC$. It might not be making money, but at least it will lose less than it would otherwise. If price is less than average variable costs and a firm produces anyway, this would be like buying ground beef for $3.00, making it into a burger and selling the burger for $2.00. Not only do you have to pay fixed costs like rent on the building but you lose money on the individual burger. But if the variable costs of a burger are $4.00 and you sell the burger for $5.00, you can use that $1.00 profit to help pay fixed costs like building rent. You might not cover all of your fixed costs, but at least you made money to help pay your fixed costs. If revenues are greater than variable costs, the firm is better off producing. It might cover all its fixed costs too and make a profit or it might not. But if revenues are less than variable costs, the firm should not produce because it will lose its fixed costs and more. A smart firm, at worse, will never lose more than its fixed costs.

Thus, if price is P_1 in Figure 13.10, the firm will not produce anything. Price is below the average variable cost line, meaning no matter how much the firm produces it will never cover its fixed costs. It is better off just not producing anything and paying the fixed cost. However, if price is P_2, the firm can at least cover its variable costs. It should produce some quantity. Specifically, it should produce q_2 units. As a rule the firm should produce another unit whenever price is greater than marginal cost. If it only costs you $100 to produce another unit and you receive a price of $150, of course you should produce another unit. Following this logic, a firm keeps increasing production until price equals marginal cost. At this point producing another unit increases costs more than price and detracts from profits. If price is P_2, the firm produces where price equals marginal cost, at q_2. Using the algebra described above, we can write profits as

$$Profit = [(P) - (AC)](q)$$

At a price of P_2 and quantity of q_2, the average cost curve is higher than price. The $[(P) - (AC)](q)$ is negative and the firm is losing money. However, because it can still cover its variable costs, it is best off cutting its losses by producing q_2 units. Next observe what happens if price is P_3. The firm again produces where price equals marginal cost (given that price is greater than the minimum average variable cost), which is at q_3. Now, the price is greater than the average cost and the firm enjoys a profit.

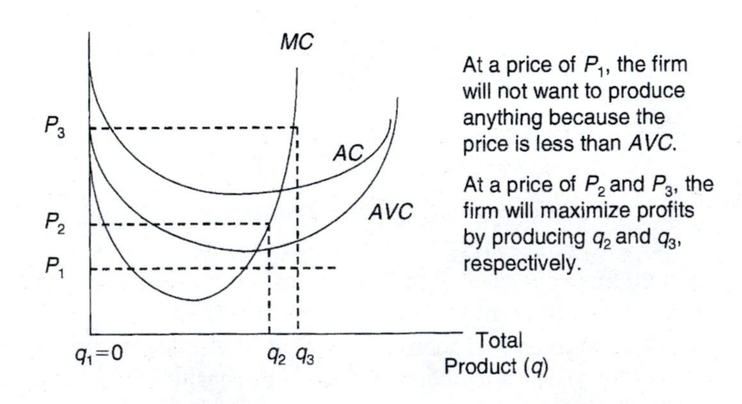

FIGURE 13.10 The Firm's Production Decision.

Towards the Long Run

In our discussion above we considered how costs changed if firms increased or decreased output. Throughout, we assumed that at least one input was being held fixed. In one example, labor could be varied but the number of tomato canning processing facilities could not. In another example, the nitrogen applied to a crop could be varied, but the number of tractors, combines, and grain bins could not. As the firm moves from the short run to the long run, inputs that were formerly fixed are now under the control of the firm. The tomato canning facility can now increase the number of processing facilities in addition to labor, and the wheat farmer can purchase more equipment in addition to nitrogen. A useful, general view of the firm is one in which raw materials and labor are variable in the short run, but capital (e.g., buildings, machinery, offices) cannot. However, in the long run, capital can be changed as well. In the long run, all inputs are variable. There are no fixed costs, so the average variable cost and average cost curve become one in the same.

For simplicity, let us assume that there are only two inputs: labor and capital. Capital is fixed in the short run, but labor is not. In the long run, both labor and capital can be varied. As you might expect, labor and capital generally complement each other. They are complements in production, meaning an increase in the use of one enhances the productivity of the other. Try painting a house with a paintbrush and then try it with a spray gun and you will understand this complementariness. Your natural painting abilities have not changed a bit, but with a spray gun in hand, your productivity soars.

The average cost curves drawn previously assume that capital is fixed. It was like assuming we varied the number of painters while holding the number of spray guns constant. But if we can increase both the number of painters and the number of spray guns, we greatly increase our painting productivity. Now we are able to paint two houses per day instead of one and our cost of painting each house declines. This is referred to as economies of scale (or increasing returns to scale)—where the firm expands its production ability by increasing the use of all inputs and finds that its average costs fall.

Economies of scale typically disappear at some production level. In Chapter 4 we discussed how beef processing firms have learned they can greatly reduce their average costs by "getting big." Building bigger processing facilities that rely more heavily on machinery and automated processes, they can transform live-cattle into beef at a lower per pound cost. Again, this is called increasing returns to scale. When increasing returns to scale exist, firms will get bigger. Their lower costs allow them to sell their output at a lower price, driving their competitors who did not "get big" out of business. However, if a single beef processing firm keeps growing and growing, they need greater and greater supplies of cattle. To induce cattle producers to increase their production, they must receive a higher price for their cattle. As the price of cattle rises, so too does the beef processing firm's costs, and average costs begin to rise. We refer to this as *diseconomies of scale* or *decreasing returns to scale*, where greater output leads to an increase in average production costs. In general, economists say that at lower levels of output there are increasing returns to scale, but as output continually expands, the firm will eventually realize decreasing returns to scale. Think of football. If you start off small, gaining weight will usually improve your playing ability. But once you hit a certain weight, you are no longer big but obese, and additional weight hurts your playing ability. As you might expect, there is an optimal weight that maximizes your playing ability. Similarly, there is one firm size that minimizes firm costs. This point is called the *minimum efficient scale*.

Between increasing and decreasing returns to scale is the *minimum efficient scale*, the point where long-run average costs are at its lowest. In Figure 13.11 there are two types of average cost curves. The long-run average costs illustrate how costs change over the long run when all inputs can be varied. The other average costs curves are short-run curves, dictating how costs change with production when at least one input is fixed. Not surprisingly, the long-run average costs are always lower. This is because in the long run the firm has control over more inputs and therefore has more options at its disposal to keep prices low.

Figure 13.11 should be interpreted in the following way. Take point A. If the firm wishes to increase production in the short run, at least one input is held fixed. This hinders the ability of the firm to expand efficiently. In a tomato canning facility there are two ways to expand production. The first is to make everyone work overtime, exerting more effort in the same factory. This is not very efficient because workers become exhausted and machinery runs longer than it is designed to run. Another is to open a new factory and hire new workers. This second method is usually more efficient (as long as the output increase is to be permanent), meaning it allows you to produce at a lower cost per unit. The first method refers to a short-run output expansion, like moving from point A to point B in Figure 13.11. Because capital (the processing facility) is being held fixed, the increase output moves along the average cost

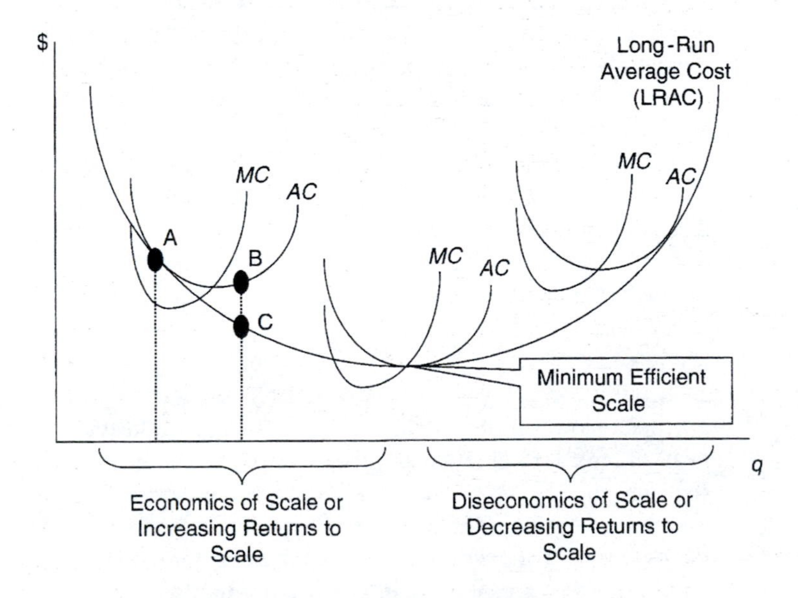

FIGURE 13.11 Firm Costs in the Long Run.

curve AC. The long-run expansion increases both hours worked by laborers and capital, and so average costs are even lower. The firm moves along the long-run average cost curve—from point A to point C. The same output expansion is achieved at a lower average cost.

The difference between the short and long run boils down to the number of options available to the firm. A firm wants to increase output, and we presume it increases output at the lowest cost possible. The more inputs the firm can vary, the more options it had to find a less expensive means for expanding output. In the short run the firm's options are limited because some inputs are held fixed. The long run gives the firm more ways to increase output at a lower cost, and if a less expensive way can be found, the firm will pursue it. Just like a consumer can be made happier by being given more choices, a firm can decrease its cost by being given more freedom to alter input usage.

These average cost curves are simply a snapshot of the firm. In reality, the curves themselves change as input prices change and with technological advancements. Consider the brewing industry between 1950 and today. Beer production is usually measured in millions of barrels, where one barrel is 31 gallons. In 1950 the minimum efficient scale for brewers was 0.1 million barrels. This was the lowest point on the long-run average cost curve for the typical firm. Since then, technological innovations have allowed brewers to realize economies of scale past 0.1 million barrels. After these technological changes, the minimum efficient size for brewing increased to 8 million barrels in 1970 and 18 million barrels in 2000. Although the demand for beer has grown since 1950, mainly since incomes and populations have risen, demand has not grown as fast as the minimum efficient scale. As a result, it takes fewer brewers to meet the market demand for beer. Some brewers had to leave the market, and it would undoubtedly be those with higher costs. As Figure 13.12 shows, the increase in the minimum efficient scale had a pronounced effect on the number of mass-producing brewers (number of brewers excluding your local microbrewery). This number has fallen from 350 in 1950 to 24 today. Today, most of the beer consumed in the United States is produced by one of three brewers: Anheuser-Busch, Miller, and Coors (Tremblay and Tremblay 2005). This is in large part due to a growing minimum efficient scale. It pays for firms to get bigger, and they do. Yet consumer demand does not grow as fast, so some firms go out of business.

Year	Minimum Efficient Scale (millions of barrels)	Number of Mass-Producing Brewing Companies
1950	0.1	350
1970	8.0	75
2000	18.0	24

FIGURE 13.12 Economies of Scale and Number of Brewers in the U.S. Brewing Industry.

Source: Tremblay and Tremblay (2005).

SUMMARY

In the introduction to this chapter we discussed how engineers build models of bridges before the real thing, and NASA simulates how space shuttles will fly in space before actually flying there. If you cannot build a model bridge, you cannot build the real thing. And if you cannot manage model or simulated businesses profitably, your chances at the real thing are small. Agribusiness managers face numerous, complex problems, such as how many inputs like fertilizer or labor should be used each week, how much output should be produced each month, and how much capital should be procured to ensure the long-run profitability of the firm.

To prepare students for decision making in these complex environments, this chapter built a "model firm." We assumed both the output price and input prices are fixed and that the firm produces a single identical product. Based on these assumptions the "model firm" became a collection of marginal product curves and cost curves. The chapter then discussed how to determine the optimal input use, production levels, and how firms' costs differ in the short and long run.

CHAPTER FOURTEEN

Agriculture and Society

It is not an exaggeration to say that all of agriculture is intrinsically a struggle *against* nature.

—*Erik Lichtengerg*, professor, University of Maryland (2004)

INTRODUCTION

In the title of this book, *Agricultural Marketing and Price Analysis*, the word *agricultural* comes first. Understanding agricultural markets requires a foundation in economics *and* agriculture. Economics is a social science with the versatility to become relevant in any area, especially agriculture. Agriculture has changed greatly since man planted his first seed, and these changes present challenges in which economics is well suited to assist. Three of these challenges are genetically modified food, antibiotic use in livestock production, and nutrient runoff from farming. These are three among many important and contentious issues. These three issues are revisited frequently throughout this book, illustrating how economics can assist governments and firms to successfully confront these challenges.

The objectives of this chapter are to understand how agriculture and society has evolved by studying the history of the following three issues:

1. genetically modified food
2. antibiotic use in livestock
3. nutrient runoff from farming

Indeed, a scholar understands both the current issues and the history behind those issues. One cannot understand modern agriculture without understanding its roots. What follows is agriculture's history, starting with the first farmers in modern-day Iraq. We will see how humans genetically altered plants and animals even before they learned to farm, which will help us put genetic modification of food into a historical perspective. Then we will study the close relationship between livestock and humans, how this relationship altered history, and why antibiotic use in agriculture is a concern to some people today. Our focus will then turn to the nutrients of life: nitrogen, phosphorus, and potassium. For thousands of years these nutrients limited human's food supply, yet today they are almost unlimited. The result has been a much larger and cheaper food supply, but greater water pollution as well.

PLANT DOMESTICATION AND GENETICALLY MODIFIED FOOD

Between 6000 and 9000 B.C., humans in the Fertile Crescent (present-day Iraq) made the first transition from hunter-gatherers to an agricultural society. Humans have always survived by consuming plants and animals. What makes agricultural societies different from hunter-gatherers is their large consumption of *domesticated* plants and animals. What makes domesticated species different from the wild counterparts is that their genes were altered by humans selecting which plants and animals would reproduce. From the Fertile Crescent, the use of agriculture to provide food would spread East and West first, and then North and South. Other regions also adopted agriculture independently, but not until later. For example, Indians in (present-day eastern United States) invented crop farming independently around 2500 B.C. (Diamond 1999).

Agriculture emerged in the Fertile Crescent first because its climate was ideal for producing a storable grain seed. Winters in the Fertile Crescent around 9000 B.C. were mild and wet. The summers were long, hot, and dry. For plant species to survive they had to produce a seed that could survive these long, hot, and dry summers. Put differently, the seeds had to be storable. This storable seed allowed humans to develop sedentary civilizations. Instead of wandering the country searching for new food, people could collect the grain seeds in one place and store them, providing food year round. Although it took humans some time to learn crop farming (*Homo sapiens* first evolved around 150,000–200,000 years ago, but only started farming around 9000 B.C.), they eventually learned that if you plant a seed, a crop usually follows (Diamond 1999).

At first, humans in the Fertile Crescent simply planted seeds from wild grains. These wild plants would later become domesticated and would evolve (with our help) into barley, peas, wheat, and lentils.[1] Domestication refers to the act of genetically altering a plant or animal to become more useful to humans, simply through the act of humans deciding which individual plant seeds to sow. Over time, humans genetically altered wild grains in such a way that they no longer resembled their ancestor. What is interesting is that we altered the genetics of plants and animals without even knowing it.

Next time you pass a wheat field, if it is close to harvest, notice that the plant does not drop the seed to the ground. It holds the seed high and erect. That is because those wheat varieties have a particular gene sequence that tells the plant to hold the

---

[1]Wild grains in China would later evolve—with our help—to rice; in Mesoamerica corn, beans, and squash; in the Andes potatoes; and in the eastern United States sunflowers.

seed high even when the seed is fully formed. In the wild, this gene sequence is fatal. Wild plants ensure the survival of their species by making sure their seeds fall to the ground to germinate. Think back to our ancestors in the Fertile Crescent around 9000 B.C. Much of their food came from the gathering of wild grain seed. Most of these wild grains will release their seed when formed, but there is genetic variation within any species, and by genetic mutation there would certainly be some wild grains that did not drop their seed.

Now here is the important part. If you are gathering seed, is it easier to pick seeds held erect above the ground or seeds you must bend down to gather? Wild grains with the genetic mutation that prevented them from dropping seeds were easier for humans to pick and so were more likely to be picked and planted later. Humans *preferred* plants with this genetic mutation, so they *preferred* to plant seeds with this genetic mutation, and over time plants with this mutation came to dominate the population. These new plants with different genes were an improvement; they produced seeds that were easier to harvest (Diamond 1999).

This was the first step towards plant domestication. Unknowingly, by picking seeds that were easiest to pick, humans altered the genetic makeup of these wild grains to make them more useful to humans. Many more genetic alterations were made, also without humans even knowing it. Imagine a tribe gathering seeds, some of which to eat now and some to store for future consumption or planting. As mentioned before, there is genetic variation across all species. Just like some people weigh more than others, some seeds are bigger than others. In the wild, both big and small seeds would form, but humans were more likely to plant big seeds because big seeds provide more food. Given humans prefer to plant big seeds, over time the plant population becomes dominated by plants with the genes producing big seeds. Over time, plants' seeds got bigger and bigger. When corn was first being domesticated in Mesoamerica around 5000 B.C., its ears were only about 2 centimeters long. By 3400–3200 B.C. it had grown to 4.3 cm. Today it is easy to find ears in excess of 6 cm, all because humans prefer more food to less.

This is yet another example of plant domestication, where humans altered the genetic makeup of wild plants to better serve their needs. This type of genetic alteration is referred to here as *genetic selection*. You plant only the crops possessing the genes you desire, and over time most all crops will possess those genes. The problem with genetic selection is that it is slow. Often one must wait for a genetic mutation (a natural but random change in the genetic makeup of a plant) to produce a better plant, and this can be quite a long wait. This wait is being reduced by sophisticated plant breeding technologies. Monsanto uses "gene markers" that identify potentially useful genes, collect data on genes from plants, and crunch these data in powerful supercomputers. Plant breeders today include statisticians and computer scientists in addition to biologists and agronomists. These technologies reduced the time it took to alter one particular soybean variety named Vistive by three years (Leonard 2006).

Instead of waiting for mutations to occur, one can force gene mutations. You may find it surprising that crop breeders around the world frequently induce genetic mutations by zapping plants with nuclear radiation or chemicals. This process has been termed *mutagenesis*. As you may suspect, this leads to the death of many plants, but mutagenesis of enough plants will eventually produce a better plant. There are over 2,252 crop varieties in over 70 countries produced using mutagenesis, and if you have ever drunk a beer or eaten pasta, you have likely consumed one of these varieties (DeGregori 2007).

An even more direct route is *genetic modification*, where a sequence of genes from a plant is directly removed and replaced with a sequence from another organism, usually a bacteria. Organisms that contain strands of genes from different species are referred to as *genetically modified organisms* (GMOs) or *transgenic organisms*. Two transgenic plants are Bt cotton and Bt corn, where genes from a soil bacteria named *Bacillus thuringiensis* are inserted into the plant, inducing it to produce its own pesticide. This pesticide is harmless to humans and lowers the farmers' cost of production because they need not apply as much pesticide. Pesticides can comprise 10% to 30% of total crop production costs, so this cost reduction can be significant and lead to lower food prices (Oklahoma State University 2005a). Moreover, pesticides pose human health risks, so Bt corn and Bt cotton may improve human health as well. Other crops have been genetically modified to become tolerant towards selected herbicides and are referred to as herbicide-tolerant varieties.[2] Weeds have not undergone this modification, so the farmer can directly apply a single herbicide on the entire field, killing everything except the crop. Many farmers have found that this lowers production costs and can even lower total herbicide use.

Genetic modification is now a common practice in seed production. As shown in Figure 14.1, almost all soybeans are genetically modified to become herbicide-tolerant,

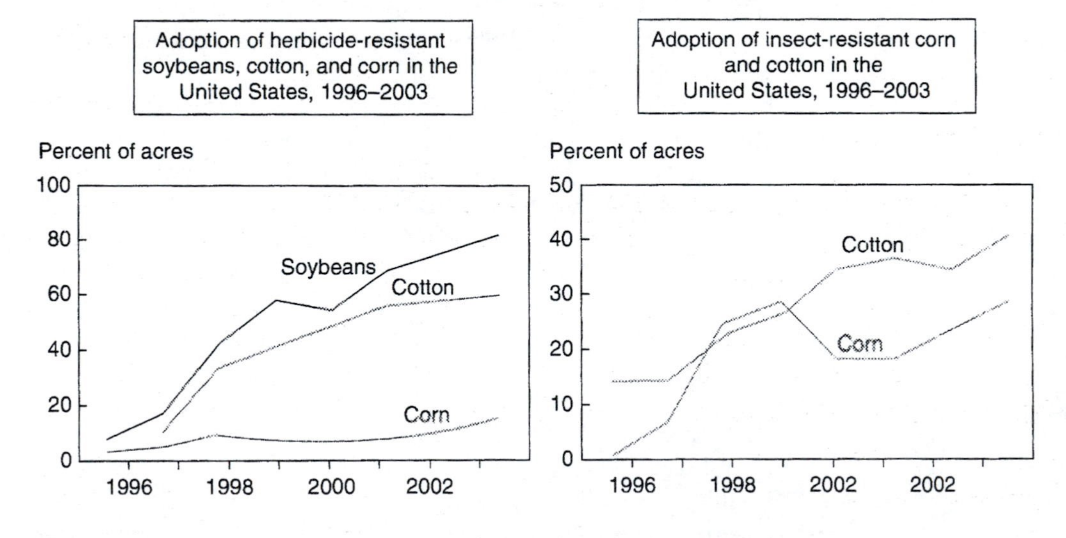

FIGURE 14.1

Source: Price, Lin, Falck-Zepeda, and Fernandex-Cornego (2003). Graphs reprinted with permission from the Economic Research Service.

---

[2]Pesticides refer to chemicals applied to any pest, whether it is an insect, weed, or disease. Herbicides are pesticides for weeds, insecticides are pesticides for insects, and fungicides are pesticides for diseases.

and almost half of all cotton is produced from the Bt variety. Widespread adoption of genetically modified crops has yielded benefits for society. Some estimate the benefits to the United States to be worth more than $800 million (Price, Lin, Flack-Zepeda, and Fernandex-Cornego 2003). Yet, the widespread adoption has been accompanied by strong opposition by some groups. There are concerns that some consumers may be allergic to genetically altered crops. Gerber Products Company produces the familiar Gerber baby foods. In 1999, it received a fax from Greenpeace asking if the company had taken steps to avoid genetically modified ingredients in their baby foods. In an effort to avoid any food safety concerns, Gerber immediately announced to the public it would limit the use of ingredients from genetically modified crops (Schweikhardt and Browne 2001).

Also, genetically modified crops may induce genetic alterations in pests and weeds that make them increasingly difficult to control. Animals and plants evolved to survive in their environment. Bt corn is genetically altered to produce its own pesticide. It is likely that pests will ultimately develop a resistance to this pesticide. Roundup Ready soybeans are resistant against the Roundup herbicide, and weeds are not. However, widespread use of Roundup Ready soybeans will likely lead to the development of resistance to Roundup in weeds also. Some fear these weeds will become immune to herbicides and become increasingly difficult to control. Finally, many fear that genetically altered crops will breed with regular crops. This fear is founded, because scientists have found DNA from Bt corn in native corn varieties in Mexico (Quist and Chapela 2001). The two crop types may exchange genes in such a way that we lose our former "natural" crops, and some are concerned this genetic contamination will prevent us from ever going back to nongenetically modified food. Pharma crops are also of concern. These are crops that are genetically modified not to produce food, but pharmaceutical and industrial chemicals. As of yet, no business actually sells chemicals produced from pharma crops. Yet the USDA has approved more than 100 field tests for pharma crops. Even though the Food and Drug Administration has yet to approve the production of pharmaceuticals from these crops, it is only a matter of time. Precautions have been taken to prevent the pharma crops from cross-pollinating with traditional crops. Some companies have gone so far as to grow the pharma crops in caves under artificial light.

Concerns over genetically modified organisms are real and understandable, and organizations opposed to GMO crops are large and well-funded. To see for yourself, visit the site of *Union of Concerned Scientists* and peruse their large library of documents against GMO food. Similar controversies have been generated in the livestock sector. Like plants, animals have undergone genetic alteration by humans through the selection process. The pioneer of livestock breeding, Robert Blackwell (1725–1795), developed a breeding program that increased the average weight of sheep from 28 lbs. to 80 lbs. and the average weight of cattle from 370 lbs. to 800 lbs. Today, most livestock are bred artificially, so that the semen from one male with desirable genes can breed many more females than possible in the wild. Ovary transplants are common on purebred beef farms, and you have probably heard of Dolly, the sheep clone. Like plants, the livestock breeding technology has gone so far as to include genetic modification. Transgenic salmon have been produced and may one day be approved for consumer sale by the Food and Drug Administration.

Hopefully, this section illustrates that genetic modification is just another form of genetic alteration and is nothing new to humans. The only thing that has changed is the rate at which we alter plant and animal genes. It is easy to think that genetically modified food is "unnatural." But recall that our food is derived from domesticated plants and animals, which means their genetics have been altered from their natural state to become more useful to humans. Our base crops like corn, wheat, and rice are not natural plants you find in the wild, but a feat of human engineering. The same can be said for our livestock. Agriculture itself is a form of engineering, where products produced by nature are tweaked to better serve humans.

We do not write this in an attempt to convince you that genetically modified food is "good" or "bad," only to instill an understanding of how the issue relates to the history of agriculture. There are reasons to fear and embrace genetic modification. But whether you decide to fear it, embrace it, or something in between, you should understand that genetic modification is just one form of genetic alteration. And genetic alteration has occurred for thousands of years and, as long as we don't nuke each other to death, will occur for thousands more.

DISEASE, CONQUESTS, AND ANTIBIOTIC USE IN LIVESTOCK

Animals with the potential for domestication must

(1) be an herbivore
(2) grow quickly
(3) be willing to breed under captivity
(4) not have a nasty disposition
(5) not have a tendency to panic
(6) live in herds
(7) have well-developed herd hierarchy
(8) be willing to share territories with other herds

Animal domestication began shortly after plant domestication (between 8000 and 2500 B.C.), and mostly took place in Eurasia. Not all wild animals can be domesticated. All animals that have potential for domestication must possess eight characteristics.

The need for each trait is straightforward. It is cheaper to feed an animal grass than meat, so the animal must be an herbivore, and to provide meat efficiently must grow quickly. If an animal cannot breed under captivity, you cannot selectively breed for better genes. Indians and Egyptians tried to domesticate cheetahs but they would rarely breed in captivity. Why were horses domesticated but not zebras? Because zebras are very ill-tempered, becoming especially dangerous as they grow older. Zebras in circuses always have a muzzle. This is because zebras have a tendency to bite and not let go. Animals that tend to panic are hard to keep together, and if animals will not live in herds, it is even more difficult to contain them. Animals with a well-developed hierarchy are willing to accept orders from a boss, allowing humans to become this boss and control the animal.

Few animals in the wild possess all eight characteristics, and most of those who did were located in Eurasia. Thus, animal domestication and livestock farming began in Eurasia. Even though other regions domesticated some animals (e.g., dogs were domesticated in North America before the arrival of Europeans), their use of livestock for food provision was minor in comparison with humans in Eurasia. It is important to note that plant and animal domestication was first adopted on a large scale in Eurasia; thus, we say agriculture was first adopted large scale in Eurasia. Agriculture can support a much larger population density than reliance on hunting-gathering. Consequently, these agricultural communities became larger and more organized than the hunters and gatherers who were scattered about. History reveals that large,

agricultural, and organized societies almost always conquer societies that are
wise (Diamond 1999).

Consider North America, which held a large population of natives and only natives prior to the sixteenth century. These natives had developed agricultural systems, but these were mostly cropping systems. The only livestock used was the dog, and it comprised only a small component of meat provision. Conversely, the Europeans had domesticated many animals for animal labor and food. Eventually, Europeans invaded North America and conquered every inch of it, and you may find it surprising that the major reason Europeans conquered the Native Americans is because Europeans relied mostly on livestock for meat (Diamond 1999).

What does livestock production have to do with war? Today, not much. But this was not always true. To see why, we must discuss epidemic diseases and their causes. Before agricultural communities were formed, epidemics did not exist. But once agriculture allowed people to live in close proximity to one another, the major killers of mankind arrived.[3] These include smallpox, flu, tuberculosis, malaria, plague, measles, and cholera. Epidemics such as these are crowd diseases. They can only spread through crowds, and they spread fast, inflicting great misery. Before humans lived in cities, crowd diseases had no human crowd. Diseases are bacteria and viruses, living things. All living things come from something. So if crowd diseases suddenly appeared when cities were formed, where did they initially come from?

The answer is livestock. All livestock are domesticated from wild animals that lived in herds. Herds are crowds of animals, and crowd diseases were prevalent in animal herds. All of our major crowd diseases are bacteria or viruses that evolved from livestock diseases. The measles, tuberculosis, and smallpox came from cattle; the flu from pigs and ducks; and malaria from chickens. Once humans started living in close proximity to livestock, these diseases evolved from animal diseases to infect humans.

The important thing to understand about crowd diseases is that even though they kill many people (the Black Death killed one quarter of Europe's population between 1346 and 1352), the crowds eventually develop immunity. Those immune crowds can still spread the disease to others who haven't developed immunity though. When Cortes arrived in present-day Mexico in 1519, he faced a large number of fierce Aztec warriors. Cortes was outnumbered but possessed one important weapon—smallpox. A slave from Spanish Cuba arrived infected with smallpox in 1520. Smallpox spread throughout the Aztecs, reducing their population from 20 million to 14.6 million. But not one of Cortes' men died from smallpox, because they were immune.

A similar story can be told for much of the world. Eurasians easily conquered all of North America and Australia (and many islands in the Pacific Ocean) by bringing their diseases with them when they invaded. Only one hundred years after Columbus arrived in North America, the native population declined dramatically, perhaps by as much as 95%, mostly due to these crowd diseases. One might wonder why the natives did not pass their diseases on to the Europeans. The reason is that they had no crowd

---

[3]Recall that hunters and gatherers cannot live close to one another because they would soon deplete the food supply. Agriculture produces much more food per acre than hunting and gathering.

diseases. Even though some lived in densely populated areas, they did not live close to livestock, and so crowd diseases never transferred from herds of animals to herds of Native Americans (Diamond 1999).

That is the past, but the fact that disease can spread from livestock to humans is important today. If you do not have an agricultural background, your image of a farm may include cows grazing in pastures, hogs playing in the mud in an open pen, and chickens scratching in the sand outside a red barn. For the most part, those days are over. Cattle still spend most of their life in pastures, but before slaughter are fed grain in densely populated pens. Hogs and chickens rarely ever see the outside of a building. Hogs are raised on a concrete slab and chickens are raised on sawdust, both inside of a densely populated building.

Living nose to nose, disease can spread quickly within these confined livestock buildings. Sickness and mortality rates can be reduced by regularly adding low doses of antibiotics to the livestock feed and water. This is a common practice in hog and poultry production. These doses are lower than that prescribed to treat an existing disease, and so this practice is referred to as *sub*therapeutic antibiotic use. Livestock are healthier and grow faster under subtherapeutic antibiotic use, ultimately leading to lower consumer prices.

Sows give birth in a farrowing crate, designed to keep the mother from lying on her farrows (baby pigs).

When raised in open lots, many farrows die from the weight of their mother.

Once fertilized eggs are lain, they are removed from the hen and hatched in an incubator. Later, they are delivered to poultry farms to be raised.

Hogs are generally raised in densely populated buildings and spend most of their lives on concrete. The buildings keep the hogs dry and at a comfortable temperature, promoting growth.

Like hogs, chickens are raised in densely populated buildings. The building floor is covered in sawdust. The chickens will never leave the building until slaughter.

FIGURE 14.2 Examples of Confined Livestock Production Facilities.

There is a potential risk to subtherapeutic antibiotic use though. These low antibiotic doses may mitigate disease, but they do not kill the bacteria. Eventually, genetic mutations within the bacterial population will emerge making the bacteria immune to the antibiotic. Bacteria with this immunity will then come to dominate the population. Colonies of antibiotic-resistant bacteria will result. This can threaten swine health but pose human health problems as well. Many of the antibiotics used in growing livestock are often used by humans. As we saw previously, disease can easily transfer from livestock to humans. Thus, it is possible that harmful antibiotic-resistant bacteria can transfer to humans, producing diseases that antibiotics no longer treat.

This has caused some groups to push for bans on subtherapeutic antibiotic use in livestock. Numerous bills have been proposed in Congress, though so far none have passed. Other countries, such as Denmark, already have bans on subtherapeutic antibiotics. In an effort to curb consumer concern, McDonald's announced in 2004 that it would no longer purchase pork from producers using the same antibiotics in livestock production as used in humans. The debate continues. The pork industry contends that the human health risk is minuscule compared to the benefits of subtherapeutic antibiotic use. Special interest groups like the Union for Concerned Scientists contend otherwise, arguing that the risk is substantial and outweighs the benefits of lower pork prices (Lusk, Norwood, and Pruitt 2007). Like we said before, we are not trying to convince you that one side is right and one side is wrong. Both sides have valid points and concerns. We merely wish to present the historical relationship between humans, their livestock, and the diseases that both sides face. It is difficult to fully understand this issue without understanding the history behind it.

SUSTAINABLE AGRICULTURE AND INDUSTRIAL FERTILIZERS

It is common these days to hear the term *sustainable agriculture* thrown around. Sustainable agriculture is often viewed as an alternative to today's common agricultural practices. Although there is no one definition of sustainable agriculture, its advocates generally see several problems with current practices. These are excessive environmental degradation, excessive depletion of nonrenewable resources, and social equity problems. To gain an understanding of these concerns, we will discuss one feature of modern agriculture that many view as an unsustainable practice: the use of industrial fertilizers.

Plants need water, carbon dioxide, and sunlight to grow. These inputs are relatively plentiful. More scarce is nitrogen, potassium, and phosphorus; nutrients also necessary for plant growth. Because animals depend on plants, they too rely on these nutrients. Phosphorus and potassium (denoted P and K, respectively) are minerals contained in the soil and can only be transported through soil erosion or animal droppings. If we take modern agriculture out of the picture, there is a cycle of P and K from the soil, to the plant, to the animal, and back to the soil. The amount of P and K is relatively fixed. Even though P and K can move from one region to another via soil erosion and animal transport (animals eating plants in one area and defecating in another), the total amount available for life varies little.

Excluding the effects of modern agriculture, the amount of nitrogen available to sustain life is also fixed. Like P and K, nitrogen (N) is recycled between plant and animal when the animal consumes plants and the nitrogen it contains, returning that nitrogen to the soil through excretion. There is also an exchange of nitrogen between the atmosphere and the soil. The atmosphere is roughly 75% nitrogen, but this nitrogen is elemental nitrogen (N_2), which is inaccessible to plants. Elemental nitrogen can be transformed to a plant-accessible form through certain atmospheric conditions like lightning, but most comes from *nitrogen fixation* by soil bacteria. Some of these nitrogen-fixing bacteria live independently of plants, and some have a symbiotic relationship with plants. Either way, these nitrogen-fixing bacteria convert atmospheric nitrogen to a molecule that can be consumed by the plant for growth and reproduction. Then, there is another class of bacteria called denitrifying bacteria, which take plant-accessible nitrogen and return it back to the atmosphere as elemental nitrogen (N_2).

Before the advent of industrial fertilizers, agriculture relied on this nitrogen cycle for plant and animal growth. For the most part, there was no addition or subtraction to the total nitrogen available for life. Although nitrogen availability certainly shifted across years and regions, on average it remained the same. Along with other essentials like sunlight, water, and carbon dioxide, these nutrients supply the world with food. Plants need them to grow and reproduce, and animals need the plants to grow and reproduce. These nutrients are also scarce, meaning that they are the main constraint on life growth in the natural world. If the supply of N, P, and K increases, nature will respond by producing more plants and animals. After humans adopted agriculture they eventually learned how to increase soil nutrients through natural fertilizers. The Chinese learned they could increase soil nutrients by treating bones with lime and placing it in the soil and by straining their urine through soil. The first pilgrims to America were taught by the Indians that placing a small fish in the same hole as a corn seed would yield a greater corn harvest. Also, soil nutrients could be replenished by leaving land fallow for a period of time.[4]

Beginning in the eighteenth century better methods of soil fertilization were found. Farmers learned that planting a legume crop left the soil more fertile,[5] and in Flanders[6] they learned that excrement from outhouses could be applied to cropland and lead to greater crop yields. They called the outhouse excrement "night soil." Ground bone meal became a popular fertilizer in the nineteenth century, because it supplies plants with phosphorus. Although this was an advancement, it is important to note that there still existed a natural balance of nutrients. Crops consume nutrients from the soil, and humans consume nutrients from the crop. Then, humans return

---

[4]Leaving land fallow means not planting anything on the land. This gives soil bacteria a chance to replenish the nitrogen removed from the previous harvest.

[5]A legume is a flowering plant family containing bacteria in its roots that convert atmospheric nitrogen to a plant accessible nitrogen. Thus, legumes will grow well even if there is no nitrogen in the soil, and after the plant dies it will release some of the nitrogen it extracted from the atmosphere into the soil for other crops to consume.

[6]A historic region eventually assimilated by Belgium, the Netherlands, and France.

those nutrients largely in the form of excrement. The total amount of nutrients did not change. However much was taken from the soil was returned (Rider 1995).

Eventually humans devised more advanced ways of fertilizing the soil, ways that altered the nutrient balance towards a steady increase in nutrients available for life. The "Father of the Fertilizer Industry" is said to be the German Justus von Liebig. He began mass-producing commercial fertilizers by mining rock phosphate. Today, we still acquire phosphorus fertilizer through mining. Scattered throughout the world are deposits of ancient marine life that became phosphate rock. Combining this rock with sulphuric acid produces a phosphorus solution that can be applied to crops. Potassium fertilizer is also mined. Most of our dry land today was covered by oceans long ago. When these oceans evaporated, they left behind large potash deposits, which are today mined and applied to land to supply crops with greater potassium (The Fertilizer Institute 2005).

World War I caused worldwide suffering but also gave rise to inexpensive nitrogen fertilizers. Many World War I bombs were nitrogen-based, and many factories were built to produce these bombs. After the war these factories were converted so they could take atmospheric nitrogen and convert it to a plant-accessible nitrogen (much like soil bacteria). Today, we produce nitrogen by mixing natural gas and air at high temperatures and high pressures. The result is anhydrous ammonia, which is a molecule with one nitrogen atom and three hydrogen atoms (NH_3). Because nitrogen fertilizer production requires the extraction of natural gas, it relies on fossil fuels.

There are two important facts to be gleaned from this. The first is that our food system is essentially based on the harvest of nonrenewable resources. All crops receive fertilizer treatments, which consist of nitrogen, phosphorus, and potassium. These crops are either consumed by humans or consumed by livestock who are later consumed by humans. Nitrogen is produced from harvesting natural gas, phosphorus from mining phosphorus rock deposits, and potassium from mining potassium salt deposits. Thus, our current agriculture system is unsustainable. We cannot continually harvest these resources at the current rate. Just like we must someday find an alternative to fossil fuels for energy production, we must someday find an alternative to fertilizer production. Without an alternative, crop yields will fall and food will become more expensive.

The second fact is that our methods of producing fertilizer are increasing the nutrients available for plant and animal growth. Industrial fertilizers make available nutrients that before were inaccessible to plants. The nitrogen in fertilizers was formerly in the air, and the phosphorus and potassium was contained in mineral deposits. The total amount of plant and animal life must increase with the discovery of industrial fertilizers. In fact, that is the point of industrial fertilizers: increasing food production. This is one of the major reasons food is so cheap compared to the past.

This sounds good. It is hard to argue that more food is bad. But there are drawbacks to constantly increasing the nutrient supply to the world. Industrial fertilizers lead to greater life, but sometimes they encourage too much growth of certain life forms that lead to environmental degradation. Not all the nutrients produced in fertilizers will be consumed by crops and livestock. Some will leave the field as nutrient runoff.[7] Livestock manure (ultimately derived from plant nutrients as the animal consumes grain) is often expensive to haul to faraway fields and therefore is frequently overapplied in fields near the livestock production facility. This overapplication means there are more nutrients than the plant can consume, increasing the potential for nutrient runoff. Many of these nutrients will enter surface waters, providing more nutrients for aquatic life. These nutrients are first consumed by algae, and when these algae die and sink to the bottom, they are consumed by bacteria. As more and more nutrients feed these bacteria, the colonies can become so large that the bacteria consume all the available oxygen, killing all aquatic life in that water body in a process referred to as eutrophication. This is not a mere possibility. Eutrophication has been documented in many cases throughout the United States, and nutrient runoff from agriculture is often the cause.

Of all the rivers and streams whose health has been studied by the Environmental Protection Agency (which is only 19% of all the river and stream miles), close to 20% have been polluted by agricultural practices. A total of 7% of these rivers and streams have undergone some degree of eutrophication, partly from agricultural nutrient runoff and partly from human waste treatment plants. Agriculture is also a leading contributor to lake pollution and a major factor of estuary pollution (Environmental Protection Agency 2002).

Concerns over pollution from nutrient runoff has led to the formation of special interest groups and regulations seeking to curb agricultural pollution. The attorney general of North Carolina forced swine producers to invest millions of dollars researching more environmentally friendly manure management practices. The attorney general of Oklahoma is suing a group of poultry producers in Arkansas whose applications of poultry litter to land has led to nutrient runoff polluting the drinking water of Tulsa, Oklahoma. Eutrophication of the Tar-Pamlico estuary in North Carolina induced the state government to impose restrictions on nutrient runoff from agricultural and nonagricultural sources. There are other undesirable effects from industrial fertilizers. Nitrogen fertilizers can lead to greenhouse gases and greater global warming. They can also contaminate drinking water, posing a particular threat to infants.

Agriculture is presented with a challenge. It cannot completely abandon the use of industrial fertilizers. Crop yields would fall and livestock would become more expensive to feed. The result would be high food prices and consumer backlash. Yet, agriculture must learn to utilize industrial fertilizers in a more sustainable manner. Its reliance on nonrenewable resources means we must discover alternatives to producing fertilizer from mineral deposits and natural gas.

Agriculture must also learn to mitigate the pollution produced from nutrient runoff. Industrial fertilizers increase the total available nutrients for life. They can

[7]Nitrogen usually leaves the field in a water-soluble form, flowing underneath the field surface. Phosphorus tends to leave the field via soil erosion.

FIGURE 14.3 Lake Eucha Polluted by Nutrient Runoff from Agriculture. This picture shows large algae blooms in Lake Eucha in eastern Oklahoma. These blooms largely stem from overapplications of poultry manure to crops. If these blooms continue, the lake will eventually suffer under eutrophication, leaving a dead lake. Because this lake provides drinking water to the city of Tulsa, the city has had to spend large amounts of money to make the water drinkable again.

increase life in a good way by producing more wheat, beef, and corn for human consumption. They can increase life in a bad way by increasing water bacteria populations to the point of eutrophication. However, eutrophication can be reduced by environmentally friendly management practices. These are often referred to as *best management practices (BMPs)*. Applying manure to land at a rate crops can consume prevents excess application of nutrients and therefore nutrient runoff. Farmers can install strips of natural vegetation around cropland, which will capture some of the excess nutrients. By diverting nutrient runoff from surface and ground waters, agriculture can enhance food production while limiting the extent of water pollution.

However, BMPs cost the farmer money. Some farmers will adopt them without encouragement out of a concern for the environment. Most need government action to encourage adoption of BMPs. In the past farmers could employ whatever production practices they desired and would receive little to no criticism from the public. Today, there are well-organized and well-funded interest groups whose sole purpose is to reduce agricultural pollution. They oppose many modern agricultural practices through lawsuits, and they support more stringent environmental laws. There is no doubt that agricultural practices must change, but there is great debate over how fast it should change. A widespread movement towards sustainable agriculture means we pay higher food costs now, with the benefits of less water pollution and a more sustainable future. The policy debate centers around whether the costs are smaller or larger than the benefits. The purpose of this section is not to advocate one position or another, but to clarify the topic of sustainable agriculture and how industrial fertilizers change the nutrient balance in the world. For those interested in the sustainable agriculture area, there are numerous other topics such as soil erosion and pesticide use that have important implications for our ability to feed people today and in the future.

SUMMARY

This textbook applies economic tools to agricultural issues. Three reoccurring issues are genetically modified food, antibiotic use in swine production, and nutrient runoff from farming. But developing a well-informed opinion on these subjects requires more than economics. Understanding the history of agriculture is a prerequisite for understanding genetically modified food. Genetic modification is less of a new thing and is more like doing the same thing faster—altering the genetic makeup of plants and animals to better feed humans. Humans have a long history of complex interactions with livestock and the pathogens that infect livestock. Most major human diseases stem from our close proximity to livestock. This is one reason among many that antibiotic use in livestock is a concern for today. Finally, we discussed nutrient runoff from pollution. All life on earth relies on nitrogen, phosphorus, and potassium. The major reason for our abundant food supply is that technologies allow us to produce these three nutrients at higher rates than what would prevail in nature. The greater nutrients lead to more food but also to more water pollution. Food production is naturally tied to certain forms of pollution, making food policy and environmental policy a joint topic.

References

Adams, W. J., and J. L. Yellen. 1976. Commodity bundling and the burden of a monopoly. *The Quarterly Journal of Economics* 90(3): 475–498.

Ajzen, I., and M. Fishbein. 1980. *Understanding attitudes and predicting social behavior*. Englewood Cliffs, NJ: Prentice Hall.

Alston, J. M., J. W. Freebairn, and J. S. James. 2001. Beggar-thy-neighbor advertising: Theory and application to generic commodity promotion programs. *American Journal of Agricultural Economics* 83 (November): 888–902.

Arnot, C., and C. Gauldin. 2006. Cargill, Iowa agree. *Feedstuffs*, February 6, 14.

Baker, G. A., and T. A. Burnham. 2001. Consumer response to genetically modified foods: Market segment analysis and implications for producers and policy makers. *Journal of Agricultural and Resource Economics* 6: 387–403.

Baron, J., and N. P. Maxwell. 1996. Cost of public goods affects willingness to pay for them. *Journal of Behavioral Decision Making* 9: 173–183.

Beghin, J. C., B. E. Osta, J. R. Cherlow, and S. Mohanty. 2001. The cost of the U.S. sugar program revisited. *Working Paper 01-WP 273,* Center for Agricultural and Rural Development, Iowa State University, March.

Bhatnager, Parija. 2005. Coke, Pepsi losing the fiz. *CNN Money*, March 8. http://money.cnn.com/2005/03/07/news/fortune500/cokepepsi_sales/ (accessed October 8, 2005).

Bhuyan, S. 2005. Does vertical integration effect market power? Evidence from U.S. food manufacturing industries. *Journal of Agricultural and Applied Economics* 37 (April): 263–276.

Brannon, Ike. 2005. What is a life worth? *Regulation* (Winter): 60–63.

Bredahl, L. 1999. Consumers' cognitions with regard to genetically modified foods. Results of a qualitative study in four countries. *Appetite* 33: 343–360.

Brown, Robert W. 1993. *Economic Inquiry* 31 (October): 671–680.

Carson, T. C. et al. 2003. Contingent valuation and lost passive use: Damages from the Exxon Valdez oil spill. *Environmental and Resource Economics* 25: 257–286.

Center for Disease Control and Prevention. 2004. Trends in intake of energy and macronutrients—United States, 1971–2000. *Morbidity and Mortality Weekly Report*. 53: 80–82. http://www.cdc.gov/mmwr/preview/mmwrhtml/mm5304a3.htm.

Center for Disease Control and Prevention. Available online at http://www.cdc.gov/nccdphp/dnpa/obesity/economic_consequences.htm.

Chaloupka, F. J., M. Grossman, and H. Saffer. 2002. The effects of price on alcohol consumption and alcohol-related problems. *Alcohol Research and Health*. (Winter).

Chouinard, Hayley H., D. E. Davis, J. T. LaFrance, and J. M. Perloff. 2001. The effects of a fat tax on dairy products. Department of Agricultural & Resource Economics, UCB. CUDARE Working Paper 1007, August 1. Available at http://repositories.cdlib.org/are_ucb/1007.

Cramer, G. L., E. J. Wailes, and S. Shangnan. 1993. Impacts of liberalizing trade in the world rice market. *American Journal of Agricultural Economics* 75 (February): 219–226.

Crespi, J. M., and R. J. Sexton. 2004. Bidding for cattle in the Texas panhandle. *American Journal of Agricultural Economics* 86 (August): 660–674.

Cobia, D., ed. 1989. *Cooperative in agriculture*. Prentice Hall.

Crutchfield, S. R., J. C. Buzby, T. Roberts, M. Ollinger, and C. Jordan. 1997. An economic assessment of food safety regulations: The new approach to meat and poultry inspection. Economic Research Service. United States Department of Agriculture. Agricultural Economic Report No. 755.

Cummings, R. G., G. W. Harrison, and E. E. Rutström. 1995. Homegrown values and hypothetical surveys: Is the dichotomous choice approach incentive-compatible? *American Economic Review* 5: 260–266.

Cutler, David M., E. L. Gleaser, and J. Shapiro. 2004. Why have Americans become more obese? *Journal of Economic Perspectives* 17 (Summer): 93–118.

Davis, George C. 2005. The significance and insignificance of demand analysis in evaluating promotion programs. *American Journal of Agricultural Economics* 87 (August): 673–688.

Dawkins, Richard. 1999. *The selfish gene*. New York: Oxford University Press.

de Ferran, F., and K. G. Grunert. In press. French fair trade coffee buyers' purchasing motives: An exploratory study using means-end chains analysis. *Food Quality and Preference*.

DeGregori, T. A Primer on genetic modification vis a vis the Southern African drought and famine. Common Wealth Partnership for Technology Management. http://www.cptm.org/Genetic%20Modification.htm (accessed August 10, 2007).

Denney, D., L. L. Byung, D. W. Noh, and V. J. Tremblay. 2004. Excise tax effects in the U.S. brewing industry. Working Paper. Department of Economics. Oregon State University, April.

Diamond, J. 1999. *Guns, germs, and steel*. New York: W. W. Norton & Company.

Dickinson, D. L., and D. Bailey. 2002. Meat traceability: Are U.S. consumers willing to pay for it? *Journal of Agricultural and Resource Economics* 27 (December): 348–364.

Dillman, D. A. 2000. *Mail and Internet surveys: The tailored design method*. 2nd ed. New York: John Wiley & Sons.

Duffy, M., and A. Holste. 2005. Estimated returns to Iowa farmland. Staff General Research Papers 12396. Iowa State University, Department of Economics.

Eales, J. S., and L. J. Unnevehr. 1988. Demand for beef and chicken products: separability and structural change. *American Journal of Agricultural Economics* 70: 521–532.

Eckholm, E. 2006. New campaign shows progress for homeless. *New York Times* June 7.

Economic Research Service. 2002. United State Department of Agriculture. Briefs: Agricultural trade. *Agricultural Outlook*, November.

Economic Research Service. 2005a. United States Department of Agriculture. Briefing Room. Food cpi, prices, and expenditures: expenditures as a share of disposable income. http://www.ers.usda.gov/Briefing/CPIFoodAndExpenditures/Data/table7.htm (accessed December 14, 2005).

Economic Research Service. 2005b. United States Department of Agriculture. Briefing Room. Marketing bill and farm value components of consumer expenditures for domestically produced farm foods. http://www.ers.usda.gov/Data/FoodMarketIndicators/default.asp?TableSet=3 (accessed December 16, 2005).

Economic Research Service. 2005c. United States Department of Agriculture. Briefing Room. Briefing Room: Food marketing and price spreads. http://www.ers.usda.gov/Briefing/FoodPriceSpreads/bill/ (accessed December 16, 2005).

Economic Research Service. 2005d. United States Department of Agriculture. Briefing Room. The 2002 Farm Bill: Title I—Commodity programs. http://www.ers.usda.gov/Features/farmbill/titles/ titleIcommodities.htm (accessed December 27, 2005).

Economic Research Service. 2006. United States Department of Agriculture. Key topics: Trade. http://www.ers.usda.gov/topics/view.asp?T=104200 (accessed February 26, 2006).

The Economist. 2005. The march of the robo-traders. *The Economist Technology Quarterly*, September 17.

Elitzak, H. 1999. Food cost review, 1950–97. Economic Research Service. United States Department of Agriculture. Agriculture Economic Report No. 780, June.

Elzinga, K. 1970. Predatory pricing: The case of the gunpowder trust. *Journal of Law and Economics* 13(1): 223–240.

Encarta Reference Suite 2000. Quotations. Economics. Microsoft Corporation. 1993–1999.

Energy Information Administration. Department of Energy. http://www.eia.doe.gov/emeu/international/petroleu.html#WorldReserves (accessed September 21, 2005).

Enterprise Corn, Soybean, and Cotton Budgets. Oklahoma State University. Department of Agricultural Economics.

Environmental Protection Agency. 2002. 2000 Water Quality Inventory. EPA-841-R-02-001, August.

Fatka, J. 2006. Antitrust suit looks at bundling. *Feedstuffs*, October 9.

Feedstuffs. 2004. European Commission imposes fine on animal feed vitamin cartel. December 20.

Feedstuffs. 2005a. Albertsons may consider end. September 12.

Feedstuffs. 2005b. Cow herd responds to 'green things.' November 14, 18.

The Fertilizer Institute. 2005. Nutrients for life. http://www.sharingcommonground.org/CFI/toolkit/default.asp (accessed September 5, 2005).

Fishbein, M., and I. Ajzen. 1975. *Belief, attitude, intention and behavior: An introduction to theory and research*. California: Addison-Wesley.

Gardner, B. 2002. *American agriculture in the twentieth century*. Cambridge, MA: Harvard University Press.

Garoyan, L. 1983. Developments in the theory of farmer cooperatives: A discussion. *American Journal of Agricultural Economics* 65 (December): 1096–1098.

Gayer, T., J. Horowitz, and J. A. List. 2005. When economists dream, they dream of clear skies. *The Economists' Voice* 2(2): Article 7.

Genescove, D., and W. P. Mullin. 1997. Predation and its rate of return: The sugar industry, 1887–1914. National Bureau of Economic Research. Working Paper 6032.

Genescove, D., and W. P. Mullin. 1998. Testing static oligopoly models: Conduct and cost in the sugar industry, 1890–1914. *The RAND Journal of Economics* 29 (Summer): 355–377.

Gladwell, M. 2005. *Blink*. New York: Time Warner Book Group.

Goodwin, B. K., and A. K. Misra. 2006. Are "Decoupled" farm program payments really decoupled? An empirical evaluation. *American Journal of Agricultural Economics* 88 (February): 73–89.

Hakes, Jahn J., and R. D. Sauer. 2006. An economic evaluation of the *Moneyball* Hypothesis. *Journal of Economic Perspectives* 20 (Summer): 173–185.

Harold and Kumar Go to White Castle. 2004. Directed by Danny Leiner. Written by Jon Hurwitz and Hayden Schlossberg. Starring John Cho and Kal Penn.

Harper, D., and P. J. Lassek. 2003. Poultry suit settled for $7.5 million. *Tulsa World,* July 17, A1.

Harris, J. M., P. R. Kaufman, S. W. Martinez, and C. Price. 2003. The U.S. food marketing system, 2002. Economic Research Service, United States Department of Agriculture. Agricultural Economic Report No. 811, June.

Hess, J. D., and E. Gerstner. 1991. Price-matching policies. *Managerial and Decision Economics* 12 (August): 305–315.

Hessel, O., R. Sloof, and G. Kuilen. 2004. Cultural differences in ultimatum game experiments: Evidence from a meta-analysis. *Experimental Economics* 72(2): 171–188.

Hirsch, W. B. 2005. The Exxon Valdez litigation justice delayed: Seven years later and no end in sight. Available at http://www.lieffcabraser.com/ wbh_exxart.htm (accessed September 7, 2005).

Huang, K. S., and B. Lin. 2000. Estimation of food demand and nutrient elasticities from household survey data. Economic Research Service. United States Department of Agriculture. Technical Bulletin No. TB1887, September.

Huck, S., H. Normann, and J. Oechssler. 2004. Two are few and four are many: Number effects in experimental oligopolies. *Journal of Economic Behavior and Organization* 53: 435–446.

Iyengar, S. S., and M. R. Lepper. 2000. When choice is demotivating: Can one desire too much of a good thing? *Journal of Personality and Social Psychology*. 70: 996–1006.

Jerardo, A. 2004. Ag trade balance . . . More than just a number. *Amber Waves* 2 (February).

Johnson, K. 2005. Spain's gory pastime creates bull market for star surgeons. *Wall Street Journal,* November 10, A1.

Jome, E. 2005. Hurricane Katrina causes lower Illinois grain prices. *Media Relations*. Illinois State University, September 8.

Kahneman, D., J. L. Knetsch, and R. Thaler. 2001. Fairness as a constraint on profit seeking: Entitlements in the market. *American Economic Review* 76(4): 728–741.

Keem I. K. 1999. Non-cooperative tacit collusion, complementary bidding, and incumbency premium. *Review of Industrial Organization* (15): 115–134.

Key, N., and W. McBride. 2003. Production contracts and productivity in the U.S. hog sector. *American Journal of Agricultural Economics,* 85 (February): 121–133.

King, R. P. 2003. Is there a future for wholesaler-supplied supermarkets? *Choices*, December.

Kinnucan, H. W., and R. N. Nelson. 1993. Vertical control and the farm-retail price spread for eggs. *Review of Agricultural Economics,* 15: 473–482.

Kirkpatrick, D. D. 2005. Storm and crisis. *New York Times*, September 7.

Knutson, R. D., J. B. Penn, and W. T. Boehm. 1990. *Agricultural and food policy*. 2nd ed. Englewood Cliffs, NJ: Prentice Hall.

Kreps, D. M. and R. Wilson. 1982. Reputation and imperfect information. *Journal of Economic Theory* 27: 253–279.

Kuchler, F., A. Tegene, and J. M. Harris. 2005. Taxing snack foods: Manipulating diet quality or financing information programs? *Review of Agricultural Economics* 27: 4–20.

Landsburg, Steven E. (1995). *The armchair economist*. New York: The Free Press.

Lawrence, J. D. 2001. Profiting from the cattle cycle: Alternative cow herd investment strategies. Iowa Beef Center. AgDM Newsletter Article, May.

Lee, I. K. 1999. Non-cooperative tacit collusion, complementary bidding, and incumbency premium. *Review of Industrial Organization* 15: 115–134.

Leonard, C. 2006. Monsanto's math reinvents soybean. *Tulsa World*, July 2, section E6.

Lewitt, E., and D. Coate. 1982. The potential for using excise taxes to reduce smoking. *Journal of Health Economics*, 1: 121–145.

Lewitt, E. M., and C. Douglas. 1981. The potential for using excise taxes to reduce smoking. NBER Working Paper No. W0764, September.

Levitt, S. D., and S. J. Dubner. 2005. *Freakonomics*. New York: HarperCollins Publishers Inc.

Lichtenberg, E. 2004. Some hard truths about agriculture and the environment. *Agricultural and Resource Economics Review* 33 (April): 24–33.

Livestock Marketing Information Center. 2005a. Weekly and monthly cattle price databases. http://www.lmic.info/ (accessed November 15, 2005).

Livestock Marketing Information Center. 2005b. Monthly feedstuffs database. http://www.lmic.info/ (accessed November 16, 2005).

Lomborg, B. 2001. *The skeptical environmentalist*. Cambridge University Press.

Lusk, J. L. 2003. Effect of cheap talk on consumer willingness-to-pay for golden rice. *American Journal of Agricultural Economics* 85: 840–856.

Lusk, J. L., and J. D. Anderson. 2004. Effects of country-of-labeling on meat producers and consumers. *Journal of Agricultural and Resource Economics* 29 (August): 185–205.

Lusk, J. L., M. S. Daniel, D. R. Mark, and C. L. Lusk. 2001. Alternative calibration and auction institutions for predicting consumer willingness-to-pay for non-genetically modified corn chips. *Journal of Agriculture and Resource Economics* 26: 40–57.

Lusk, J. L., R. Little, A. Williams, J. Anderson, and B. McKinley. 2003. Utilizing ultrasound technology to improve livestock marketing decisions. *Review of Agricultural Economics* 25 (Spring/Summer): 203–217.

Lusk, J. L., F. B. Norwood, and J. R. Pruitt. 2006. Consumer demand for a ban on antibiotic drug use in pork production. *American Journal of Agricultural Economics*. 88: 1015–1033.

MacDonald, J., J. Perry, M. Ahearn, D. Banker, W. Chambers, C. Dimitri, N. Key, K. Nelson, and L. Southerland. 2004. Contracts, markets, and prices: Organizing the production and use of agricultural commodities. Economic Research Service. United States Department of Agriculture, Agricultural Economic Report Number 837, November.

MacDonald, J. M., and M. E. Ollinger. 2000. Scale economies and consolidation in hog slaughter. *American Journal of Agricultural Economics* 82 (May): 334–346.

MacDonald, J. M., and M. E. Ollinger. 2005. Technology, labor wars, and producer dynamics: Explaining consolidation in beefpacking. *American Journal of Agricultural Economics* 87 (November): 1020–1033.

Marsh, J. M. 2003. Impacts of declining U.S. retail beef demand on farm-level beef prices and production. *American Journal of Agricultural Economics* 85 (November): 902–913.

Martinez, S. W. 1999. Vertical coordination in the pork and broiler industries: Implications for pork and chicken products. Economic Research Service, United States Department of Agriculture. Agricultural Economic Report No. 777, April.

Martinez, S. W. 2002. Vertical coordination of marketing systems: Lessons from the poultry, egg, and pork industries. Economic Research Service, United States Department of Agriculture. Agricultural Economic Report No. 807, April.

Maslow, A. 1954. *Motivation and personality*. New York: Harper.

McClure, S. M., J. Li, D. Tomlin, K. S. Cypert, L. M. Montague, and P. R. Montague. 2004. Neural correlates of behavioral preference for culturally familiar drinks. *Neuron* 44: 379–87.

McGee, J. S. 1958. Predatory price cutting: The Standard Oil (N.J.) case. *Journal of Law and Economics* 1: 137–169.

McMahon, K. 1998. Four packers kill 57% of hogs. *National Hog Farmer*, March 1.

Milgrom, P., and J. Roberts. 1982. Limit pricing and entry under incomplete information: An equilibrium analysis. *Econometrica* 50(2): 443–459.

Muren, A., and R. Pyddoke. 1999. Does collusion without communication exist? Research Papers in Economics. Number 1999:11. Department of Economics. Stockholm University.

Murphy, E. 2004. Economic impact of the Woody Adelgid on residential property values. Ph.D. Dissertation. North Carolina State University.

MSN Money. 2005. 33 states top $3 a gallon. CNBC Market Dispatches. http://moneycentral.msn.com (accessed September 7, 2005).

National Agricultural Statistics Service. 2005. United States Department of Agriculture. Online Historical Database. www.nass.usda.gov (accessed November 16, 2005).

New York Times. 1995. Court backs Wal-Mart on pricing. January 10.

Nichols, D. 2005. Economic outlook for late 2005 and 2006. Prepared for the Economic Outlook Conference. The Management Institute. School of Business. University of Wisconsin at Madison, September 16.

Norton, R. 2005. Unintended consequences. *The Concise Encyclopedia of Economics*. http://www.econlib.org/library/Enc/UnintendedConsequences.html (accessed September 13, 2005).

Norwood, F. B. 2001. Pesticide productivity bias due to unobserved variables. Dissertation. Department of Economics. North Carolina State University, November 12.

Norwood, F. B. 2006. Less choice is better, sometimes. *Journal of Agricultural and Food Industrial Organization* 4(1): Article 3.

Norwood, F. B., and J. Chvosta. 2005. Phosphorus based applications of livestock manure and the law of unintended consequences. *Journal of Agricultural and Applied Economics* 37 (April).

Oklahoma State University. 2005a. Enterprise corn, soybean, and cotton budgets. Department of Agricultural Economics. http://agecon.okstate.edu/budgets/sample_pdf_files.asp (accessed November 15, 2005).

Oklahoma State University. 2005b. Cow-calf sample enterprise budget. Department of Agricultural Economics. http://agecon.okstate.edu/budgets/sample_pdf_files.asp (accessed November 15, 2005).

Oklahoma State University. 2006. Experiment 502: Wheat grain yield response to nitrogen. http://nue.okstate.edu/Long_Term_Experiments/E502.htm (accessed February 28, 2006).

Ollinger, M., J. M. MacDonald, and M. Madison. Technological change and economies of scale in U.S. poultry processing. *American Journal of Agricultural Economics*. 87(1) (February 2005): 116–129.

Palmquist, R. B., F. M. Roka, and T. Vukina. 1997. Hog operations, environmental effects, and residential property values. *Land Economics* 73 (February): 114–124.

Pasour, E. C., and R. R. Rucker. 2005. *Plowshares and pork barrels*. Oakland, CA: The Independent Institute.

Powell, J. 2003. *FDR's folly*. New York: Crown Forum.

Prewitt, M. 1992. Reduced-fat burgers bomb with diners; chilly response prompts fast feeders to desert high-tech alternative. *Nation's Restaurant News*, November 16.

Preston, J. 1996. Mexico's political inversion: The city that can't fix the air. *New York Times*, February 4.

Price, G. K., W. Lin, J. B. Falck-Zepeda, and J. Fernandex-Cornego. 2003. Size and distribution of market benefits from adopting biotech crops. Economic Research Service. United States Department of Agriculture. Technical Bulletin Number 1906, November.

Qaim, M., and A. De Janvry. 2003. Genetically modified crops, corporate pricing strategies, and farmers' adoption: The case of Bt cotton in Argentina. *American Journal of Agricultural Economics* 85(November): 814–828.

Quist, D., and I. H. Chapela. 2001. Transgenic DNA introduced into traditional maize landraces in Oaxaca, Mexico. *Nature* 414: 541–543.

Rider, C. 1995. *An introduction to economic history*. Cincinnati, OH: South Western College Publishing.

Rokeach, M. 1968. *Beliefs, attitudes, and values*. San Francisco: Jossey-Bass.

Rokeach, M. 1973. *The nature of human values*. New York: Free Press.

Rosen, Ellen. 2005. Taking on credit card fees, with allies. *New York Times*, October 6.

Saba, A., and M. Vassallo. 2002. Consumer attitudes toward the use of gene technology in tomato production. *Food Quality and Preference* 13: 13–21.

Sachs, J. D. 2005. Can extreme poverty be eliminated? *Scientific American*, September, 56–65.

Schuff, S. 2005a. Big ag wants CAFTA; U.S. sugar fights back. *Feedstuffs*, April 11.

Schuff, S. 2005b. Grain to benefit through CAFTA. *Feedstuffs*, April 18.

Schroeder, T. C., C. E. Ward, J. Lawrence, and D. Feuz. 2002. Cattle marketing trends and concerns: Cattle feeder survey results. Kansas State University, June.

Schroeter, C. 2005. Determining the impact of food price and income changes on body weight. Ph.D. Dissertation, Purdue University.

Schweikhardt, D., and W. Browne. 2001. Politics by other means: The emergence of a new politics of food in the United States. *Review of Agricultural Economics* 23 (Fall/Winter): 302–318.

Sen, B. 2006. The relationship between beer taxes, other alcohol policies, and child homicide deaths. *The Berkeley Electronic Journals in Economic Analysis and Policy* 6(1): Article 9.

Shepherd, W. G. 1997. *The Economics of industrial organization*. 4th ed. Upper Saddle River, NJ: Prentice Hall.

Shor, M. 2006. Game Theory.net. Available at http://www2.owen.vanderbilt.edu/mike.shor/ (accessed March 22, 2006).

Sorenson, T. L. 2004. Limit pricing with incomplete information: Answers to frequently asked questions. *Journal of Economic Education* 35(1): 62–78.

Spar, D. L. 2006. Continuity and change in the international diamond market. *Journal of Economic Perspectives* 20(Summer): 195–208.

Sterns, L. D., and T. A. Petry. 1996. Hog market cycles. North Dakota State University, North Dakota State Extension Service. EC-1101, January.

Stumborg, B. E., K. A. Baerenklau, and R. C. Bishop. 2001. Nonpoint source pollution and present values: A contingent valuation study of Lake Mendota. *Review of Agricultural Economics* 23 (Spring/Summer): 120–132.

Surowiecki, J. 2002. *The wisdom of crowds*. Doubleday Publishers.

Towler, G., and R. Shepherd. "Modification of Fishbein and Ajzen's Theory of Reasoned Action to Predict Chip Consumption." *Food Quality and Preference*. 3 (1992): 37–45.

Tracy, J., and H. Schneider. 2001. Stocks in the household portfolio: A look back at the 1990s. *Current Issues in Economics and Finance* 7 (April).

Tulsa World. U.S. trade deficit hits all-time record. Associated Press Wire Service, February 11, E6.

Tremblay, V. J., and C. H. Tremblay. 2005. The U.S. brewing industry: Data and economic analysis. Cambridge, MA: The MIT Press.

Waldman, D. E. 2004. *Microeconomics*. Boston, MA: Pearson Addison Wesley.

Ward, C. Beef, pork, and poultry industry considerations. Oklahoma State University Cooperative Extension Service. F-552.

Ward, C. Captive supply trends since mandatory price reporting. Oklahoma State University Cooperative Extension Service. F-597.

Ward, C., T. C. Schroeder, A. P. Barkley, and S. R. Koontz. 1996. Role of captive supplies in beef packing. Grain Inspection, Packers and Stockyards Administration. United States Department of Agriculture. GIPSA-RR 96-3, May.

Wheelan, C. 2002. *Naked economics*. New York: W. W. Norton & Company.

The World Bank. 2005. World development indicators 2005. http://www.worldbank.org (accessed September 24, 2005).

World Trade Organization. 2005. Understanding the WTO: Basics. http://www.wto.org/english/hewto_e/whatis_e/ tif_e/tif_e.htm (accessed August 19, 2005).

Wossink, A., and C. Gardebroek. 2006. Environmentally policy uncertainty and marketable permit systems: The Dutch phosphate quota program. *American Journal of Agricultural Economics* 88 (February): 16–27.

Xia, T., and R. J. Sexton. 2004. The competitive implications of top-of-the-market and related contract-pricing clauses. *American Journal of Agricultural Economics* 86 (February): 124–138.

Zaichkowsky, J. L. The personal involvement inventory: Reduction, revision, and application to advertising. *Journal of Advertising* 23(1994): 59–71.

Answers to End-of-Chapter Crossword Puzzles and Study Questions

CHAPTER 1

Crossword Puzzle

Study Questions

1. Cattle are owned; they are private property, unlike the buffalo. One could not consume cattle today so that they can breed to provide beef in the future. This is like an investment. If the animal is not owned, there is little incentive to reserve animals for breeding because someone else may come along and profit from them. Thus, there was no incentive to make this investment, and buffalo were slaughtered in large numbers, leaving very few for breeding.
2. Macroeconomics: large economies, economic growth, inflation

   Microeconomics: individual markets, collections of markets, individual firms, and consumers

   Environmental and Resource Economics: the environment and natural resources

   Behavioral Economics: the psychology of individuals in economic decisions

   Labor Economics: labor markets

   Agricultural Economics: agricultural markets
3. The price will be higher than the old price, such that it costs the farmer almost \$15 more per acre to use the new fertilizer. At the least it would cost $(2/3)\$15 = \$10$ more.
4. The price in western Kansas should be at least $\$3.25 - \$0.15 = \$3.10$ and no more than $\$3.25 + \$0.15 = \$3.40$.
5. The storage cost is about \$0.50/4 months = \$0.125 per bushel.
6.

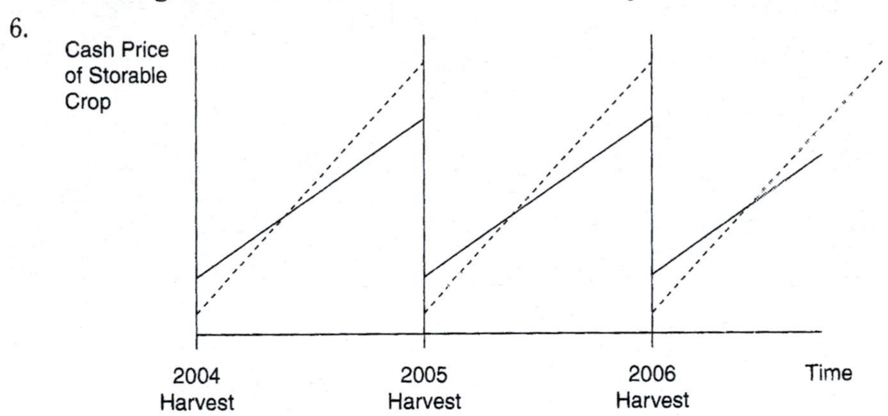

FIGURE 1.12

7. The price difference should fall. Prices will be higher in April and lower in May.
8. Because the amount of toys one dollar can buy was more in 1960 than in 2005.
9. The government could use the value of a statistical life and estimates on the number of lives saved from the ban to calculate the monetary benefits of the ban. This could then be compared to the \$20 million cost of the ban to see if the benefits outweigh the costs.

10. The value of the next best alternative, which is planning canola and earning \$60,000. Economic profits from corn are then \$80,000 − \$60,000 = \$20,000.

11. $\$125{,}000 \times (1.08)^{-1} = \$115{,}741$

    $\$125{,}000 \times (1.08)^{-2} = \$107{,}167$

    $\$125{,}000 \times (1.08)^{-3} = \$99{,}229$

    $\$125{,}000 \times (1.08)^{-4} = \$91{,}179$

    $\$125{,}000 \times (1.08)^{-5} = \$85{,}073$

    Present value of extra profits sum to \$499,089. Compared to the present costs of \$575,000, the upgrade is not profitable.

CHAPTER 2

Crossword Puzzle

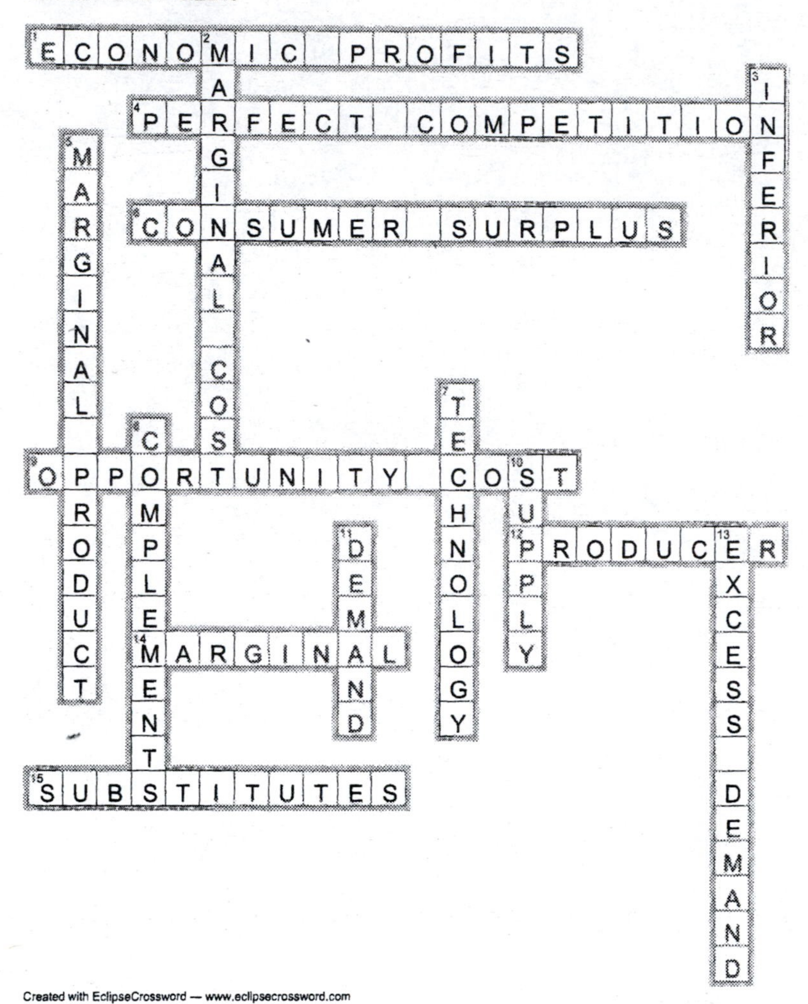

Created with EclipseCrossword — www.eclipsecrossword.com

Study Questions

1. The opportunity cost of selling is $4,000 and the value placed on the car is $5,000. Thus, the two should be able to strike a price somewhere between $4,000 and $5,000 for the El Camino.

2. Regular or accounting profits equal total revenues minus accounting costs, and accounting costs do not account for opportunity costs. Economic profits equal total revenues minus opportunity costs. Earning zero economic profits only indicates that Ashley is no better off running her advertising agency than she would be in her next best alternative. It does not indicate that she is in trouble

financially. In fact, she could be making lots of money but still earning zero economic profits as long as her next best alternative employment also paid lots of money. For example, if her next best alternative is to work for a rival company at a $50,000 salary, then zero economic profits implies that she is currently earning $50,000 running her company.

3. *Fill in the blanks*. Economists assume that the marginal opportunity cost of production is __increasing__ in quantity produced and marginal consumer value is __decreasing__ in quantity consumed.
4. a
5. Consumer surplus = ($5 − $2) + ($3 − $2) = $3 + $1 = $4.
6. a
7. decreases
8. increases
9. increases
10. decreases
11. decreases
12. a
13. a
14. b
15. a
16. c
17. b
18. a
19. c
20. c
21. b
22. 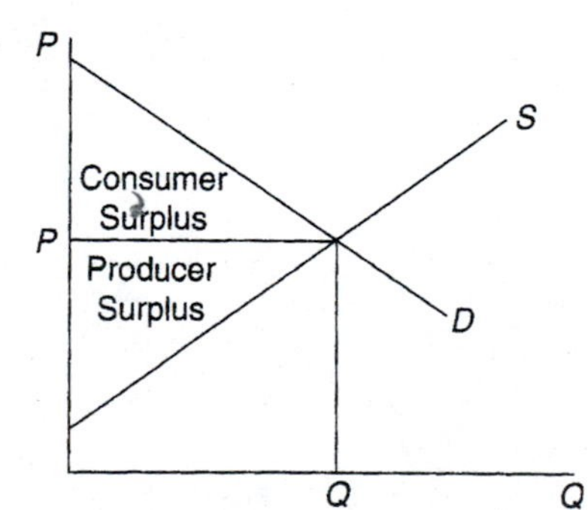

FIGURE 2.13

23.

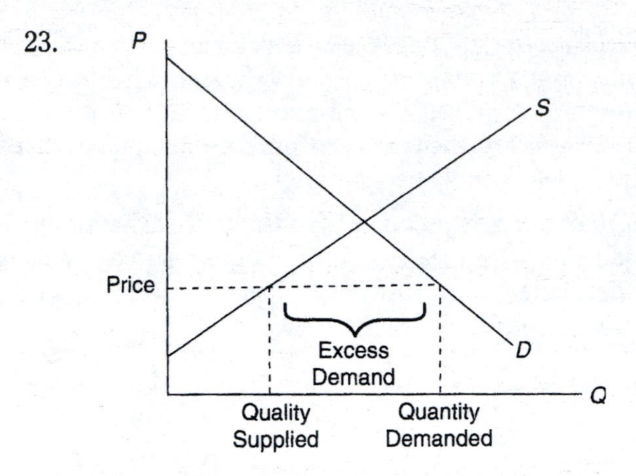

FIGURE 2.14

24. $Q = 26, P = 410$

CHAPTER 3

Crossword Puzzle

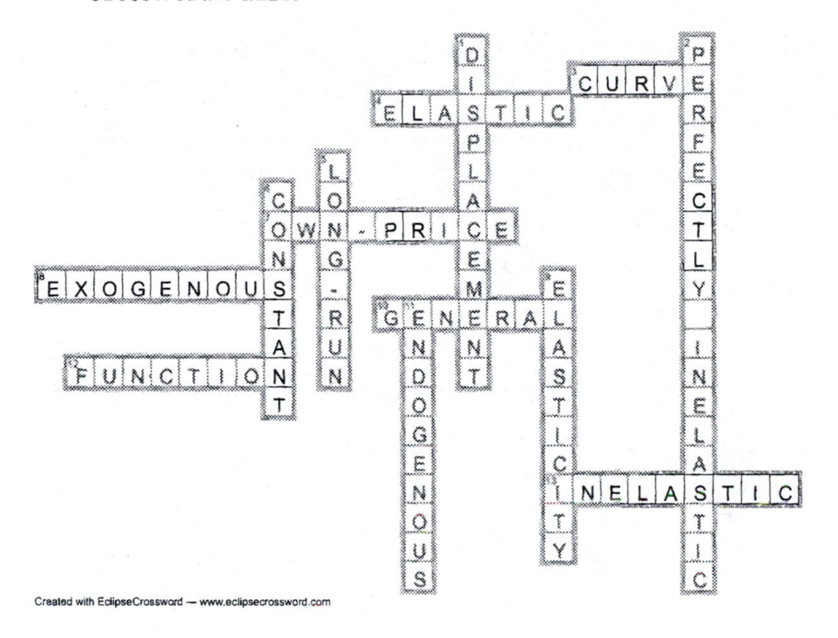

Study Questions

1. 16/–8 = –2, elastic
2. 4/8 = 0.5, inelastic
3. substitutes
4. Pork is a more threatening competitor. A 1% fall in the pork price steals more market share of beef than a 1% fall in poultry or fish prices. Thus, pork is a closer substitute for beef, making it a more threatening competitor.
5. Rice is more of a staple food in Vietnam than the U.S. Their culture involves greater use of rice in their recipes. Thus, there are fewer substitutes for rice in Vietnam than the U.S., making Vietnamese less sensitive to changes in the rice price than the U.S., and making their demand less elastic (smaller in absolute value).
6. Much of the arable land in Vietnam is more suited to rice production, and rice being more of a staple crop in Vietnam, producers have more experience and human capital in rice production than other crops. Thus, there are fewer substitutes for rice production, fewer alternative crops Vietnamese farmers can raise. Thus, they are less sensitive to changes in the rice price than U.S. farmers, and their supply is less elastic (smaller in value).
7. $\%\Delta QS = \%\Delta QD$

   $E_S(\%\Delta P) + S_S = E_D(\%\Delta P) + S_D$

   $0.15(\%\Delta P) + 0.15(-15) = -0.56(\%\Delta P) + 0$

$$0.15(\%\Delta P) + 0.56(\%\Delta P) = 0.15(15)$$
$$[0.15 + 0.56](\%\Delta P) = 0.15(15)$$
$$(\%\Delta P) = [0.15(15)]/[0.15 + 0.56]$$
$$(\%\Delta P) = [0.15(15)]/[0.15 + 0.56] = 3.17\%$$
$$\%\Delta QS = E_S(\%\Delta P) + S_S = 0.15(3.17) + 0.15(-15) = -1.78\%$$
$$\%\Delta QD = E_D(\%\Delta P) + S_D = -0.56(3.17) + 0 = -1.78\%$$

8. $\%\Delta QS = \%\Delta QD$
$$E_S(\%\Delta P) + S_S = E_D(\%\Delta P) + S_D$$
$$0.15(\%\Delta P) + 0 = -0.56(\%\Delta P) + 0.10(-25)$$
$$0.15(\%\Delta P) + 0.56(\%\Delta P) = 0.10(-25)$$
$$[0.15 + 0.56](\%\Delta P) = 0.10(-25)$$
$$(\%\Delta P) = 0.10(-25)/[0.15 + 0.56] = -3.52$$
$$\%\Delta QS = E_S(\%\Delta P) + S_S = 0.15(-3.52) + 0 = -0.528\%$$
$$\%\Delta QD = E_D(\%\Delta P) + S_D = -0.56(-3.52) + 0.10(-25) = -0.528\%$$

9. $Q_t = 367 - 3.12(P_t^{\text{Pork}}) + 1.35(P_t^{\text{Beef}}) + 1.69(P_t^{\text{Poultry}}) + 0.12(I_t) + 0.52(Q_{t-1})$
$$Q_t = 367 - 3.12(P_t^{\text{Pork}}) + 1.35(P_t^{\text{Beef}}) + 1.69(P_t^{\text{Poultry}}) + 0.12(I_t) + 0.52(Q_{t-1} = Q_t)$$
$$Q_t - 0.52(Q_t) = 367 - 3.12(P_t^{\text{Pork}}) + 1.35(P_t^{\text{Beef}}) + 1.69(P_t^{\text{Poultry}}) + 0.12(I_t)$$
$$Q_t\,(1 - 0.52) = 367 - 3.12(P_t^{\text{Pork}}) + 1.35(P_t^{\text{Beef}}) + 1.69(P_t^{\text{Poultry}}) + 0.12(I_t)$$
$$Q_t = [367 - 3.12(P_t^{\text{Pork}}) + 1.35(P_t^{\text{Beef}}) + 1.69(P_t^{\text{Poultry}}) + 0.12(I_t)]/(1 - 0.52)$$
$$Q_t = 765.58 - 6.50(P_t^{\text{Pork}}) + 2.81(P_t^{\text{Beef}}) + 3.52(P_t^{\text{Poultry}}) + 0.25(I_t)$$

10. $-3.12 \times (172/1325) = -0.405$, inelastic

11. $2.81 \times (226/1325) = 0.479$, substitutes

12. $1.69 \times (78/1325) = 0.099$, substitutes

13. $0.25 \times (3069/1325) = 0.579$, normal

14. $Q_t = 765.58 - 6.50(P_t^{\text{Pork}}) + 2.81(P_t^{\text{Beef}}) + 3.52(P_t^{\text{Poultry}}) + 0.25(I_t)$
$$Q_t = 765.58 - 6.50(P_t^{\text{Pork}}) + 2.81(226) + 3.52(78) + 0.25(3069)$$
$$Q_t = 765.58 - 6.50(P_t^{\text{Pork}}) + 635 + 275 + 767$$
$$Q_t = 2443 - 6.50(P_t^{\text{Pork}})$$
$$Q_t - 2443 = -6.50(P_t^{\text{Pork}})$$
$$P_t^{\text{Pork}} = -2443/-6.5 + Q_t/-6.5$$
$$P_t^{\text{Pork}} = 376 - 0.1538(Q_t)$$
For example, if $Q_t = 1000$, $P_t^{\text{Pork}} = 376 - 0.1538(1000) = 222.2$.

CHAPTER 4

Crossword Puzzle

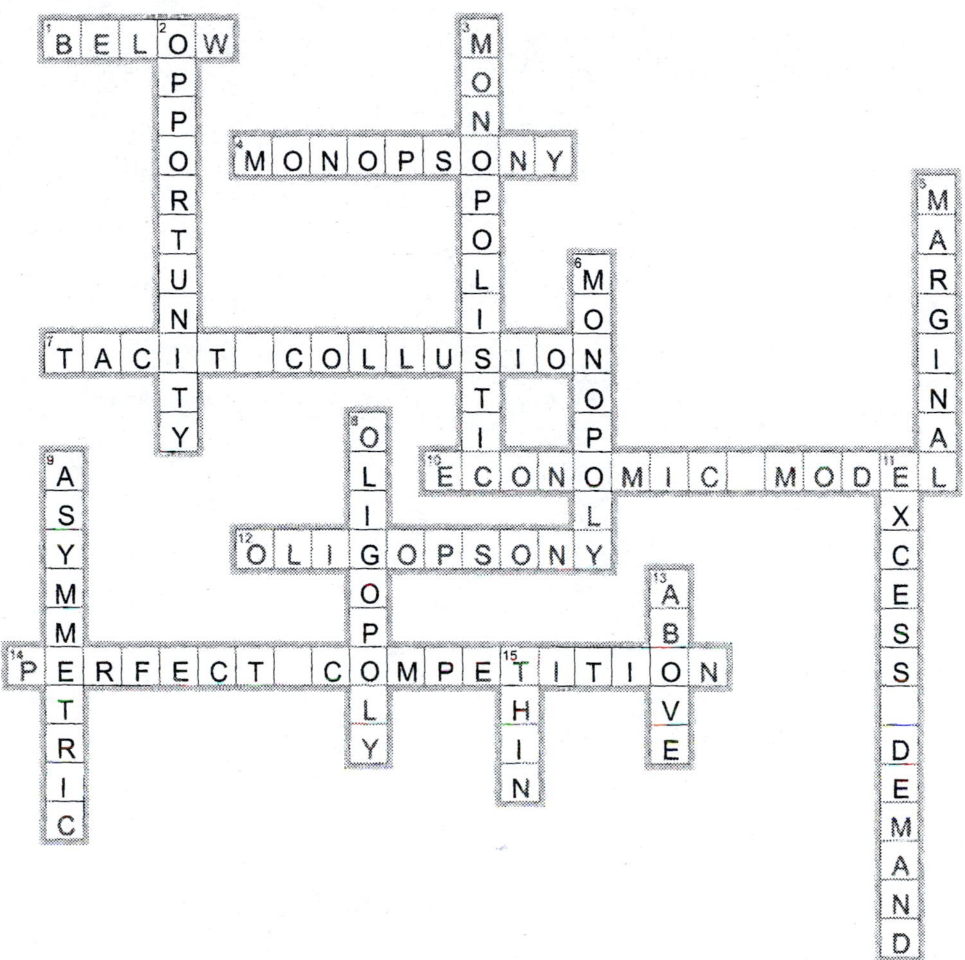

Study Questions

1.

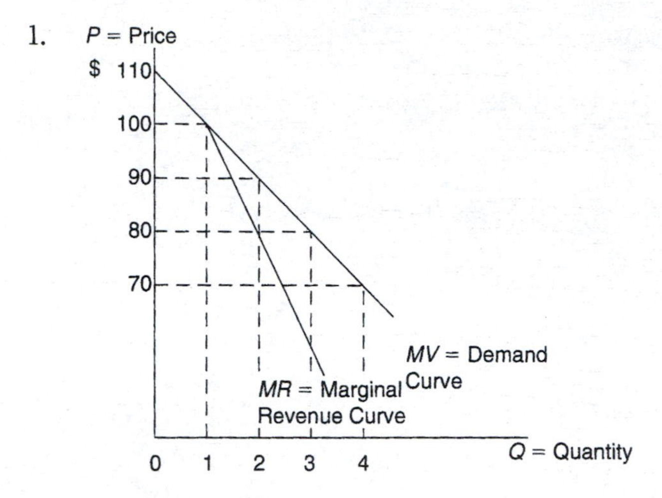

FIGURE 4.15

Quantity Demanded	Price	Total Revenues	Marginal Revenue
0	\$110	\$110 × 0 = \$0	
1	\$100	\$100 × 1 = \$100	\$100 − \$0 = \$100
2	\$90	\$90 × 2 = \$180	\$180 − \$100 = \$80
3	\$80	\$80 × 3 = \$240	\$240 − \$180 = \$60
4	\$70	\$70 × 4 = \$280	\$280 − \$240 = \$40

2.

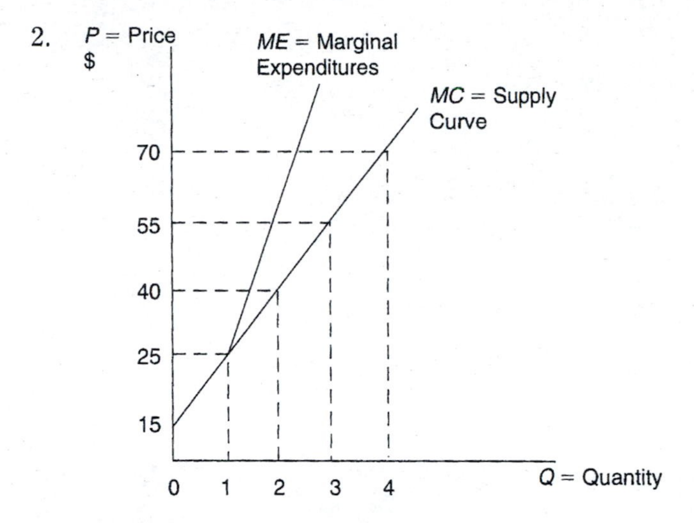

FIGURE 4.16

Quantity Supplied	Price	Total Expenditures	Marginal Expenditures
0	\$15	\$15 × 0 = \$0	
1	\$25	\$25 × 1 = \$25	\$25 − \$0 = \$25
2	\$40	\$40 × 2 = \$80	\$80 − \$25 = \$55
3	\$55	\$55 × 3 = \$165	\$165 − \$80 = \$85
4	\$70	\$70 × 4 = \$280	\$280 − \$165 = \$115

3. higher, lower
4. lower, lower
5. monopoly
6. monopsony
7. $P = \$410$
   $Q = 26$
8. $P = \$556.25$
   $Q = \$16.25$
9. $P = \$335.70$
   $Q = \$18.57$
10. See Chapter 2 where this question is directly addressed.

CHAPTER 5

Crossword Puzzle

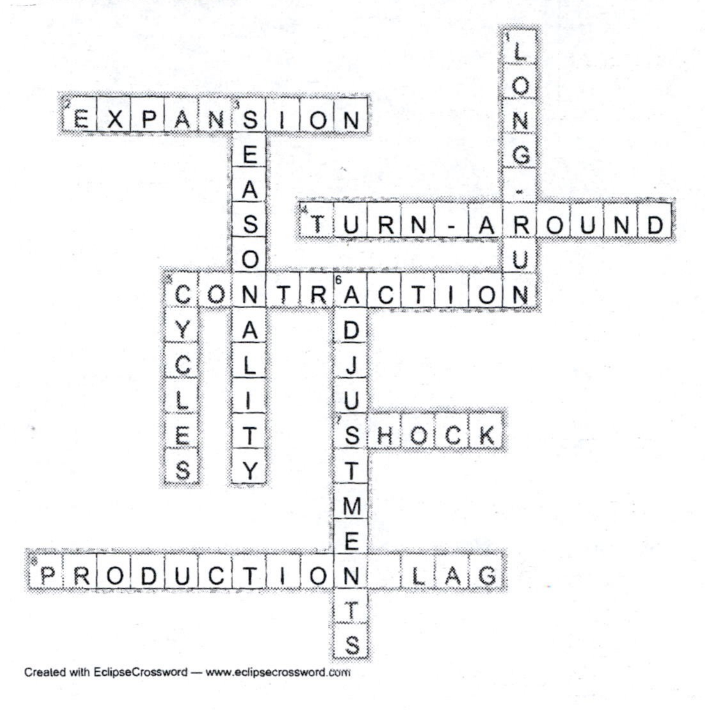

Study Questions

1.

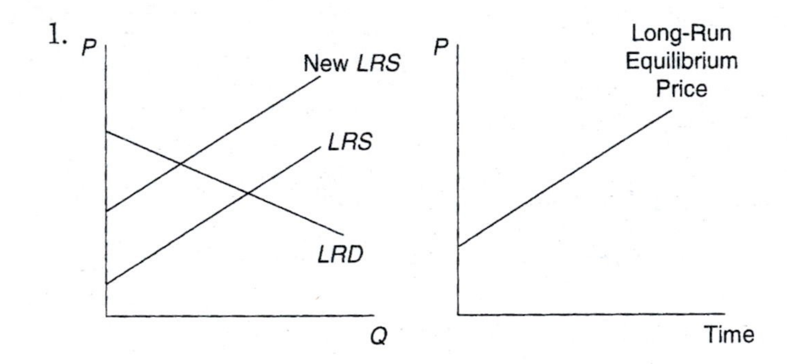

FIGURE 5.19

2. 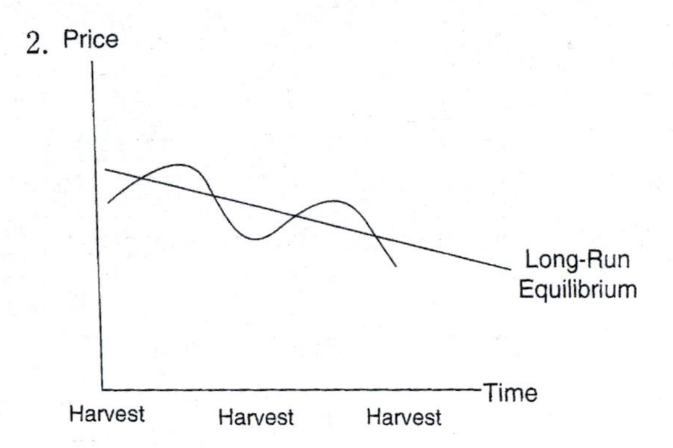

FIGURE 5.20

3. 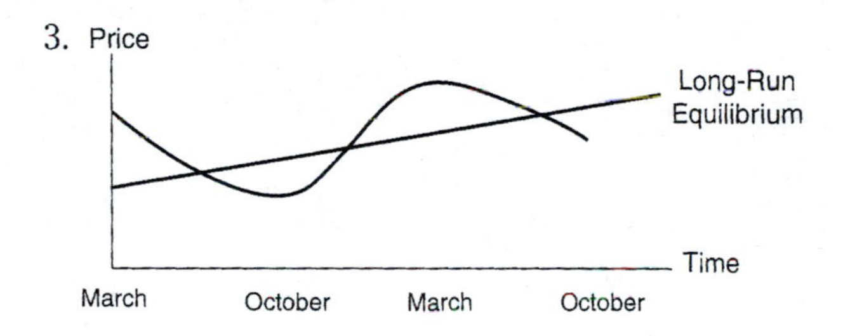

FIGURE 5.21

4. The difference between the high March price and the low October price would fall, thus, the effects of seasonality would dampen and actual prices would come closer to the long-run equilibrium price.

5. 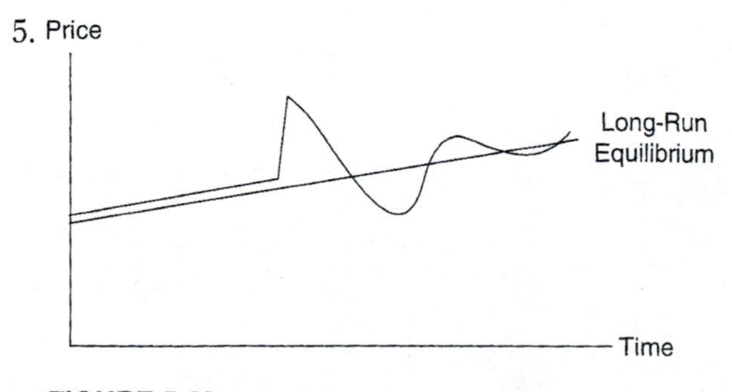

FIGURE 5.22

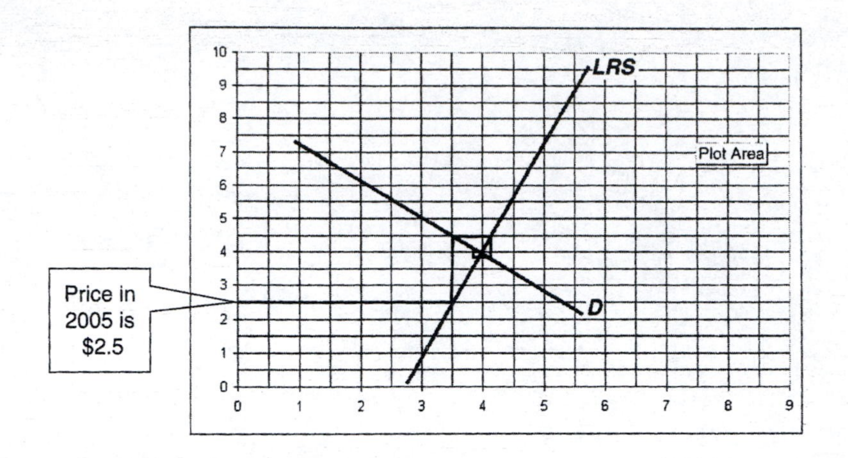

FIGURE 5.23

6. How much pork will be produced in 2006? 3.5

   What will be the pork price in 2006? 4.5

   How much pork will be produced in 2007? 4.1

7.

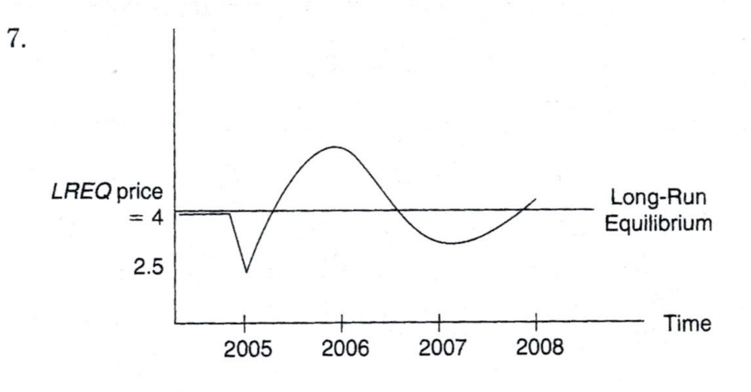

FIGURE 5.24

CHAPTER 6

CROSSWORD PUZZLE

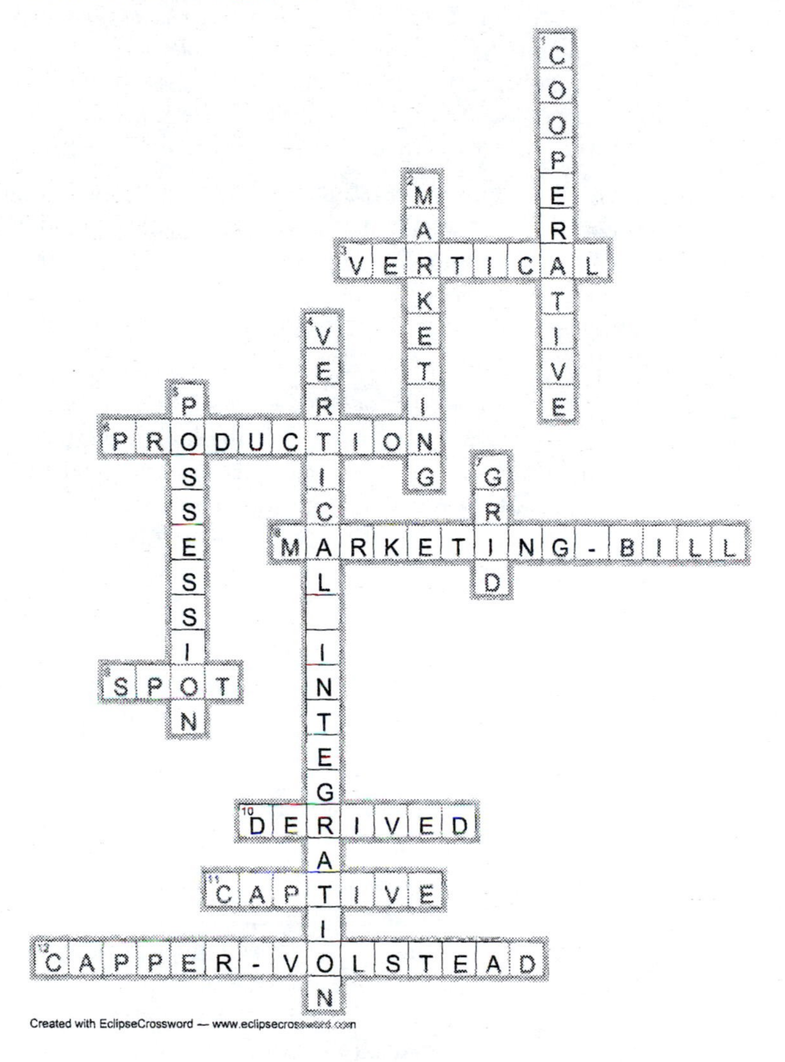

Study Questions

1. This farmer's market is concentrating on the provision of form utility. It provides "organic food," which is food in a particular form not found anywhere else in town. By locating miles from town, only accepting cash as payment, and only opening one day per week, the market has a disadvantage at providing place, time, and possession utility.

2. With average pricing, cattle are sold in groups for a lump sum, which is equivalent to saying that each cattle receives the same price regardless of its quality. Also, average pricing typically assigns a price to the live animal before it is slaughtered, prohibiting one from assigning higher or lower prices based on the meat tenderness. Buyers are forced to guess at the meat tenderness and will not be willing to offer higher prices unless they get some confirmation of meat tenderness. Thus, cattle with more tender meat receive little to no premium over cattle with tough meat, which discourages a producer from trying to produce tender beef.
3. The grid pricing system assigns each head of cattle a unique price based on its evaluated carcass quality. The grid system could be modified from that in Figure 5.7 to provide premiums for tender meat. Seeing carcasses with tender meat bringing greater revenues, producers will attempt to target cattle with tender meat genes in their breeding decisions, ultimately leading to a greater percentage of cattle possessing tender beef.
4. Tender beef can be sold for a higher price in grocery stores, so beef processors have an incentive to acquire cattle with tender beef. Instead of offering financial incentives to the producer for providing cattle with tender beef genes, the processor can purchase the cattle with these genes themselves, breed those cattle, and then pay farmers to raise the offspring. This amounts to a production contract, where the processor owns the cattle and the farmer is paid to raise the animal until ready for slaughter. By owning the animal, the processor is ensured the animal's beef with be tender.
5. Following from Question 4, the processor could not only purchase and breed cattle with tender beef genes, but they can build their own farms and raise the offspring themselves. This is referred to as vertical integration, where the beef processor owns two parts of the food marketing channel. Thus, by owning and raising cattle with tender beef genes, the processor assures itself that it can process and sell tender beef.
6. If the for-profit beef processors' success is due to their market power, then a farmer-owned processing facility (i.e., a cooperative) removes some of their market power, allowing farmers to receive higher cattle prices. The cooperative may also receive a tax advantage. Conversely, if the processors' success is due to excellent management or the ownership of patented process, the cooperative will not be able to compete unless it can also acquire managers and production processes similar in ability and efficiency. This may be too costly to acquire.
7. For each pound of live-hog produced, $250/125 = \frac{1}{2}$ pounds of retail pork is made. Thus, the retail equivalent of 2,000 lbs. of hogs is 1,000 lbs.
8. 120,000,000,000 lbs. wheat $\times$ 0.7 $=$ 84,000,000,000 lbs. flour

9.

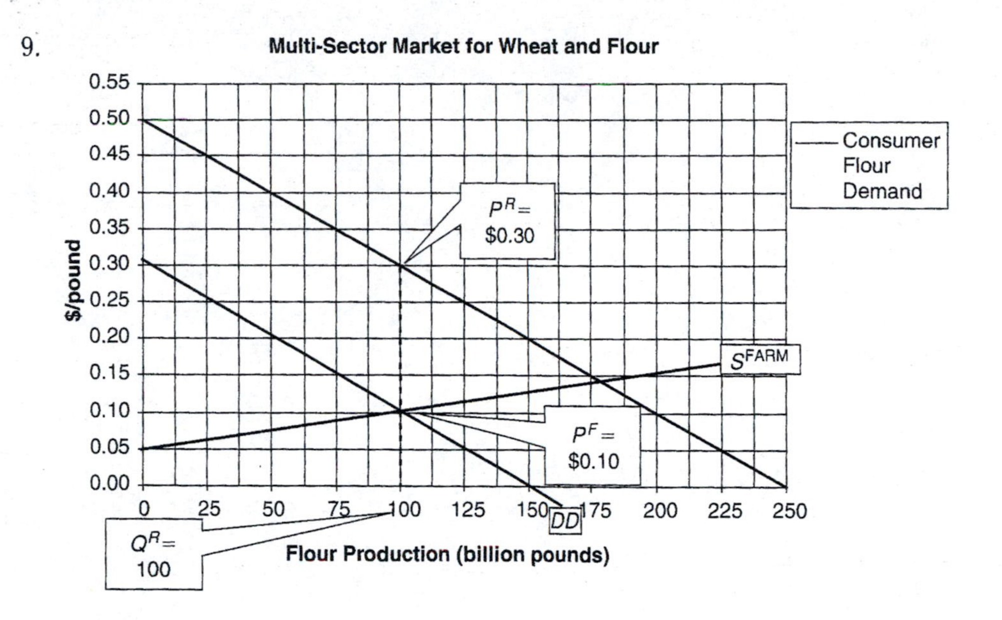
Multi-Sector Market for Wheat and Flour
$/pound
0.55
0.50
0.45
0.40
0.35
0.30
0.25
0.20
0.15
0.10
0.05
0.00
0
25
50
75
100
125
150
175
200
225
250
Flour Production (billion pounds)
Consumer Flour Demand
$P^R =$ $0.30
S^{FARM}
$P^F =$ $0.10
$Q^R =$ 100
DD

FIGURE 6.16

CHAPTER 7

Crossword Puzzle

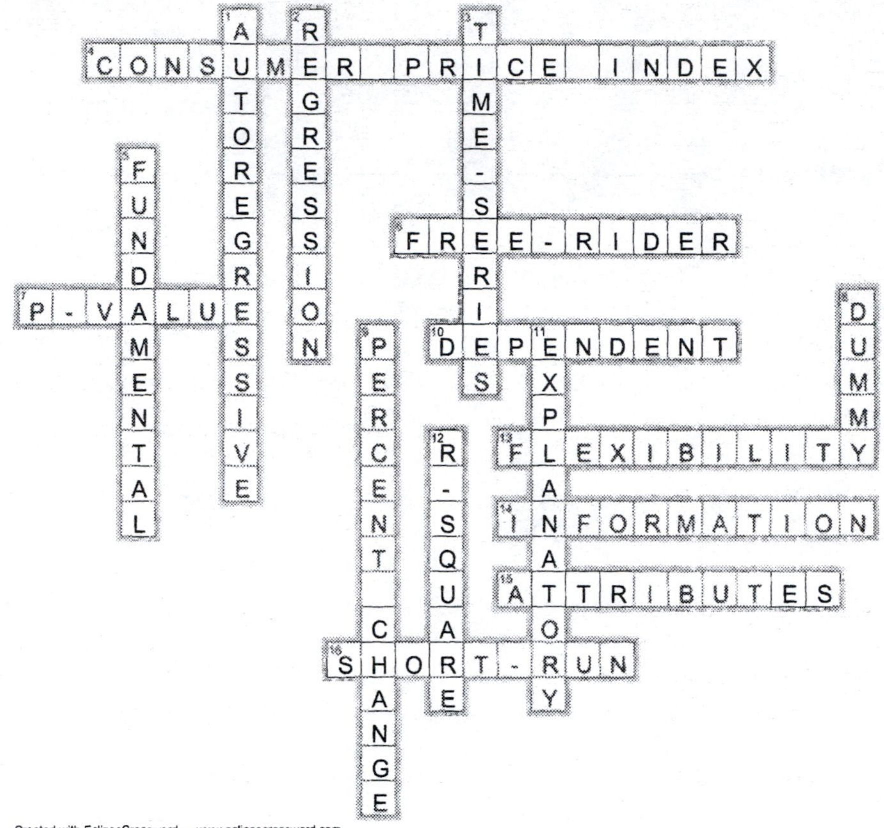

Study Questions

1. We estimate the regression Bid = $a_0 + a_1$ (Female) and get the estimates Bid = 0.45 − 0.02 (Female), where the p-value for the Female coefficient is 0.79. The p-value is much higher than 0.10, thus, we conclude Female has no real impact on Bid and the willingness-to-pay for males and females are the same.

2. a. Long-Run Supply: $Q_t = -200 + 4(P_{t-5}) + 0.5(C_{t-5}) + 0.2(Q_{t-5})$

$$Q_t = -200 + 4(P_{t-5} = P_t) + 0.5(C_{t-5} = C_t) + 0.2(Q_{t-5} = Q_t)$$
$$Q_t - 0.2(Q_t) = -200 + 4(P_t) + 0.5(C_t)$$
$$0.8(Q_t) = -200 + 4(P_t) + 0.5(C_t)$$
$$Q_t = [-200 + 4(P_t) + 0.5(C_t)]/0.8$$
$$Q_t = -250 + 5(P_t) + 0.625(C_t)$$

Long-Run Demand: $P_t = 1000 - 5(Q_t) + 0.5(S_t) + 0.05(Q_{t-5})$
$P_t = 1000 - 5(Q_t) + 0.05(Q_{t-5} = Q_t) + 0.5(S_t)$
$P_t = 1000 - 4.95(Q_t) + 0.5(S_t)$

b. Long-Run Supply: $Q_t = -250 + 5(P_t) + 0.625(C_t)$

$Q_t = -250 + 5(P_t) + 0.625(C_t = 20)$

$Q_t = -250 + 5(P_t) + 12.5$

$Q_t = -237.5 + 5(P_t)$

$-5(P_t) = -237.5 - Q_t$

$P_t = 237.5/5 + Q_t/5$

Long-Run Supply Curve: $P_t = 47.5 + 0.2(Q_t)$

Long-Run Demand: $P_t = 1000 - 4.95(Q_t) + 0.5(S_t)$
$P_t = 1000 - 4.95(Q_t) + 0.5(S_t = 100)$
$P_t = 1000 - 4.95(Q_t) + 50$

Long-Run Demand Curve: $P_t = 1050 - 4.95(Q_t)$

c. $47.5 + 0.2(Q_t) = 1050 - 4.95(Q_t)$
$5.15(Q_t) = 1002.5$
$Q_t = 1002.5/5.15 = 194.66$

Long-Run Supply Curve: $P_t = 47.5 + 0.2(194.66) = 86.43$

Long-Run Demand Curve: $P_t = 1050 - 4.95(194.66) = 86.43$

3. a. a_1

   b. b_1

4. a. The estimate is *Yield* = 312 + 6.957 (*Time Trend*).

   b. The p-value on the time trend variable is 0.00; so yes, cotton yields do increase over time.

   c. Each year cotton yields increase 6.957 pounds per acre.

   d. Notice that the time trend variable is calculated as the year minus 1952. In 2010, we would expect cotton yields to equal *Yield* = $312 + 6.957(2010-1952 = 58) = 312 + 6.957 \times 58 = 715.51$ pounds per acre.

CHAPTER 8

Crossword Puzzle

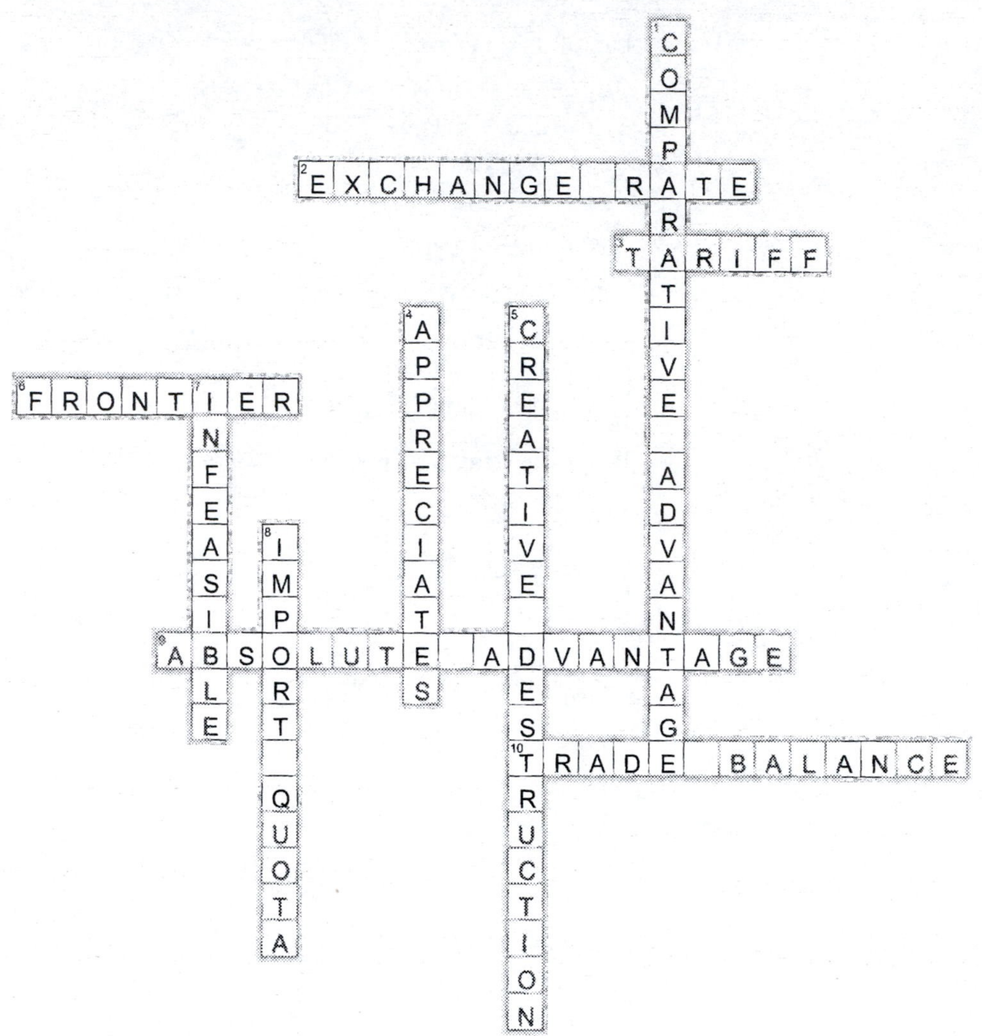

Study Questions

1.

	Opportunity Cost of Beer	Opportunity Cost of Cigars
U.S.	1/2 cigars	2 beers
Cuba	1 cigars	1 beers

2. Beer—United States

   Cigars—Cubas

3.

	Peer Production (# beers produced)	Cigar Production (# cigars produced)
U.S.	10	0
Cuba	0	10

4. d
5. The trade balance, if investments are counted, measure the amount of goods and services we export to other countries minus the amount we import from others. If no country gives anything away for free or at reduced prices, then clearly exports must equal imports. But foreign aid is the act of giving countries free goods and services, which would make the trade balance positive.
6. Protectionist policies tend to provide large benefits to small, well-organized groups and harm consumers as a whole. Thus, the small groups who reap large benefits are more likely to lobby for protectionist policies than consumers who are harmed only slightly from protectionist policies of one particular good.
7. Cuba will begin exporting cigars to the United States where the price is higher and the United States will begin importing cheaper cigars. As long as Cuban cigars are cheaper, U.S. consumers will continue to bid up the price of Cuban cigars until it equals the U.S. price. At this point, consumers are indifferent between U.S. and Cuban cigars.
8. World Price = 12

   U.S. Imports = 4

   Chinese Exports = 4
9. rises, rises
10. falls, falls
11. **United States**

    Consumer Surplus Before Trade A (use letters in Figure 8.18.)

    Producer Surplus Before Trade D + B (use letters in Figure 8.18.)

    Total Surplus Before Trade A + D + B (use letters in Figure 8.18.)

    Consumer Surplus After Trade A + B + C (use letters in Figure 8.18.)

    Producer Surplus After Trade D (use letters in Figure 8.18.)

    Total Surplus After Trade A + B + C + D (use letters in Figure 8.18.)

    Japan

    Consumer Surplus Before Trade E + F (use letters in Figure 8.18.)

    Producer Surplus Before Trade H (use letters in Figure 8.18.)

Total Surplus Before Trade ___E + F + H___ (use letters in Figure 8.18.)

Consumer Surplus After Trade ___E___ (use letters in Figure 8.18.)

Producer Surplus After Trade ___H + F + G___ (use letters in Figure 8.18.)

Total Surplus After Trade ___E + H + F + G___ (use letters in Figure 8.18.)

Final Welfare Analysis

Total Welfare (Total Surplus) Change for United States increased by C. (Use letters in Figure 8.18.)

Total Welfare (Total Surplus) Change for Japan increased by G. (Use letters in Figure 8.18.)

CHAPTER 9

Crossword Puzzle

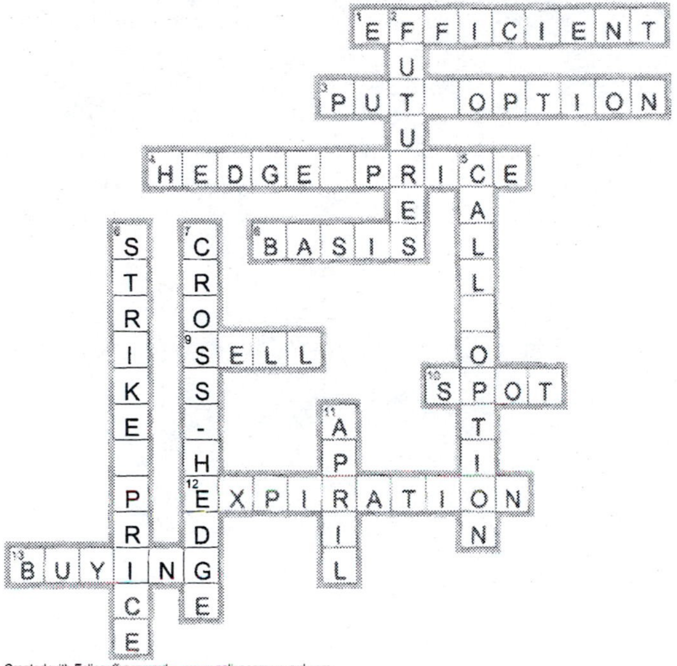

Study Questions

1. b
2. spot − futures price = 7.35 − 7.50 = −0.15
3. profits = (selling price − buying price)(# contracts)(units per contract)

   = (6.23 − 7.50)(3)(5000) = −$19,050
4. selling, buying, spot market
5. hedge price = futures price at hedge execution + basis

   = 6.23 + (7.35 − 7.50) = $6.08
6. 85.97 cents per lb
7. expected hedge price = futures price at hedge execution + expected basis

   = 84.42 + 2.25 = $86.67/cwt
8. efficient
9. She can cross-hedge by selling November corn futures in November, offsetting by buying the same number of November corn futures in November, and selling her sorghum in the spot market. A hedge is simply a futures transaction that makes you money when prices are unfavorable and losing you money when prices are

favorable. Sorghum and corn are close substitutes in animal feed, so their prices rise and fall together. If sorghum prices fall, corn prices will also fall and the futures transaction will make money, partially offsetting the losses from low sorghum prices. If sorghum prices rise, corn prices will also rise and the futures transaction will lose money, at a time when high sorghum prices favor the producer.

10. a, e
11. False. In February, the spot price refers to the price of corn to be exchanged in February, whereas the price of March corn futures contract refers to the price of corn to be exchanged in March. Because the supply and demand for corn differ in February and March, two prices do not have to be similar.
12. a
13. a
14. \$2.38/bushel
15. futures profits = (selling price − buying price)(# contracts)(units per contract)
    = (\$3.10 − \$3.41)(1)(5000) = −\$1,550
16. expected hedge price = futures price at hedge execution + expected basis
    = \$3.30 − \$0.10 = \$3.20
17. hedge price = futures price at hedge execution + basis
    = \$3.30 + (\$3.45 − \$3.50) = \$3.25
18. minimum expected price = strike price + expected basis − option premium
    = \$4.00 + \$0.25 − \$0.40 = \$3.85

CHAPTER 10

Crossword Puzzle

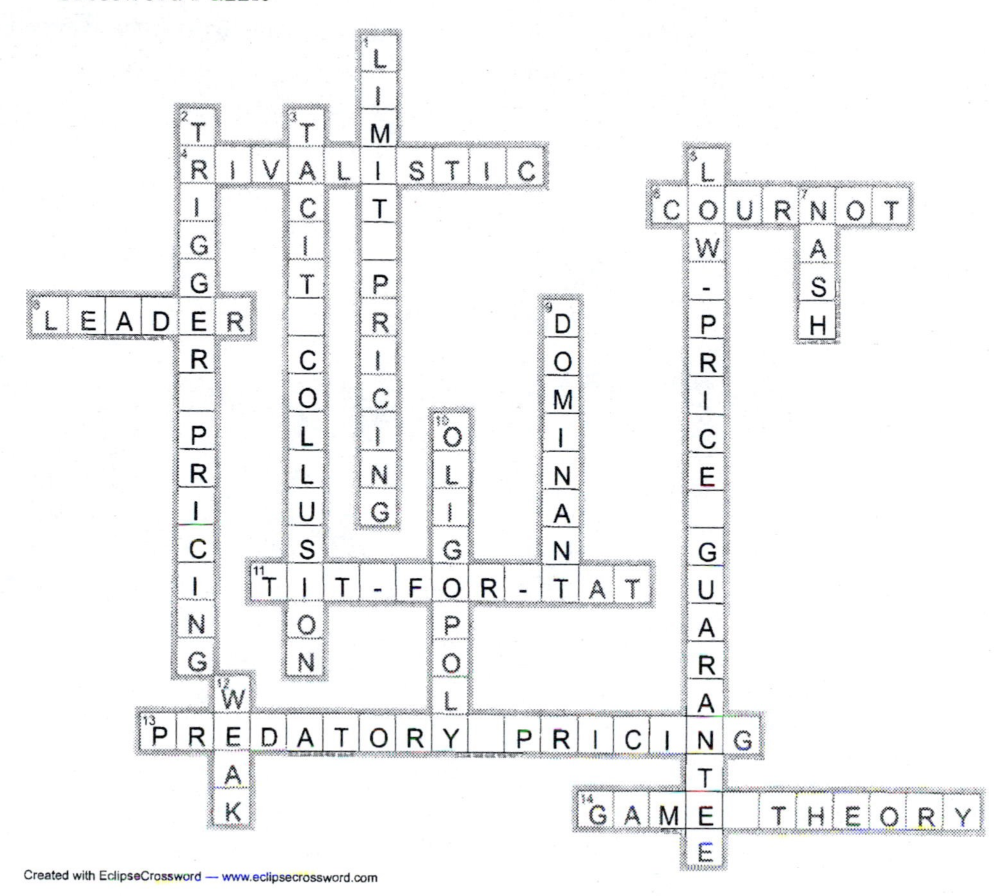

Study Questions

1. dominant
2. Nash
3. see Chapter 10
4. tit-for-tat
5. asymmetric, asymmetric
6. strong, weak
7. advertise, advertise
8. trigger pricing

9. firm homogeneity
10. A weak monopolist has the same costs of production as its competitors. It engages in predatory pricing trying to fool others that it is a strong monopolist with a lower cost of production, hoping to convince the competitors to leave the market.

CHAPTER 11

Crossword Puzzle

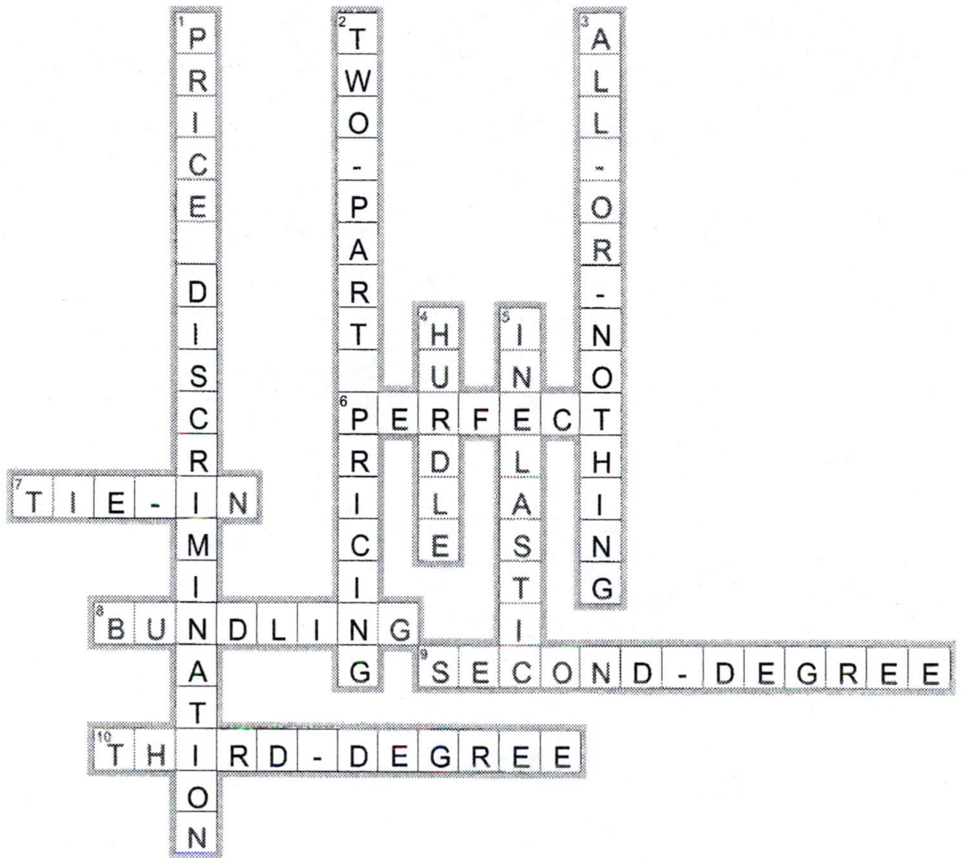

Study Questions

1. $(1/2)(50 - 20)(100) = \$1{,}500$
2. $$\begin{aligned} \text{bundle} &= 100 \text{ units} \\ \text{bundle price} &= (1/2)(50 - 20)(100) + (100)(20) = \$1{,}500 + \$2{,}000 \\ &= \$3{,}500; \text{ profits} = \$1{,}500 \end{aligned}$$
3. $$\begin{aligned} \text{fixed fee} &= \$1{,}500 \\ \text{per unit price} &= \$20 \\ \text{profits} &= \$1{,}500 \end{aligned}$$
4. The demand for Group A is less elastic.

CHAPTER 12

Crossword Puzzle

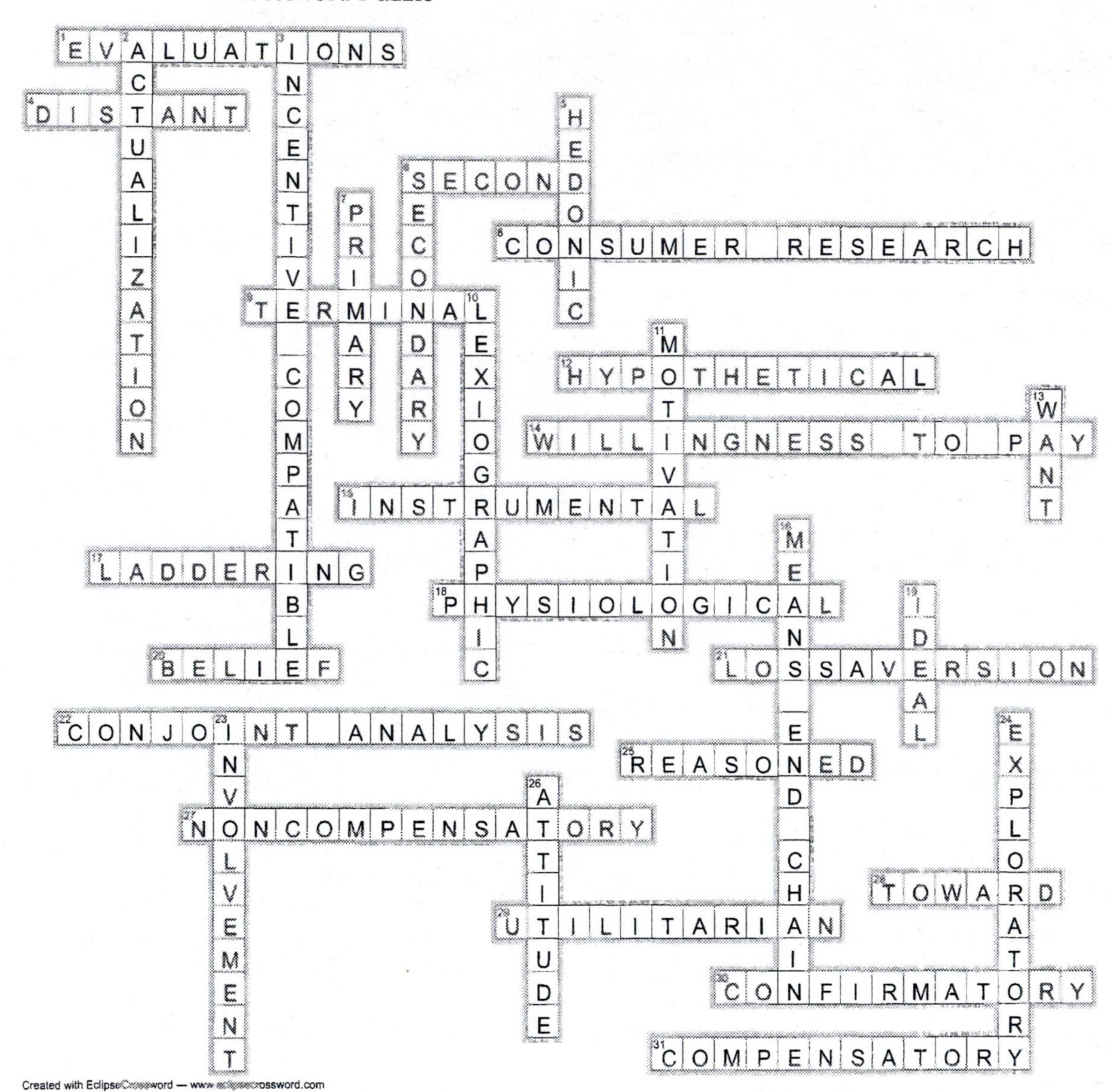

Study Questions

1. The main advantage of primary data is that the researcher can ensure that the collected data are able to adequately answer the question of interest. However, such data are often time intensive to obtain and expensive to collect. Secondary

data, on the other hand, are often freely available and can be rapidly used and analyzed. The primary disadvantages of secondary data are that they may not yield ideal tests of the hypotheses of interest and often come in aggregated form, making it difficult to uncover consumer heterogeneity, a key determinant of creating profitable segmentation strategies.

2. Define the Problem; Determine Research Design; Analyze and Interpret Data; Present Results

3. There are three attributes: origin label, organic label, and price. Suppose each attribute is varied at two levels: origin label (United States or Mexico), organic label (Organic or Nonorganic), and price (\$4.00 or \$8.00). This means there are $2^3 = 8$ possible beef steak descriptions, as shown below.

	Steak Attribute		
Steak	Origin	Organic	Price
1	United States	Yes	\$4.00
2	United States	No	\$4.00
3	United States	Yes	\$2.00
4	United States	No	\$2.00
5	Mexico	Yes	\$4.00
6	Mexico	No	\$4.00
7	Mexico	Yes	\$2.00
8	Mexico	No	\$2.00

Survey respondents would be asked to rate each of the 8 steaks on a scale of 1 to 7 where 7 is very desirable and 1 is very undesirable. Then, a regression would be run to estimate the coefficients of the following model:

$\text{Rating} = \beta_0 + \beta_1\text{U.S.} + \beta_2\text{Organic} + \beta_3\text{Price}$

If $\beta_1 > \beta_2$, then origin is more important than organic. Willingness-to-pay for U.S. origin versus Mexican origin is $-\beta_1/\beta_3$ and willingness-to-pay for organic versus nonorganic is $-\beta_2/\beta_3$. Assuming each person only purchases one steak during the time period of interest, a demand curve for organic beef could be constructed by estimating the ratings model for each respondent, calculating WTP for organic for each person, sorting the WTP values from highest to lowest, and plotting the sorted WTP values. Alternatively, if one wanted to consider the price of substitutes, then assume each person can buy two portions organic at one price and nonorganic at another. Assume the person will choose the product with the highest predicted rating. Then, a demand curve could be constructed by determining how the percentage of people choosing organic could change as the price of organic changes.

4. (a) Change people's current beliefs about purchasing the product (b) add additional positive or negative beliefs not currently present, (c) change the evaluation of beliefs, or (d) affect social norms by altering the overall social image of GM.

5. A means-end chain could be constructed as follows. Prompt the respondent to think about the attributes that are important in choosing between 2% milk,

whole milk, and milk in general. This might be done by asking a respondent to rank four different products according to their preference. After the products are ranked, ask respondents to give reasons for their ranking. Typically, the answers will be something like "this product has attribute *X* whereas the other product had attribute *Y*." Once a set of attributes is elicited from the respondent, ask questions like "why is that important to you?" After an answer is given, again ask "why is that important to you?" to encourage the respondent to give more and more abstract explanations until they can go no further or until they provide an explanation that resembled one of the values in Figure 12.3. Repeat the process for each attribute. Through this series of questioning, you should be able to construct a chain from each concrete attributes to functional and psychosocial consequences to instrumental and terminal values. There will be as many chains as there are attributes.

6. Attitudes can be measured using multiple or single-item measures but should probably utilize an interval scale. An example single-item attitude measure would be "On a scale of 1 to 5 where 1 is strongly disagree and 5 is strongly agree, please respond to the following statement: I have a favorable opinion of purchasing genetically modified food." Attitudes toward purchasing GM food would be expected to be influenced by the product of beliefs about purchasing GM food and the evaluations of those beliefs. Attitudes would also be expected by social norms and the individual's willingness to comply with those norms.

7. A disadvantage of experimental auctions is that unlike stated preference data, consumers cannot be asked to evaluation *any* product—the goods must be deliverable. Another disadvantage is that relative to survey methods, it is relatively more difficult and costly to obtain data from a large and representative sample of consumers. The primary advantage of experimental auctions is that they involve real choices with real money and real goods and provide incentives for people to truthfully reveal their preferences. Further, unlike discrete choice questions, auctions provide an exact measure of willingness-to-pay for each individual.

Index